Lecture Notes in Mathematics

Edited by A. Dold, B. Eckmann and F. Takens

1404

M.-P. Malliavin (Ed.)

Séminaire d'Algèbre Paul Dubreil et Marie-Paul Malliavin

Proceedings, Paris 1987–1988
(39ème Année)

Springer-Verlag
Berlin Heidelberg New York London Paris Tokyo Hong Kong

Editeur

Marie-Paule Malliavin
Université Pierre et Marie Curie, Mathématiques
4, place Jussieu, 75252 Paris Cedex 02, France

Mathematics Subject Classification (1980): 14C20, 14F05, 14L30, 16A03, 16A20, 16A33, 16A46, 16A48, 16A54, 16A61, 16A62, 16A64, 16A89, 20G05, 32M99, 47F05

ISBN 3-540-51812-6 Springer-Verlag Berlin Heidelberg New York
ISBN 0-387-51812-6 Springer-Verlag New York Berlin Heidelberg

Printing and binding: Druckhaus Beltz, Hemsbach/Bergstr.
2146/3140-543210 – Printed on acid-free paper

LISTE DES AUTEURS

* *
*

TABLE DES MATIÈRES

PREVIOUS VOLUMES OF THE "SEMINAIRE Paul DUBREIL" WERE PUBLISHED IN THE LECTURE NOTES, VOLUMES 586 (1976), 641 (1977), 740 (1978), 795 (1979), 867 (1980), 924 (1981), 1029 (1982), 1146 (1983-84), 1220 (1985) and 1296 (1986).

ALGEBRES PRE-INCLINEES ET CATEGORIES DERIVEES

Ibrahim Assem et Andrzej Skowronski

[*Cet article apparaît dans sa forme définitive et aucune autre version de ce travail ne sera soumise ailleurs pour publication*].

Soient k un corps algébriquement clos et A une k-algèbre (associative, unifère) de dimension finie. Sans perte de généralité, on peut supposer A sobre et connexe. Tous nos modules sont à droite et de k-dimension finie. On notera mod A la catégorie des A-modules et $D^b(A)$ la catégorie dérivée des complexes bornés sur mod A [V] . La structure de $D^b(A)$ est connue dans les cas où A est une algèbre héréditaire de représentation finie ou docile, ou bien A est une algèbre tubulaire canonique (au sens de Ringel [R1]) [H2] [HR2] . Dans ces deux cas, tout cycle de $D^b(A)$ se trouve entièrement dans un tube.

Nous prouvons que, réciproquement, si tout cycle de $D^b(A)$ se trouve dans un tube, alors $D^b(A)$ est équivalente, en tant que catégorie triangulée, à $D^b(C)$, où C est soit héréditaire de représentation finie ou docile, soit une algèbre tubulaire canonique. En outre, c'est le cas si et seulement si A s'obtient de C au moyen d'une suite finie d'inclinaisons et de co-inclinaisons. Si C est héréditaire, A est en fait une algèbre pré-inclinée [AH] . La démonstration de notre résultat (section (5)) nous permet d'obtenir une classification des algèbres pré-inclinées de type Dynkin ou Euclidien. Nous introduisons une notion de connexité simple pour les algèbres triangulaires qui ne sont pas nécessairement de représentation finie et montrons que la classification se scinde en deux cas, le cas où l'algèbre est simplement connexe et celui où elle est pré-inclinée de type $\tilde{\mathbb{A}}_m$ (section (4)). Ce dernier cas est résolu dans la section (3), alors que le cas simplement connexe se divise encore en deux parties: si l'algèbre est de représentation infinie, nous donnons une description complète par carquois liés (section (2)) et si elle est de représentation finie, nous donnons un critère pratique en termes de la forme quadratique de l'algèbre (section (6)). Enfin, dans une dernière section, nous présentons une classification détaillée par carquois liés des algèbres pré-inclinées de type $\mathbb{D}_n$. Ce résultat, compte tenu de [AH] et [H1] , achève la classification des algèbres pré-inclinées de type Dynkin. Il a été utile dans les parties combinatoires des démonstrations des résultats précédents, ainsi que dans d'autres problèmes de classification d'algèbres dociles (voir, par exemple, [ANS] et [NS]). Sauf dans la dernière partie, la plupart des démonstrations ne sont qu'esquissées. Les détails paraîtront dans [AS1] [AS2] [AS3] et [AS4] . Dans la section (7), par contre, nous donnons une démonstration complète de notre résultat.

Ces résultats ont été présentés par le premier auteur au Séminaire Malliavin en mars 1987. Ils ont été obtenus alors que les deux auteurs visitaient l'Université de Bielefeld en tant que boursiers Alexandre von Humboldt. Ils voudraient remercier C. M. Ringel pour son hospitalité.

1. Préliminaires

1.1 On rappelle qu'un carquois Q est défini par la donnée d'un ensemble de points Q_o et d'un ensemble de flèches Q_1. Une relation d'un point x à un point y est une combinaison linéaire $\rho = \sum_{j=1}^{m} \lambda_j w_j$ où, pour chaque $1 \le j \le m$, λ_j est un scalaire non-nul et w_j un chemin de x à y de longueur au moins deux (nous distinguons soigneusement entre un chemin de Q, orienté par définition, et une allée, qui ne l'est pas). Une relation ρ est une relation-zéro (respectivement, une relation de commutativité) si $m = 1$ (respectivement, $m = 2$). Un ensemble de relations engendre un idéal bilatère I de l'algèbre des chemins kQ de Q. La paire (Q,I) est alors appelée un carquois lié. Si $I = 0$, le carquois est dit libre. On sait que, pour toute k-algèbre A (localement) de dimension finie, sobre (c'est à dire telle que $A/\text{rad}\, A \tilde{\to} k \times \ldots \times k$) et connexe (c'est à dire dont les seuls idempotents centraux sont 0 et 1), il existe un carquois lié connexe (Q_A,I) tel qu'il existe un isomorphisme $A \tilde{\to} kQ_A/I$ (appelé une présentation de A) [G1]. L'algèbre A est dite triangulaire si son carquois Q_A n'a pas de cycles orientés. Pour chaque $i \in (Q_A)_o$, on notera e_i l'idempotent primitif correspondant de A, $S(i)$ le A-module simple correspondant et $P(i)$ (respectivement, $I(i)$) la couverture projective (respectivement, l'enveloppe injective) de $S(i)$. Le support d'un A-module M est l'ensemble $\{i \in (Q_A)_o \mid \text{Hom}_A(P(i),M) \neq 0\}$. L'algèbre d'un carquois lié $A = kQ/I$ peut aussi être considérée comme une k-catégorie dont l'ensemble d'objets A_o est Q_o et l'ensemble $A(x,y)$ des morphismes de x à y est le quotient de l'espace vectoriel $kQ(x,y)$ des combinaisons linéaires des chemins de x à y par le sous-espace $I(x,y) = I \cap kQ(x,y)$ [BoG].

1.2 Soit K une catégorie de Krull-Schmidt, c'est à dire une k-catégorie où les idempotents scindent. Un cycle de K est une suite de morphismes non-nuls et non-inversibles:

$$M_o \xrightarrow{f_1} M_1 \xrightarrow{f_2} M_2 \to \ldots \xrightarrow{f_m} M_m = M_o$$

où les M_i sont des objets indécomposables de K [R1]. Pour une algèbre A, la catégorie mod A est dite dirigée si elle ne contient pas de cycles. Si c'est le cas, alors A est nécessairement de représentation finie [R1]. Le carquois $\Gamma(K)$ d'une catégorie de Krull-Schmidt K a pour points les classes d'isomorphisme $[M]$ des objets indécomposables M de K et il existe une flèche $[M] \to [N]$ s'il existe un morphisme irréductible $M \to N$ dans K. Si $K = \text{mod}\, A$ ou $D^b(A)$, alors $\Gamma(K)$ est un carquois de translation. On notera τ la translation de ce carquois. Le carquois $\Gamma(\text{mod}\, A)$, noté plus brièvement Γ_A, est le carquois d'Auslander-Reiten de A [G1][R1][H2]. Un carquois de translation Γ sans flèches multiples est appelé un tube [R1] s'il contient un chemin cyclique et si sa réalisation topologique $|\Gamma| = S^1 \times \mathbb{R}_0^+$ (où S^1 est le cercle unité et $\mathbb{R}_0^+$ l'ensemble des réels non-négatifs).

Une catégorie de Krull-Schmidt K sera dite de cycles finis [AS3] si, pour chaque cycle $M_0 \to M_1 \to \dots \to M_m = M_0$ de K , les objets M_i se trouvent dans un tube de $\Gamma(K)$.

1.3 Soit A une algèbre. Un module T_A est dit inclinant (respectivement, co-inclinant) [HR1] si les conditions suivantes sont satisfaites:

(T1) $\mathrm{Ext}^2_A(T,-) = 0$ (respectivement, (T1') $\mathrm{Ext}^2_A(-,T) = 0$)

(T2) $\mathrm{Ext}^1_A(T,T) = 0$

(T3) Le nombre de facteurs directs indécomposables non-isomorphes de T_A égale le rang du groupe de Grothendieck $K_0(A)$ de A .

Soit $B = \mathrm{End}\, T_A$. On dit alors que $(B, {}_BT_A, A)$ est un triplet inclinant (respectivement, co-inclinant). Un module inclinant T_A est dit séparant si, pour tout A-module indécomposable M , on a soit $\mathrm{Hom}_A(T,M) = 0$, soit $\mathrm{Ext}^1_A(T,M) = 0$ [A2] . Des exemples de modules inclinants séparants sont les modules inclinants d'APR [APR] définis comme suit: à chaque puits i de Q_A , on associe le module inclinant

$$T_A^{(i)} = \tau^{-1}P(i) \oplus (\bigoplus_{j \neq i} P(j)) \quad .$$

Deux algèbres A et B sont dites équivalentes pour les inclinaisons et les co-inclinaisons (ou de la même classe d'inclinaison) [AS2] s'il existe une suite d'algèbres $A = A_0, A_1, \dots, A_m, A_{m+1} = B$ et une suite de modules $T^i_{A_i}$ $(0 \leq i \leq m)$ inclinants ou co-inclinants telles que $A_{i+1} = \mathrm{End}\, T^i_{A_i}$ pour tout i . Etant donné un carquois fini connexe $\vec{\Delta}$ sans cycles orientés, une algèbre A est dite pré-inclinée de type $\vec{\Delta}$ [AH] si A est équivalente pour les inclinaisons et les co-inclinaisons à $B = k\vec{\Delta}$ et, en outre, chaque $T^i_{A_i}$ est un module inclinant séparant. Si $m \leq 1$, on dit que A est inclinée [HR1] . Par exemple, si H est une algèbre héréditaire docile et T_H est un module inclinant dont tous les facteurs directs indécomposables sont pré-projectifs (ou pré-injectifs), alors l'algèbre $A = \mathrm{End}\, T_H$, appelée docile dérobée, est inclinée. Rappelons qu'une algèbre héréditaire est de représentation finie (respectivement, docile) si et seulement si le graphe sous-jacent de son carquois est un graphe de Dynkin (respectivement, un graphe Euclidien).

Il suit de [H2] que, si (B,T,A) est un triplet inclinant, alors $D^b(A) \xrightarrow{\sim} D^b(B)$, en tant que catégories triangulées. En particulier, si A est pré-inclinée de type $\vec{\Delta}$, alors $D^b(A) \xrightarrow{\sim} D^b(k\vec{\Delta})$, en tant que catégories triangulées. Happel, Rickard et Schofield [HRS] ont prouvé que pour un carquois fini connexe $\vec{\Delta}$ sans cycles orientés et une algèbre A , les conditions suivantes sont équivalentes:

(i) $D^b(A) \xrightarrow{\sim} D^b(k\vec{\Delta})$, en tant que catégories triangulées.

(ii) A est pré-inclinée de type $\vec{\Delta}$.

(iii) A et $k\vec{\Delta}$ sont de la même classe d'inclinaison.

1.4 Soient A une algèbre de dimension finie et $D = \mathrm{Hom}_k(-,k)$ la dualité usuelle sur mod A. <u>L'algèbre répétitive</u> $\hat{A}$ de A est l'algèbre:

$$\hat{A} = \begin{bmatrix} \ddots & & & & 0 \\ & A_{-1} & & & \\ & M_0 & A_0 & & \\ & & M_1 & A_1 & \\ 0 & & & \ddots & \ddots \end{bmatrix}$$

où les matrices n'ont qu'un nombre fini de coefficients non-nuls, $A_p = A$, $M_p = {}_A(DA)_A$ pour tout $p \in \mathbb{Z}$, tous les autres coefficients sont nuls. L'addition est l'addition usuelle des matrices et la multiplication est induite des applications canoniques $A \otimes_A DA \tilde{\to} DA$, $DA \otimes_A A \tilde{\to} DA$ et l'application nulle $DA \otimes_A DA \to 0$ [HW]. C'est une algèbre associative, non unifère, localement de dimension finie et auto-injective.

Happel a démontré que, si A est de dimension globale finie, alors $D^b(A)$ est équivalente, en tant que catégorie triangulée, à la catégorie stable $\underline{\mathrm{mod}}\,\hat{A}$ associée à mod $\hat{A}$ [H2]. D'autre part, si (B,T,A) est un triplet inclinant (et A est de dimension globale arbitraire), alors $\underline{\mathrm{mod}}\,\hat{A} \tilde{\to} \underline{\mathrm{mod}}\,\hat{B}$ [TW1][W] (voir aussi [H2]).

1.5 Rappelons que l'<u>extension</u> (respectivement, la <u>co-extension</u>) ponctuelle d'une algèbre A par un A-module M est l'algèbre

$$A[M] = \begin{bmatrix} A & 0 \\ M & k \end{bmatrix} \quad \text{(respectivement,} \quad [M]A = \begin{bmatrix} k & 0 \\ DM & A \end{bmatrix})$$

avec l'addition et la multiplication ordinaire des matrices. Son carquois contient Q_A en tant que sous-carquois plein et un point supplémentaire, dit d'extension, qui est une source (respectivement, de co-extension, qui est un puits). Soient A une algèbre triangulaire et $i \in (Q_A)_0$ un puits. La <u>réflexion</u> S_i^+A de A au puits i [HW] est le quotient de $A[I(i)_A]$ par l'idéal bilatère engendré par e_i. Il est clair que les algèbres répétitives de A et S_i^+A sont isomorphes. En outre A et S_i^+A sont de la même classe d'inclinaison [TW2]. Dualement, partant d'une source j de Q_A, on définit la réflexion S_j^-A de A en j.

1.6 On considère le carquois suivant, où $n_1 \leq n_2 \leq \ldots \leq n_t$:

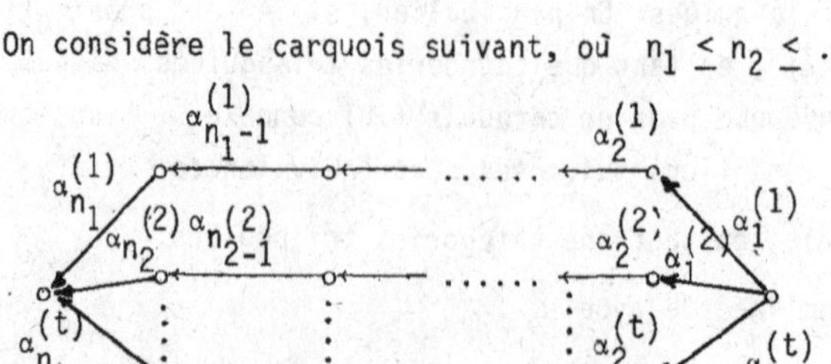

Notons $\alpha^{(s)}$ le chemin $\alpha_1^{(s)}\alpha_2^{(s)}\ldots\alpha_{n_s}^{(s)}$ $(1 \le s \le t)$ et U l'espace vectoriel de base $<\alpha^{(1)},\ldots,\alpha^{(t)}>$. Soit I un sous-espace de dimension t-2 dont l'intersection avec tout sous-espace $<\alpha^{(s)},\alpha^{(s')}>$ $(s \neq s')$ de U est nulle. L'algèbre $A = kQ/I$ est dite canonique de type $(n_1,n_2,\ldots,n_t)$ [R1] . Ringel a démontré que les algèbres dociles canoniques sont les suivantes: l'algèbre de type (2,2,2,2) où $I = <\alpha^{(1)} + \alpha^{(2)} + \alpha^{(3)}, \alpha^{(1)} + \lambda\cdot\alpha^{(2)} + \alpha^{(4)}>$ $(\lambda \in k \setminus \{0,1\})$, les algèbres de type (p,q,r) avec $\frac{1}{p} + \frac{1}{q} + \frac{1}{r} \ge 1$ où $I = <\alpha^{(1)} + \alpha^{(2)} + \alpha^{(3)}>$ et l'algèbre de type (p,q) où I = 0 [R1] . Les algèbres canoniques des types (p,q), (2,2,r), (2,3,3), (2,3,4) et (2,3,5) sont dites domestiques. Elles sont inclinées de types respectifs $\tilde{\mathbb{A}}_{p+q+1}$, $\tilde{\mathbb{D}}_{r+2}$, $\tilde{\mathbb{E}}_6$, $\tilde{\mathbb{E}}_7$ et $\tilde{\mathbb{E}}_8$. Les autres algèbres dociles canoniques, c'est à dire des types (3,3,3), (2,4,4), (2,3,6) et (2,2,2,2) sont dites tubulaires.

Soit C une algèbre docile canonique. La catégorie dérivée $D^b(C)$ est décrite dans [H2][HR2] . Si C est domestique, le carquois de $D^b(C)$ est formé de composantes transjectives $\mathcal{C}_i$ $(i \in \mathbb{Z})$ et de composantes régulières $\mathcal{R}_i$ $(i \in \mathbb{Z})$. Les composantes transjectives sont isomorphes à la bande infinie $\mathbb{Z}\Delta$, où Δ est le graphe Euclidien correspondant à C . Chaque composante régulière est formée d'une famille de tubes indexés par la droite projective $\mathbb{P}_1(k)$. En outre, si $\mathrm{Hom}_{D^b(C)}(M^\cdot,N^\cdot) \neq 0$ pour $M^\cdot$, $N^\cdot$ indécomposables et $M^\cdot$ dans $\mathcal{C}_i$ (respectivement, $\mathcal{R}_i$), alors $N^\cdot$ appartient soit à $\mathcal{C}_i$, soit à $\mathcal{R}_i$, soit à $\mathcal{C}_{i+1}$ (respectivement, soit à $\mathcal{R}_i$, soit à $\mathcal{C}_{i+1}$, soit à $\mathcal{R}_{i+1}$). Pour une algèbre tubulaire canonique C , le carquois de $D^b(C)$ est formé de familles $\mathcal{R}_q$, $q \in \mathbb{Q}$, où chaque $\mathcal{R}_q$ est une famille de tubes indexés par $\mathbb{P}_1(k)$. En outre, si $\mathrm{Hom}_{D^b(C)}(M^\cdot,N^\cdot) \neq 0$ pour $M^\cdot$, $N^\cdot$ indécomposables et $M^\cdot$ dans $\mathcal{R}_q$, alors $N^\cdot$ appartient à $\mathcal{R}_p$ pour un $q \le p \le q+3$. Dans les deux cas, la catégorie $D^b(C)$ est de cycles finis. Un autre cas où $D^b(C)$ est (trivialement) de cycles finis est le suivant: soit C une algèbre héréditaire de représentation finie et Δ le graphe sous-jacent de son carquois, alors le carquois de $D^b(C)$ est isomorphe à $\mathbb{Z}\Delta$ [H2] . En particulier, $D^b(C)$ est dirigée.

2. Agrandissement d'une algèbre par branches:

2.1 Le fil conducteur à travers les résultats de classification est la notion d'agrandissement d'une algèbre par branches. La branche complète est un carquois lié infini (Q,I) défini comme suit. L'ensemble de points Q_0 est l'ensemble des mots sur l'alphabet formé de deux lettres α et β :
$Q_0 = \{x = \alpha^{n_1}\beta^{m_1}\ldots\alpha^{n_t}\beta^{m_t} \mid t \in \mathbb{N},\ n_i,m_j \in \mathbb{Z},\ 1 \le i,j \le t\}$. Pour chaque $x \in Q_0$, on définit deux flèches de source x : $\alpha_x : x \to x\alpha$ et $\beta_x : x \to x\beta$ (cela équivaut

à dire que tout $x \in Q_o$ est le but de deux flèches $\alpha_{x\alpha^{-1}} : x\alpha^{-1} \to x$ et $\beta_{x\beta^{-1}} : x\beta^{-1} \to x$). Enfin, I est l'idéal de kQ engendré par tous les mots des formes $\alpha_x\beta_{x\alpha}$ et $\beta_x\alpha_{x\beta}$ (pour $x \in Q_o$). Les flèches de la forme α_x (respectivement, β_x) seront simplement appelées des α-flèches (respectivement, des β-flèches).

Une branche est par définition un sous-carquois lié plein, fini et connexe de la branche complète. Il résulte de [AH] qu'une algèbre A est pré-inclinée de type $\mathbb{A}_n$ si et seulement si $A \xrightarrow{\sim} kQ/I$, où le carquois lié (Q,I) est une branche. Une racine d'extension (respectivement, de co-extension) dans une branche est un sommet a qui n'est pas la source (respectivement, pas le but) d'une α-flèche. La branche est alors appelée une branche d'extension (respectivement, de co-extension) en a . Ainsi une branche d'extension en a est un sous-carquois lié, plein, fini et connexe, contenant a , de l'arbre infini suivant, lié par toutes les relations possibles des formes $\alpha\beta = 0$ et $\beta\alpha = 0$:

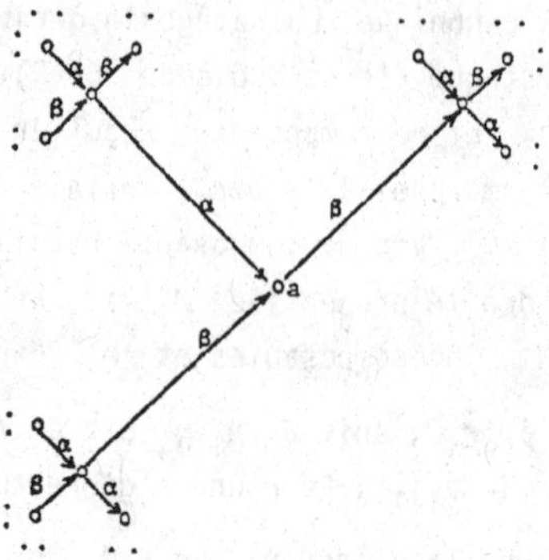

Une branche d'extension de la forme $\ldots \circ \xrightarrow{\alpha} \circ \xrightarrow{\alpha} \ldots \xrightarrow{\alpha} \circ\ a$ est appelée une droite dirigée d'extension. On définit dualement une droite dirigée de co-extension. Le nombre de points d'une branche K est appelé sa longueur et noté $|K|$. On conviendra de considérer le carquois vide comme une branche de longueur zéro.

Soient $A = kQ/I$ l'algèbre d'un carquois lié (Q,I) et (Q',I') un sous-carquois lié plein de (Q,I) contenant une source a . On dit que A s'obtient à partir de kQ'/I' en enracinant une branche d'extension (Q",I") en a si (Q",I") est un sous-carquois lié plein de (Q,I) tel que:

(1) $Q'_0 \cap Q''_0 = \{a\}$, $Q'_0 \cup Q''_0 = Q_0$

(2) I est engendré par I', I" et tous les chemins $\beta\gamma$, où $\beta \in Q''_1$ a pour but a , et $\gamma \in Q'_1$ a pour source a .

On définit dualement l'enracinement de branches de co-extension.

Soient C une algèbre, et $E_1,\ldots,E_t$ des C-modules indécomposables deux à deux non-isomorphes. Pour $1 \leq i \leq t$, soient K_i une branche d'extension en a_i et

K_i' une branche de co-extension en a_i' (K_i ou K_i' peut être vide). On définit par récurrence l'agrandissement A de C aux modules E_i par les branches d'extension K_i et de co-extension K_i' comme suit. L'algèbre $C[E_1,K_1]$ s'obtient de l'extension ponctuelle $C[E_1]$ en enracinant la branche K_1 au point d'extension a_1. Pour $1 < j \leq t$, $C[E_i,K_i]_{i=1}^{j}$ s'obtient de l'extension ponctuelle $(C[E_i,K_i]_{i=1}^{j-1})[E_j]$ en enracinant la branche K_j au point d'extension a_j. Alors $B = C[E_i,K_i]_{i=1}^{t}$ est appelée l'extension de C aux modules E_i par les branches d'extension K_i. On définit de même la co-extension $B' = {}_{i=1}^{t}[E_i,K_i']C$. Notons maintenant, pour $1 \leq i \leq t$, E_i' l'unique B-module indécomposable dont la restriction à C égale E_i et la restriction à K_i est l'unique K_i-module indécomposable de support le chemin maximal non-nul de but a (formé de α-flèches). Alors $[E_1',K_1']B$ s'obtient de la co-extension ponctuelle $[E_1']B$ en enracinant K_1' au point de co-extension a_1'. Pour $1 < j \leq t$, ${}_{i=1}^{j}[E_i',K_i']B$ s'obtient de $[E_j']({}_{i=1}^{j-1}[E_i',K_i']B)$ en enracinant K_j' au point de co-extension a_j'. Alors $A = {}_{i=1}^{t}[E_i',K_i']B$ est l'agrandissement de C. On dit que C est le coeur de A. C'est une sous-catégorie pleine et convexe de A.

La conjecture générale est que toute algèbre pré-inclinée ou équivalente pour les inclinaisons et les co-inclinaisons à une algèbre tubulaire canonique est un agrandissement par branches de certaines algèbres plus élémentaires.

2.2 Dans cette section, nous nous limiterons au cas suivant. Soit C une algèbre docile dérobée, de famille tubulaire $(\mathcal{T}_\lambda)_{\lambda \in \mathbb{P}_1(k)}$. Nous noterons r_λ le rang du tube (stable) $\mathcal{T}_\lambda$ ($\lambda \in \mathbb{P}_1(k)$). Soient $E_1,E_2,\ldots,E_t$ des C-modules simples réguliers deux à deux non-isomorphes, $K_1,K_2,\ldots,K_t$ des branches d'extension et $K_1',K_2',\ldots,K_t'$ des branches de co-extension. On définit le type tubulaire $n_A = (n_\lambda)_{\lambda \in \mathbb{P}_1(k)}$ de l'agrandissement $A = {}_{i=1}^{t}[E_i',K_i']\,C\,[E_i,K_i]_{i=1}^{t}$ par:

$$n_\lambda = r_\lambda + \sum_{E_i \in \mathcal{T}_\lambda} (|K_i| + |K_i'|) \quad .$$

On conviendra d'écrire, au lieu de $(n_\lambda)_{\lambda \in \mathbb{P}_1(k)}$, la suite finie ayant au moins deux n_λ et comprenant tous ceux qui sont plus grands que 1, ordonnés suivant un ordre non-décroissant. Un agrandissement A d'une algèbre docile dérobée C en des modules simples réguliers est dit domestique (respectivement, tubulaire) si son type tubulaire est une des suites suivantes: (p,q), $(2,2,r)$, $(2,3,3)$, $(2,3,4)$ ou $(2,3,5)$ (respectivement, $(3,3,3)$, $(2,4,4)$, $(2,3,6)$ ou $(2,2,2,2)$).

2.3 Ces concepts généralisent ceux de [R1]: une branche tronquée en a (branche dans la terminologie de [R1]) est un sous-carquois lié plein, fini et connexe, contenant a, de l'arbre infini suivant, lié par toutes les relations de la forme $\alpha\beta = 0$:

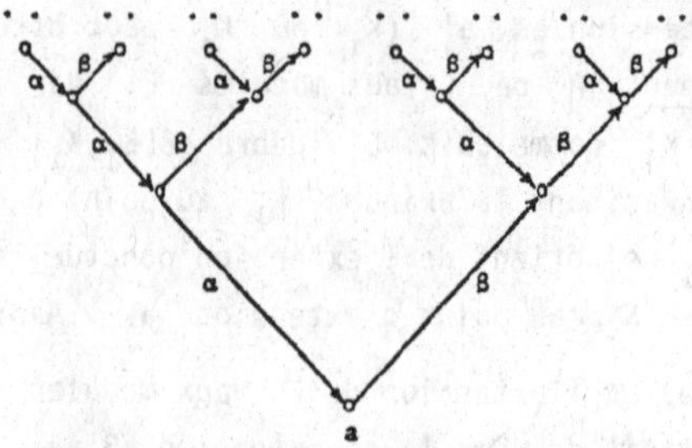

En d'autres termes, une branche tronquée est une branche telle que le but d'une β-flèche n'est pas la source d'une α-flèche. Si C est une algèbre docile dérobée, $E_1,\ldots,E_t$ des C-modules simples réguliers deux à deux non-isomorphes et $K_1,\ldots,K_t$ des branches tronquées, l'extension par branches $B = C[E_i,K_i]_{i=1}^{t}$ est une extension tubulaire au sens de [R1](4.7). Ringel a démontré que, si A est une extension (respectivement, co-extension) domestique d'une algèbre docile dérobée en des modules simples réguliers par des branches tronquées, alors A est une algèbre inclinée de type Euclidien ayant une tranche complète dans sa composante pré-injective (respectivement, pré-projective) et aucun projectif (respectivement, injectif) dans cette composante. Réciproquement, toute algèbre inclinée de représentation infinie de type Euclidien est d'une de ces formes [R1](4.9).

<u>2.4 THEOREME</u> [AS3] . Une algèbre A est pré-inclinée de type Euclidien (respectivement, équivalente pour les inclinaisons et les co-inclinaisons à une algèbre tubulaire canonique) et de représentation infinie si et seulement si A est isomorphe à un agrandissement domestique (respectivement, tubulaire) d'une algèbre docile dérobée en des modules simples réguliers. En outre, dans ce cas, n_A égale le type tubulaire de l'algèbre héréditaire (respectivement, tubulaire canonique) correspondante.

<u>Démonstration</u>. La nécessité suivra de (3.1) si A est pré-inclinée de type $\tilde{A}_m$ et de (5.3) dans les autres cas. Nous démontrons ici la suffisance. Soit A un agrandissement domestique (respectivement, tubulaire) de l'algèbre docile dérobée C en des modules simples réguliers. Il suffit (d'après (1.3)) de prouver que A est de la classe d'inclinaison d'une algèbre héréditaire de type Euclidien (respectivement, d'une algèbre tubulaire canonique). Pour commencer, on trouve une algèbre B de la classe d'inclinaison de A , qui est un agrandissement de C par des droites dirigées et telle que $n_B = n_A$. En effet, on applique l'algorithme de [AH](2.3) pour effacer successivement les relations sur les branches. Comme à chaque étape les calculs ont lieu à l'intérieur d'une branche donnée, le type tubulaire n'est pas affecté.

Nous montrons maintenant qu'il existe une algèbre D , de la classe d'inclinaison de B (donc de A), qui est une extension de C par des droites dirigées, et telle que $n_D = n_B$. En effet, soit K' une droite de co-extension de B et i son puits. Pour réfléchir en i , on remarque que, d'après (2.1), le support de la restriction de $I(i)_B$ à K' égale K' , le support de sa restriction à C est un C-module simple régulier E et, s'il existe aussi une droite d'extension K correspondant à E , alors la restriction de $I(i)$ à K est K . Enfin, ce support ne coupe pas les autres droites d'extension et de co-extension. Ainsi, dans l'algèbre S_i^+B , le puits i de B est remplacé par une source, qui est point d'extension de $B/\langle e_i\rangle$ et le module d'extension est dans le tube de $B/\langle e_i\rangle$ contenant E . Par conséquent, $n_B = n_{S_i^+B}$. Si on applique successivement ce processus aux points de K' , on obtient une algèbre de la même classe d'inclinaison que B , de même type tubulaire, mais dans laquelle K' est remplacée par une droite dirigée d'extension. On obtient D en appliquant ce processus successivement à toutes les droites de co-extension.

Supposons que A est domestique. Comme les droites dirigées sont des branches tronquées, il résulte de (2.3) que D est une algèbre inclinée de type Euclidien ayant une tranche complète dans sa composante pré-injective. Par contre, si A est tubulaire, D est, par définition [R1](5), une algèbre tubulaire. Mais alors il suit de [HR2](1) qu'elle est équivalente , pour les inclinaisons et les co-inclinaisons, à une algèbre tubulaire canonique. Dans les deux cas, le type tubulaire est préservé. Cela achève la démonstration.

<u>Exemple</u>: Soit C l'algèbre docile dérobée de type $\tilde{\mathbb{E}}_6$ donnée par le carquois:

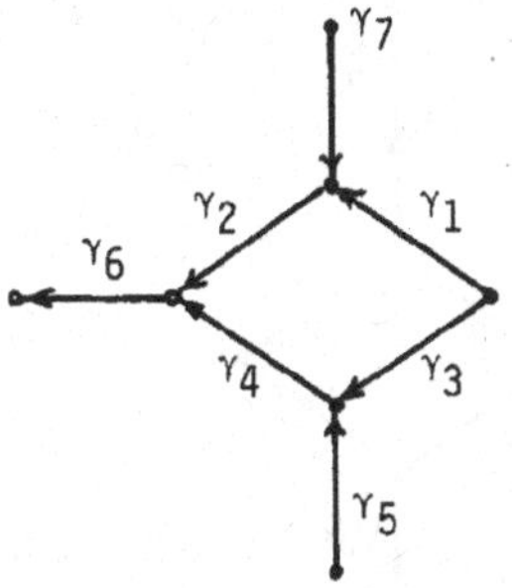

lié par $\gamma_1\gamma_2 = \gamma_3\gamma_4$. On a $n_C = (2,3,3)$. Soit E_C le module simple régulier de vecteur-dimension $\begin{smallmatrix} & & 0 & \\ & & 1 & \\ 0 & 1 & & 0 \\ & & 0 & \\ & & 0 & \end{smallmatrix}$. On enracine en E la branche d'extension de longueur un, et la branche de co-extension de longueur deux formée d'une β-flèche. Cela donne l'algèbre A du carquois:

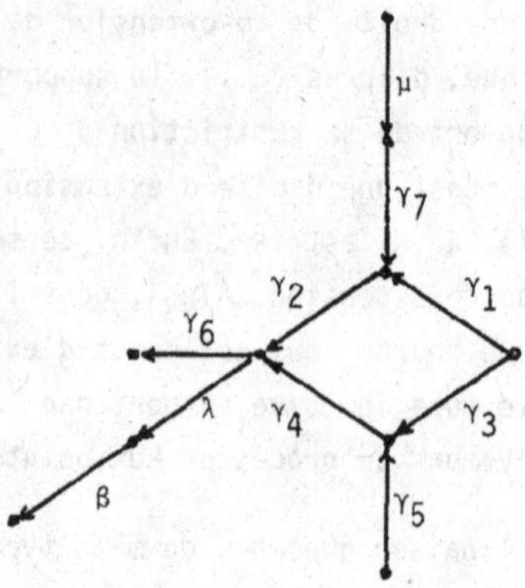

lié par $\gamma_1\gamma_2 = \gamma_3\gamma_4$, $\gamma_4\lambda = 0$, $\lambda\beta = 0$, $\mu\gamma_7\gamma_2\gamma_6 = 0$. C'est un agrandissement de C de type tubulaire $n_A = (2,3,6)$. Donc A est dans la classe d'inclinaison de l'algèbre tubulaire canonique de ce type. Afin de voir ceci, on commence par appliquer un module inclinant d'APR correspondant au but de β . Cela donne une algèbre B , de même carquois que A , lié seulement par $\gamma_1\gamma_2 = \gamma_3\gamma_4$, $\gamma_4\lambda = 0$ et $\mu\gamma_7\gamma_2\gamma_6 = 0$. On applique ensuite successivement des réflexions correspondant aux buts de β et λ . On obtient l'algèbre D du carquois:

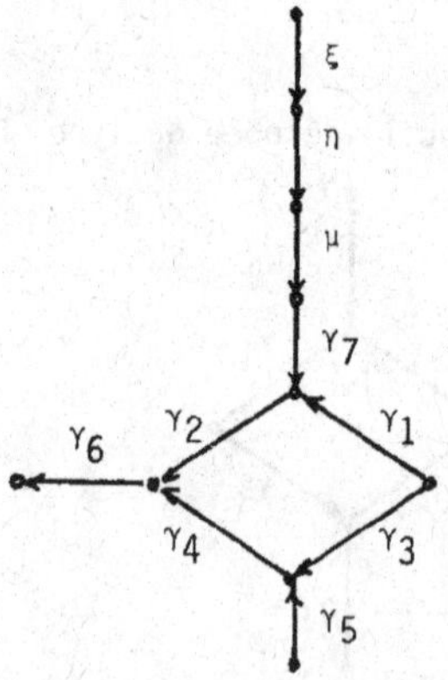

lié par $\gamma_1\gamma_2 = \gamma_3\gamma_4$ et $\mu\gamma_7\gamma_2\gamma_6 = 0$. C'est bien une algèbre tubulaire de type (2,3,6).

3. Algèbres pré-inclinées de type $\tilde{\mathbb{A}}_m$:

3.1 Dans cette section, nous exposons la classification des algèbres pré-inclinées de type $\tilde{\mathbb{A}}_m$ $(m \geq 1)$ [AS1] :

<u>THEOREME</u>. Une algèbre A est pré-inclinée de type $\tilde{\mathbb{A}}_m$ si et seulement s'il existe une présentation $A \tilde{\to} kQ/I$ de A par un carquois lié (Q,I) de m+1 points

tel que:

(R1) Le nombre de flèches de source ou de but donné est au plus égal à deux.

(R2) Pour chaque flèche α, il existe au plus une flèche β et une flèche γ telles que $\alpha\beta$ et $\gamma\alpha$ n'appartiennent pas à I.

(R3) I est engendré par un ensemble de chemins (relations-zéro) de longueur deux.

(R4) Pour chaque flèche α, il existe au plus une flèche ξ et une flèche η telles que $\alpha\xi$ et $\eta\alpha$ appartiennent à I.

(R5) Q contient un cycle unique (non-orienté) C.

(R6) Le nombre de relations-zéro de C dans le sens des aiguilles d'une montre égale le nombre de relations-zéro dans le sens inverse.

En d'autres termes, les algèbres pré-inclinées de type $\tilde{\mathbb{A}}_m$ sont des agrandissements par branches des algèbres dont le carquois ordinaire est un cycle non-orienté, lié par des relations-zéro de longueur deux et satisfaisant la condition (R6).

Exemple: On considère l'algèbre A du carquois:

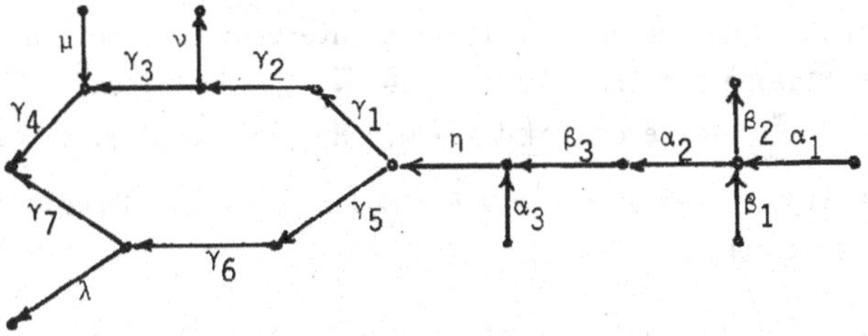

lié par $I = \langle\gamma_1\gamma_2, \gamma_2\gamma_3, \mu\gamma_4, \gamma_5\gamma_6, \eta\gamma_5, \beta_3\eta, \alpha_2\beta_3, \beta_1\alpha_2, \alpha_1\beta_2\rangle$. On voit de suite que A satisfait les conditions (R1) à (R6) et donc est une algèbre pré-inclinée de type $\tilde{\mathbb{A}}_{16}$. Observons que A est de représentation finie.

3.2 Il suit immédiatement de l'énoncé du théorème que les algèbres pré-inclinées de type $\tilde{\mathbb{A}}_m$ sont bisérielles. Rappelons qu'une k-catégorie localement bornée Λ est dite bisérielle si le radical de tout Λ-module projectif indécomposable (à droite ou à gauche) qui n'est pas unisériel est la somme de deux sous-modules unisériels dont l'intersection est simple ou nulle. Λ est dite bisérielle spéciale [SW] si elle admet une présentation $\Lambda \xrightarrow{\sim} kQ/I$ telle que (Q,I) satisfait les conditions (R1) et (R2). Toute algèbre bisérielle spéciale est bisérielle, et toute algèbre bisérielle de représentation finie est bisérielle spéciale [SW]. Rappelons aussi que Λ est dite satisfaire $\alpha'(\Lambda) \leq 2$ si, pour tout $x \in (\Gamma_\Lambda)_0$, il existe au plus deux σ-orbites de flèches ayant x pour source ou pour but [H1](5). On sait qu'une algèbre Λ est pré-inclinée de type $\mathbb{A}_n$ si et seulement si elle est simplement connexe de représentation finie, bisérielle spéciale et satisfait $\alpha'(\Lambda) \leq 2$. On en déduit:

<u>LEMME</u>. Une catégorie localement bornée Λ est bisérielle spéciale avec $\alpha'(\Lambda) \leq 2$ si et seulement si elle admet une présentation $\Lambda \tilde{\to} kQ/I$ où (Q,I) satisfait les conditions (R1) (R2) (R3) et (R4).

<u>Démonstration</u>: Si Λ est bisérielle spéciale et $\alpha'(\Lambda) \leq 2$, alors, par définition, $\Lambda \tilde{\to} kQ/I$, où (Q,I) satisfait (R1) et (R2). Comme il est clair que tout indécomposable projectif-injectif est unisériel, alors I est engendré par un ensemble de chemins. On considère un revêtement galoisien $F : \tilde{\Lambda} \to \Lambda$ avec le groupe fondamental libre de Q [G2]. Le carquois $\tilde{Q}$ de $\tilde{\Lambda}$ est un arbre lié par l'ensemble $\tilde{I}$ des chemins w de $\tilde{Q}$ tels que $F(w) \in I$. Comme le foncteur de rabaissement $F_\lambda : \mathrm{mod}\,\tilde{\Lambda} \to \mathrm{mod}\,\Lambda$ associé à F préserve les suites d'Auslander-Reiten, on a $\alpha'(\tilde{\Lambda}) \leq 2$. Par conséquent $(\tilde{Q},\tilde{I})$ satisfait (R3) et (R4). Il en est donc de même de (Q,I) . Réciproquement, si (Q,I) satisfait ces conditions, alors Λ est bisérielle spéciale et toute sous-catégorie finie K du revêtement universel $\tilde{\Lambda}$ de Λ est pré-inclinée de type $\mathbb{A}_n$. En particulier, $\alpha'(K) \leq 2$. Il suit alors de [DS][WW] que $\alpha'(\Lambda) \leq 2$.

3.3 La démonstration du théorème (3.1) fait intervenir les notions précédentes. Soit en effet A une algèbre pré-inclinée de type $\tilde{\mathbb{A}}_m$. Alors il existe une algèbre héréditaire H de type $\tilde{\mathbb{A}}_m$ telle que $\underline{\mathrm{mod}}\,\hat{A} \tilde{\to} \underline{\mathrm{mod}}\,\hat{H}$ (voir (1.3)). On démontre:

(a) Soit A une algèbre telle que $\underline{\mathrm{mod}}\,\hat{A} \tilde{\to} \underline{\mathrm{mod}}\,\hat{H}$ pour H héréditaire de type $\tilde{\mathbb{A}}_m$. Alors $\hat{A}$ est bisérielle.

Il suit de [H2](7.3) et [S] que A est triangulaire. On prouve:

(b) Soit A une algèbre triangulaire telle que $\hat{A}$ est bisérielle. Alors A est bisérielle spéciale avec $\alpha'(A) \leq 2$.

Remarquons qu'il s'ensuit immédiatement que $\hat{A}$ est en fait bisérielle spéciale. Nous avons montré que A admet une présentation $A \tilde{\to} kQ/I$ où le carquois lié (Q,I) satisfait les conditions (R1) à (R4). En outre, Q n'est pas un arbre puisque sinon A serait pré-inclinée de type $\mathbb{A}_n$ et alors $\hat{A}$ serait localement de représentation finie [AHR], une absurdité, puisque $\underline{\mathrm{mod}}\,\hat{A} \tilde{\to} \underline{\mathrm{mod}}\,\hat{H}$ avec H de représentation infinie. Par conséquent, Q contient au moins un cycle non-orienté. D'autre part, comme $D^b(H) \tilde{\to} \underline{\mathrm{mod}}\,\hat{H}$ est de cycles finis, il en est de même de $D^b(A)$. On démontre:

(c) Soit A une algèbre triangulaire telle que $D^b(A)$ est de cycles finis. Alors toute sous-catégorie pleine K de A qui est bisérielle spéciale et telle que $\alpha'(K) \leq 2$ contient au plus un cycle.

En particulier, Q contient exactement un cycle non-orienté C . Il résulte alors de [S] que la condition (R6) est satisfaite. Enfin, comme A est pré-inclinée de type $\tilde{\mathbb{A}}_m$, Q a $m+1$ points. Cela achève la démonstration de la nécessité. La suffisance se démontre directement, en construisant une suite de modules inclinants séparants aboutissant à une algèbre héréditaire de type $\tilde{\mathbb{A}}_m$.

3.4 Nous caractérisons également les algèbres pré-inclinées de type $\tilde{\mathbb{A}}_m$ de représentation finie. Un chemin $\gamma: x_1 \to \ldots \to x_t$ de Γ_A est dit <u>sectionnel</u> [B](2.1). si $x_i \neq \tau x_{i+2}$ pour $1 \leq i \leq t-2$. Un chemin sectionnel γ est dit <u>maximal</u> s'il n'est pas un sous-chemin d'un chemin sectionnel $\gamma' \neq \gamma$. Nous pouvons énoncer:

<u>THEOREME</u> [AS1]. Soit A une algèbre. Les conditions suivantes sont équivalentes:

(i) A est pré-inclinée de type $\tilde{\mathbb{A}}_m$ de représentation finie

(ii) A satisfait les conditions (R1) à (R6) et:

(R7) Le cycle C est lié par au moins une relation-zéro

(iii) A satisfait les conditions suivantes:

(a) mod A est dirigée avec $m+1$ projectifs

(b) $\alpha'(A) \leq 2$

(c) Le groupe d'homotopie $\pi_1(\Gamma_A)$ de Γ_A (voir [BoG] (1.2)) est isomorphe à $\mathbb{Z}$.

(d) Pour chaque représentant w d'un générateur de $\pi_1(\Gamma_A)$, le nombre de chemins sectionnels maximaux de w dans le sens des aiguilles d'une montre égale le nombre de chemins sectionnels maximaux dans le sens inverse.

3.5 Soit A une algèbre telle que mod A est dirigée. Alors A admet une présentation $A \xrightarrow{\sim} kQ/I$ telle que le carquois lié (Q,I) satisfait les conditions (R1) à (R5) si et seulement si $\alpha'(A) \leq 2$ et $\pi_1(\Gamma_A) \xrightarrow{\sim} \mathbb{Z}$. En effet, supposons que $A \xrightarrow{\sim} kQ/I$ où (Q,I) satisfait (R1) à (R5). Comme mod A est dirigée, A est de représentation finie. Il suit du lemme (3.2) que A est bisériélle spéciale avec $\alpha'(A) \leq 2$. Enfin, (R5) et [BrG](1.2) [MP](4.3) entraînent que $\pi_1(\Gamma_A) \xrightarrow{\sim} \mathbb{Z}$. Réciproquement si $\alpha'(A) \leq 2$ et $\pi_1(\Gamma_A) \xrightarrow{\sim} \mathbb{Z}$, il suit encore de [BrG](1.2) [MP](4.3) que le carquois Q de A a exactement un cycle non-orienté. Cela donne (R5). Comme mod A est dirigée, A est de représentation finie. Il résulte de [SW](1.2) que A est bisérielle spéciale. D'après le lemme (3.2), il existe une présentation $A \xrightarrow{\sim} kQ/I$ de A telle que (Q,I) satisfait (R1) à (R4).

Soient A une algèbre satisfaisant les conditions équivalentes précédentes et w une allée de Γ_A. On notera $\mu_-(w)$ (respectivement, $\mu_+(w)$) le nombre de chemins sectionnels maximaux de w dans le sens des aiguilles d'une montre (respectivement, dans le sens inverse). On prouve qu'alors, si w est un représentant d'un générateur du groupe cyclique infini $\pi_1(\Gamma_A)$, la différence entre le nombre de relations-zéro sur le cycle unique de (R5) dans le sens des aiguilles d'une montre et dans le sens inverse égale, en valeur absolue, $|\mu_+(w) - \mu_-(w)|$.

Le théorème suit directement de ces remarques. Il est évident que (i) implique (ii), et il résulte de [SW] que (ii) implique (i). Soit A une algèbre pré-inclinée de type $\tilde{\mathbb{A}}_m$ de représentation finie. Il suit alors de [H2](7.6) que $\operatorname{mod} A$ est dirigée. Le théorème (3.1) et le lemme (3.2) entraînent $\alpha'(A) \leq 2$. Les conditions (c) et (d) suivent alors des remarques précédentes. Donc (i) implique (iii). Enfin, (iii) implique (ii) résulte aussi des remarques précédentes.

4. Algèbres simplement connexes:

4.1 Dans [BoG], Bongartz et Gabriel définissent une notion de connexité simple pour les algèbres de représentation finie. Dans cette section, nous introduisons une généralisation naturelle de cette notion pour les algèbres triangulaires qui ne sont pas nécessairement de représentation finie. Rappelons d'abord la notion de groupe fondamental d'un carquois lié [Gr][MP]. Soit (Q,I) un carquois lié connexe. Une relation $\rho = \sum_{j=1}^{m} \lambda_j w_j \in I(x,y)$ est dite minimale si $m \geq 2$ et si, pour tout sous-ensemble propre J de $\{1,2,\ldots,m\}$, $\sum_{j \in J} \lambda_j w_j \notin I$. Notons $m(I)$ l'ensemble des relations minimales de l'idéal I , et $\pi_1(Q,x_o)$ le groupe fondamental du carquois Q au point x_o . Soit $N(Q,m(I),x_o)$ le sous-groupe distingué de $\pi_1(Q,x_o)$ engendré par tous les éléments de la forme $[\gamma v u^{-1} \gamma^{-1}]$ où γ est une allée de x_o à x et u,v sont des chemins de x à y tels qu'il existe un élément $\sum_{j=1}^{m} \lambda_j w_j \in m(I)$ avec $u = w_j$ et $v = w_l$ pour $1 \leq j,l \leq m$. Le groupe fondamental $\pi_1(Q,I)$ de (Q,I) est le groupe quotient:

$$\pi_1(Q,I) = \pi_1(Q,x_o)/N(Q,m(I),x_o) .$$

Une algèbre triangulaire A est dite simplement connexe [AS2] si, pour toute présentation $A \cong kQ/I$ de A , le groupe fondamental $\pi_1(Q,I)$ du carquois lié (Q,I) est trivial.

Il suit directement de [BrG](1.2) [MP](4.3) que, si A est triangulaire et de représentation finie, alors, pour toute présentation $A \cong kQ/I$, on a $\pi_1(Q,I) \cong \pi_1(\Gamma_A)$. Par conséquent une algèbre de représentation finie est simplement connexe si et seulement si elle l'est au sens de [BoG].

Exemple : L'algèbre A du carquois:

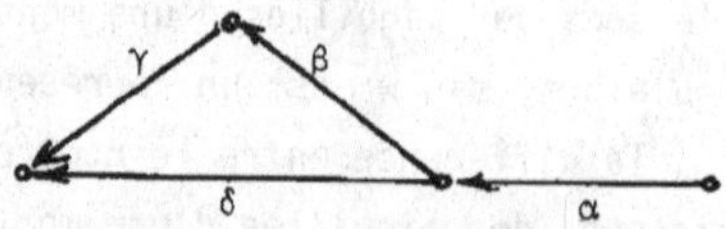

lié par $\alpha\beta\gamma = \alpha\delta$ n'est pas simplement connexe. En effet, on obtient une nouvelle présentation $A \tilde{\to} kQ/I$ de A en remplaçant δ par $\delta' = \delta - \beta\gamma$. Alors $I = \langle\alpha\delta'\rangle$ et $\pi_1(Q,I) \tilde{\to} \mathbb{Z}$. Par contre, l'algèbre B du carquois:

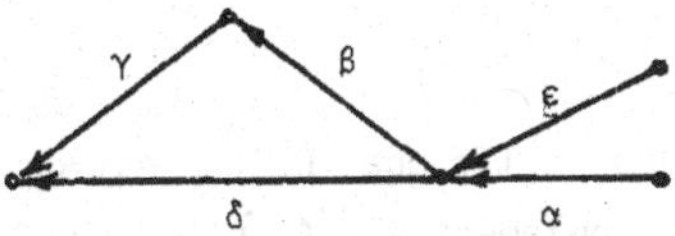

lié par $\alpha\beta = \alpha\beta\gamma$, $\varepsilon\delta = 0$ est simplement connexe.

4.2 Un critère pratique de connexité simple pour les algèbres de représentation finie est la condition de séparation de Bautista, Larriôn et Salmerôn [BLS]. Un A-module indécomposable projectif $P(i)$ est dit de <u>radical séparé</u> si les supports de deux facteurs indécomposables non-isomorphes quelconques de rad $P(i)$ sont contenus dans deux composantes connexes distinctes du sous-carquois plein de Q_A obtenu en omettant tous les points j tels qu'il existe un chemin de j à i. Si chaque indécomposable projectif est de radical séparé, A est dite satisfaire la <u>condition de séparation</u>. Une algèbre triangulaire de représentation finie satisfait la condition de séparation si et seulement si elle est simplement connexe [BLS].

Rappelons qu'une algèbre A est dite <u>Schurienne</u> si, pour toute paire $x,y \in A_0$, $\dim_k A(x,y) \leq 1$. Une algèbre Schurienne A est dite <u>sans $\tilde{\mathbb{A}}$</u> si A ne contient pas de sous-catégorie pleine $B \tilde{\to} kQ$, où le graphe sous-jacent de Q est $\tilde{\mathbb{A}}_m$ (pour un $m \geq 1$).

<u>LEMME</u>. Si A est une algèbre Schurienne, sans $\tilde{\mathbb{A}}$ et simplement connexe, alors elle satisfait la condition de séparation.

<u>Démonstration</u>. En effet, A étant Schurienne et simplement connexe, il résulte de [BrG](2.3) que le premier groupe d'homologie de A s'annule. Mais alors, puisque A est sans $\tilde{\mathbb{A}}$, il suit de [Bol](2.3) [BrG](2.9) que A satisfait la condition de séparation.

Il existe des algèbres simplement connexes qui ne satisfont pas la condition de séparation. C'est le cas par exemple de l'algèbre du carquois:

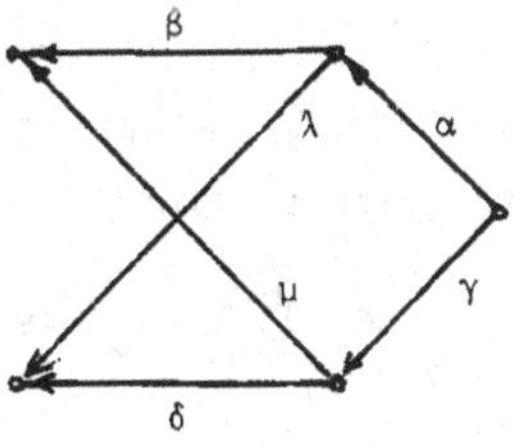

lié par $\alpha\beta = \gamma\mu$, $\alpha\lambda = \gamma\delta$. Cet exemple montre aussi qu'il n'est pas toujours vrai qu'une sous-catégorie pleine et convexe d'une algèbre simplement connexe est simplement connexe (contrairement au cas de représentation finie [BrG](2.8)).

4.3 Nous donnons un critère de connexité simple pour les algèbres dont la catégorie dérivée est de cycles finis:

THEOREME [AS2]. Soit A une algèbre telle que $D^b(A)$ est de cycles finis. Alors A est simplement connexe si et seulement si A n'est pas pré-inclinée de type $\tilde{\mathbb{A}}_m$ $(m \geq 1)$.

Rappelons que toute algèbre pré-inclinée de type Dynkin est simplement connexe de représentation finie [Al](3.5). Une conséquence immédiate du théorème précédent est que toute algèbre pré-inclinée de type Euclidien $\tilde{\mathbb{D}}_n$ $(n \geq 4)$ ou $\tilde{\mathbb{E}}_p$ $(p = 6, 7$ ou $8)$ est simplement connexe. De même, toute algèbre équivalente pour les inclinaisons et les co-inclinaisons à une algèbre tubulaire canonique est simplement connexe.

La démonstration utilise essentiellement le théorème (3.1). En effet, il suit immédiatement de ce théorème que, si A est pré-inclinée de type $\tilde{\mathbb{A}}_m$, alors il existe une présentation $A \cong kQ/I$ telle que $\pi_1(Q,I) \cong \mathbb{Z}$. Réciproquement, supposons que A n'est pas simplement connexe. Comme $D^b(A)$ est de cycles finis, A est triangulaire (voir (5.1)). Soit $A \cong kQ/I$ une présentation de A telle que $\pi_1(Q,I) \neq 0$. Alors (Q,I) contient une allée fermée w de longueur minimale qui n'est pas contractile. Soit C la sous-catégorie de A formée par les objets et les morphismes de w. Au moyen de calculs combinatoires longs et fastidieux, on prouve que C est une sous-catégorie pleine de A, liée par des relations-zéro de longueur deux. On montre ensuite que le nombre de relations-zéro dans le sens des aiguilles d'une montre égale le nombre de relations-zéro dans le sens inverse. On prouve alors que C est convexe dans A et enfin que A est un agrandissement de C par des branches, ce qui achève la démonstration.

Notre théorème montre que la classification des algèbres dont la catégorie dérivée est de cycles finis se scinde en deux cas: celui où l'algèbre est pré-inclinée de type $\tilde{\mathbb{A}}_m$, déjà résolu, celui où l'algèbre est simplement connexe, qui sera résolu dans les deux sections suivantes.

5. Catégories dérivées de cycles finis:

5.1 Notre objectif est maintenant de caractériser les algèbres A dont la catégorie dérivée $D^b(A)$ est de cycles finis (voir (1.6)).

LEMME. Soit A une algèbre telle que $D^b(A)$ est de cycles finis. Alors A est triangulaire et $\operatorname{mod} \hat{A}$ est de cycles finis.

Démonstration. Si A n'est pas triangulaire, elle contient un cycle orienté

$a_0 \to a_1 \to \dots \to a_t = a_0$ dans son carquois ordinaire. Pour chaque $1 \leq i \leq t$, soit M_i le module unisériel de longueur deux dont la coiffe est $S(a_{i-1})$ et le socle $S(a_i)$.

La longueur d'un complexe indécomposable borné $X^{\cdot} = (X^{\ell}, d^{\ell})_{\ell \in \mathbb{Z}}$ est par définition la cardinalité de l'ensemble fini $\{\ell \in \mathbb{Z} \mid d^{\ell} \neq 0\}$. Soient maintenant $j \geq 1$ et $0 \leq i < t$ tels que $j \equiv i$ (modulo t). Il existe un complexe indécomposable $X_j^{\cdot}$ de longueur $j-1$ de la forme:

$$\dots \to 0 \to M_i \to M_{i-1} \to \dots \to M_1 \to 0 \to \dots$$

où M_1 est en degré zéro, et les différentielles sont les morphismes évidents. Il est clair que l'on a une suite de monomorphismes $g_j^{\cdot} : X_j^{\cdot} \to X_{j+1}^{\cdot}$. Notons aussi $X^{\cdot}$ le complexe indécomposable de longueur nulle ayant M_t en degré zéro. L'homomorphisme $M_1 \to M_t$ d'image $S(a_0)$ induit un morphisme $f^{\cdot} : X_1^{\cdot} \to X^{\cdot}$. D'autre part, pour tout $j \geq 1$, il existe un morphisme $h_j^{\cdot} : X_j^{\cdot} \to X^{\cdot}$ tel que l'on a $f^{\cdot} = h_{j+1}^{\cdot} g_j^{\cdot} \dots g_1^{\cdot}$. Comme $f^{\cdot}$ se trouve sur le cycle orienté de complexes indécomposables de longueur nulle définis par les modules M_i $(1 \leq i \leq t)$, on en déduit une contradiction à l'hypothèse que $D^b(A)$ est de cycles finis.

Par conséquent A est triangulaire et en particulier de dimension globale finie. Donc $D^b(A) \xrightarrow{\sim} \underline{\text{mod}}\, \hat{A}$ (1.4) et $\underline{\text{mod}}\, \hat{A}$ est de cycles finis. Soit maintenant $N_0 \to N_1 \to \dots \to N_s = N_0$ un cycle de $\underline{\text{mod}}\, \hat{A}$. Il suit de [R2](3), Lemme, qu'il existe un cycle correspondant $N_0 \to N_1' \to N_1 \to N_2' \to N_2 \to \dots N_s' \to N_s = N_0$ de $\text{mod}\, \hat{A}$. Ce cycle se trouve dans un tube du carquois d'Auslander-Reiten stable de $\hat{A}$. Donc le cycle original se trouve dans un tube de $\Gamma_{\hat{A}}$.

5.2 Notre résultat principal est le suivant:

THEOREME [AS3]. Soit A une algèbre telle que $D^b(A)$ est de cycles finis. Alors A est soit pré-inclinée de type Dynkin ou Euclidien, soit équivalente pour les inclinaisons et les co-inclinaisons à une algèbre tubulaire canonique.

Avant de donner une esquisse de la démonstration, indiquons quelques conséquences:

COROLLAIRE (a): Soit A une algèbre. Les conditions suivantes sont équivalentes:

(i) $D^b(A)$ est de cycles finis

(ii) Il existe une algèbre C héréditaire de type Dynkin ou Euclidien, ou tubulaire canonique telle que $D^b(A) \xrightarrow{\sim} D^b(C)$, en tant que catégories triangulées.

(iii) A est pré-inclinée de type Dynkin ou Euclidien, ou équivalente pour les inclinaisons et les co-inclinaisons à une algèbre tubulaire canonique.

(iv) $\underline{\text{mod}}\, \hat{A}$ est de cycles finis.

Démonstration: (i) ⇒ (iv) suit de (5.1), alors que (iv) ⇒ (i) est immédiat. D'autre part, (i) ⇒ (iii) est le théorème, alors que (iii) ⇒ (ii) ⇒ (i) suivent de (1.3).

COROLLAIRE (b): Soit C une algèbre tubulaire canonique. Alors $D^b(A) \simeq D^b(C)$, en tant que catégoriestriangulées, si et seulement si A est équivalente à C pour les inclinaisons et les co-inclinaisons.

Démonstration: On utilise l'équivalence (ii) ⇔ (iii) du corollaire (a) et le théorème de Happel, Rickard et Schofield (voir (1.3)).

COROLLAIRE (c): Soient A une algèbre pré-inclinée de type Dynkin (respectivementm pré-inclinée de type Euclidien, équivalente pour les inclinaisons et les co-inclinaisons à une algèbre tubulaire canonique) et B une algèbre telle que $\underline{\text{mod}}\,\hat{A} \simeq \underline{\text{mod}}\,\hat{B}$. Alors B est aussi une algèbre pré-inclinée de type Dynkin (respectivement, pré-inclinée de type Euclidien, équivalente pour les inclinaisons et les co-inclinaisons à une algèbre tubulaire canonique).

Démonstration: En effet, il suit de l'hypothèse faite sur A que mod $\hat{A}$ est de cycles finis et de l'équivalence stable que mod $\hat{B}$ est aussi de cycles finis.

5.3 Passons maintenant à la démonstration du théorème. Soit A une algèbre telle que $D^b(A)$ est de cycles finis. Il résulte de (5.1) que A est triangulaire et mod $\hat{A}$ est de cycles finis. Si A n'est pas simplement connexe, alors il résulte du théorème (4.3) que A est pré-inclinée de type $\tilde{\mathbb{A}}_m$. On peut donc supposer que A est simplement connexe. Si A est pré-inclinée de type Dynkin, il n'y a rien à démontrer. Sinon, il existe une suite d'algèbres $A = A_0, A_1, \ldots, A_{m+1} = B$ et une suite de modules inclinants d'APR, $T^i_{A_i}$ $(0 \leq i \leq m)$ telles que $A_{i+1} = \text{End}\, T^i_{A_i}$ et B est de représentation infinie (voir [A2] ou [TW2] et [AHR]). Comme $D^b(A) \simeq D^b(B)$, mod $\hat{B}$ est de cycles finis. D'autre part, B n'est pas pré-inclinée de type $\tilde{\mathbb{A}}_m$ puisque sinon on aurait $\underline{\text{mod}}\,\hat{A} \simeq \underline{\text{mod}}\,\hat{B} \simeq \underline{\text{mod}}\,\hat{H}$ pour H héréditaire de type $\tilde{\mathbb{A}}_m$ et ceci entraînerait,d'après (3.3), que A est pré-inclinée de type $\tilde{\mathbb{A}}_m$, une contradiction. Donc B est simplement connexe. On peut donc supposer dès le départ que A est simplement connexe de représentation infinie. Compte tenu du théorème (2.4), il suffira de montrer que A est un agrandissement par branches d'une algèbre docile dérobée C en des modules simples réguliers, et que le type tubulaire n_A de A est domestique ou tubulaire mais pas de la forme (p,q) . Cela donnera en même temps la réciproque du théorème (2.4). Nous indiquons les étapes du travail.

On commence par prouver que A contient une sous-catégorie pleine et convexe C qui est une algèbre docile dérobée. On considère ensuite les extensions ponctuelles de C. On démontre successivement:

1) Soit $B = C[M]$ une extension ponctuelle de C qui est une sous-catégorie pleine de $\hat{A}$. Alors M est un C-module simple régulier.

2) Soit $B = C[M]$ une extension ponctuelle de C par un C-module simple régulier. Alors $B[M]$ n'est pas une sous-catégorie pleine de $\hat{A}$.

3) Soient $B = C[M]$ une extension ponctuelle de C par un C-module simple régulier M, avec point d'extension a, et $D = B[X]$ une extension ponctuelle de B, avec point d'extension b. Supposons que D est une sous-catégorie pleine de A et N est un facteur direct indécomposable de X contenant $S(a)$ dans sa coiffe. Alors $N \simeq P(a)_B$ ou $N \simeq S(a)_B$.

On prouve ensuite que trois sortes de configurations ne peuvent apparaître:

4) A ne contient pas de sous-catégorie pleine de la forme:

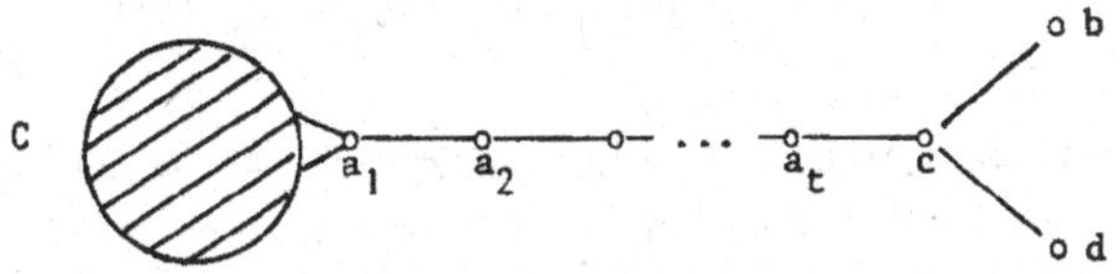

où les arêtes peuvent être orientées arbitrairement, et la sous-catégorie pleine formée par les points a_t, b, c, d est héréditaire.

5) A ne contient pas de sous-catégorie pleine de la forme:

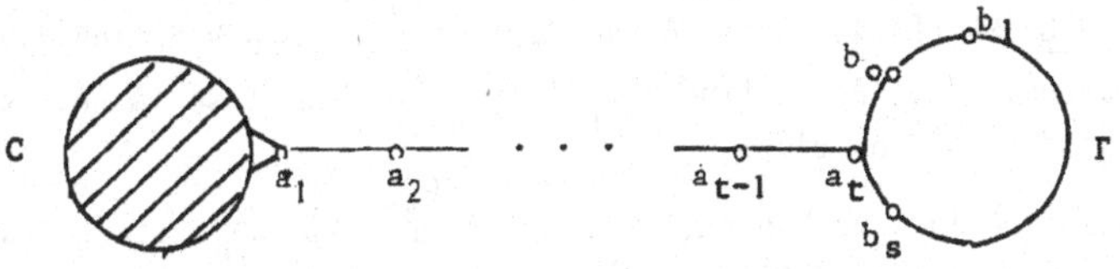

où les arêtes peuvent être orientées arbitrairement, et Γ est un cycle non-commutatif.

6) Soient a et b deux objets de A n'appartenant pas à C, tous deux reliés à C par (au moins) une arête. Alors toute allée de A passant par a et b doit couper C.

Les deux dernières assertions se démontrent au moyen de techniques de revêtement. Il s'ensuit facilement que A est un agrandissement de C par branches en des modules simples réguliers. Enfin, pour montrer que n_A est domestique ou tubulaire, on prouve que, si ce n'est pas le cas, mod $\hat{A}$ contient une sous-catégorie de la forme mod K, pour K héréditaire sauvage, et cela contredit le fait que mod $\hat{A}$ est de cycles finis.

6. Formes quadratiques:

6.1. Nous avons obtenu une description complète des algèbres pré-inclinées de type Euclidien de représentation infinie (théorème (2.4)) ainsi qu'une classification des algèbres pré-inclinées de type $\tilde{\mathbb{A}}_m$ (théorèmes (3.1) et (3.4)). Il reste donc à caractériser les algèbres pré-inclinées de représentation finie, de type Dynkin ou Euclidien $\neq \tilde{\mathbb{A}}_m$. On sait déjà d'après le théorème (4.3) qu'elles sont simplement connexes. Nous donnons maintenant un critère pratique permettant d'identifier ces algèbres au moyen de leurs formes quadratiques. Rappelons la définition. Soient A une algèbre de dimension globale finie et S(1),... S(n) un ensemble complet de représentants des classes d'isomorphisme des A-modules simples. La forme quadratique (homologique) q_A de A est la forme sur le groupe de Grothendieck $K_0(A)$ de A dont la matrice $\kappa_A = [\kappa_{ij}]_{1 \leq i,j \leq n}$ est donnée par:

$$\kappa_{ij} = \sum_{s \geq o} (-1)^s \dim_k \mathrm{Ext}_A^s(S(i),S(j)).$$

Cette somme est finie puisque l'on a supposé dim.gl. $A < \infty$. Il résulte de [HR1] (3.2) que, si (B,T,A) est un triplet inclinant, alors les formes q_A et q_B ou, ce qui revient au même, leurs matrices κ_A et κ_B sont $\mathbb{Z}$-congruentes, c'est à dire qu'il existe une matrice $\chi \in GL_n(\mathbb{Z})$ telle que $\kappa_A = \chi^t \kappa_B \chi$. En particulier, si A est pré-inclinée de type Dynkin (respectivement, Euclidien) alors q_A est définie positive (respectivement, semi-définie positive de corang un).

6.2. THÉORÈME [AS4]. Soit A une algèbre de représentation finie.
(i) A est pré-inclinée de type Dynkin si et seulement si A est simplement connexe et q_A est définie positive.
(ii) A est pré-inclinée de type Euclidien $\tilde{\mathbb{D}}_n$ ou $\tilde{\mathbb{E}}_p$ si et seulement si A est simplement connexe et q_A est semi-définie positive de corang un.

La première partie du théorème a également été obtenue par Happel (communication privée). Nous en obtenons une démonstration simple, qui découle d'ailleurs de la démonstration de la seconde partie. Cette dernière utilise le théorème (2.4). En effet, la nécessité étant évidente, nous devons démontrer la suffisance. Si A est simplement connexe avec q_A semi-définie positive, A ne peut être une algèbre pré-inclinée de type Dynkin. On démontre que cela entraîne l'existence d'une suite de réflexions de A telle que $B = S_{i_t}^+ \dots S_{i_1}^+ A$ est de représentation infinie mais que, pour tout $j < t$, $S_{i_j}^+ \dots S_{i_1}^+ A$ est de représentation finie [ANS] (3.4). Il faut ensuite prouver que B est un agrandissement par branches d'une algèbre docile dérobée en des modules simples réguliers. Les étapes sont semblables à celles de la démonstration du théorème (5.3). On déduit du théorème la proposition:

PROPOSITION. Soit A une algèbre pré-inclinée de type Dynkin (respectivement, Euclidien) et B une sous-catégorie pleine et convexe de A. Alors B est une algèbre pré-inclinée de type Dynkin (respectivement, Dynkin ou Euclidien).

7. Algèbres pré-inclinées de type $\mathbb{D}_n$:

7.1. Dans cette section, nous classifions les algèbres pré-inclinées de type $\mathbb{D}_n$ par leurs carquois liés. Dans [Y], Yamagata avait donné une caractérisation, en termes de leurs carquois d'Auslander-Reiten, des algèbres dont l'extension triviale est de représentation finie et de classe de Cartan $\mathbb{D}_n$ (c'est à dire, par [AHR], des algèbres pré-inclinées de type $\mathbb{D}_n$). D'autre part, les algèbres inclinées de type $\mathbb{D}_n$ ont été classifiées par Conti [C] au moyen de la théorie des arbres gradués de [BoG]. Nous allons montrer que les algèbres pré-inclinées de type $\mathbb{D}_n$ sont des agrandissements par branches. Afin de définir les coeurs de ces agrandissements, nous utiliserons les deux lemmes suivants de D. Happel [H1] (6).

LEMME (a) Soit A une algèbre pré-inclinée de type $\mathbb{D}_n$. Si le carquois de A n'est pas un arbre, il contient un sous-carquois lié plein de la forme:

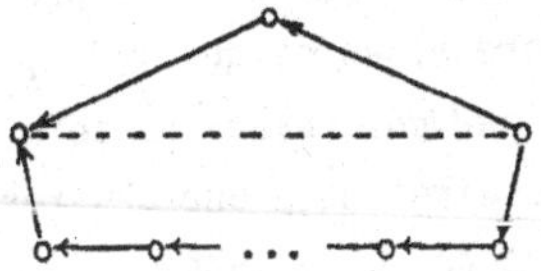

Ici, et dans toute cette section, des lignes en pointillé dans un carquois indiquent que la somme des chemins de la source au but est nulle. Une arête non-orientée peut être orientée arbitrairement.

LEMME (b) Soit A une algèbre pré-inclinée de type $\mathbb{D}_n$. Si A n'est pas donnée par le carquois lié du lemme (a), son carquois contient une source ou un puits i n'ayant qu'un seul voisin et tel que $A/\langle e_i \rangle$ est pré-inclinée de type $\mathbb{A}_{n-1}$ ou $\mathbb{D}_{n-1}$.

Nous définissons un $\mathbb{D}_n$-cadre comme étant le carquois lié (Q,I) d'une algèbre pré-inclinée A de type $\mathbb{D}_n$ tel que, pour toute source ou tout puits $i \in Q_0$, $A/\langle e_i \rangle$ est pré-inclinée de type $\mathbb{A}_{n-1}$. Il est évident que toute algèbre pré-inclinée de type $\mathbb{D}_n$ contient un sous-carquois lié plein qui est un $\mathbb{D}_m$-cadre pour un $m \leq n$.

<u>PROPOSITION</u>. Les $\mathbb{D}_n$-cadres sont les carquois liés

(F1) 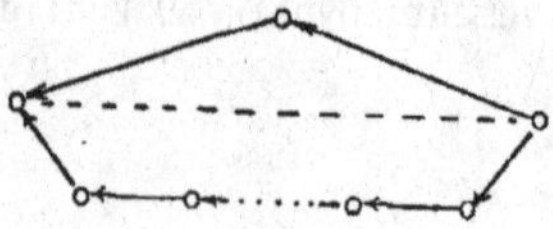 (F2) 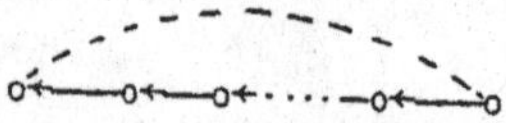

(F3) 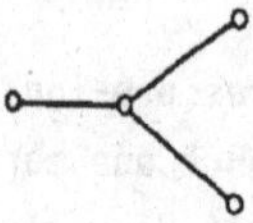 (F4) 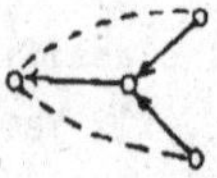 (F5) 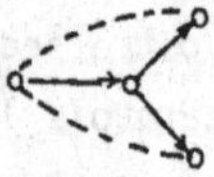

(Dans (F1) et (F2), on suppose que l'on a au moins 4 points).

<u>Démonstration</u>: Comme il est évident que les carquois liés de l'énoncé sont des $\mathbb{D}_n$-cadres, il faut prouver que ce sont les seuls. Soit donc (Q,I) un $\mathbb{D}_n$-cadre arbitraire.

Si (Q,I) n'est pas un arbre, il contient un sous-carquois lié plein (Q',I') de la forme (F1). Mais alors Q = Q'. En effet, si ce n'est pas le cas, et si on enlève une source ou un puits de Q qui n'appartient pas à Q', le carquois lié résultant n'est évidemment pas celui d'une algèbre pré-inclinée de type $\mathbb{A}_{n-1}$. Supposons que (Q,I) est un arbre et aussi que (Q,I) n'est pas de la forme (F2). En particulier, I est engendré par des relations-zéro de longueur deux. Nous montrerons que (Q,I) est nécessairement d'une des formes (F3)(F4) ou (F5). Il est facile de voir que tout point de Q a au plus 5 voisins et que, si c'est le cas, (Q,I)est décrit localement par le carquois lié:

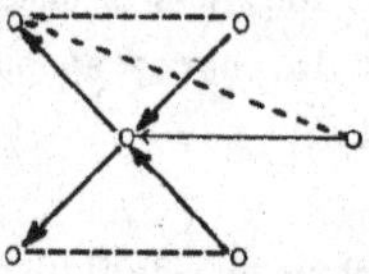

ou son opposé [H1] (6). Ainsi (Q,I) contient deux sous-carquois liés pleins des formes (F3) et (F4) (dualement, (F3) et (F5)). Cela entraîne, comme plus haut, que (Q,I) est en fait égal à l'un de ces cadres. Si tous les points de (Q,I) ont au plus 4 voisins, il est tout aussi facile de voir que (Q,I) est décrit localement par un des carquois liés:

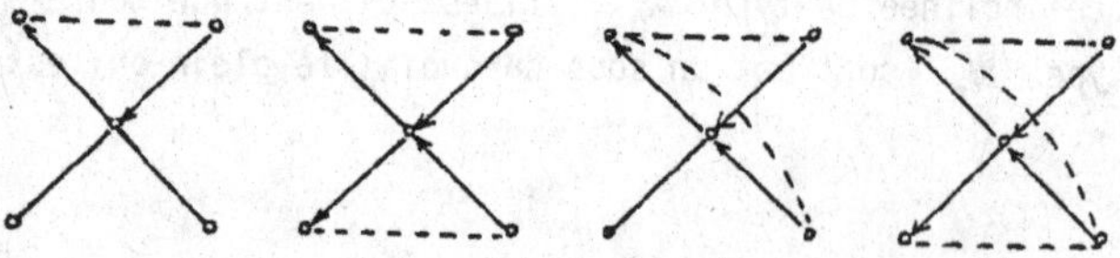

ou des carquois opposés. Le même raisonnement entraîne que (Q,I) est égal à un cadre d'une des formes (F3)(F4) ou (F5). De même, si chaque point de (Q,I) a au plus 3 voisins, (Q,I) est décrit localement par un des carquois liés:

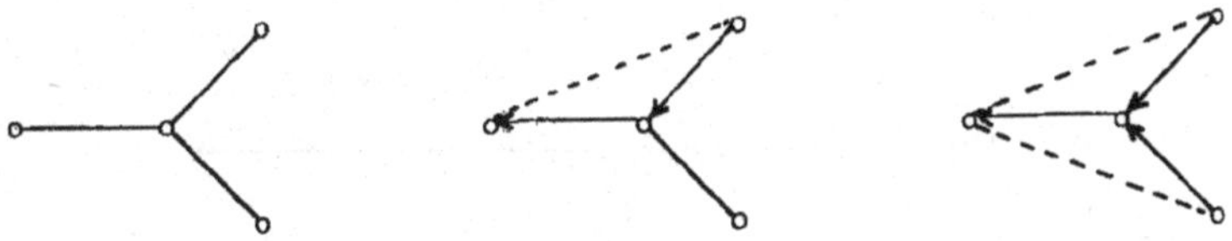

ou des carquois opposés. Donc (Q,I) est d'une des formes (F3)(F4) ou (F5). Enfin, si chaque point a au plus deux voisins, (Q,I) étant un $\mathbb{D}_n$-cadre doit contenir des relations-zéro de longueur au moins trois, une contradiction qui achève la démonstration.

Remarque: Observons que chaque $\mathbb{D}_n$-cadre contient une sous-catégorie pleine qui est un $\mathbb{D}_4$-cadre.

7.2. Soit (Q,I) un carquois lié, (Q',I') un sous-carquois lié plein et i un point de Q ayant au plus deux voisins dans Q'. Notons Q_i le sous-carquois lié plein de Q formé par i et ses voisins dans Q'. On dit que (Q,I) est un agrandissement de (Q',I') en i si le sous-carquois lié plein formé par la composante connexe de $(Q\setminus Q') \cup Q_i$ contenant i est une branche (voir (2.1)). Nous pouvons énoncer:

THÉORÈME. Une algèbre A est pré-inclinée de type $\mathbb{D}_n$ si et seulement si $A \cong kQ/I$, où (Q,I) est un sous-carquois lié plein et connexe, contenant au moins un cadre, d'un agrandissement d'un des carquois liés suivants aux points marqués par des astérisques:

(D1)

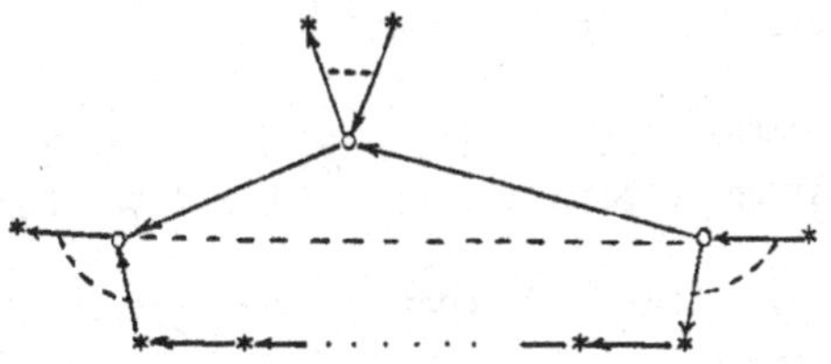

(D2)

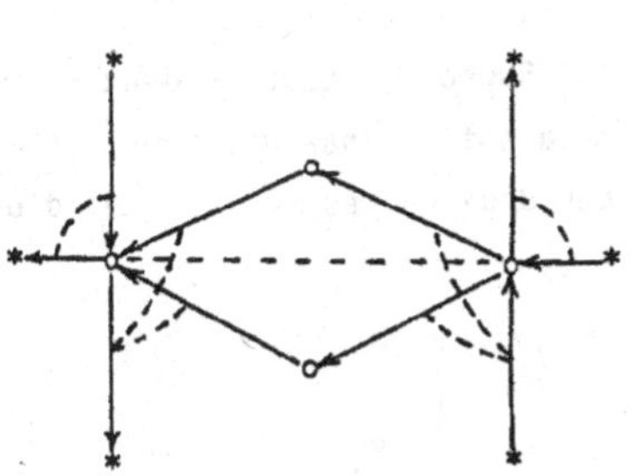

(D3) 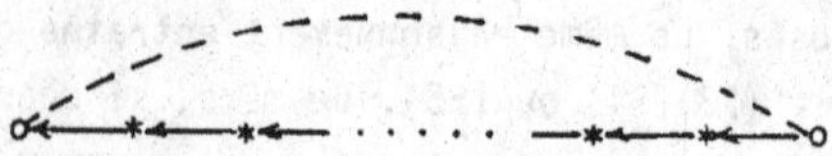

(D4) 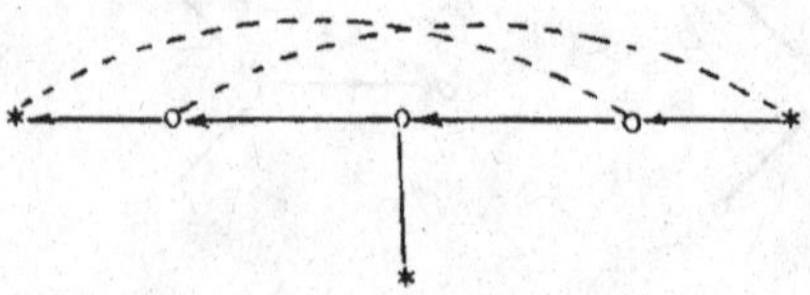

(D5) 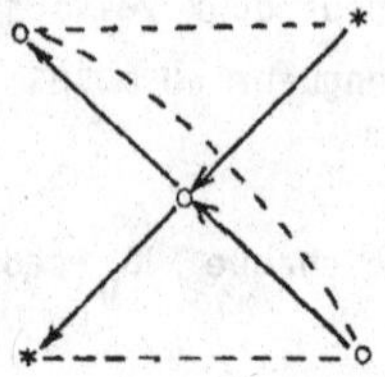

Dans (D1) (respectivement, (D3)), la relation de commutativité (respectivement, la relation-zéro) est supposée comprendre au moins 4 points. Remarquons que les carquois liés (D1) à (D5) sont auto-duals, ce qui facilite la vérification. Celle-ci se fait de la manière suivante. Si (Q,I) est un carquois lié donné, on commence par identifier le ou les $\mathbb{D}_n$-cadres qu'il contient, puis on essaie de faire coïncider ces cadres avec des cadres isomorphes apparaissant dans un des carquois liés (D1) à (D5). Il ne reste plus qu' à vérifier que le reste du carquois lié (Q,I) est une union disjointe de branches. On voit, par exemple, que l'on obtient toutes les algèbres héréditaires de type $\mathbb{D}_n$ à partir de (D1) et (D2) (et aussi de (D4)).

7.3. Démonstration de la nécessité: Dans tout ce qui suit, on supposera que $A = kQ/I$ est une algèbre pré-inclinée de type $\mathbb{D}_n$.

LEMME (a). Si (Q,I) contient deux cadres, ces deux cadres ont au moins deux flèches en commun.

Démonstration: (i) Supposons que ces deux cadres C' et C" n'ont aucune flèche en commun. Remplaçant A par une sous-catégorie pleine convenable, on peut supposer A formée des deux cadres C', C" et d'une allée de radical carré nul joignant C' et C"

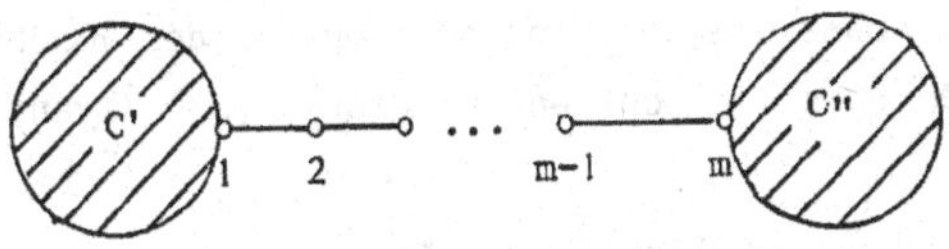

Si on applique des réflexions convenables aux puits et aux sources de C' et C", on trouve une algèbre contenant une sous-catégorie pleine B (telle que $\hat{B}$ est une sous-catégorie pleine de $\hat{A}$) dont le carquois lié est de la forme:

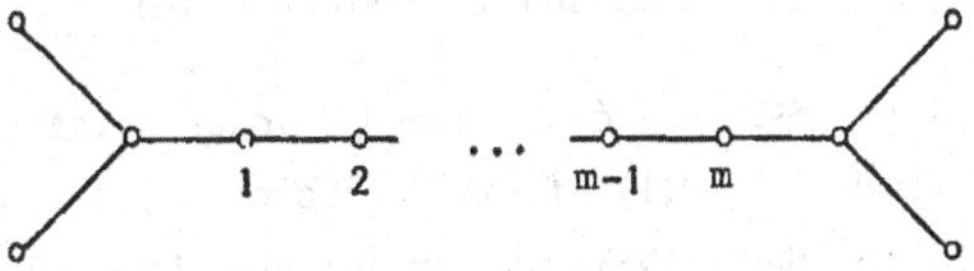

avec la même allée $1 - 2 - .. - m\text{-}1 - m$ que A et deux cadres de type (F3). Appliquant, si nécessaire, des modules inclinants d'APR correspondant aux extrémités des cadres, on trouve une algèbre C de radical carré nul, de même graphe sous-jacent que B et telle que $\underline{\text{mod}}\ \hat{C} \not\simeq \underline{\text{mod}}\ \hat{B}$. Mais $\hat{C}$ n'est pas localement de représentation finie et cela contredit l'hypothèse que A est pré-inclinée de type Dynkin.

(ii) Supposons maintenant que C' et C" ont exactement une flèche en commun. Remplaçant éventuellement A par une sous-catégorie pleine, on peut supposer A formée par la réunion de deux $\mathbb{D}_4$-cadres. On a alors, à dualité près, les possibilités suivantes:

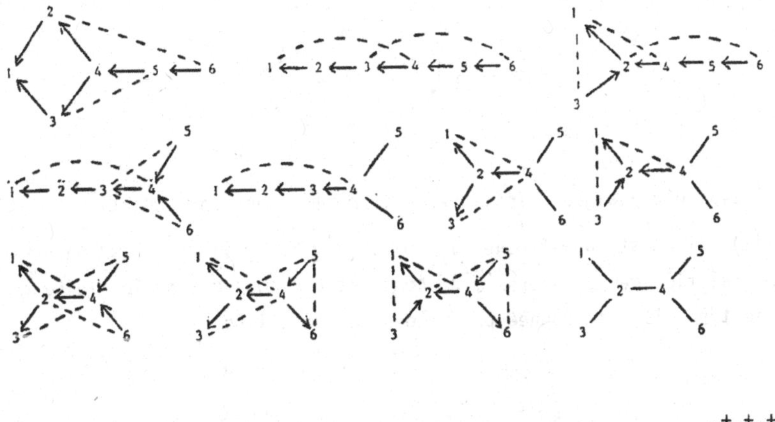

Dans le premier cas, A est pré-inclinée de type $\mathbb{E}_6$. Dans les autres, $S_3^+S_2^+S_1^+A$, $S_6^-S_1^+A$, $S_6^-S_5^-S_1^+A$, S_1^+A, $S_3^+S_1^+A$, S_1^+A, $S_6^-S_5^-S_3^+S_1^+A$, $S_5^-S_3^+S_1^+A$, $S_5^-S_1^+A$ et A respectivement sont des algèbres héréditaires de type $\tilde{\mathbb{D}}_5$. Cette contradiction achève la démonstration.

Le lemme suivant donne une description complète des algèbres pré-inclinées de type $\mathbb{D}_n$ dont le carquois lié contient un cadre (F2) ayant au moins 5 points.

LEMME (b). Si (Q,I) contient un cadre C

(F2) $1 \leftarrow 2 \leftarrow 3 \leftarrow \cdots \leftarrow m-1 \leftarrow m$

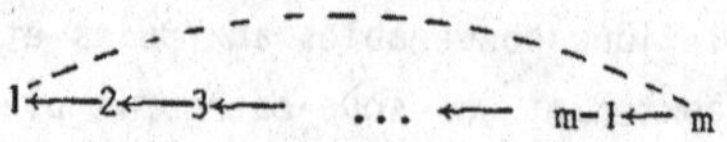

avec $m \geq 5$, alors (Q,I) est un agrandissement du cadre C aux points $2,\ldots,m-1$.

Démonstration: Soit $m+1$ un voisin dans Q d'un point de C et B l'algèbre du sous-carquois lié plein de (Q,I) formé des points $1,2\ldots m,m+1$. On considèrera, à dualité près, toutes les façons possibles de joindre $m+1$ au cadre.

(i) Il existe une arête $m - m+1$ qui n'est pas contenue dans une relation de B. Alors le carquois lié de $S_1^+ B$ contient un sous-carquois plein héréditaire de type $\mathbb{E}_6$:

$$\begin{array}{c} 1 \\ \downarrow \\ m-3 \leftarrow m-2 \leftarrow m-1 \leftarrow m \leftarrow m+1 \end{array}$$

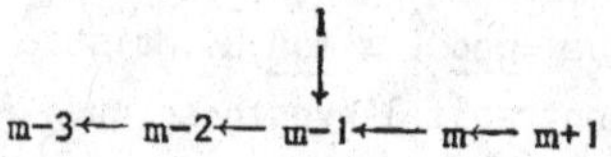

Cela contredit le fait que le carquois de $\underline{\mathrm{mod}}\,\hat{A}$ est isomorphe à $\mathbb{Z}\mathbb{D}_n$.

(ii) Il existe une flèche $m-1 \leftarrow m+1$ qui n'est pas contenue dans une relation de B. Alors le carquois lié de $S_2^+ S_1^+ B$ contient le sous-carquois lié plein:

$$m-2 \leftarrow m-1 \cdots \overset{m}{\underset{m+1 \leftarrow 1}{\diamond}} 2$$

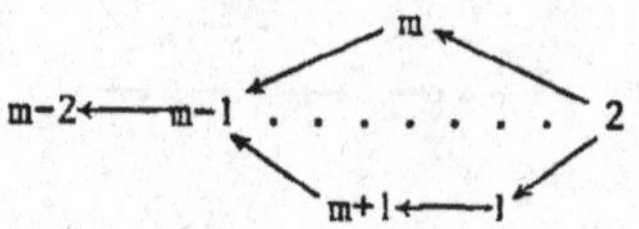

qui est celui d'une algèbre inclinée de type $\mathbb{E}_6$, une contradiction.

(iii) Il existe une flèche $i \leftarrow m+1$ $(1 < i < m-1)$ qui n'est pas contenue dans une relation de B. Alors $S_1^+ B$ contient une sous-catégorie pleine qui est inclinée de type $\mathbb{E}_6$ et donnée par un des carquois liés:

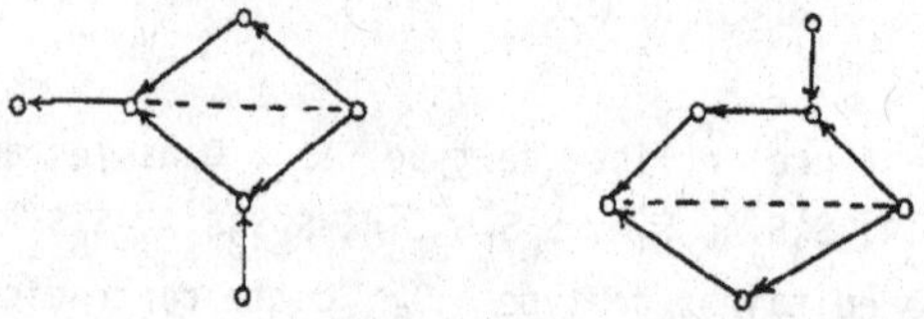

On en déduit encore une contradiction.

(iv) Il existe une flèche $m \leftarrow m+1$ contenue dans une relation-zéro de longueur $m-1$. Dans ce cas, le carquois lié de $S_3^+ S_2^+ S_1^+ B$ contient le carquois lié de l'algèbre inclinée de type $\mathbb{E}_6$:

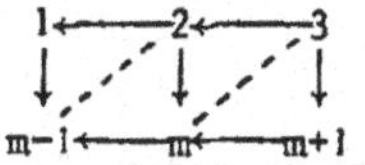

(v) Il existe une flèche $m \leftarrow m+1$ et $m+1$ est la source d'une relation-zéro de longueur $\ell < m-1$ dans B. Dans ce cas, il existe une sous-catégorie pleine de $S_{m-1}^+ \dots S_1^+ B$ qui est héréditaire de type $\mathbb{E}_6$ ou bien inclinée de ce type donnée par le carquois lié:

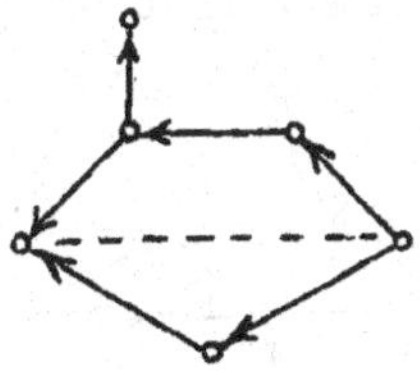

(vi) Il existe une flèche $i \leftarrow m+1$, $i \leq m-2$ et une relation-zéro dans B de source $m+1$ et de longueur au moins trois. Alors le carquois lié de $S_1^+ B$ contient le carquois héréditaire de type $\tilde{\mathbb{D}}_p$:

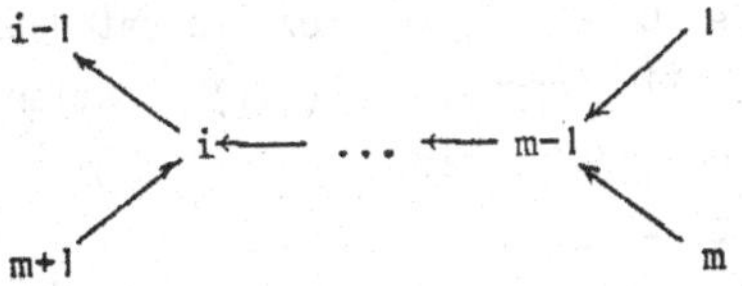

ce qui contredit le fait que $\hat{A}$ est localement de représentation finie. De même si $i = m-1$.

Enfin, supposons que (Q,I) contient trois relations-zéro $m+2 \leftarrow i \leftarrow m+1$, $m+2 \leftarrow i \leftarrow i+1$ et $i-1 \leftarrow i \leftarrow m+1$ (pour $2 \leq i \leq m-1$) et soit D l'algèbre du sous-carquois lié plein de (Q,I) formé par les points $1,2 \dots m$, $m+1$, $m+2$. Alors le carquois lié de $S_{m+2}^+ S_1^+ D$ contient le carquois héréditaire:

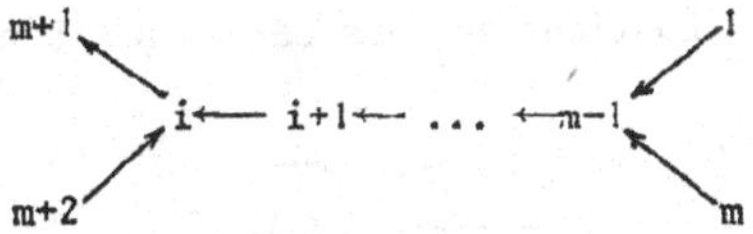

ce qui contredit le fait que $\hat{A}$ est localement de représentation finie. Les calculs précédents montrent aussi qu'un cadre de type (F2) (ayant au moins 5 points) ne partage aucune flèche avec un cadre de type (F1). Le résultat suit alors du lemme (a).

<u>COROLLAIRE</u>. Deux cadres de (Q,I) ayant une flèche en commun ont exactement deux flèches en commun.

<u>LEMME</u> (c). Si (Q,I) contient le cadre C:

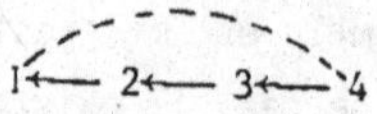

alors (Q,I) est soit un agrandissement de C en 2 et 3, soit un sous-carquois lié d'un agrandissement de (D4):

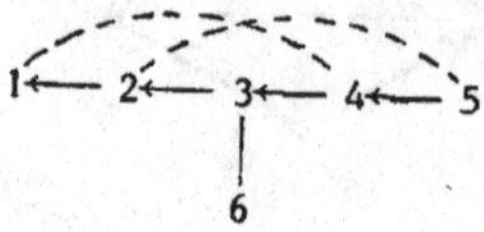

en 1, 5 et 6.

<u>Démonstration</u>: Supposons d'abord que (Q,I) contient un cadre C' ≠ C. Il suit du corollaire précédent que C et C' ont exactement deux flèches en commun. C' n'est pas de la forme (F1) car sinon (Q,I) contient, à dualité près, un sous-carquois lié plein d'une des formes:

B':

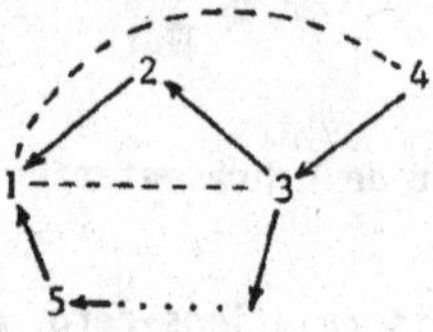

B":

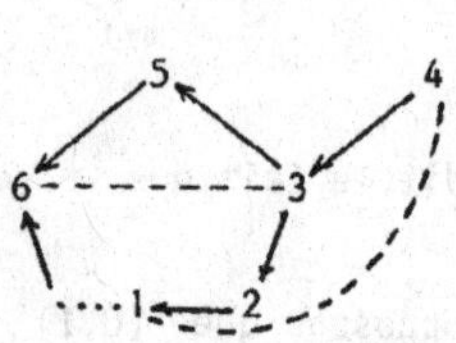

et S_1^+B' (respectivement, S_6^+B'') contient un sous-carquois plein héréditaire de type $\tilde{\mathbb{D}}_4$. Le lemme (b) entraîne que C' est d'une des formes (F2) (avec 4 points) ou (F3).

Supposons que (Q,I) contient un sous-carquois lié plein D de la forme:

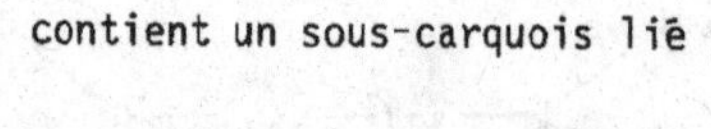

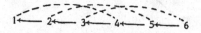

Alors $S_3^+S_2^+S_1^+D$ est inclinée de type $\mathbb{E}_6$ donnée par le carquois lié:

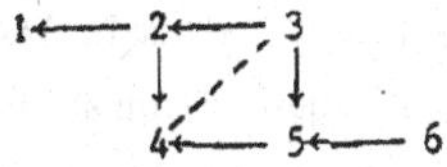

ce qui contredit le fait que le carquois de <u>mod</u> $\hat{A}$ est isomorphe à $\mathbb{Z}\mathbb{D}_n$. En outre, (Q,I) ne contient pas de sous-carquois lié plein E de la forme:

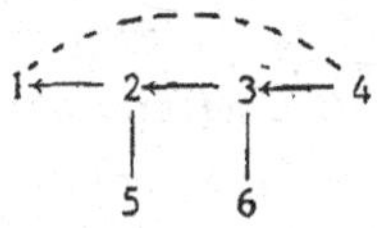

En effet, pour toute orientation des deux arêtes libres, E est une algèbre inclinée de type $\mathbb{E}_6$, encore une contradiction. Par conséquent, il suit du lemme (a) que (Q,I) est un agrandissement de C, ou un sous-carquois lié plein d'un agrandissement de (D4). Il est facile de voir que l'on ne peut agrandir C qu'en 2 et 3 et (D4) en 1, 5 et 6.

Nous avons prouvé la nécessité dans le cas où (Q,I) contient un cadre (F2). Dans les autres cas, elle résulte du lemme suivant:

<u>LEMME</u> (d). Si (Q,I) ne contient pas de cadre de type (F2), alors (Q,I) est un sous-carquois lié, contenant au moins un cadre, d'un agrandissement de (D1)(D2) ou (D5) aux points marqués d'un astérisque (voir (7.2)).

<u>Démonstration</u>: On suppose d'abord que (Q,I) contient un cadre (F1) ayant au moins 5 points. Alors les lemmes (b) et (c) et le corollaire du lemme (b) entraînent que le carquois lié (Q,I) est un sous-carquois lié plein d'un agrandissement de (D1) (si ce n'est pas le cas, (Q,I) contient un sous-carquois lié plein qui est incliné de type $\mathbb{E}_6$). Supposons maintenant que (Q,I) contient un $\mathbb{D}_4$-cadre (F1). Nous montrerons que (Q,I) est un sous-carquois lié plein d'un agrandissement de (D1) ou (D2). En effet, supposons que ce n'est pas le cas. Comme (Q,I) ne contient pas de sous-carquois lié plein héréditaire de type $\tilde{\mathbb{D}}_4$ ou incliné de type $\mathbb{E}_6$, alors (Q,I) contient, à dualité près, un des sous-carquois liés pleins suivants:

C: 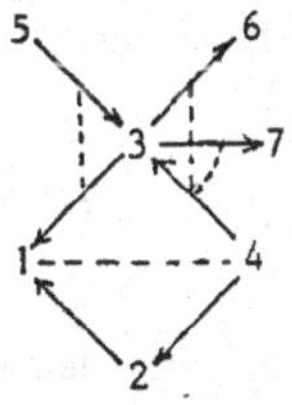

D: 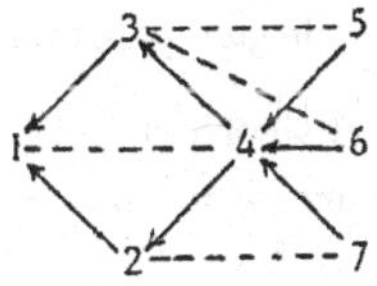

Or, S_1^+D contient un sous-carquois héréditaire de type $\tilde{\mathbb{D}}_4$ et $S_7^+S_6^+S_1^+C$ est héréditaire de type $\tilde{\mathbb{D}}_6$, ce qui contredit le fait que $\hat{A}$ est localement de représentation finie. Ainsi (Q,I) est un sous-carquois lié plein d'un agrandissement de (D1) ou (D2). On montre facilement que l'agrandissement ne peut avoir lieu qu'aux points marqués d'un astérisque. Enfin, une analyse semblable à partir du cadre (F3) (respectivement, (F4), (F5)) donne un agrandissement de (D1)(D2) ou (D4) (respectivement, (D2) ou (D5) dans les deux cas) aux points marqués d'un astérisque. Cela achève la démonstration.

7.4. Démonstration de la suffisance: Soit C la classe des algèbres données par un carquois lié satisfaisant les conditions du théorème. On doit montrer que, si A appartient à C, alors A est pré-inclinée de type $\mathbb{D}_n$. Il suffit de prouver que C est fermée sous l'action des modules inclinants d'APR. En effet, si A dans C n'est pas héréditaire, il existe une suite d'algèbres $A = A_o, A_1, \ldots A_t$ et une suite de modules inclinants d'APR $T^i_{A_i}$ $(o \leq i < t)$ telles que $A_{i+1} = \operatorname{End} T^1_{A_i}$ et $n(A_t) > n(A)$ (où $n(C)$ désigne le nombre de classes d'isomorphisme de C-modules indécomposables) [A2][TW2]. Or, après un certain nombre de telles étapes, on aboutit à une algèbre héréditaire de la classe C (sinon il existerait un nombre infini d'algèbres de C, en particulier simplement connexes de représentation finie, deux à deux non-isomorphes et ayant le même nombre de modules simples, une absurdité). Et il est évident qu'une algèbre de C est héréditaire si et seulement si elle est l'algèbre des chemins d'un carquois de graphe sous-jacent $\mathbb{D}_n$.

Pour montrer que C est fermée sous l'action des modules inclinants d'APR, on remarque qu'il suffit d'étudier ceux qui correspondent aux puits qui sont les buts d'au moins une relation (sinon l'action du module consiste simplement à inverser l'orientation des flèches passant par ce puits, sans affecter les relations). L'assertion (et donc le théorème) suivent du lemme suivant dont la démonstration est simple et peut être laissée au lecteur.

LEMME. Soient A une algèbre de C, a un puits de Q_A qui est le but d'au moins une relation, $T_A^{(a)}$ le module inclinant d'APR correspondant et $B = \operatorname{End} T_A^{(a)}$. Notons b^* (pour $b \neq a$) le point de Q_B correspondant au facteur indécomposable $P(b)_A$ de $T^{(a)}$, et a^* le point correspondant à $\tau^{-1}P(a)$. Alors $B \simeq kQ_B/I_B$ où (Q_B, I_B) est obtenue comme suit:

(i) A chaque flèche $b \to a$ de Q_A correspond une flèche $a^* \to b^*$.

(ii) A chaque relation de c vers a correspond une flèche $c^* \to a^*$.

(iii) A chaque flèche $b \to c$ $(c \neq a)$ telle qu'il n'existe pas de relation $b \to c \to a$ correspond une flèche $b^* \to c^*$. Si une telle relation existe, on a des flèches $b^* \to a^* \to c^*$.

(iv) Toute flèche de Q_B est d'une des formes précédentes.

(v) Toute relation de I_A de but distinct de a induit, de façon évidente, une relation de I_B.

(vi) Soit w un chemin de Q_B de b^* vers c^* de longueur au moins deux, avec $a \neq b$ et $\mathrm{Hom}_A(P(c), P(b)) = 0$. Si $a \neq c$, alors w est une relation-zéro. Si $a = c$, alors w est une relation-zéro si et seulement s'il existe un unique voisin ℓ de a se trouvant sur un chemin de b vers a et en outre $\mathrm{Hom}_A(P(\ell), P(b)) = 0$.

(vii) Si w_1 et w_2 sont deux chemins de Q_B de memes extrémités et de longueur au moins deux, alors il existe une relation de commutativité liant w_1 et w_2.

(viii) Les relations précédentes donnent un système minimal de générateurs de I_B.

BIBLIOGRAPHIE:

[A1] ASSEM, I.: Iterated tilted algebras of types $\mathbb{B}_n$ and $\mathbb{C}_n$, J. Algebra 84, No. 2 (1983), 361-390.

[A2] ASSEM, I.: Separating splitting tilting modules and hereditary algebras, Bull. Can. Math. 30 (2) (1987).

[AH] ASSEM, I. et HAPPEL, D.: Generalized tilted algebras of type $\mathbb{A}_n$, Comm. Algebra 9 (1981), 2101-2125.

[AHR] ASSEM, I., HAPPEL, D. et ROLDÁN, O.: Representation-finite trivial extension algebras, J. Pure Appl. Algebra 33 (1984), 235-242.

[ANS] ASSEM, I., NEHRING, J. et SKOWROŃSKI, A.: Domestic trivial extensions of simply connected algebras, à paraître

[AS1] ASSEM, I. et SKOWROŃSKI, A.: Iterated tilted algebras of type $\tilde{\mathbb{A}}_n$, Math. Z., Band 195, Heft 2 (1987), 269-290.

[AS2] ASSEM, I. et SKOWROŃSKI, A.: On some classes of simply connected algebras, à paraître

[AS3] ASSEM, I. et SKOWROŃSKI, A.: Algebras with cycle-finite derived categories, à paraître dans Math. Annalen.

[AS4] ASSEM, I. et SKOWROŃSKI, A.: Quadratic forms and iterated tilted algebras, à paraître.

[APR] AUSLANDER, M., PLATZECK, M.-I. et REITEN, I.: Coxeter functors without diagrams, Trans. Amer. Math. Soc. 250 (1979), 1-46.

[B] BAUTISTA, R.: Sections in Auslander-Reiten quivers, Proc. ICRA II (Ottawa, 1979), Springer Lecture Notes No. 832 (1980), 74-96.

[BLS] BAUTISTA, R., LARRIÓN, F. et SALMERÓN, L.: On simply connected algebras, J. London Math. Soc. (2) 27 (1983), No. 2, 212-220.

[Bo] BONGARTZ, K.: A criterion for finite representation type, Math. Annalen, Band 269, Heft 1 (1984), 1-12.

[BoG] BONGARTZ, K. et GABRIEL, P.: Covering spaces in representation theory, Invent. Math. 65 (1981/82), No. 3, 331-378.

[BrG] BRETSCHER, O. et GABRIEL, P.: The standard form of a representation-finite algebra, Bull. Soc. Math. France 111 (1983), 21-40.

[C] CONTI, B.: Simply connected algebras of tree class $\mathbb{A}_n$ and $\mathbb{D}_n$, Proc. ICRA IV (Ottawa, 1984), Springer Lecture Notes No. 1177 (1986), 60-90.

[DS] DOWBOR, P. et SKOWROŃSKI, A.: Galois coverings of representation-infinite algebras, Comment. Math. Helv. Vol. 62, No. 2 (1987), 311-337.

[G1] GABRIEL, P.: Auslander-Reiten sequences and representation-finite algebras, Proc. ICRA II (Ottawa, 1979), Springer Lecture Notes No. 831 (1980), 1-71.

[G2] GABRIEL, P.: The universal cover of a representation-finite algebra, Proc. ICRA III (Puebla, 1980), Springer Lecture Notes No. 903 (1981), 65-105.

[Gr] GREEN, E.L.: Graphs with relations, coverings and group-graded algebras, Trans. Amer. Math. Soc., Vol. 279 (1983), 297-310.

[H1] HAPPEL, D.: Tilting sets on cylinders, Proc. London Math. Soc (3), 51 (1985), 21-55.

[H2] HAPPEL, D.: On the derived category of a finite-dimensional algebra, à paraître dans Comment. Math. Helv.

[HRS] HAPPEL, D., RICKARD, J. et SCHOFIELD, A.: Piecewise hereditary algebras, à paraître.

[HR1] HAPPEL, D. et RINGEL, C.M.: Tilted algebras, Trans. Amer. Math. Soc. 274 (1982), No. 2, 399-443.

[HR2] HAPPEL, D. et RINGEL, C.M.: The derived category of a tubular algebra, Proc. ICRA IV (Ottawa, 1984), Springer Lecture Notes No. 1177 (1986), 156-180.

[HW] HUGHES, D. et WASCHBOSCH, J.: Trivial extensions of tilted algebras, Proc. London Math. Soc. 46 (3) (1983), 347-364.

[MP] MARTINEZ-VILLA, R. et DE LA PEÑA, J.A.: The universal cover of a quiver with relations, J. Pure Appl. Algebra 30 (1983), 277-292.

[NS] NEHRING, J. et SKOWROŃSKI, A.: Polynomial growth trivial extensions of simply connected algebras, à paraître.

[R1] RINGEL, C.M.: Tame algebras and integral quadratic forms, Springer Lecture Notes No. 1099 (1984).

[R2] RINGEL, C.M.: Representation theory of finite-dimensional algebras, dans: Representations of Algebras, Proc. Durham 1985, Cambridge University Press (1986), 7-80.

[S] SKOWROŃSKI, A.: Generalization of Yamagata's theorem on trivial extensions, Archiv. Math. 48 (1987), 68-76.

[SW] SKOWROŃSKI, A. et WASCHBOSCH, J.: Representation-finite biserial algebras, J. Reine Angew. Math. 345 (1983), 172-181.

[TW1] TACHIKAWA, H. et WAKAMATSU, T.: Tilting functors and stable equivalences for self-injective algebras, à paraître dans J. Algebra.

[TW2] TACHIKAWA, H. et WAKAMATSU, T. Applications of reflection functors for self-injective algebras, Proc. ICRA IV (1984) Springer Lecture Notes No. 1077 (1986), 308-327.

[V] VERDIER, J.-L.: Catégories dérivées, état 0, Springer Lecture Notes No. 569 (1977), 262-311.

[WW] WALD, B. et WASCHBÜSCH, J.: Tame biserial algebras, J. Algebra 95 (1985), 480-500.

[W] WAKAMATSU, T.: Stable equivalence between universal covers of trivial extension self-injective algebras, Tsukuba J. Math. Vol. 9, No. 2 (1985), 299-316.

[Y] YAMAGATA, K.: On algebras whose trivial extensions are of finite representation type II. Preprint (1983).

Ibrahim Assem
Département de Mathématiques et de Statistiques
Université Carleton
Promenade Colonel By
Ottawa, Ontario
Canada, K1S 5B6

Andrzej Skowroński
Institut de Mathématiques
Université Nicolas Copernic
Chopina 12/18
87-100 Toruń
Pologne.

STABLE CALCULUS OF THE MIXED TENSOR CHARACTER I

Ranee Kathryn Brylinski(*)(**)
Brown University
Department of Mathematics
Providence, R.I. 02912
USA

[This paper is in final form and no version of it will be submitted for publication else-where].

§1. INTRODUCTION.

The general linear group $GL_n=GL(n,\mathbb{C})$ acts by matrix conjugation on its Lie algebra $\underline{gl}_n=\underline{gl}(n,\mathbb{C})$, the space of all nXn matrices. In terms of Lie theory, $\underline{gl}_n$ is the adjoint representation of the complex reductive Lie group GL_n.

Given a rational representation V of a reductive Lie group G, determining the decomposition of symmetric, exterior, and tensor powers of V into a direct sum of irreducible G-representations is a central problem in the representation theory of G. Indeed, these decompositions are the key for many problems of algebra (e.g., double-centralizer theorems) and geometry (e.g., properties of coordinate rings) of G-spaces and G-varieties.

The case where V is the adjoint representation $\underline{g}$ is especially rich – cf. Kostant's fundamental work on the exterior algebra $\underline{E}(\underline{g})$ and differential forms on G/P, and with the symmetric algebra $\underline{S}(\underline{g})$ and functions on adjoint orbits. Write Irred(G) for the set of (equivalence classes) of irreducible rational (so finite-dimensional algebraic) representations of G.

(*) Research Supported by the NSF

(**) *Current address:* The Pennsylvania State University, Department of Mathematics, 312 McAllister Building, University Park, Pennsylvania 16802 USA

This paper is in final form, and no version of it will be submitted for publication elsewhere.

For $G=GL_n$, the adjoint representation $\underline{gl}_n$ is the space M_n of all nXn matrices, with GL_n acting by matrix conjugation. As the center acts trivially, this action descends naturally to an action of PGL_n on M_n.

Our purpose here is to study the PGL_n-decomposition of the spaces $\underline{E}(M_n)$, for any polynomial functor $\underline{E}$. Such functors $\underline{E}$ are known to be sums of Schur functors $\underline{S}^{\pi}$ (π a partition); in particular, the pth symmetric power $\underline{S}^p$, the pth exterior power $\underline{E}^p$, and the pth tensor power $\underline{T}^p$, are all polynomial functors.

In this paper, we produce a way of separating out the dependence on n from the dependence on $\underline{E}$, for the character of the representation $\underline{E}(M_n)$ of PGL_n. We do this in §3 by introducing a <u>mixed-tensor parametrization</u> of PGL_n-irreducibles, and a <u>mixed tensor symbol mts V</u> of any PGL_n-representation V. The latter is a generating function for the multiplicities of PGL_n-irreducibles in V, where we use the mixed tensor parametrization to concoct the generating function. Then given $\underline{E}$, we can study the representation theoretic structure of $\underline{E}(M_n)$ as <u>a function of n</u> by studying the algebraic function mts $\underline{E}(M_n)$ as a function of n.

An immediate question is how $\underline{E}(M_n)$ behaves in this scheme as $n \to \infty$. In §4 and §5, we prove that $\underline{E}(M_n)$ <u>stabilizes</u> in a precise sense for n large, so that a limiting mixed tensor symbol $\lim_{n\to\infty}$ mts $\underline{E}(M_n)$ exists; we call this limit the <u>stable limit</u> or the <u>stable mixed tensor symbol</u> mts $\underline{E}(M_\infty)$. In fact, the limit is achieved for $n \geq 2p$.

The way this theory works is that mts $\underline{E}(M_n)$ and the stable limit take values in the ring $\Lambda(x;y)$ of symmetric functions in two infinite variable sets x and y; i.e., the ring of formal power series of bounded degree in $(x_1,x_2,\ldots,y_1,y_2,\ldots)$ which are symmetric in $x_1,x_2,\ldots$ and symmetric in $y_1,y_2,\ldots$.

In the case $\underline{E}$ is graded in a variable set w, e.g., w=(q) with $\underline{E}$ the graded symmetric algebra $\underline{E}=\underline{S}_q := \sum_{p\geq 0} \underline{S}^p q^p$, then these results carry through by dint of analyzing each graded component separately. But then mts $\underline{E}(M_n)$ and mts $\underline{E}(M_\infty)$ all take values in the larger ring $(\Lambda(x;y))[[w]]$. This brings in a key difference between the ungraded and graded cases: if $\underline{E}$ is an ordinary polynomial functor $\underline{E}$, then $\underline{E}$ has some degree p, and mts $\underline{E}(M_n)$ = mts $\underline{E}(M_\infty)$ for all $n \geq 2p$, while if $\underline{E}$ is graded, then $\underline{E}$ may have unbounded degree, so that mts $\underline{E}(M_\infty)$ is achieved by mts $\underline{E}(M_n)$ for no value of n.

In §6, we prove (Hidden Symmetry Theorem 6.3) that the stable limit always has a remarkable symmetry property which we now describe. For every value of n, mts $\underline{E}(M_n)$ has the same total degree in the x and y variables, so it, and its stable limit, can be expressed as a power series in the products $x_i y_j$ with integral coefficients. Theorem 6.3 says that the stable limit mts $\underline{E}(M_\infty)$ can in fact be uniquely expressed as a <u>fully symmetric function in the products</u> $x_i y_j$ (again with integral coefficients). This full symmetry is a very strong property; it fails for mts $\underline{E}(M_n)$ for n small compared to the degree of $\underline{E}$.

In §7, we give a more comprehensive theory of hidden symmetries, and we define resulting operations of <u>higher order skew-division</u> on symmetric functions.

In §8, we use the hidden symmetries to compute the stable mixed tensor characters of all symmetric powers of M_n, and then to study the generalized exponents of PGL_n. Other applications of hidden symmetries will be developed in our sequel paper [B3].

In some sense, the idea of a mixed tensor parametrization of irreducible GL_n-representations goes back to Littlewood, for he ([L]) investigated in 1944 the relevant formulae in the context of symmetric functions. Starting in the late 1960's, R.C. King and others studied mixed tensor representations in a series of papers (see, e.g., [Ki] and [Ki-P1).

Most recently, Kazuhiko Koike [Kk] has done a lot of work decomposing characters of classical groups using mixed tensor representations, and his "universal characters" have been used by Jun-ichi Matsuzawa ([Mt]) to study generalized exponents of classical groups.

The stable theory of the mixed tensor character turns out to carry a great deal of information on (i) the finite characters mts $\underline{E}(M_n)$, and hence on the usual characters ch $\underline{E}(M_n)$, and also on (ii) the structure of these spaces $\underline{E}(M_n)$. This will be demonstrated fully in two subsequent papers - [B2] (see [G] for an announcement of the results) and [B3], respectively.

We now sketch some key ideas and constructions from §3.

The essential difficulty in studying $\underline{E}(M_n)$ as function of n is already apparent in the case $\underline{E} = \underline{S}^1$ by considering limiting characters. The character of this representation, the trace on $\underline{S}^1(M_n)$ of a diagonal matrix with entries $z_1,\ldots,z_n$, is, for all $n\geq 1$,

(1.1)
$$\begin{aligned} ch_{Gl_n} M_n &= (z_1 +..+ z_n)(1/z_1 +..+ 1/z_n) \\ &= n + \sum_{i\neq j} z_i/z_j . \\ &= (n-1 + \sum_{i\neq j} z_i/z_j) + 1 \\ &= ch_{GL_n} \underline{pgl}_n + ch_{GL_n} \mathbb{C} . \end{aligned}$$

So while the splitting of this character into two pieces, the PGL_n-adjoint character of plus the trivial character, is constant as a function of n, the character itself has no limit as $n\to\infty$ (see second line). This means the ordinary character does not behave as well as the representation behaves as a function of n.

In general, we prove (Theorem 5.7) that the decomposition of $\underline{E}(M_n)$ is a stable function of n, i.e., constant for n large. This means that (i) the number of irreducible components of $\underline{E}(M_n)$, counted with multiplicity, is constant for n large, and that (ii) we give an abstract scheme (independent of choice of $\underline{E}$) of matching irreducibles of PGL_n with certain irreducibles of PGL_{n+1} so that, for n large, the PGL_n-multiplicities of $\underline{E}(M_n)$ exactly equal the PGL_{n+1}-multiplicities of $\underline{E}(M_{n+1})$. If $\underline{E}$ has degree p, then $n\geq 2p$ is large enough.

To find where such a "matching" might come from, we looked at the overall behavior of $\underline{E}(M_n)$ as n gets large.

The problem of the blowing-up of the character of $\underline{E}(M_n)$ as $n\to\infty$ is clearly caused by cancelling in the product (1.1) of z_i with $1/z_i$. To get around this, we extend the representation $M_n=\mathbb{C}^n\otimes(\mathbb{C}^n)^*$ from GL_n to the product group $GL_n \times GL_n$ in the obvious way: we embed GL_n as the diagonal subgroup $\{(g,g)\mid g\in GL_n\}$ inside $GL_n\times GL_n$, so that M_n is the restriction of the product representation $\mathbb{C}^n\otimes(\mathbb{C}^n)^*$ of $GL_n\times GL_n$. Then we have the following character expansions, when we evaluate (cf. §3) characters under the left GL_n in the variable set $(x_1,\ldots,x_n)$, and under the right GL_n in the variable set $(1/y_1,\ldots,1/y_n)$:

(1.2)
$$ch_{Gl_n\times Gl_n} \mathbb{C}^n\otimes(\mathbb{C}^n)^* = (x_1+..+x_n)(y_1+..+y_n) = \sum_{1\leq i,j\leq n} x_i y_j ,$$

and for any polynomial functor $\underline{E}$,

(1.3) $$\mathrm{ch}_{GL_n \times GL_n}\ \underline{E}(\mathbb{C}^n \otimes (\mathbb{C}^n)^*) = F(x_i y_j)_{1\le i,j\le n},$$

where F is the characteristic of $\underline{E}$, i.e., the function satisfying

(1.4) $$F(z_1,\ldots,z_m) = \mathrm{ch}_{Gl_m}(\underline{E}(\mathbb{C}^m)),$$

for all m.

As $n\to\infty$, (1.3) has a limit, namely $F(xy)=F(x_i y_j)_{i,j\ge 1}$, in the ring $\Lambda(x;y)$. Moreover, (1.3) and (1.4) separate out the n-dependence from the $\underline{E}$-dependence.

So behavior of $\underline{E}(M_n)$ under $GL_n \times GL_n$ is more controlled than under PGL_n. To relate the two, we can follow either (or both) of two routes.

Along the first route, we decompose $\underline{E}(M_n)$ into irreducible spaces under $GL_n \times GL_n$, and then to break these spaces into PGL_n-irreducibles. Indeed, the classical results (see §2) on intertwining of representations of symmetric groups and general linear groups give the exact decomposition of $\underline{E}(\mathbb{C}^n \otimes (\mathbb{C}^n)^*)$ into a direct sum of irreducibles of the form $(\underline{S}^\alpha \mathbb{C}^n)\otimes(\underline{S}^\beta \mathbb{C}^n)^*$; here α and β are partitions of equal size. Then the splitting of these spaces under PGL_n comes from "contractions" of cogredient and contragredient tensors.

So this first route suggested to us that we index PGL_n-irreducibles by how they sit inside these tensor product spaces $(\underline{S}^\alpha \mathbb{C}^n)\otimes(\underline{S}^\beta \mathbb{C}^n)^*$: we write $V^{(n)}_{\alpha\beta}$ for the Cartan piece (the highest piece) in $(\underline{S}^\alpha \mathbb{C}^n)\otimes(\underline{S}^\beta \mathbb{C}^n)^*$ as long as $\ell(\alpha)+\ell(\beta)\le n$. This gives a new way (Lemma 3.4) to index the irreducibles of PGL_n: $\{V^{(n)}_{\alpha\beta} \mid |\alpha|=|\beta|, \ell(\alpha)+\ell(\beta)\le n\}$ is the mixed tensor parametrization.

In this way, for each pair α,β of equal size, we construct a family $\{V^{(n)}_{\alpha\beta}\}_{n\ge 1}$ of representations of of PGL_n with n varying. For example, if $\alpha=\beta=(1)$, then $V^{(n)}_{\alpha\beta}$ is the adjoint representation of PGL_n. Then, to achieve the matching described above, we simply match $V^{(n)}_{\alpha\beta}$ to $V^{(n+1)}_{\alpha\beta}$.

The mixed tensor symbol of a PGL_n-representation U is a natural generating function inside $\Lambda(x;y)$ for the irreducible multiplicities of U, but constructed with regard to the mixed tensor parametrization:

(1.5) $\quad \text{mts } U := \sum \langle V^{(n)}_{\alpha\beta}, U \rangle s_{\alpha}(x) s_{\beta}(y),$

with summation over the set of pairs α,β satisfying the two conditions above (see Definition 3.7).

Along the second route, we introduce a linear injection of representation rings, the <u>extended character</u> ec: $\underline{R}(PGL_n) \to \underline{R}(GL_n \times GL_n)$ which associates a $GL_n \times GL_n$-character to each PGL_n-representation in the same spirit as (1.2) and (1.3). We compute the extended character of PGL_n-irreducibles V_λ, and then express ec $\underline{E}(M_n)$ as a linear combination of the ec V_λ.

In §3, we construct such a map ec by first proving (Theorem 3.9) that the set $\{ (\underline{S}^{\alpha}\mathbb{C}^n) \otimes (\underline{S}^{\beta}\mathbb{C}^n)^* \mid |\alpha| = |\beta|, \ell(\alpha) + \ell(\beta) \leq n \}$ can "play the role" in the representation ring of the set of irreducibles, in that this set of reducibles constitutes a linear basis of $\underline{R}(PGL_n)$. Then, in $\underline{R}(PGL_n)$, the irreducible classes V_λ are certain alternating sums of these reducible basis elements; this alternating sum formula (Theorem 4.10) actually goes back to Littlewood in the context of symmetric function expansions. With this basis in hand, we construct our map ec; note that the classes ec V_λ are then <u>virtual classes</u> in $\underline{R}(GL_n \times GL_n)$.

It turns out that for n large ($n \geq 2\deg(F)$), the extended representation of $F(M_n)$ has no virtual components, and is just the obvious candidate $F(\mathbb{C}^n \otimes (\mathbb{C}^n)^*)$ discussed above. So for n large, (1.3) gives the extended character ec $F(M_n)$, and the problem of decomposing $\underline{E}(M_n)$ into PGL_n-irreducibles becomes one of expressing the relatively simple $GL_n \times GL_n$-character (1.3) as a linear combination of the much more complicated alternating sums ec V_λ, V_λ irreducible. In §4, we develop the calculus for doing this in the stable range.

The interlacing of these two routes is evident throughout the paper. In particular, the blatant symmetry of (1.3) in the product variable set forces the rather subtle hidden symmmetry in xy of the mts.

<u>Acknowledgments.</u> The fundamental part of this work was carried out while I was an NSF Mathematical Sciences Postdoctoral Research Fellow (1981-82) at MIT and then a Member (1982-83) at the Institute for Advanced Studies. I express sincere thanks to the NSF and the IAS for their support.

It is a pleasure to thank Professor Marie-Paule Malliavin for her invitations to speak in the Algebra Seminar on this work and its sequels, and then to contribute to this volume.

Remark. This work was circulated in earlier forms (in particular, my 1983 handwritten manuscript "A stability theory for polynomial functions on gl_n"), has been refered to in the literature, and part of it was announced in [G]. All this was done under my former name "R.K. Gupta".

§2. SYMMETRIC FUNCTIONS AND INTERTWINING OF REPRESENTATIONS.

In this section, we set notations and sketch the theories of symmetric functions and representations of symmetric groups and general linear groups, together with their fundamental intertwining. We often refer to Macdonald's book [M].

All vector spaces and representations are complex linear ones. For any vector space V, we write $V^{[r]}$ for the r-fold tensor product $V^{\otimes r}$.

Throughout the paper, the symbols x,y,z, and $x^{(i)},y^{(i)},z^{(i)}$, will mean infinite variable sets, while $u,v,u^{(i)},v^{(i)}$ will mean arbitrary variable sets, for $i \geq 1$.

This section has five subsections I-V. Their dependence is sequential, except that II and III can be read in any order. These five subsections recall definitions and results, and we make some definitions in a new way. In III, we also give some proofs, to hopefully dispel the apparently prevalent notion that the theory of intertwining of symmetric group and general linear group representations lacks plain statements and straightforward proofs. This point is important here, because that theory is the bedrock of this paper.

In some sense, the material in II (symmetric functions) is equivalent to that in III (representations), via IV (characters). We adopt the view that II can be established on its own, or as a corollary of III and IV, while III is a theory on representation rings and is stronger than II. Concretely, this means that we prove the basic results of III in the calculus of symmetric group representations, rather than in the calculus of symmetric functions. However, in applications of III (such as this paper), symmetric functions often provide the most convenient language, and they support operations which are rather unnatural in the representation rings. In II, we present a basis-free sketch of symmetric functions, and in particular we construct Littlewood's internal product map abstractly.

I. Partitions and Tableaux.

We mainly follow the notations set out in [M,I,1] for partitions. A partition π is a non-increasing sequence $\pi=(\pi_1,\pi_2,..)$ of non-negative integers, with the π_i called the parts of π. The size or magnitude $|\pi|$ of π is the sum p of its parts; we often write "$\pi \vdash p$" to say "π is a partition of (size) p". The length $l(\pi)$ is the number of non-zero parts. In writing a partition, we often ignore some or all of trailing zeroes. The zero partition (0) is distinguished from the empty partition $\emptyset$.

The diagram, or Young diagram, or shape, of π is the diagram of $l(\pi)$ rows with π_i boxes in the ith row (the rows are justified to the left). We often confuse a partition with its diagram. The transpose or conjugate partition π' of π is obtained by interchanging the rows and columns of π. We write $\alpha\subseteq\beta$, iff the diagram of α fits inside the diagram of β, so iff $\alpha_i\leq\beta_i$ for all $i\geq 1$.

The row-sum $\alpha+\beta$ of two non-empty partitions α and β is the new partition $(\alpha_1+\beta_1,\alpha_2+\beta_2,..)$. If α or β is empty, then $\alpha+\beta$ is empty.

We define, for each positive integer n and partition μ, the partition $(n\backslash\mu)$ as follows: $(n\backslash\mu)$ is empty unless $l(\mu)\leq n$ and μ is non-empty, in which case

$$(n\backslash\mu) := (\mu_1-\mu_n,\ \mu_1-\mu_{n-1},\dots,\ \mu_1-\mu_1).$$

Tableaux of shape π are obtained by putting numbers (positive integers) in the boxes of the diagram of π, with one number per box, in such a way that all rows (read left to right) and all columns (read top to bottom) are non-decreasing sequences. The adjectives column-strict and row-strict for tableaux are then self-explanatory. (Here, we differ from Macdonald's terminology in [M], where all tableaux are required to be column-strict.)

Each partition π of p defines (up to conjugacy) a subgroup $\mathcal{S}_\pi := \mathcal{S}_{\pi_1}\times\mathcal{S}_{\pi_2}\cdots\times\mathcal{S}_{\pi_{l(\pi)}}$ of the symmetric group $\mathcal{S}_p$.

II. Symmetric Functions - Formal Properties.

Given a set $u=(u_1,\ldots,u_n)$ of n variables, let $\Lambda(u)=\Lambda(u_1,\ldots,u_n)$ be the ring of symmetric polynomials in u, i.e., the ring of polynomials in $\mathbb{Z}[u_1,\ldots,u_n]$ invariant under the natural permutation action of the symmetric group $\mathcal{S}_n$. Then $\Lambda(u) = \bigoplus_{p\geq 0} \Lambda^p(u)$ is graded by polynomial degree deg.

Write $x\langle n\rangle$ for the truncated variable set $(x_1,\ldots,x_n)$. The <u>product</u> variable set of u and v is $uv:=(u_i v_j)_{i,j\geq 1}$; the <u>sum</u> variable set is the concatenation $u+v:=(u_1,u_2,\ldots,v_1,v_2,\ldots)$. We consider a set u of n variables as the same thing as an infinite variable set u with $u_{n+1}=u_{n+2}=\ldots=0$. Write $|w|$ for the cardinality of a set w.

The collection of rings $(\Lambda(x\langle n\rangle))_{n\geq 1}$, along with the restriction homomorphisms $\Lambda(x\langle n+1\rangle)\rightarrow\Lambda(x\langle n\rangle)$ sending $f(x_1,\ldots,x_{n+1})$ to $f(x_1,\ldots,x_n,0)$, forms an inverse system. The inverse limit $\Lambda(x)=\Lambda(x_1,x_2,\ldots)$ (respectively, $\Lambda((x))=\Lambda((x_1,x_2,\ldots))$) in the category of graded rings (resp., rings) is the ring of symmetric functions (resp., symmetric power series) in x. Note symmetric functions are not polynomials, but instead are power series of bounded degree.

A useful feature of the graded decomposition of $\Lambda(x)$ is that if $n\geq p\geq 0$, then the restriction map from $\Lambda^p(x)$ to $\Lambda^p(x\langle n\rangle)$ is a linear isomorphism. This says: symmetric functions of degree p are completely determined in p variables. The inverse to this restriction isomorphism is the <u>full symmetrization</u> of a symmetric function $f(x_1,\ldots,x_n)$ into x.

Usually, one works with symmetric functions in infinitely many variables, and then obtains the results for finite variable sets by restriction.

Whenever possible, we suppress the variable set so that we have "abstract symmetric functions". We then write $\Lambda_n = \bigoplus_{p\geq 0} \Lambda_n^p$ for $\Lambda(x\langle n\rangle)$ and $\Lambda = \bigoplus_{p\geq 0} \Lambda^p$ and $\Lambda(x)$. Given $f\in\Lambda$, f can be evaluated to $f(u)\in\Lambda(u)$ for any u. In general, "symmetric function" means a symmetric function in infinitely many variables, unless otherwise specified.

We define the <u>index</u> ind(f) of a symmetric function $f\in\Lambda$ to be the smallest value of r such that the full symmetrization of $f(x_1,\ldots,x_r)$ from $\Lambda(x\langle r\rangle)$ to $\Lambda(x)$ is equal to $f(x)$. So ind(f) is the smallest number of variables needed "to see f", and $\mathrm{ind}(f) \leq \deg(f)$ always. The classical stability property of multiplication of symmetric functions is: if $f, g \in \Lambda$, then $\mathrm{ind}(fg) = \mathrm{ind}(f) + \mathrm{ind}(g)$, so that the product $f(x)g(x)$ can be computed by taking the full symmetrization of $f(x\langle n\rangle)g(x\langle n\rangle)$, as long as $n \geq \mathrm{ind}(f) + \mathrm{ind}(g)$.

The classical bases of Λ^p (resp., Λ^p_n) are naturally indexed by partitions π of size p (resp., and length n or less). In each of these bases $(f_\pi)_\pi$ by Schur functions, power-sum symmetric functions, etc., the zero partition gives $f_{(0)}=1$, while the empty partition gives $f_\emptyset=0$.

Define in $(\Lambda(u))[[q]]$, with q an indeterminate, the <u>complete homogeneous series</u>

$$H_q(u) := \prod_{i\geq 1} (1-qu_i)^{-1} = \sum_{p\geq 0} h_p(u)q^p,$$

and the <u>elementary series</u>

$$A_q(u) := \prod_{i\geq 1} (1+qu_i) = \sum_{p\geq 0} a_p(u)q^p.$$

The degree p components h_p and a_p are the pth <u>complete homogeneous</u> and pth <u>elementary symmetric</u> function, respectively. Obviously, we have the identity $H_q(u)A_{-q}(u)=1$. If q is assigned an integer value t, then $H_t(u)$ and $A_t(u)$ lie in $\Lambda\langle\langle u\rangle\rangle$. Set $H(u):=H_1(u)$.

For any two finite variable sets u and v, set $\Lambda(u;v):=\Lambda(u)\otimes\Lambda(v)$. Let $\Lambda(x;y)$ (resp., $\Lambda\langle\langle x;y\rangle\rangle$) be the inverse limit in the category of graded rings (resp., of rings) of the inverse system $(\Lambda(x\langle n\rangle;y\langle n\rangle))_{n\geq 1}$. This is the ring of bounded degree power series (resp., of all power series in x and y) which are symmetric in the individual variable sets x and y. Then $\Lambda(x;y):=\Lambda(x)\otimes\Lambda(y)$, and $\Lambda(x;y)$ is our canonical model for $\Lambda\otimes\Lambda$. Write elements of $\Lambda(x;y)$ as $t(x;y)$; these are "symmetric functions of x and y". Note $\Lambda(x+y)$, the ring of symmetric functions of x+y, is a proper subring of $\Lambda(x;y)$.

These considerations for two variable sets extend in the obvious manner to any finite collection of variable sets. In particular, we define the <u>index</u> of $t(x^{(1)};..;x^{(k)}) \in \Lambda(x^{(1)};..;x^{(k)})$ to be the smallest value of r such that the symmetrization of $t(x^{(1)}\langle r\rangle;..;x^{(k)}\langle r\rangle)$ to $\Lambda(x^{(1)};..;x^{(k)})$ is equal to $t(x^{(1)};..;x^{(k)})$.

The classical scalar product $\langle\ ,\ \rangle_\Lambda$ on Λ is determined by the symmetric series $H(xy)$ as follows. Recall that a non-degenerate bilinear form $k(\ ,\ )$ on a finite-rank free abelian group M is the same thing as a $\mathbb{Z}$-bilinear map $k: M\times M \to \mathbb{Z}$ inducing an isomorphism between M and its dual. In particular, k is an element of $(M\otimes M)^*$, so that identifying $(M\otimes M)^*$ with $M\otimes M$ via k will identify k to element $|k|$ of $M\otimes M$. We call $|k|$ the <u>tensor invariant</u> of k. This is, in fact, the unique element in $M\otimes M$ with the property that two bases $(e_1,\dots,e_r)$ and $(f_1,\dots,f_r)$ of M are dual under k iff $\sum_{i=1}^{r} e_i\otimes f_i = |k|$.

The <u>natural scalar product</u> $\langle\ ,\ \rangle_{\Lambda(u)}$, or simply $\langle\ ,\ \rangle_u$, is the bilinear form on $\Lambda(u)$ with tensor invariant equal to $H(uv)$, where v is another variable set of the same cardinality as u, so that $\Lambda(u;v) \cong \Lambda(u)\otimes\Lambda(u)$. Under $\langle\ ,\ \rangle_u$, homogeneous components $\Lambda^p(u)$ of different degree are orthogonal, and each Λ^p has a (canonical, up to multiplication by ± 1) orthonormal basis.

In this paper, "$\langle f,g\rangle$", with $f,g\in\Lambda$, means $\langle f,g\rangle_\Lambda$.

The <u>natural scalar product</u> extends naturally via the product to rings of symmetric functions in several variable sets; for two sets we obtain $\langle\ ,\ \rangle_{\Lambda(u;v)} = \langle\ ,\ \rangle_{u;v}$ on $\Lambda(u;v)$ given by

$$\langle d(u)e(v),\ f(u)g(v)\rangle_{u;v} = \langle d(u),\ f(u)\rangle_u \langle e(v),\ g(v)\rangle_v.$$

For the rest of this subsection, let $\langle\ ,\ \rangle_k$ be the natural scalar product on $\Lambda(x^{(1)};\dots;x^{(k)})$.

From now on, all adjoints of maps between rings of symmetric functions will be constructed with respect to these natural scalar products.

The adjoint of the map $\Lambda(z)\to\Lambda(x;y)$ sending $f(z)$ to $f(x+y)$ is the multiplication map $M: \Lambda(x;y)\to\Lambda(z)$ sending $g_1(x)g_2(y)$ to $g_1(z)g_2(z)$. The adjoint of the map $\Lambda(z)\to\Lambda(x;y)$ sending $f(z)$ to $f(xy)$ is the <u>internal product</u> map $I: \Lambda(x;y)\to\Lambda(z)$; we write $(g_1*g_2)(z)$ for $I(g_1(x)g_2(y))$.

So for all $e,f,g\in\Lambda$,

(2.1) $$\langle e(x+y),\ f(x)g(y)\rangle_{x;y} = \langle e,\ fg\rangle;$$

(2.2) $$\langle e(xy),\ f(x)g(y)\rangle_{x;y} = \langle e,\ f*g\rangle.$$

These identities extend to more variable sets. For $e, f_1, \ldots, f_k \in \Lambda$,

(2.1') $\langle e(x^{(1)}+..+x^{(k)}),\ f_1(x^{(1)})\cdot\cdot f_k(x^{(k)})\rangle_k = \langle e,\ f_1\cdot\cdot f_k\rangle$;

(2.2') $\langle e(x^{(1)}\cdot\cdot x^{(k)}),\ f_1(x^{(1)})\cdot\cdot f_k(x^{(k)})\rangle_k = \langle e,\ f_1*\cdot\cdot*f_k\rangle$.

From these identities, we get expansion formulae. For $i=1,\ldots,k$, let $\{b_{i;\lambda}\}_\lambda$ be a basis of Λ, with $\{\bar{b}_{i;\lambda}\}_\lambda$ the dual basis under $\langle\ ,\ \rangle$. Then for all $e\in\Lambda$,

(2.1") $e(x^{(1)}+..+x^{(k)}) = \sum \langle e,\ \bar{b}_{1;\lambda_1}\cdot\cdot\bar{b}_{k;\lambda_k}\rangle b_{1;\lambda_1}(x^{(1)})\cdot\cdot b_{k;\lambda_k}(x^{(k)})$,

(2.2") $e(x^{(1)}\cdot\cdot x^{(k)}) = \sum \langle e,\ \bar{b}_{1;\lambda_1}*\cdot\cdot*\bar{b}_{k;\lambda_k}\rangle b_{1;\lambda_1}(x^{(1)})\cdot\cdot b_{k;\lambda_k}(x^{(k)})$,

with both summations over k-tuples $\lambda_1,\ldots,\lambda_k$ of partitions.

If we replace $x^{(1)},\ldots,x^{(k)}$ by arbitrary variable sets $u^{(1)},\ldots,u^{(k)}$, then (2.1") and (2.2") remain valid, while (2.1') holds if $|u^{(i)}|\geq \mathrm{ind}(e)$ for all i, and (2.2') holds if $|u^{(i)}|\geq \deg(e)$ for all i.

For all $f,g\in\Lambda$, we have $\langle f,\ g\rangle = \langle f(x)g(y),\ H(xy)\rangle_{x;y} = (f*g)(1)$.

For each $f\in\Lambda$, <u>skew-division</u> by f is the linear map $D(\ /f)\colon \Lambda\to\Lambda$, sending e to $D(e/f)$, which is adjoint to multiplication by f. We write also $D_x(\ /f(x))$ for skew-division on $\Lambda(x)$ by $f(x)$. Explicitly, the defining equation for the <u>skew-quotient</u> $D(e/f)$ is $\langle e,fg\rangle=\langle D(e/f),g\rangle$, for all $e,f,g\in\Lambda$. Note that $D(D(e/f)/g)=D(e/(fg))$.

Similarly, define skew-division $D_{x;y}(\ /t(x;y))\colon \Lambda(x;y)\to\Lambda(x;y)$, for $t(x;y)\in\Lambda(x;y)$, as the adjoint of multiplication by $t(x;y)$.

On the other hand, for all $f\in\Lambda$, the internal product with f gives a self-adjoint map $\Lambda\to\Lambda$. Note that for $e\in\Lambda$, $D(e/f)$ and $e*f$ make sense even if f is a symmetric power series. Then $e*H=e$.

III. <u>Representations of Symmetric and General Linear Groups.</u>

Let G be a complex reductive algebraic group G. A <u>rational representation</u> of G is a locally finite representation of G whose matrix coefficients are regular functions on G. In this paper, we will generally assume (without explicit comment) that representations are rational <u>with (each graded component having) finite multiplicities</u>.

Let $\underline{R}(G)$ be the rational representation ring of G, the ring (under direct sum and tensor product) of isomorphism classes of (virtual) finite-dimensional rational G-representations. Also write $\underline{CH}(G)$ for the corresponding character ring. The natural scalar product on $\underline{R}(G)$ is $\langle\ ,\ \rangle_G = \dim \mathrm{Hom}_G(\ ,\ )$.

The irreducible representations of $\mathcal{S}_p$ are naturally labeled, as say $(J_\pi)_{\pi\vdash p}$, by the partitions π of p, in a number of equivalent ways. In $\mathbb{Z}[x_1,\ldots,x_p]$, let $d(x_{i_1},\ldots,x_{i_t})$ be the Vandermonde determinant in the t variables. Then define J_π to be the representation of $\mathcal{S}_p$ spanned in $\mathbb{Z}[x_1,\ldots,x_p]$ by the product

$$d(x_1,\ldots,x_{t_1})d(x_{t_1+1},\ldots,x_{t_1+t_2})\ldots,$$

where $t_1,t_2,\ldots$ are the parts of π', the column lengths of π. It is easy to show that the J_π are a complete set of inequivalent irreducibles. To do this, one considers anti-invariant vectors under products of symmetric groups inside $\mathcal{S}_p$. An anti-invariant vector in a representation space of $\mathcal{S}_t$ is one which transforms under $\mathcal{S}_t$ according to the sign character.

Note J_π is the trivial representation $triv_p$ when π is the row partition (p), and J_π is the sign representation sgn_p when π is the column partition (1^p). For all α, $J_{\alpha'} = J_\alpha \otimes sgn_p$.

The outer product J#K of a representation J of $\mathcal{S}_p$ with one K of $\mathcal{S}_q$ is the new representation of $\mathcal{S}_{p+q}$ obtained by inducing up the product $J\otimes K$ from $\mathcal{S}_p \times \mathcal{S}_q$. Outer product induces a bilinear map $R(\mathcal{S}_p)\times R(\mathcal{S}_q)\longrightarrow R(\mathcal{S}_{p+q})$. Let $Res_{p,q}$ be the restriction operator mapping the opposite way. Define coefficients

$$(2.3) \qquad m^\pi_{\alpha\beta} := \langle J_\pi, J_\alpha \# J_\beta \rangle_{\mathcal{S}_{p+q}}$$

for partitions α,β,π of p,q,and p+q. Set $m^\pi_{\alpha\beta}:=0$ if $|\alpha|+|\beta|\neq|\pi|$. By Frobenius Reciprocity,

$$(2.3') \qquad m^\pi_{\alpha\beta} = \langle Res_{p,q}(J_\pi), J_\alpha \otimes J_\beta \rangle_{\mathcal{S}_p\times\mathcal{S}_q}.$$

By construction, then, J_π is a component of

$$sgn_\pi := sgn_{t_1} \# sgn_{t_2} \# \ldots$$

where $\pi' = (t_1,t_2,\ldots)$.

Define the <u>index</u> of a symmetric group representation J to be the largest value of t such that J has non-trivial anti-invariants upon to restriction to $\mathcal{S}_t \times \mathcal{S}_1 \times \cdots \times \mathcal{S}_1$. Then the index of J_π is easily seen to be just the length $l(\pi)$.

As symmetric group representations are self-dual, we have

(2.4) $$\langle J_1, J_2 \rangle_{\mathcal{S}_p} = \dim (J_1 \otimes J_2)^{\mathcal{S}_p},$$

for $J_1, J_2 \in R(\mathcal{S}_p)$. For partitions $\alpha_1, \ldots, \alpha_k$, define coefficients by

(2.5) $$c_{\alpha_1 \cdots \alpha_k} := \dim (J_{\alpha_1} \otimes \cdots \otimes J_{\alpha_k})^{\mathcal{S}_p}$$

if $\alpha_1, \ldots, \alpha_k$ are partitions of a common size p, and $c_{\alpha_1 \cdots \alpha_k} := 0$ otherwise. Then

(2.5') $$c_{\alpha\beta\gamma} = \langle J_\alpha, J_\beta \otimes J_\gamma \rangle = \langle J_\beta, J_\alpha \otimes J_\gamma \rangle = \langle J_\gamma, J_\alpha \otimes J_\beta \rangle,$$

where $\langle \, , \, \rangle = \langle \, , \, \rangle_{\mathcal{S}_p}$.

The direct sum $R(\mathcal{S}) = \bigoplus_{p \geq 0} R(\mathcal{S}_p)$ of the representation rings is a graded commutative ring under the direct sum and outer product operations. The sum of the usual scalar products $\langle \, , \, \rangle_{\mathcal{S}_p}$ gives a natural scalar product $\langle \, , \, \rangle_{\mathcal{S}}$ on $R(\mathcal{S})$. And tensor product of symmetric group representations gives still another operation on $R(\mathcal{S})$, which we will call <u>internal tensor product</u> in $R(\mathcal{S})$.

For $G = GL_n$, we distinguish rational representations from <u>polynomial representations</u>, those whose matrix coefficients are polynomials in the matrix entries of GL_n. Then rational and polynomial representations of GL_n differ only in that the former admit negative powers of the 1-dimensional determinant representation $\det_n$. Moreover, restriction identifies the rational representation ring of SL_n with the quotient of the polynomial representation ring of GL_n by its ideal generated by $(\det_n - 1)$. Write $\underline{R}_0(GL_n)$ and $\underline{CH}_0(GL_n)$ for the polynomial representation and character rings.

Given a GL_n-irreducible U, its restriction to SL_n and its descent (if it exists) to PGL_n remain irreducible, as the center of GL_n "acts by scalars" on U. Every irreducible of SL_n or PGL_n "comes from" (by restriction or descent) an irreducible of GL_n.

A fact from the representation theory of GL_n which we shall freely use is that every irreducible polynomial GL_n-representation occurs in some tensor power of $\mathbb{C}^n$.

The degree deg(U) of a polynomial representation U of GL_n or of a rational representation U of SL_n or PGL_n is the smallest non-negative integer r such that each irreducible component of U occurs in $(\mathbb{C}^n)^{[r]}$. The degree under GL_n is just determined by the action of the center of GL_n: for the scalar matrix $tI_{n,n}$ acts on U by some polynomial g(t), and the degree of g is then the degree of U.

With this notion of degree, $\underline{R}(GL_n)$ becomes a graded commutative ring $\bigoplus_{p\geq 0} \underline{R}^p(GL_n)$ with scalar product.

Via lifting of actions, we may regard irreducible rational PGL_n-representations as the same thing as irreducible rational representations of GL_n or of SL_n on which the center acts trivially. This in turn identifies the former with irreducible rational representations of SL_n whose degree is an integral multiple of n.

The essential link between the representation theories of the symmetric and general linear groups is forged by the joint decomposition of the pth tensor power of $\mathbb{C}^n$ under the actions of $\mathcal{S}_p$ (by permutation of tensor position) and GL_n. The fundamental result is this (see [W]):

Theorem 2.6. Let p be a non-negative integer and V a complex vector space of dimension n. Then the $\mathcal{S}_p$-isotypic components in $V^{[p]}$ coincide with the GL(V)-isotypic components, and $V^{[p]}$ is multiplicity free under the joint action of $\mathcal{S}_p \times GL(V)$. Furthermore, an $\mathcal{S}_p$-irreducible J_π occurs in $V^{[p]}$ iff $\ell(\pi)\leq n$; in particular, all $\mathcal{S}_p$-irreducibles occur iff $n\geq p$.

Proof. The first assertion is formally equivalent to the fact that the subrings A and B of End $V^{[p]}$ generated by $\mathcal{S}_p$ and GL(V) are full commutants of each other. This is what we will prove.

Identifying End $V^{[p]}$ with $(\text{End } V)^{[p]}$, we see that B is the span of the "pth power" tensors $x\otimes..\otimes x$, $x\in GL(V)$, hence B is the pth symmetric power $\underline{S}^p(\text{End } V)$. Now the action of $\mathcal{S}_p$ on End $V^{[p]}$ by conjugation identifies with the action of $\mathcal{S}_p$ on $(\text{End } V)^{[p]}$ by permutation of the p tensor positions. So the fact that $\underline{S}^p(\text{End } V)$ is the fixed space in

$(\mathrm{End}\ V)^{[p]}$ under the permutation action means that B is the commuting ring of A.

On the other hand, A is semi-simple, so A is its own double commutant. Thus A is the commuting ring of B.

To prove the second assertion of the Theorem, we first observe that the space of anti-invariants in $V^{[p]}$ under the action of $H_t = \mathcal{S}_t \times \mathcal{S}_1 \cdots \mathcal{S}_1$, for $1 \le t \le p$, is the subspace $\underline{E}^t(V) \otimes V^{[p-t]}$. Thus $V^{[p]}$ has non-trivial H_t-anti-invariants only if $t \le n$. So if J_π occurs in $V^{[p]}$, then $\ell(\pi) \le n$. Conversely, if $\ell(\pi) \le n$, then $V^{[p]}$ contains the $\mathcal{S}_p$-representation sgn_π, and hence contains J_π. []

Define the map $\theta_{p,V}$ on representations J of $\mathcal{S}_p$ by:

(2.7) $$\theta_{p,V}(J) := \mathrm{Hom}_{\mathcal{S}_p}(J, V^{[p]}) = (J \otimes V^{[p]})^{\mathcal{S}_p}.$$

Extending linearly, we get an induced map (of the same name)

(2.8) $$\theta_{p,V}\colon \underline{R}(S_p) \to \underline{R}_0^p(GL(V))$$

on representation rings, and

<u>Corollary 2.9.</u> For all p and V, $\theta_{p,V}$ is a linear surjection; if $n = \dim(V) \ge p$, then $\theta_{p,V}$ is an isomorphism which sends irreducibles to irreducibles and respects the scalar products.

For each partition π, the <u>Schur functor</u> $\underline{S}^\pi$ on vector spaces is given by

(2.10) $$\underline{S}^\pi(V) := \theta_{p,V}(J_\pi).$$

So $\underline{S}^{(p)}$ is the pth symmetric power $\underline{S}^p$, $\underline{S}^{(1^p)}$ is the pth exterior power $\underline{E}^p$, and $\underline{S}^\pi(V) = 0$ unless $\dim(V) \ge \ell(\pi)$. We may write $\underline{S}^J(V)$ for $\theta_{p,V}(J)$.

The Theorem says that the decomposition of $V^{[p]}$ produces a pairing between certain irreducibles of $\mathcal{S}_p$ and certain irreducibles of GL(V); explicitly, this is

(2.11) $$V^{[p]} = \bigoplus_{\ell(\pi) \le n} J_\pi \otimes \underline{S}^\pi(V)$$

Each space (2.10) is an irreducible representation of GL(V), of degree $|\pi|$ if $\ell(\pi)\leq n$, and also an irreducible representation of SL(V), of degree $|\pi|$ if $\ell(\pi)<n$. Moreover, the theorem implies that $(\underline{S}^{\pi}(\mathbb{C}^n))_{\ell(\pi)\leq n}$ is an exhaustive, repetition free list of the irreducible polynomial representations of GL_n, and $(\underline{S}^{\pi}(\mathbb{C}^n))_{\ell(\pi)<n}$ is the same sort of list of the irreducible rational representations of SL_n.

The dual GL_n-representation to $\underline{S}^{\pi}(\mathbb{C}^n)$ is $(\det_n)^{-m}\,\underline{S}^{(n\setminus\pi)}(\mathbb{C}^n)$ if $\ell(\pi)\leq n$ and $m=\pi_1$.

<u>Proposition 2.12.</u> Let U and V be any two vector spaces. Then the decomposition under GL(U)XGL(V) of $\underline{S}^{\pi}(U\otimes V)$, for any partiton π of p is (see 2.5)

$$(2.13)\qquad \underline{S}^{\pi}(U\otimes V) = \bigoplus_{\alpha,\beta\vdash p} c_{\pi\alpha\beta}\,\underline{S}^{\alpha}(U)\otimes\underline{S}^{\beta}(V).$$

<u>Proof.</u> (2.11) gives

$$(2.14)\qquad (U\otimes V)^{[p]} = \bigoplus_{\alpha,\beta\vdash p} J_{\alpha}\otimes J_{\beta}\otimes S^{\alpha}(U)\otimes S^{\beta}(V),$$

so that

$$\underline{S}^{\pi}(U\otimes V) = \mathrm{Hom}_{\mathcal{S}_p}(J_{\pi},(U\otimes V)^{[p]}) = \bigoplus_{\alpha,\beta\vdash p} \mathrm{Hom}_{\mathcal{S}_p}(J_{\pi},J_{\alpha}\otimes J_{\beta})\underline{S}^{\alpha}(U)\otimes\underline{S}^{\beta}(V).$$

[]

As a special case, we get the <u>Cauchy Decomposition</u> for each p:

$$(2.15)\qquad \underline{S}^{p}(U\otimes V) = \bigoplus_{\alpha\vdash p} \underline{S}^{\alpha}(U)\otimes\underline{S}^{\alpha}(V),$$

and also

$$(2.16)\qquad \underline{E}^{p}(U\otimes V) = \bigoplus_{\alpha\vdash p} \underline{S}^{\alpha}(U)\otimes\underline{S}^{\alpha'}(V)$$

A more general statement than (2.12), with the same proof, is

$$(2.17)\qquad \underline{S}^{J}(U\otimes V) = \bigoplus_{\alpha,\beta\vdash p} (J,J_{\alpha}\otimes J_{\beta})_{\mathcal{S}_p}\,\underline{S}^{\alpha}(U)\otimes\underline{S}^{\beta}(V),$$

for any representation J of $\mathcal{S}_p$.

The second main decomposition concerns, simultaneously, tensor products of GL(V)-representations and restrictions from GL(U⊗V) to GL(U)XGL(V).

<u>Proposition 2.18.</u> Let π, α, β be partitions, and U and V vector spaces. Then

(i) $\underline{S}^{\pi}(V)$ occurs (as a non-zero component) in the GL(V)-representation $\underline{S}^{\alpha}(V) \otimes \underline{S}^{\beta}(V)$ only if $|\pi| = |\alpha| + |\beta|$. We have (see 2.3)

$$\underline{S}^{\alpha}(V) \otimes \underline{S}^{\beta}(V) = \bigoplus_{\alpha\beta} m^{\pi}_{\alpha\beta}\, \underline{S}^{\pi}(V). \tag{2.19}$$

(ii) We have the decomposition under GL(U)XGL(V),

$$\underline{S}^{\pi}(U \oplus V) = \bigoplus_{\alpha\beta} m^{\pi}_{\alpha\beta}\, \underline{S}^{\alpha}(U) \otimes \underline{S}^{\beta}(V). \tag{2.20}$$

<u>Proof.</u> (i) The first statement is forced by degree considerations. Let $p = |\alpha|$, $q = |\beta|$. The second statement follows from (2.3):

$$\begin{aligned} \underline{S}^{\alpha}(V) \otimes \underline{S}^{\beta}(V) &= (J_{\alpha} \otimes V^{[p]})^{\mathcal{S}_p} \otimes (J_{\beta} \otimes V^{[q]})^{\mathcal{S}_q} \\ &= [(J_{\alpha} \otimes J_{\beta}) \otimes \mathrm{Res}_{p,q}(V^{[p]} \otimes V^{[q]})]^{\mathcal{S}_p \times \mathcal{S}_q} \\ &= [(J_{\alpha} \# J_{\beta}) \otimes V^{[p+q]}]^{\mathcal{S}_{p+q}}. \end{aligned}$$

This last equality of spaces is given by the Frobenous Reciprocity isomorphism. (ii) Follows similarly from (2.3'). □

(2.19) and (2.20) have immediate generalizations (with the same proof): the identities hold if we replace $\underline{S}^{\pi}$ with $\underline{S}^{J}$, for $J \in R(\mathcal{S}_p)$, and $m^{\pi}_{\alpha\beta}$ with $\langle J, J_{\alpha} \# J_{\beta} \rangle_{\mathcal{S}_p}$. We have also

<u>Corollary 2.21.</u> Let V be a vector space of dimension n, and let π, α, β be partitions of lengths strictly less than n. Then $\underline{S}^{\pi}(V)$ occurs in the SL(V)-representation $\underline{S}^{\alpha}(V) \otimes \underline{S}^{\beta}(V)$ only if $|\alpha| + |\beta| - |\pi| = rn$ for some non-negative integer r, in which case

$$\langle \underline{S}^{\pi}(V), \underline{S}^{\alpha}(V) \otimes \underline{S}^{\beta}(V) \rangle_{SL(V)} = m^{\mu}_{\alpha\beta}, \tag{2.22}$$

where μ is the row-sum of π with r columns of length n.

Using indices one can check

<u>Lemma 2.23.</u> For any partitions π, α, β, $m^{\pi}_{\alpha\beta} = 0$ unless $l(\pi) \leq l(\alpha) + l(\beta)$.

On the other hand, also: $m^{\pi}_{\alpha\beta} = 0$ unless $\alpha \subseteq \pi$ and $\beta \subseteq \pi$.

Finally, let us give the Fundamental Isomorphism. The family of representation rings $(\underline{R}_0(GL_n))_{n\geq 1}$, together with the graded homomorphisms $\underline{R}_0(GL_{n+1}) \longrightarrow \underline{R}_0(GL_n)$ induced by the natural inclusions of GL_n into GL_{n+1} (i.e., the ones corresponding to the splittings $\mathbb{C}^{n+1}=\mathbb{C}^n \oplus \mathbb{C}$), forms an inverse system of graded rings. Let $\underline{R}_0(GL)$ denote the inverse limit, with $\langle\ ,\ \rangle_{GL}$ the induced scalar product. Similarly (and equivalently), we have an inverse limit $\underline{CH}_0(GL)$ of the character rings.

Theorem 2.6 gives

<u>Theorem 2.24.</u> The direct sum of the maps $\theta_{p,V}$ induces an isomorphism θ: $\underline{R}(\mathcal{S}) \longrightarrow \underline{R}_0(GL)$, sending J to $\underline{S}^J$, of graded rings which also respects the scalar products.

A (covariant) functor $\underline{F}$ from the category of complex vector spaces to itself is a <u>polynomial functor</u> iff linear maps between vector spaces transform in a polynomial manner under the application of $\underline{F}$ (see [M,I,A1.1]). Schur functors are polynomial, and in fact all polynomial functors are $\mathbb{Z}$-linear combinations of Schur functors (see [M,I,A5.4]).

We may regard $\underline{R}_0(GL)$ as the ring of polynomial functors, with each polynomial functor $\underline{F}$ identified to the coherent sequence $(\underline{F}(\mathbb{C}^n))_{n\geq 1}$. Tensor product of general linear group representations goes over into product of polynomial functors. Given a polynomial functor $\underline{F}$, <u>skew-division</u> by $\underline{F}$ is the linear map $D(\ /\underline{F})$: $\underline{R}_0(GL) \longrightarrow \underline{R}_0(GL)$, sending $\underline{E}$ to $D(\underline{E}/\underline{F})$, which is adjoint (with respect to $\langle\ ,\ \rangle_{GL}$) to multiplication by $\underline{F}$.

On the other hand, the internal tensor product operation in $R(\mathcal{S})$ translates under Theorem 2.4 into the <u>internal product</u> operation $*$ on polynomial functors, and we have

$$\underline{S}^{\alpha} * \underline{S}^{\beta} = \sum_{\pi} c_{\pi\alpha\beta}\, \underline{S}^{\pi}. \tag{2.25}$$

Finally, let us note that the generalizations to several spaces of the decomposition results of this subsection follow at once. Let $U_1,\ldots,U_k$ be vector spaces, and let $\underline{F}=\underline{S}^J$ be a polynomial functor. Then we have the decompositions under $GL(U_1)\times\ldots\times GL(U_k)$:

(2.26) $\underline{F}(U_1 \otimes .. \otimes U_k)$

$$= \oplus \langle J, J_{\alpha(1)} \otimes .. \otimes J_{\alpha(k)} \rangle_S \underline{S}^{\alpha(1)}(U_1) \otimes .. \otimes \underline{S}^{\alpha(k)}(U_k)$$

$$= \oplus \underline{S}^{\alpha(1)}(U_1) \otimes .. \otimes \underline{S}^{\alpha(k-1)}(U_{k-1}) \otimes (\underline{F} * \underline{S}^{\alpha(1)} * .. * \underline{S}^{\alpha(k-1)})(U_k),$$

with the first summation over k-tuples $\alpha(1),\ldots,\alpha(k)$ of partitions and the second over (k-1)-tuples $\alpha(1),\ldots,\alpha(k-1)$; and

(2.27) $\underline{F}(U_1 \oplus .. \oplus U_k)$

$$= \oplus \langle J, J_{\alpha(1)} \# .. \# J_{\alpha(k)} \rangle_S \underline{S}^{\alpha(1)}(U_1) \otimes .. \otimes \underline{S}^{\alpha(k)}(U_k)$$

$$= \oplus \underline{S}^{\alpha(1)}(U_1) \otimes .. \otimes \underline{S}^{\alpha(k-1)}(U_{k-1}) \otimes D(\underline{F}/(\underline{S}^{\alpha(1)} .. \underline{S}^{\alpha(k-1)}))(U_k),$$

with the same summation indices.

IV. Characters.

The character ch_G U of a representation U of $G=GL_n$, SL_n, or PGL_n is the trace of a general diagonal matrix on U. If we take the diagonal coefficients to be variables $x_1,\ldots,x_n$, then the character mapping ch_{GL_n} identifies $\underline{R}_0(GL_n)$ to $\Lambda(x_1,\ldots,x_n)$, and $\underline{R}(SL_n)$ to $\Lambda(x_1,\ldots,x_n)/(x_1 \cdots x_n=1)$. These identifications of rings respect the scalar products, and, in the case of GL_n, also the gradings and the inverse system structure.

Thus we get an isomorphism $\underline{R}_0(GL) \rightarrow \Lambda$ of graded rings with scalar product. We call the limit of the character mappings ch_{GL_n} the characteristic mapping ch_{GL}, or simply ch. So just as representations have a character, polynomial functors $\underline{F}$ have a characteristic, given by (1.4). The isomorphism carries the skew-quotient of polynomial functors to the skew-quotient of their characteristics. In addition, the isomorphism carries internal product of polynomial functors to internal product of symmetric functions.

The characteristic of the Schur functor $\underline{S}^{\pi}$ is the Schur function $s_{\pi} \in \Lambda$. In particular, the character of $\underline{S}^{\pi}(\mathbb{C}^n)$ is $s_{\pi}(x_1,\ldots,x_n)$. These classical symmetric functions were originally studied by Jacobi, who defined them as quotients of alternating functions. They have purely combinatorial definitions as generating functions of tableaux (see [M,I,3,5]). The index of s_{π} is $l(\pi)$. Note $s_{(p)}=h_p$, and $s_{(1^p)}=a_p$.

Given partitions α and β, the skew Schur function $s_{\alpha/\beta}$ is defined to be the skew-quotient $D(s_\alpha/s_\beta)$. The <u>skew-shape</u>, or <u>skew-diagram</u>, α/β is obtained by cutting out the shape of β from the upper left corner of the shape of α (see [M,I,pg. 4]). A key feature of the theory is that all formulae (both algebraic and combinatorial) describing ordinary Schur functions by means of their shapes pass to formulae for skew Schur functions in terms of their skew shapes.

We write $\underline{S}^{\alpha/\beta}$ for the <u>skew Schur functor</u> $D(\underline{S}^\alpha/\underline{S}^\beta)$.

The results on decompositions of the last subsection translate into special cases of the bases expansions (2.1')-(2.2'') in the orthonormal bases of products of Schur functions.

In particular, we emphasize the following. As usual, π,α,β are partitions, and u and v sets of variables. Then

(2.28) $$s_\pi(uv) = \sum_{\alpha\beta} c_{\pi\alpha\beta}\, s_\alpha(u)s_\beta(v) = \sum_\alpha s_\alpha(u)(s_\pi * s_\alpha)(v),$$

so that

(2.29) $$h_p(uv) = \sum_{\alpha\vdash p} s_\alpha(u)s_\alpha(v); \quad a_p(uv) = \sum_{\alpha\vdash p} s_\alpha(u)s_{\alpha'}(v).$$

Also

(2.30) $$s_\alpha(u)s_\beta(u) = \sum_\pi m^\pi_{\alpha\beta}\, s_\pi(u),$$

and

(2.31) $$s_\pi(u+v) = \sum_{\alpha\beta} m^\pi_{\alpha\beta}\, s_\alpha(u)s_\beta(v) = \sum_\alpha s_\alpha(u)s_{\pi/\alpha}(v).$$

The complete homogeneous series gives

(2.32) $$H(x^{(1)}\cdot\cdot x^{(k)}) = \sum c_{\alpha_1\cdot\cdot\alpha_k} s_{\alpha_1}(x^{(1)})\cdot\cdot s_{\alpha_k}(x^{(k)})$$

$$= \sum s_{\alpha_1}(x^{(1)})\cdot\cdot s_{\alpha_{k-1}}(x^{(k-1)})[s_{\alpha_1} *\cdot\cdot* s_{\alpha_{k-1}}](x^{(k)}),$$

with the first summation over partitions $\alpha_1,\ldots,\alpha_k$ and second over $\alpha_1,\ldots,\alpha_{k-1}$.

V. Highest Weights for SL_n.

In this subsection, we translate some results from highest weight theory for representations of SL_n and PGL_n into the language of partitions. We often write $V^{(n)}_\pi$ for the GL_n-representation $\underline{S}^\pi(\mathbb{C}^n)$, to emphasize that π "combinatorially plays the role of the highest weight" when we restrict to SL_n. Note that $V^{(n)}_{(0)}$ is the trivial 1-dimensional representation $\mathbb{C}$, while $V_{\emptyset}^{(n)}$ is the zero space.

In any tensor product $V_{\alpha}^{(n)} \otimes V_{\beta}^{(n)}$ of non-zero SL_n-representations, the representation $V_{\alpha+\beta}^{(n)}$ (recall $\alpha+\beta$ is the row-sum of partitions), always occurs exactly once; this irreducible component is the one called the <u>Cartan piece</u>, or <u>highest piece</u>, of the tensor product.

One can somewhat characterize the set of partitions μ such that $V_{\mu}^{(n)}$ occurs in $V_{\alpha}^{(n)} \otimes V_{\beta}^{(n)}$. The <u>dominance partial order</u> on the set of all partitions is given by: γ dominates δ iff for all $i \geq 1$, we have $\gamma_1+\ldots+\gamma_i \geq \delta_1+\ldots+\delta_i$. We may extend dominance to a partial order on non-increasing sequences of integers with finitely many non-zero parts. For each positive integer n, define the <u>n-dominance partial order</u> on the set of partitions of length n or less by: γ n-dominates δ iff $|\gamma|-|\delta|=rn$ for some integer r, and γ dominates the sequence $(r+\delta_1,\ldots,r+\delta_n)$.

<u>Proposition 2.33.</u> Fix a positive integer n, and let λ,α,β be partitions of length n or less. Then $V_{\lambda}^{(n)}$ occurs in the tensor product $V_{\alpha}^{(n)} \otimes V_{\beta}^{(n)}$ of SL_n-representations only if the row-sum of α and β n-dominates λ.

This is the specialization to SL_n of a general fact from highest weight theory for finite-dimensional representations of a complex semi-simple Lie algebra $\underline{g}$: the highest weight of any irreducible component in a tensor product of two irreducibles is inferior in the usual partial order to the sum of the highest weights of the two factors. Here, the "usual partial order" is the one given by addition of positive roots. For $\underline{g}=\underline{sl}_n$, this translates into this n-dominance partial order on partitions.

<u>§3. MIXED TENSOR REPRESENTATIONS AND CHARACTERS.</u>

<u>Definition 3.1.</u> To each pair (α,β) of partitions, we attach an irreducible representation $V_{\alpha\beta}^{(n)}$ of SL_n as follows. If $l(\alpha)+l(\beta) \leq n$, then $V_{\alpha\beta}^{(n)}$ is the Cartan piece in the tensor product $V_{\alpha}^{(n)} \otimes (V_{\beta}^{(n)})^*$. If $l(\alpha)+l(\beta) > n$, then $V_{\alpha\beta}^{(n)} := 0$.

We call a representation obtained in this way a <u>mixed tensor representation</u>. Clearly, $V_{\beta\alpha}^{(n)}$ is the dual of $V_{\alpha\beta}^{(n)}$.

Since it is irreducible, $V_{\alpha\beta}^{(n)}$ is equal to $V_{\lambda}^{(n)}$ for a unique partition λ of less than n rows. Let us write $\mathrm{prt}_n(\alpha,\beta)$ for this λ. Then λ is just the row-sum $\alpha+(n\backslash\beta)$ (see §2), as long as $\ell(\alpha)+\ell(\beta)\leq n$. Writing this out, for $\alpha=(\alpha_1,\dots,\alpha_s)$ and $\beta=(\beta_1,\dots,\beta_t)$ with $s+t\leq n$, we have a component-wise sum

$$(3.2)\qquad \mathrm{prt}_n(\alpha,\beta) = (\alpha_1,\dots,\alpha_s,\underbrace{0,\dots,0}_{n-s-t},-\beta_t,\dots,-\beta_1) + (\beta_1,\dots,\beta_1).$$

So $|\mathrm{prt}_n(\alpha,\beta)|=|\alpha|-|\beta|+n\beta_1$. This partition has diagram

$\mathrm{prt}_n(\alpha,\beta) =$

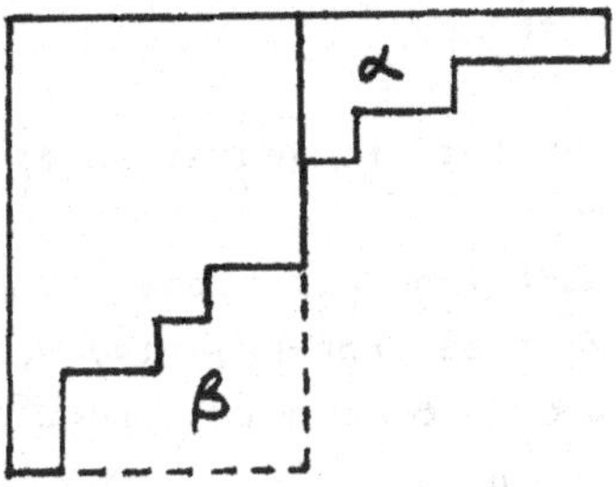

For example, $V_{(0),(0)}^{(n)}=V_{(0)}^{(n)}$ is $\mathbb{C}$, and $V_{(1)(1)}^{(n)}=V_{(2,1,\dots,1)}^{(n)}$ is the adjoint representation.

Note $\mathrm{prt}_n(\alpha,\beta)$ is the empty partition if $\ell(\alpha)+\ell(\beta)>n$.

A useful fact is

<u>Lemma 3.3.</u> Let λ,μ,α,β be partitions with $|\lambda|-|\mu|=|\alpha|-|\beta|$. Assume n satisfies $\ell(\alpha)+\ell(\beta)\leq n$, and $\lambda+(n\backslash\mu)$ n-dominates $\mathrm{prt}_n(\alpha,\beta)$. Then λ dominates α and μ dominates β.

<u>Proof.</u> By the assumptions, the sequence $(\lambda_1-\mu_n,\dots,\lambda_n-\mu_1)$ dominates the sequence $(\alpha_1,\dots,\alpha_{\ell(\alpha)},0,\dots,0,-\beta_{\ell(\beta)},\dots,-\beta_1)$, so the first $\ell(\alpha)$ dominance inequalities tell us that λ dominates α. Since the two sequences have the same sum, the last $t=\ell(\beta)$ dominance inequalities are precisely the dominance inequalities of the sequence $(\mu_1-\lambda_n,\dots,\mu_t-\lambda_{n-t})$ over $(\beta_1,\dots,\beta_t)$. So μ dominates β. []

If $|\alpha|=|\beta|$, then $V_{\alpha\beta}^{(n)}$ descends to a representation of PGL_n.

<u>Lemma 3.4.</u> Fix $n\geq 1$. Then the $V_{\alpha\beta}^{(n)}$, where α and β satisfy $\ell(\alpha)+\ell(\beta)\leq n$ and $|\alpha|=|\beta|$, form an exhaustive, repetition-free list of the irreducible finite-dimensional representations of PGL_n.

Proof. The irreducible finite-dimensional representations of PGL_n are the $V_\lambda^{(n)}$, λ a partition with $|\lambda|$ a multiple of n and $\ell(\lambda)<n$. So we just need to check that the list of partitions $prt_n(\alpha,\beta)$, with $\ell(\alpha)+\ell(\beta)\le n$ and $|\alpha|=|\beta|$, make up an exhaustive, repetition free list of such λ. Clearly, each partition $prt_n(\alpha,\beta)$ has the desired form.

Suppose λ is a partition with $|\lambda| = dn$, some integer d, and length $<n$. Then the sequence $(\lambda_1-d,\ldots,\lambda_n-d)$ is a non-increasing sequence of integers with zero-sum. It must, unless $\lambda=(0)$, involve both positive and negative integers. Thus it has the form of (3.2), for some uniquely determined α and β. □

The mixed tensor analogy to degree is given by

Definition 3.5. Let V be a rational PGL_n-representation. The depth dep(V) is the smallest non-negative value of r such that each irreducible component of V occurs in $(\mathrm{End}\ \mathbb{C}^n)^{\otimes r}$. If no such r exists, we say the depth is infinite.

Proposition 3.6. Fix $r\ge 0$. The irreducible PGL_n-representations of depth r are exactly the non-zero $V_{\alpha\beta}^{(n)}$ with $|\alpha|=|\beta|=r$.

Proof. From the decomposition (2.14), we get

$$(\mathrm{End}\ \mathbb{C}^n)^{\otimes r} = \bigoplus_{\alpha,\beta\vdash r} J_\alpha\otimes J_\beta\otimes V_\alpha^{(n)}\otimes (V_\beta^{(n)})^*,$$

which tells us that all irreducibles of depth r are found among the components of the $V_\alpha^{(n)}\otimes(V_\beta^{(n)})^*$ with $|\alpha|=|\beta|=r$, and each such component has depth at most r. So we need only check that a non-zero $V_{\alpha\beta}^{(n)}$ with $|\alpha|=|\beta|=r$ cannot occur in $V_\gamma^{(n)}\otimes(V_\delta^{(n)})^*$ if $|\gamma|=|\delta|<r$.

Proposition 2.33 says that $V_{\alpha\beta}^{(n)}$ occurs in $V_\gamma^{(n)}\otimes(V_\delta^{(n)})^*$ only if the row-sum $\gamma+(n\backslash\delta)$ n-dominates $prt_n(\alpha,\beta)$. Adding the sequences $(\gamma_1,\ldots,\gamma_n)$ and $(-\delta_n,\ldots,-\delta_1)$, we see that the row-sum is a partition $prt_n(\bar\gamma,\bar\delta)$ with $\bar\gamma\subseteq\gamma$ and $\bar\delta\subseteq\delta$. (If $n\ge\ell(\gamma)+\ell(\delta)$, then $\bar\gamma=\gamma$ and $\bar\delta=\delta$.) But $prt_n(\bar\gamma,\bar\delta)$ n-dominates $prt_n(\alpha,\beta)$ only if $\bar\gamma$ dominates α and $\bar\delta$ dominates β (Lemma 3.3), so only if γ dominates α and δ dominates β, so only if $|\gamma|\ge|\alpha|$ and $|\delta|\ge|\beta|$. □

Mixed tensor representations are a powerful tool for studying families of represesentations of PGL_n as n varies. The generating function for the decomposition of a PGL_n-representation into irreducibles parametrized via Lemma 3.4 is the following symmetric function in two sets of variables, a linear combination of products of Schur functions.

Definition 3.7. Let x and y be two infinite variable sets. The mixed tensor symbol mts U of a rational PGL_n-representation U is the generating function

$$\text{mts } U := \sum_{\alpha\beta} \langle V_{\alpha\beta}^{(n)}, U\rangle s_\alpha(x)s_\beta(y) \in \Lambda((x;y)),$$

summation over all partition pairs α,β satisfying $l(\alpha)+l(\beta)\leq n$ and $|\alpha|=|\beta|=r$.

If U is finite-dimensional, then mts U $\in \Lambda(x;y)$.

In view of Lemma 3.4, two (virtual) PGL_n-representations are isomorphic iff they have the same mixed tensor symbol.

Our overall program is to compute, for each polynomial functor $\underline{E}$, the mixed tensor symbols of the representations $\underline{E}(M_n)$ as functions of n. To do this, we first develop the notion of the "mixed tensor character", which keeps track of the decomposition of $\underline{E}(M_n)$ under the natural action of $GL_n \times GL_n$. The basis of this theory is the following surprising fact: in the reprepresentation ring, all PGL_n-representations are virtual linear combinations of some special reducible PGL_n-representations. These reducible spaces moreover arise from certain irreducible $GL_n \times GL_n$-representations.

For each pair α,β of partitions, set

$$(3.8) \qquad T_{\alpha\beta}^{(n)} := V_\alpha^{(n)} \otimes (V_\beta^{(n)})^*.$$

These are (except for rather trivial cases) reducible representations of GL_n. On the other hand, each space (3.8) can be viewed in the natural way as an irreducible representation $\bar{T}_{\alpha\beta}^{(n)}$ of $GL_n \times GL_n$. Then $T_{\alpha\beta}^{(n)}$ is the restriction from $GL_n \times GL_n$ to its diagonal subgroup GL_n of $\bar{T}_{\alpha\beta}^{(n)}$. In the case $|\alpha|=|\beta|$, $T_{\alpha\beta}^{(n)}$ has degree 0, and hence we may regard it as a representation of PGL_n (by descent). From now on, we consider only this case.

__Theorem 3.9.__ Fix $n\geq 1$. Then the set

(3.10) $\quad \{ T^{(n)}_{\alpha\beta} \mid |\alpha|=|\beta| \text{ and } \ell(\alpha)+\ell(\beta)\leq n \}$

is a linear basis of the representation ring of PGL_n. We call this the __bivariant basis__.

__Proof.__ Partially order the PGL_n-irreducibles as listed in Lemma 3.4 as follows: $V^{(n)}_{\alpha\beta}$ is above $V^{(n)}_{\gamma\delta}$ iff α dominates γ and β dominates δ. Then by the proof of Proposition 3.6, each tensor product space $T^{(n)}_{\alpha\beta}$ in (3.10) contains a unique maximal element, namely $V^{(n)}_{\alpha\beta}$. As each irreducible has only finitely many others below it, the proof follows by induction. []

The proof of also shows

__Corollary 3.11.__ (i) The depth of $T^{(n)}_{\alpha\beta}$ is the common size $|\alpha|=|\beta|$.
(ii) Each PGL_n-irreducible of depth r is, in the representation ring $\underline{R}(PGL_n)$, a (unique) linear combination of bivariant basis elements $T^{(n)}_{\alpha\beta}$ of depth less than or equal to r.

Actually, much more is true:

__Theorem 3.12.__ Let α and β be partitions and n a positive integer, such that $\ell(\alpha)+\ell(\beta)\leq n$. Then, for some integers $i_{\alpha\beta\gamma\delta}$ and $j_{\alpha\beta\gamma\delta}$, we have the equations in the representation ring of PGL_n:

(i) $\quad T^{(n)}_{\alpha\beta} = V^{(n)}_{\alpha\beta} + \sum_{\substack{\gamma\subset\alpha,\gamma\neq\alpha \\ \delta\subset\beta,\delta\neq\beta}} i_{\alpha\beta\gamma\delta} V^{(n)}_{\gamma\delta},$

(ii) $\quad V^{(n)}_{\alpha\beta} = T^{(n)}_{\alpha\beta} + \sum_{\substack{\gamma\subset\alpha,\gamma\neq\alpha \\ \delta\subset\beta,\delta\neq\beta}} j_{\alpha\beta\gamma\delta} T^{(n)}_{\gamma\delta}$

__Proof.__ Deferred to §4. []

The expansion of a (virtual) PGL_n-representation U in terms of the bivariant basis determines integral coefficients $e_U(\alpha,\beta)$. We have the character expansion

(3.13) $\quad ch_{PGL_n}(U) = \sum_{\alpha\beta} e_U(\alpha,\beta) ch_{PGL_n}(T^{(n)}_{\alpha\beta}),$

summation over pairs α,β satisfying the two conditions in (3.10). Set $e_U(\alpha,\beta) := 0$ if the pair α,β fails to satisfy the two conditions.

Now the right side of (3.13) is the diagonal restriction of the virtual character of $GL_n \times GL_n$ obtained by replacing each $T^{(n)}_{\alpha\beta}$ by $\bar{T}^{(n)}_{\alpha\beta}$.

Let us introduce a duality involution $\mathcal{I}$ on representations of $GL_n \times GL_n$ given by

$$\mathcal{I}(U \otimes V) = U \otimes V^*,$$

for any two GL_n-representations U and V. Given a $GL_n \times GL_n$-representation W, it will be more convenient for us to deal with the character of $\mathcal{I}(W)$ rather than that of W; there is no harm in this as the two carry the same information.

We then say the extended character ec U of a PGL_n-representation U is the (virtual) character of $GL_n \times GL_n$ given by

$$(3.14) \qquad \text{ec } U := \sum_{\alpha\beta} e_U(\alpha,\beta) \text{ch}_{GL_n \times GL_n}(\mathcal{I}(\bar{T}^{(n)}_{\alpha\beta}))$$

$$= \sum_{\alpha\beta} e_U(\alpha,\beta) \text{ch}_{GL_n}(V^{(n)}_\alpha) \text{ch}_{GL_n}(V^{(n)}_\beta).$$

Extending linearly, we get a linear injection

$$\text{ec} : \underline{R}(PGL_n) \longrightarrow \underline{CH}(GL_n) \otimes \underline{CH}(GL_n);$$

in particular, two (virtual) PGL_n-representations are isomorphic iff they have the same extended character.

Next take variable sets $\bar{x}$ and $\bar{y}$ of n variables. Then (see §2) $\underline{CH}(GL_n) \otimes \underline{CH}(GL_n)$ identifies with $\Lambda_n \otimes \Lambda_n = \Lambda(\bar{x};\bar{y})$, and the character of any $V^{(n)}_\lambda$ is the Schur function $s_\lambda(\bar{x})$. So (3.14) becomes

$$(3.15) \qquad \text{ec } U = \sum_{\alpha\beta} e_U(\alpha,\beta)\, s_\alpha(\bar{x}) s_\beta(\bar{y}),$$

and the extended character gives a linear injection

$$\text{ec} : \underline{R}(PGL_n) \longrightarrow \Lambda(\bar{x};\bar{y}).$$

Definition 3.16. Let x and y be two infinite variable sets. The mixed tensor character mtc U of a PGL_n-representation U is the generating function

$$\text{mtc } U = \sum_{\alpha\beta} e_U(\alpha,\beta)s_\alpha(x)s_\beta(y) \in \Lambda(x;y),$$

summation over all partition pairs α,β.

So mtc U is just the full symmetrization of ec U into infinite variable sets, and thus two (virtual) PGL_n-representations are isomorphic iff they have the same mixed tensor character.

A key point is that while the mixed tensor character of $T^{(n)}_{\alpha\beta}$ is equal to $s_\alpha(x)s_\beta(y)$ whenever $l(\alpha)+l(\beta)\leq n$, its value is much more subtle when the length sum is bigger than n, as then $T^{(n)}_{\alpha\beta}$ is then no longer a bivariant basis element.

Examples 3.17. (i) Suppose α and β are partitions of common size with $l(\alpha)+l(\beta)>n$. Then it can happen that $T^{(n)}_{\alpha\beta}$ coincides with some bivariant basis element $T^{(n)}_{\gamma\delta}$, $|\gamma|=|\delta|$ and $l(\gamma)+l(\delta)\leq n$, so that

$$\text{mtc } T^{(n)}_{\alpha\beta} = s_\gamma(x)s_\delta(y) \neq s_\alpha(x)s_\beta(y).$$

For instance, if $\alpha=\beta=(1^k)$ with $k>n/2$, then

$$T^{(n)}_{\alpha\beta} = \text{End}(\underline{E}^k(\mathbb{C}^n)) = \text{End}(\underline{E}^{n-k}(\mathbb{C}^n)) = T^{(n)}_{\gamma\delta}$$

with $\gamma=\delta=(1^{n-k})$, so that $\text{mtc } T^{(n)}_{\alpha\beta} = a_{n-k}(x)a_{n-k}(y)$.

In the general situation, what happens is that $T^{(n)}_{\alpha\beta}$ coincides with no bivariant basis elements, so that we must express $T^{(n)}_{\alpha\beta}$ as a linear combination of basis elements to get its mtc. For instance, if n=3 and $\alpha=\beta=(2,1)$, then in the representation ring of PGL_3,

$$T^{(3)}_{\alpha\beta} = V^{(3)}_{(4,2)} + V^{(3)}_{(3,3)} + V^{(3)}_{(4,1,1)} + 2V^{(3)}_{(3,2,1)} + V^{(3)}_{(2,2,2)}$$

$$= V^{(3)}_{(4,2)} + V^{(3)}_{(3,3)} + V^{(3)}_{(3)} + 2V^{(3)}_{(2,1)} + V^{(3)}_{(0)}$$

$$= T^{(3)}_{(2),(2)} + T^{(3)}_{(2),(1,1)} + T^{(3)}_{(1,1),(2)} - T^{(3)}_{(1),(1)} + T^{(3)}_{(0),(0)}.$$

So

$$\text{mtc } T^{(3)}_{\alpha\beta} = h_2(x)h_2(y) + h_2(x)a_2(y) + a_2(x)h_2(y) - h_1(x)h_1(y) + 1.$$

(ii) Let us compare ec $\underline{S}^2(M_n)$ with $\text{ch}_{GL_n\times GL_n} \underline{S}^2(\mathbb{C}^n\otimes\mathbb{C}^n)$. The latter is equal to $h_2(\overline{xy})=h_2(\bar{x})h_2(\bar{y})+a_2(\bar{x})a_2(\bar{y})$ for all $n\geq 1$. The former has this same value as long as $n \geq 4$; but has different values at $n=2,3$: ec $\underline{S}^2(M_2)=h_2(\bar{x})h_2(\bar{y})+1$ and ec $\underline{S}^2(M_3)=h_2(\bar{x})h_2(\bar{y})+h_1(\bar{x})h_1(\bar{y})$.

The linear mapping

$$\text{mtc}: \underline{R}(PGL_n) \longrightarrow \Lambda(x;y)$$

is not actually a character, i.e., tensor product of representations does not go over to multiplication of symmetric functions. One reason for this is that the image of mtc is not multiplicatively closed. However, mtc is multiplicative "stably" in the following manner, so that we feel the term "character" is justified. The result is best phrased using the following notion.

Definition 3.18. The index $\text{ind}(V_{\alpha\beta}^{(n)})$ of a non-zero irreducible $V_{\alpha\beta}^{(n)}$ is the maximum of the lengths of the partitions α and β. The index ind(U) of a PGL_n-representation is the maximum of the indices of its irreducible components.

So the index of a bivariant basis element $T_{\alpha\beta}^{(n)}$ is the maximum of $\ell(\alpha)$ and $\ell(\beta)$. Observe that if $T_{\alpha\beta}^{(n)}$ has index $\leq n/2$, then it must be a bivariant basis element, which implies $\text{mtc}\ T_{\alpha\beta}^{(n)}=s_\alpha(x)s_\beta(y)$.

Proposition 3.19. Suppose U and V are PGL_n-representations of indices r and s. If $n \geq 2r+2s$, then $\text{mtc}\ U\otimes V=(\text{mtc}\ U)(\text{mtc}\ V)$.

Proof. By Theorem 3.12(ii), every representation of index p is a linear combination (in $\underline{R}(PGL_n)$) of bivariant basis elements of index p or less. So it suffices to take $U=T_{\alpha\beta}^{(n)}$ and $V=T_{\gamma\delta}^{(n)}$, where $\alpha,\beta,\gamma,\delta,n$ satisfy $|\alpha|=|\beta|$, $|\gamma|=|\delta|$, $\ell(\alpha),\ell(\beta)\leq r$, $\ell(\gamma),\ell(\delta)\leq s$, and $n \geq 2r+2s$. Then

$$(3.20) \qquad (\text{mtc}\ U)(\text{mtc}\ V) = s_\alpha(x)s_\beta(y)s_\gamma(x)s_\delta(y) = \sum_{\lambda\mu} m_{\alpha\gamma}^{\lambda}\, m_{\beta\delta}^{\mu}\, s_\lambda(x)s_\mu(y).$$

The first equality follows as $n \geq 2r \geq \ell(\alpha)+\ell(\beta)$ and $n \geq 2s \geq \ell(\gamma)+\ell(\delta)$, so that U and V are bivariant basis elements.

On the other hand,

$$(3.21) \qquad U\otimes V = V_\alpha^{(n)} \otimes V_\beta^{(n)} \otimes (V_\gamma^{(n)})^* (V_\delta^{(n)})^* = \bigoplus_{\lambda\mu} m_{\alpha\gamma}^{\lambda}\, m_{\beta\delta}^{\mu}\, T_{\lambda\mu}^{(n)}.$$

If $m_{\alpha\gamma}^{\lambda}, m_{\beta\delta}^{\mu} \neq 0$, then by Lemma 2.23,

$$\ell(\lambda) \leq \ell(\alpha)+\ell(\gamma) \leq r+s \text{ and } \ell(\mu) \leq \ell(\beta)+\ell(\delta) \leq r+s,$$

so that $\ell(\lambda)+\ell(\mu) \leq n$, and hence $T^{(n)}_{\lambda\mu}$ is a bivariant basis element. Thus the mixed tensor character of $U \otimes V$ is given by (3.20). []

The mtc is extremely easy to compute for the representations we have set out to study, <u>as long as certain minimal bounds on n are satisfied</u>.

Given a symmetric function f and partitions α, β, set

$$c^{f}_{\alpha\beta} := \langle f, s_\alpha * s_\beta \rangle .$$

<u>Proposition 3.22.</u> Let $\underline{F}$ be a homogeneous polynomial functor of degree p, with F its characteristic function. Then for all $n \geq 2p$,

$$\text{(3.23)} \qquad \text{mtc } \underline{F}(M_n) = F(xy) = \sum_{\alpha,\beta \vdash p} c^{F}_{\alpha\beta} s_\alpha(x) s_\beta(y),$$

with x and y infinite sets of variables.

<u>Proof.</u> Restricting the Cauchy decomposition under $GL_n \times GL_n$ to PGL_n gives

$$\text{(3.24)} \qquad \underline{F}(M_n) = \underline{F}(\mathbb{C}^n \otimes (\mathbb{C}^n)^*) = \bigoplus_{\alpha,\beta \vdash p} c^{F}_{\alpha\beta} T^{(n)}_{\alpha\beta}.$$

For each space $T^{(n)}_{\alpha\beta}$, we have $\ell(\alpha)+\ell(\beta) \leq |\alpha|+|\beta| = 2p$. So for $n \geq 2p$, the mtc $\underline{F}(M_n)$ is equal to the sum in (3.23), and hence equal to F(xy). []

<u>Remark 3.25.</u> This proof is best understood in the following terms. Let $\bar{x}$ and $\bar{y}$ be sets of n variables. From (3.24), the extended character of $\underline{F}(M_n)$ is equal to $F(\bar{x}\bar{y})$ as long as $n \geq 2p$. Since $\bar{x}$ and $\bar{y}$ have size $\geq p$, the full symmetrization of $F(\bar{x}\bar{y})$ is equal to F(xy).

The last two Propositions show that the mixed tensor character behaves much better when n is comparatively large; in fact they suggest a <u>stable</u> behavior. In the next section, we formulate the stable behavior of mixed tensor characters and symbols, and explain how to pass formally between the two.

§4. STABLE CALCULUS OF MIXED TENSOR REPRESENTATIONS.

In this section we give transition formulae between the two bases of the representation ring of PGL_n -- the basis of Lemma 3.4 and the basis of Theorem 3.9. In particular, we give the stable decomposition of tensor products. The main goal of this section is Theorem 4.24 below, which gives relatively simple formula for moving back and forth between the stable mtc and the stable mts.

The transition formulae (Theorem 4.10, but without size equalities- cf. Remark 4.20) were stated by Littlewood in [L]; he justified them in examples, but did not prove them in general. Recently, Koike ([Kk]) gave complete proofs, by determinental arguments. We point out below that the two transition formulae are easily deducible from each other. We prove in Proposition 4.1 a stronger version of the first formula, by showing the irreducible multiplicities in all tensor products $T^{(n)}_{\alpha\beta}$ are non-decreasing as functions of n.

As usual, x and y will be infinite variable sets.

Proposition 4.1. Let α,β,λ,μ be partitions with $|\alpha|=|\beta|=r$ and $|\lambda|=|\mu|=p$. Then $V^{(n)}_{\alpha\beta}$ occurs (as a non-zero component) in $T^{(n)}_{\lambda\mu}$ only if $\alpha\subseteq\lambda$ and $\beta\subseteq\mu$. The multiplicity

$$(4.2)\qquad m^{\alpha\beta}_{\lambda\mu}(n) := \langle V^{(n)}_{\alpha\beta},\ T^{(n)}_{\lambda\mu}\rangle$$

is a non-decreasing function of n, constant for $n\geq\min(\ell(\lambda)+\ell(\beta),\ell(\alpha)+\ell(\mu))$, so in particular for $n\geq p+r$. Call this constant the stable multiplicity $m^{\alpha\beta}_{\lambda\mu}$. Then

$$(4.3)\qquad m^{\alpha\beta}_{\lambda\mu} = \langle s_{\lambda/\alpha},\ s_{\mu/\beta}\rangle.$$

Proof. Suppose $V^{(n)}_{\alpha\beta}$ occurs in $T^{(n)}_{\lambda\mu}$. Then the row-sum $\lambda+(n\backslash\mu)$ n-dominates $prt_n(\alpha,\beta)$; i.e., $|\lambda+(n\backslash\mu)|=\mu_1 n$ is larger than or equal to $|prt_n(\alpha,\beta)|=\beta_1 n$, and $\lambda+(n\backslash\mu)$ dominates $\theta_n=(\mu_1-\beta_1,\ldots,\mu_1-\beta_1)+prt_n(\alpha,\beta)$.

Now $\theta=\theta_n$ was obtained from $prt_n(\alpha,\beta)$ by adding enough columns of length n to make θ the same size as $\lambda+(n\backslash\mu)$. So (see §2)

$$(4.4)\qquad \langle V^{(n)}_{\alpha\beta},\ T^{(n)}_{\lambda\mu}\rangle = \langle V^{(n)}_{\theta},\ V^{(n)}_{\lambda}\otimes V^{(n)}_{(n\backslash\mu)}\rangle = \langle s_\theta,\ s_\lambda s_{(n\backslash\mu)}\rangle$$
$$= \langle s_{\theta/(n\backslash\mu)},\ s_\lambda\rangle.$$

So in particular, the skew-shape $\theta/(n\backslash\mu)$ is non-empty, or equivalently, θ contains $(n\backslash\mu)$. Now

$$\theta=(\mu_1,\ldots,\mu_1)+(\alpha_1,\ldots,\alpha_a,0,\ldots,0,-\beta_b,\ldots,-\beta_1);\ (n\backslash\mu)=(\mu_1-\mu_n,\ldots,\mu_1-\mu_1),$$

where $a=\ell(\alpha)$, $b=\ell(\beta)$, and all sequences have length n. So the inequalities $\theta_i\geq(n\backslash\mu)_i$ force the inequalities $\beta_i\leq\mu_i$, which mean $\beta\subseteq\mu$.

Moreover, for $n-\ell(\mu)\geq\ell(\alpha)$, the skew-shape $\theta/(n\backslash\mu)$ separates into a disjoint union of two shapes, namely α and the image under a half-turn of the skew-shape μ/β. Pictorially, the separaration is quite striking

(4.5)

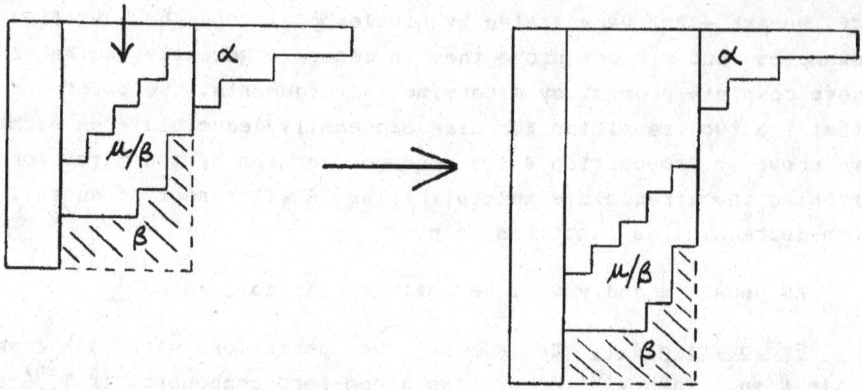

This separation of shapes implies the factorization in Λ (see [M]): $s_{\theta/(n\backslash\mu)}=s_\alpha s_{\mu/\beta}$. Thus, for $n\geq\ell(\mu)+\ell(\alpha)$, the value of (4.4) is independent of n and is equal to

(4.6) $$\langle s_\alpha s_{\mu/\beta},\ s_\lambda\rangle=\langle s_\alpha,\ s_{\mu/\beta}s_\lambda\rangle=\langle s_{\alpha/\lambda},\ s_{\mu/\beta}\rangle.$$

Furthermore, for each n, $m^{\alpha\beta}_{\lambda\mu}(n)=m^{\beta\alpha}_{\mu\lambda}(n)$, as

$$\langle V^{(n)}_{\alpha\beta},\ T^{(n)}_{\lambda\mu}\rangle=\langle (V^{(n)}_{\alpha\beta})^*,\ (T^{(n)}_{\lambda\mu})^*\rangle=\langle V^{(n)}_{\beta\alpha},\ T^{(n)}_{\mu\lambda}\rangle.$$

So also $\alpha\subseteq\lambda$ for all n, and the value of $m^{\alpha\beta}_{\lambda\mu}(n)$ stabilizes for $n\geq\ell(\lambda)+\ell(\beta)$.

It remains to prove $m^{\alpha\beta}_{\lambda\mu}(n)$ is non-decreasing as a function of n. Here we need a finer analysis of the Schur function multiplicity (4.4), considering not just the skew-shapes, but also looking at the skew-tableaux.

The Littlewood-Richardson rule computes $\langle s_{\theta/(n\backslash\mu)},\ s_\lambda\rangle$ as the number of row-strict skew-tableaux T of weight λ' and shape $\theta/(n\backslash\mu)$

whose word $w(T)$ is a lattice permutation. Here "words" are sequences of letters from the ordered alphabet $(1,2,..)$, and the word of T is obtained by reading the tableau column by column, beginning at the leftmost column, and progressing to the right. Each row is from bottom to top. A word $t_1t_2..t_m$ is a lattice permutation iff in each subword $t_1..t_j$, the letter i occurs no more times than the letter i-1, for all integers $i>1$. So in particular, the first number of a lattice permutation must be 1. (This form of the L-R rule is the *transpose* of the usual form as given in, for instance, [M]. The transpose is valid as for any partitions α,β,γ, we have $\langle s_\alpha, s_\beta s_\gamma\rangle = \langle s_{\alpha'}, s_{\beta'} s_{\gamma'}\rangle$.)

We will prove the inequality

$$\langle s_{\theta_n/(n\backslash\mu)}, s_\lambda\rangle \le \langle s_{\theta_{n+1}/((n+1)\backslash\mu)}, s_\lambda\rangle$$

by producing an appropriate injection of tableaux. Fix n and consider the skew shapes $\theta_n/(n\backslash\mu)$ and $\theta_{n+1}/((n+1)\backslash\mu)$. Pushing down each of the first μ_1 columns by one block transforms the first skew shape to the second. Clearly, this transfomation sends, in a one to one manner, row-strict tableaux of shape $\theta_n/(n\backslash\mu)$ and weight λ' to row-strict tableaux of shape $\theta_{n+1}/((n+1)\backslash\mu)$ and weight λ'. Moreover, the word of the tableaux is left unchanged. So we are finished. []

For future use in §8, we record the folowing offshoot of the proof:

<u>Corollary 4.7.</u> With notation as in Proposition 4.1, suppose also that β is the partition (1^r) and that $\lambda=\mu$. Set $m=\ell(\mu)$. Then for $n\geq m+r$, the multiplicity of $V^{(n)}_{\alpha\beta}$ in $T^{(n)}_{\mu\mu}$ is the number of pairs (S,P), where S is row-strict tableau of shape α, P is a row tableau of length m-r, and the entries of S and P taken together are precisely the parts $\mu_1,\ldots,\mu_m$ of μ.

<u>Proof.</u> Proceeding under the Littlewood-Richardson rule as in the last part of the proof of 4.1, we see the desired multiplicity is the number of row-strict tableaux T of weight μ' and skew-shape $\theta/(n\backslash\mu)$ whose word is a lattice permutation. Since the skew-shape splits into the shapes α and μ/β, with the latter rotated through a half-turn, and the entries of T are $1,\ldots,\mu_1,1,\ldots,\mu_2,\ldots,1,\ldots,\mu_m$, clearly T must be of the form

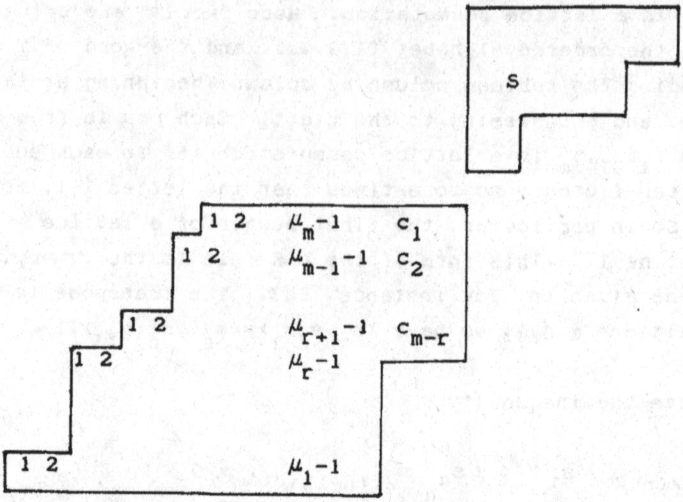

where S is a row-strict tableaux of shape α, $(c_1,\ldots,c_{m-r})$ is a non-decreasing sequence P, and the entries of S and P are $\mu_1,\ldots,\mu_m$. On the other hand, every pair (S,P) satisfying these three conditions yields an allowable tableau T. []

Let us extend some of our notations to skew-partitions. Given partitions $\alpha,\beta,\gamma,\delta$, set

$$(4.8)\qquad V^{(n)}_{\alpha/\gamma,\beta/\delta} := \bigoplus_{\lambda\mu} \langle s_{\alpha/\gamma}, s_\lambda\rangle\langle s_{\beta/\delta}, s_\mu\rangle V^{(n)}_{\lambda\mu}$$

$$= \bigoplus_{\lambda\mu} \langle s_{\alpha/\gamma}(x)s_{\beta/\delta}(y),\ s_\lambda(x)s_\mu(y)\rangle_{x;y} V^{(n)}_{\lambda\mu} \ ;$$

$$(4.9)\qquad T^{(n)}_{\alpha/\gamma,\beta/\delta} := \bigoplus_{\lambda\mu} \langle s_{\alpha/\gamma}, s_\lambda\rangle\langle s_{\beta/\delta}, s_\mu\rangle T^{(n)}_{\lambda\mu}$$

So if $l(\alpha)-l(\gamma)+l(\beta)-l(\delta)\leq n$, then

$$(4.8')\qquad \text{mts } V^{(n)}_{\alpha/\gamma,\beta/\delta} = s_{\alpha/\gamma}(x)s_{\beta/\delta}(y),$$

$$(4.9')\qquad \text{mtc } T^{(n)}_{\alpha/\gamma,\beta/\delta} = s_{\alpha/\gamma}(x)s_{\beta/\delta}(y).$$

<u>Theorem 4.10. (Littlewood [L]).</u> We have the following equations in the representation ring of PGL_n. Suppose α and β are partitions of common size r and $n\geq l(\alpha)+l(\beta)$. Then

$$(4.11)\qquad T^{(n)}_{\alpha\beta} = \sum_{\pi} V^{(n)}_{\alpha/\pi,\beta/\pi} \ ;$$

$$(4.12)\qquad V^{(n)}_{\alpha\beta} = \sum_{\theta} (-1)^{|\theta|} T^{(n)}_{\alpha/\theta,\beta/\theta'}.$$

<u>Proof.</u> Let us first show that, on account of the identity $H(xy)A_{-1}(xy)=1$ in $\Lambda(x;y)$, (4.11) and (4.12) are equivalent. Assuming (4.11) is true, let us compute the right side of (4.12) by computing its mixed tensor symbol. Write D for $D_{x;y}$. Then

$$(4.13)\qquad \sum_{\theta} (-1)^{|\theta|} \text{mts } T^{(n)}_{\alpha/\theta,\beta/\theta'}$$

$$= \sum_{\theta\pi} (-1)^{|\theta|} \langle s_{\alpha/\theta}(x)s_{\beta/\theta'}(y),\ s_{\lambda}(x)s_{\mu}(y)\rangle_{x;y} s_{\lambda/\pi}(x)s_{\mu/\pi}(y)$$

$$= \sum_{\theta\pi} (-1)^{|\theta|} D([s_{\alpha/\theta}(x)s_{\beta/\theta'}(y)]/[s_{\pi}(x)s_{\pi}(y)])$$

$$= D([s_{\alpha}(x)s_{\beta}(y)]/[\sum_{\pi\theta} (-1)^{|\theta|} s_{\theta}(x)s_{\theta'}(y)s_{\pi}(x)s_{\pi}(y)])$$

The inside sum is just $A_{-1}(xy)H(xy)=1$, so (4.13) is equal to

$$(4.14)\qquad D([s_{\alpha}(x)s_{\beta}(y)]/1) = s_{\alpha}(x)s_{\beta}(y) = \text{mts } V^{(n)}_{\alpha\beta}.$$

This proves that (4.11) implies (4.12). Similarly, we prove the converse, using the mtc.

Now by Proposition 4.1, the coefficient of $V^{(n)}_{\gamma\delta}$ in $T^{(n)}_{\alpha\beta}$ is equal to

$$(4.15)\qquad \langle s_{\alpha/\gamma},\ s_{\beta/\delta}\rangle = \langle s_{\alpha/\gamma}(x)s_{\beta/\delta}(y),\ H(xy)\rangle_{x;y}$$

$$= \langle s_{\alpha/\gamma}(x)s_{\beta/\delta}(y),\ \sum_{\pi} s_{\pi}(x)s_{\pi}(y)\rangle_{x;y}$$

$$= \langle s_{\alpha}(x)s_{\beta}(y),\ \sum_{\pi} s_{\gamma}(x)s_{\delta}(y)s_{\pi}(x)s_{\pi}(y)\rangle_{x;y}$$

$$= \langle \sum_{\pi} s_{\alpha/\pi}(x)s_{\beta/\pi}(y),\ s_{\gamma}(x)s_{\delta}(y)\rangle_{x;y}$$

So we obtain (4.11). []

The proof yields the alternate formulae

<u>Corollary 4.16.</u> Keep the notations of 4.10. Then

$$(4.17)\qquad T^{(n)}_{\alpha\beta} = \sum_{\gamma\delta} \langle s_{\alpha/\gamma},\ s_{\beta/\delta}\rangle V^{(n)}_{\gamma\delta}.$$

$$(4.18)\qquad V^{(n)}_{\alpha\beta} = \sum_{\gamma\delta} (-1)^{r-|\gamma|} \langle s_{\alpha/\gamma},\ s_{(\beta/\delta)'}\rangle T^{(n)}_{\gamma\delta}.$$

<u>Remarks 4.19.</u> (1) That transition equations exist is clear from

the last section (see Cor. 3.11(ii)). But the form of them, the independence of n for large enough values, is much stronger.
(ii) The behavior of a tensor product of a cogredient space with a contragredient space is essentially different from that of a tensor product of two cogredient spaces, in than,e.g., non-zero multiplicities can be non-stable. For the equation (4.11) is not valid for small values of n. Essentially, the components are right, but the multiplicities can be wrong. For instance, if $\lambda=\mu=(2,1)$, then the coefficient of $V^{(n)}_{(1)(1)}$ in $T^{(n)}_{\lambda\mu}$ is equal to 1 if n=2, and 2 if $n\geq 3$.

Remark 4.20. Proposition 4.1 and the formulae of Theorem 4.10 hold without the conditions on equality of sizes. The same proofs go through.

An immediate consequence of 4.10 is Theorem 3.12. Also the calculations in the proof of 4.10 give

Corollary 4.21. Take α,β,n as in Th. 4.10. Then

$$\text{(4.22)} \qquad \text{mts } T^{(n)}_{\alpha\beta} = D_{x;y}(s_\alpha(x)s_\beta(y)/H(xy));$$

$$\text{(4.23)} \qquad \text{mtc } V^{(n)}_{\alpha\beta} = D_{x;y}(s_\alpha(x)s_\beta(y)/A_{-1}(xy)).$$

These last equations give at once the transformation relations between the stable mixed tensor character and symbol.

Theorem 4.24. If U is a rational finite-dimensional PGL_n-representation of index (or depth) d with $n\geq 2d$, then we have the two identities in $\Lambda(x;y)$

$$\text{(4.25)} \qquad \text{mts } U = D_{x;y}([\text{mtc } U]/H(xy)),$$

$$\text{(4.26)} \qquad \text{mtc } U = D_{x;y}([\text{mts } U]/A_{-1}(xy)).$$

As before, the two formulae are transparently equivalent on account of the fact $A_{-1}(xy)H(xy)=1$.

§5. STABILITY OF POLYNOMIAL POWERS OF M_n.

We will now formulate a precise notion of stability, and apply it to the representations $\underline{E}(M_n)$, for $\underline{E}$ a (graded) polynomial functor.

We fix for this section a family $(L_n)_{n\geq k}$ of vector spaces, with each L_n a rational finite-dimensional representation of PGL_n.

Let x and y be infinite variable sets. By the limit of a sequence $(t_n(x;y))_{n\geq k}$ in $\Lambda(x;y)$ we mean the limit in the natural inverse limit topology. So this sequence has limit g in $\Lambda(x;y)$ iff for each $p\geq 1$, there exists $N_p\geq 1$ such that for all $n\geq N_p$, $g(x\langle p\rangle;y\langle p\rangle)=t_n(x\langle p\rangle;y\langle p\rangle)$. In this case, we say the limits exists, and write $g=\lim_{n\to\infty} t_n$ or $g=t_\infty$.

Lemma 5.1. Let $(t_n(x;y))_{n\geq k}$ be a sequence in $\Lambda(x;y)$. Then given $g\in\Lambda(x;y)$, there exists N such that $g=t_n$ for all $n\geq N$ iff $\lim_{n\to\infty} t_n = g$ and $(t_n)_{n\geq k}$ has bounded index.

Proof. Immediate, as symmetric functions of index p are completely determined in p variables. []

In the case $(t_n)_{n\geq k}$ satisfies the equivalent conditions of the Lemma, we call it a strongly stable sequence, with stable bound equal to the smallest possible value of N, and with stable index equal to the index of g.

Remark 5.1'. The sequence $(t_n)_{n\geq 1}$ given by $t_n(x;y)=a_1(x)a_1(y)$ if n is odd, and $t_n(x;y)=a_1(x)a_1(y)+a_{n/2}(x)a_{n/2}(y)$ if n is even, is stable, but not strongly stable.

Lemma 5.2. The sequence $(\text{mtc } L_n)_{n\geq k}$ is strongly stable in $\Lambda(x;y)$ iff the sequence $(\text{mts } L_n)_{n\geq k}$ is strongly stable in $\Lambda(x;y)$. In this case the two sequences have the same stable bound and stable index.

Proof. If either sequence in $\Lambda(x;y)$ is stongly stable then it has bounded index, so the sequence of representations $(L_n)_{n\geq k}$ has bounded index, say d. Then for all $n\geq 2d$, mtc L_n and mts L_n are computable from each other by the formulae of Theorem 4.24. For the second statement, use Littlewood's formulae. []

Definition 5.3. The sequence $(L_n)_{n\geq k}$ is a strongly stable sequence of PGL_n-representations with stable bound N and stable index d iff $(\text{mts } L_n)_{n\geq k}$ is a strongly stable sequence in $\Lambda(x;y)$ with stable bound N and stable index d. If also $(\text{mts } L_n)_{n\geq k}$ has bounded bidegree (r,r), then we say that r is the stable depth of $(L_n)_{n\geq k}$.

Clearly, $(L_n)_{n\geq k}$ is strongly stable with stable bound N or less iff we have (i) for all $n\geq N$, L_n has index N-1 or less, and also $V^{(n)}_{\alpha\beta}$ occurs in L_n only if $\ell(\alpha)+\ell(\beta)\leq N$, and (ii) for each pair α,β of equal size partitions with $\ell(\alpha),\ell(\beta)\leq N$, the multiplicity of $V^{(n)}_{\alpha\beta}$ in L_n is constant for $n\geq N$. Let us call $(L_n)_{n\geq k}$ <u>purely strongly stable</u> iff in addition, these multiplicities are non-decreasing functions of n.

The proof of Lemma 5.2 shows also that if $(L_n)_{n\geq k}$ is a strongly stable sequence of representations, then the <u>stable mixed tensor character</u>, or simply <u>stable character</u>,

(5.4) $$\text{mtc } L_\infty := \lim_{n\to\infty} \text{mtc } L_n ,$$

and the <u>stable mixed tensor symbol</u>, or simply <u>stable symbol</u>,

(5.5) $$\text{mts } L_\infty := \lim_{n\to\infty} \text{mts } L_n$$

satisfy

(5.6) $$\text{mts } L_\infty = D_{x;y}([\text{mtc } L_\infty]/H(xy)) ;$$

$$\text{mtc } L_\infty = D_{x;y}([\text{mts } L_\infty]/A_{-1}(xy)) .$$

In this language, we have

<u>Theorem 5.7.</u> Let $\underline{F}$ be a polynomial functor of degree p with characteristic function F. Then the sequence of PGL_n-representations $(\underline{F}(M_n))_{n\geq 1}$ is purely strongly stable with stable bound not more than 2p and stable depth p. The stable mixed tensor character and symbol both have bidegree (p,p), and are given by

(5.8) $$\text{mtc } \underline{F}(M_\infty) = F(xy), \quad \text{mts } \underline{F}(M_\infty) = D_{x;y}(F(xy)/H(xy)) .$$

Moreover, for each $r\geq 0$, the family of index r components in $(\underline{F}(M_n))_{n\geq 1}$ is stable with stable bound not more than p+r.

<u>Proof.</u> The results about $(\underline{F}(M_n))_{n\geq k}$ are immediate from Proposition 3.22. The last statement follows from the proof of 3.22 in view of Proposition 4.1. □

Often it is natural to consider, for A an abelian semi-group, the more general case of A-graded polynomial functors $\underline{E}$, i.e., polynomial functors $\underline{E}$ from the category of vector spaces to the category of A-graded vector spaces with finite-dimensional components. Let us suppose that A is just the semi-group of monomials in a variable set $w=(w_1,w_2,\ldots)$ and then say "w-graded" instead of A-graded. For instance, $\underline{\mathrm{Sym}}_q := \sum_{p\geq 0} \underline{S}^p q^p$ and $\sum_{p\geq 0} \underline{E}^p q^p$ are q-graded polynomial functors.

Also, families of graded PGL_n-representations arise naturally from geometry. For instance, if $(X_n)_{n\geq k}$ is a family of varieties with X_n carrying an action of $PGL_n \times C^*$, then we can look at the family of coordinate rings $(R(X_n))_{n\geq k}$, a family of q-graded representations.

We decompose a w-graded polynomial functor $\underline{E}$ into a sum $\underline{E} = \sum_M \underline{E}_M$ over the set mon(w) of monomials M in w, with each $\underline{E}_M$ a polynomial functor. Then the <u>characteristic</u> F of $\underline{E}$ is the sum $\sum_M F_M M \in \Lambda[[w]]$, where $F_M \in \Lambda$ is the characteristic of $\underline{E}_M$.

Similarly, a <u>w-graded vector space</u> U is a direct sum $\bigoplus_M U_M$ over mon(w) of finite-dimensional vector spaces U_M. For our present purposes, a <u>w-graded PGL_n-representation</u> is a w-graded vector space U such that each component U_M carries a rational representation of PGL_n.

In general, when forming any sort of characters of w-graded representations, we take a weighted sum over mon(w) of the characters of the components. So if U is a w-graded PGL_n-representation, then $\mathrm{mts}\ U = \sum_M (\mathrm{mts}\ U_M) M \in (\Lambda(x;y))[[w]]$, etc.

<u>Definition 5.9.</u> Fix a variable set w. A family $(U_n)_{n\geq k}$ of w-graded PGL_n- representations is <u>stable</u> iff for each monomial M on w, the family $((U_n)_M)_{n\geq k}$ is strongly stable.

We say $(U_n)_{n\geq k}$ is <u>strongly stable</u> if in addition the family of stable bounds of the $((U_n)_M)_{n\geq k}$ is bounded as M varies. This notion will be useful in the sequel paper.

If $(U_n)_{n\geq k}$ is a stable family of PGL_n-representations, then we have the <u>stable mtc</u> and <u>stable mts</u>

(5.10) $\quad \mathrm{mtc}\, U_\infty := \sum_M (\mathrm{mtc}\, U_\infty)M,$

(5.11) $\quad \mathrm{mts}\, U_\infty := \sum_M (\mathrm{mts}\, U_\infty)M.$

These are elements of $(\Lambda(x;y))[[w]]$; they of course obey the formulae (5.6).

Let $\underline{F}$ be a w-graded polynomial functor with characteristic F. Theorem 5.7 gives

Corollary 5.12. The family $(\underline{F}(M_n))$ of w-graded PGL_n-representations is stable, with stable character and symbol given by (5.8).

Extend $\langle\ ,\ \rangle_{x;y}$ $\mathbb{Z}[[w]]$-linearly to a bilinear form

$$\Lambda(x;y)\times(\Lambda(x;y))[[w]] \longrightarrow \mathbb{Z}[[w]].$$

Then for each value of n, we have a linear map

(5.13) $$g_n^F\colon\ \Lambda(x;y) \longrightarrow \mathbb{Z}[[w]]$$

given by

(5.14) $$g_n^F(t(x;y)) := \langle t(x;y),\ \mathrm{mts}\ \underline{F}(M_n)\rangle_{x;y}\ ,$$

for every $t(x;y)\in\Lambda(x;y)$. In addition, we write

(5.15) $$g_n^F(\alpha,\beta) := g_n^F(s_\alpha(x)s_\beta(y)).$$

These definitions carry through to give "g_∞^F" if we set "$n=\infty$". Then the stable multiplicities of $(\underline{F}(M_n))_{n\geq 1}$ are

(5.16) $$g_\infty^F(\alpha,\beta) = g_\infty^F(s\alpha(x)s\beta(y)) = \lim_{n\to\infty} \sum_M \langle V_{\alpha\beta}^{(n)},\ \underline{F}_M(M_n)\rangle M \in \mathbb{Z}[[w]],$$

summation over all monomials M on w.

§6. HIDDEN STABLE SYMMETRIES.

The key property of stable multiplicities is a very surprising and powerful one. Let x and y be infinite variable sets. Recall we write $x\langle n\rangle$ for the truncation $(x_1,\ldots,x_n)$.

We will use the following notation for evaluating functions given in one variable set in terms of another variable. For any variable sets u,v, set $f(x)|_{x\to u} := f(u)$, $t(x;y)|_{x\to u;y\to v} := t(u;v)$, etc.

We say $t(x;y)\in\Lambda(x;y)$ is <u>fully symmetric in the product set xy</u> iff $t(x;y)$ satisfies $t(x;y)=f(xy)$ for some symmetric function f. This condition completely determines f.

<u>Lemma 6.1.</u> If e and f are symmetric functions such that e(xy) is equal to f(xy) as elements of $\Lambda(x;y)$, then e and f are the same symmetric function.

<u>Proof.</u> Evaluate f(xy) on y=(1,0,0,..) to get f(x). []

So the map

$$\Lambda \longrightarrow \Lambda(x;y), \quad f \mapsto f(xy), \tag{6.2}$$

is a linear embedding; <u>we denote its image by</u> $\Lambda(xy)$.

<u>Hidden Symmetry Theorem 6.3.</u> Let $\underline{E}$ be a w-graded polynomial functor (w a variable set). Then the stable mixed tensor symbol mts $\underline{E}(M_\infty)$ is fully symmetric in the product variable set $xy=(x_iy_j)_{i,j\geq 1}$, i.e. mts $\underline{E}(M_\infty)$ lies in the subring $(\Lambda(xy))[[w]]$ of $(\Lambda(x;y))[[w]]$.

<u>Proof.</u> By linearity, it is enough to prove the theorem in the case where $\underline{E}$ is a Schur functor $\underline{S}^\pi$, for π a partition of p. Then $\underline{E}$ has degree p, so we know by 5.7 that for $n\geq 2p$, $\underline{E}(M_n)$ stabilizes and mts $\underline{E}(M_n)$ = mts $\underline{E}(M_\infty)$ has bidegree (p,p). So mts $\underline{E}(M_\infty)$ has index at most p, which means that mts $\underline{E}(M_\infty)$ is equal to the full symmetrization, from $\Lambda(x\langle p\rangle;y\langle p\rangle)$ to $\Lambda(x;y)$, of its truncation mts $\underline{E}(M_\infty)|_{x\to x\langle p\rangle;y\to y\langle p\rangle}$.

Thus the full symmetry of mts $\underline{E}(M_\infty)$ in the infinite variable set xy is equivalent to the full symmetry of mts $\underline{E}(M_\infty)|_{x\to x\langle p\rangle;y\to y\langle p\rangle}$ in the finite variable set $x\langle p\rangle y\langle p\rangle$. We will prove the latter. From now on, we write $\bar{x}$ and $\bar{y}$ for the truncations $x\langle p\rangle$ and $y\langle p\rangle$.

Fix p-dimensional vector spaces U and V; identify the polynomial character rings of GL(U) and GL(V) with $\Lambda[\bar{x}]$ and $\Lambda[\bar{y}]$, respectively.

We will construct a representation W of GL(U)XGL(V) such that in $\Lambda(\bar{x};\bar{y})$ we have

$$(6.4)\qquad ch_{GL(U)\times GL(V)}\, W = \text{mts}\, \underline{S}^{\pi}(M_{\infty})|_{x\to\bar{x};y\to\bar{y}}.$$

Then we will construct an action of $GL(U\otimes V)$ on W which extends the action of $GL(U)\times GL(V)$. Having such a $GL(U\otimes V)$ action implies that (6.4) is the restriction of a $GL(U\otimes V)$-character, so that (6.4) is indeed a symmetric function of $\bar{x}_i\bar{y}_j$, $1\leq i,j\leq p$.

To construct W, we first form the space

$$L := (U\oplus(\mathbb{C}^p)^*)\otimes(V\oplus\mathbb{C}^p).$$

L carries natural commuting actions of $GL(U)$, $GL(V)$, and $GL(p)$, where $GL(p)$ acts simultaneously on both $(\mathbb{C}^p)^*$ and $\mathbb{C}^p$. Thus L, and hence any image $\underline{E}(L)$, $\underline{E}$ a polynomial functor, carries an action, the <u>canonical action</u>, of $GL(U)\times GL(V)\times GL(p)$.

Define W to be the space of invariants

$$(6.5)\qquad W := [\underline{S}^{\pi}(L)]^{GL(p)}.$$

To find the character of W, we first decompose using the standard rules (see §2):

$$\underline{S}^{\pi}(L) = \bigoplus_{\alpha\beta} c_{\pi\alpha\beta}\, \underline{S}^{\alpha}(U\oplus(\mathbb{C}^p)^*)\otimes\underline{S}^{\beta}(V\oplus\mathbb{C}^p)$$

$$(6.6)\qquad = \bigoplus_{\alpha\beta\gamma\delta} c_{\pi\alpha\beta}\, \underline{S}^{\alpha/\gamma}(U)\otimes\underline{S}^{\gamma}(\mathbb{C}^p)^*\otimes\underline{S}^{\beta/\delta}(V)\otimes\underline{S}^{\delta}(\mathbb{C}^p).$$

Now if a term in (6.6) contributes to the invariants under $GL(p)$, then we must have $\gamma=\delta$. On the other hand, if $\alpha,\beta,\gamma,\delta$ satisfy $\gamma=\delta$ and $c_{\pi\alpha\beta}\,\underline{S}^{\alpha/\gamma}(U)\otimes\underline{S}^{\beta/\delta}(V)\neq 0$, then the corresponding term of (6.6) contributes to the $GL(p)$-invariants. This is because the conditions $c_{\pi\alpha\beta}\neq 0$, $\alpha/\gamma\neq\emptyset$, and $\beta/\delta\neq\emptyset$ imply, respectively, that $|\alpha|=|\beta|=p$, $\gamma\subseteq\alpha$, and $\delta\subseteq\beta$, so that they cumulatively imply that $\ell(\gamma),\ell(\delta)\leq p$. As the contribution is 1-dimensional, this implies

$$(6.7)\qquad W = \bigoplus_{\alpha\beta\gamma} c_{\pi\alpha\beta}\, \underline{S}^{\alpha/\gamma}(U)\otimes\underline{S}^{\beta/\gamma}(V),$$

so that

(6.8) $$\operatorname{ch}_{GL(U)\times GL(V)} W = \sum_{\alpha\beta\gamma} c_{\pi\alpha\beta}\, s_{\alpha/\gamma}(\bar{x}) s_{\beta/\gamma}(\bar{y})$$

$$= D_{x;y}\left([\sum_{\alpha\beta} c_{\pi\alpha\beta} s_{\alpha}(x) s_{\beta}(y)] / [\sum_{\gamma} s_{\gamma}(x) s_{\gamma}(y)]\right)\Big|_{x\to\bar{x};y\to\bar{y}}$$

$$= D_{x;y}(s_{\pi}(xy)/H(xy))\Big|_{x\to\bar{x};y\to\bar{y}}$$

$$= \text{mts } \underline{S}^{\pi}(M_{\infty})\Big|_{x\to\bar{x};y\to\bar{y}}.$$

To construct a $GL(U\otimes V)$-action on W extending the $GL(U)\times GL(V)$-action, we first locate W as a component of $(L^{\otimes p})^{GL(p)}$. The action of the symmetric group $\mathcal{S}_p$ on $L^{\otimes p}$ by permutation of tensor position commutes with the action $GL(U)\times GL(V)\times GL(p)$. The classical decomposition of $L^{\otimes p}$ under $\mathcal{S}_p\times GL(L)$ is (2.11)

(6.9) $$L^{\otimes p} = \bigoplus_{\pi\vdash p} J_{\pi} \otimes \underline{S}^{\pi}(L),$$

This tells us the J_{π}-isotypic component of $L^{\otimes p}$ is isomorphic under $GL(L)$ to $\dim(J_{\pi})$ copies of $\underline{S}^{\pi}(L)$, hence that the J_{π}-isotypic component of $(L^{\otimes p})^{GL(p)}{}^{\pi}$ is isomorphic under $GL(U)\times GL(V)$ to $\dim(J_{\pi})$ copies of W.

Thus, to construct on W an action of $GL(U\otimes V)$ which extends the action of $GL(U)\times GL(V)$, it suffices to construct on $(L^{\otimes p})^{GL(p)}$ an action of $GL(U\otimes V)$ which <u>extends the action of</u> $GL(U)\times GL(V)$ <u>and commutes with the action of</u> $\mathcal{S}_p$.

Let us expand

$$L^{\otimes p} = [(U\oplus (\mathbb{C}^p)^*)\otimes (V\oplus \mathbb{C}^p)]^{\otimes p}$$

$$= [(U\otimes V) \oplus (U\otimes \mathbb{C}^p) \oplus ((\mathbb{C}^p)^*\otimes V) \oplus ((\mathbb{C}^p)^*\otimes \mathbb{C}^p)]^{\otimes p}.$$

So $L^{\otimes p}$ is the direct sum of the 4^p p-fold tensor products $N_u = N_{i_1}\otimes\ldots\otimes N_{i_p}$, $u=(i_1,\ldots,i_p)$ with $1\le i_1,\ldots,i_p\le 4$, where

(6.10) $$N_1=U\otimes V,\quad N_2=U\otimes\mathbb{C}^p,\quad N_3=(\mathbb{C}^p)^*\otimes V,\quad N_4=(\mathbb{C}^p)^*\otimes\mathbb{C}^p.$$

These 4^p components are permuted by the action of $\mathcal{S}_p$ on $L^{\otimes p}$.

Given positive integers s and t both less than or equal to p, the space $((\mathbb{C}^p)^*)^{\otimes s}\otimes(\mathbb{C}^p)^{\otimes t}$ has non-zero $GL(p)$-invariants iff $s=t$. So

$(L^{\otimes p})^{GL(p)}$ is a direct sum of those components $[N_u]^{GL(p)}$, for which 2 and 3 occur equally often in the sequence u.

Let us take then a general such component $[N_u]^{GL(p)}$ of $(L^{\otimes p})^{GL(p)}$, in which the multiplicities of 1,2,3,4 in u are respectively, a,b,b,d. So a+2b+d=p. Then as a representation of GL(U)XGL(V)XGL(p),

$$N_u = U^{\otimes(a+b)} \otimes V^{\otimes(a+b)} \otimes (\mathrm{End}\ \mathbb{C}^p)^{\otimes(b+d)}.$$

So as a representation of GL(U)XGL(V),

(6.11) $$[N_u]^{GL(p)} = U^{\otimes(a+b)} \otimes V^{\otimes(a+b)} \otimes [(\mathrm{End}\ \mathbb{C}^p)^{\otimes(b+d)}]^{GL(p)}.$$

Since $(b+d) \leq p$, the space of GL(p)-invariants in

$$(\mathrm{End}\ \mathbb{C}^p)^{\otimes(b+d)} = ((\mathbb{C}^p)^* \otimes \mathbb{C}^p)^{\otimes(b+d)},$$

identifies naturally with the group ring $\mathbb{C}[\mathcal{S}_{b+d}]$, so that this space of invariants breaks up into a direct sum of (b+d)! 1-dimensional spaces A_P, each one spanned by an invariant f_P built from a permutation P matching the factors $(\mathbb{C}^p)^*$ with the factors $\mathbb{C}^p$.

For example, the matching P drawn below,

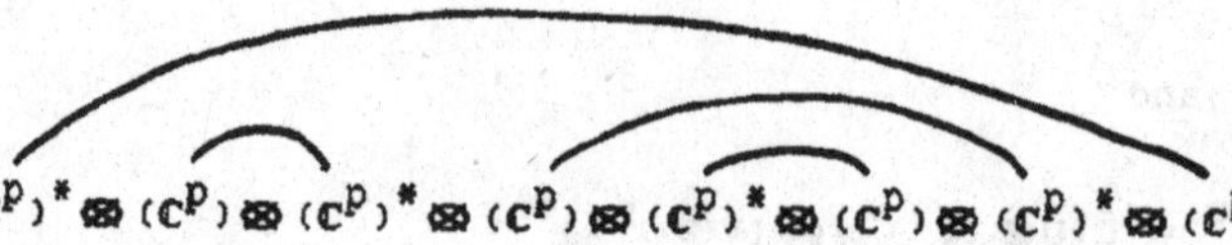

$$(\mathbb{C}^p)^* \otimes (\mathbb{C}^p) \otimes (\mathbb{C}^p)^* \otimes (\mathbb{C}^p) \otimes (\mathbb{C}^p)^* \otimes (\mathbb{C}^p) \otimes (\mathbb{C}^p)^* \otimes (\mathbb{C}^p)$$

gives rise to the invariant

$$f_P = \sum_{1 \leq i,j,k,l \leq p} (\nu_i \otimes e_j \otimes \nu_j \otimes e_l \otimes \nu_k \otimes e_k \otimes \nu_l \otimes e_i),$$

where $(\nu_i)_{1 \leq i \leq p}$ and $(e_i)_{1 \leq i \leq p}$ are dual bases of $(\mathbb{C}^p)^*$ and $\mathbb{C}^p$. So

(6.12) $$[(\mathrm{End}\ \mathbb{C}^p)^{\otimes(b+d)}]^{GL(p)} = \bigoplus_{P \in S_{b+d}} A_P,$$

The substitution of (6.12) into (6.11) effects a decomposition of $[N_u]^{GL(p)}$ into a direct sum of (b+d)! spaces.

$$(6.13)\qquad [N_u]^{GL(p)} = \bigoplus_P ([N_u]^{GL(p)})_P$$

with P ranging over the (b+d)! matchings between the factors $(\mathbb{C}^p)^*$ and the factors $\mathbb{C}^p$ occuring in N_u.

Recall each factor N_{i_j} in N_u is a tensor product $N_{i_j}^{(1)} \otimes N_{i_j}^{(2)}$. We will say the two factors $N_{i_j}^{(1)}$ and $N_{i_j}^{(2)}$ are "mated" in couples and call each factor the "mate" of the other. Examining (6.10), we see that the mate of a copy of U can only be a copy of V or $\mathbb{C}^p$, while the mate of a copy of V can only be a copy of U or $(\mathbb{C}^p)^*$.

Now to build an action of $GL(U \otimes V)$ on each space (6.13), it suffices to construct, given the matching P, a matching Q of the (a+b) factors U with the (a+b) factors V occuring in N_u. Given P, construct Q as follows. First, we decree Q will match together the a copies of U and V which are already mated in couples. We complete the definition of Q by an iterative proceedure.

(i) Pick any copy $N_{i_j}^{(1)}$ of U which has not yet been matched to a copy of V by Q.

(ii) By hypothesis, its mate $N_{i_j}^{(2)}$ is not a copy of V, so it is a copy of $\mathbb{C}^p$. So P matches this $N_{i_j}^{(2)}$ to some copy $N_{i_k}^{(1)}$, $k \neq j$, of $(\mathbb{C}^p)^*$. Look at its mate $N_{i_k}^{(2)}$.

If $N_{i_k}^{(2)}$ is a copy of V, then have Q match $N_{i_j}^{(1)}$ to $N_{i_k}^{(2)}$, and return to step (i). Otherwise, $N_{i_k}^{(2)}$ is a copy of $\mathbb{C}^p$, and we repeat the work just done, with $N_{i_k}^{(2)}$ replacing $N_{i_j}^{(2)}$. Eventually, we must arrive at some copy $N_{i_m}^{(2)}$ of V, which we then match to $N_{i_j}^{(1)}$. Then we return to step (i).

Since there are only b copies of U not mated with copies of V, the definition of the matching Q is complete after performing step (i) b times. So in this way we construct on each space $([N_u]^{GL(p)})_P$ an action of $GL(U \otimes V)$ extending the $GL(U) \times GL(V)$-action.

For example, figure (6.14) shows some typical components N_u, with a given matching P drawn above, and the resulting matching Q drawn below. In the figure, $E:=\mathbb{C}^p$.

FIGURE (6.14)

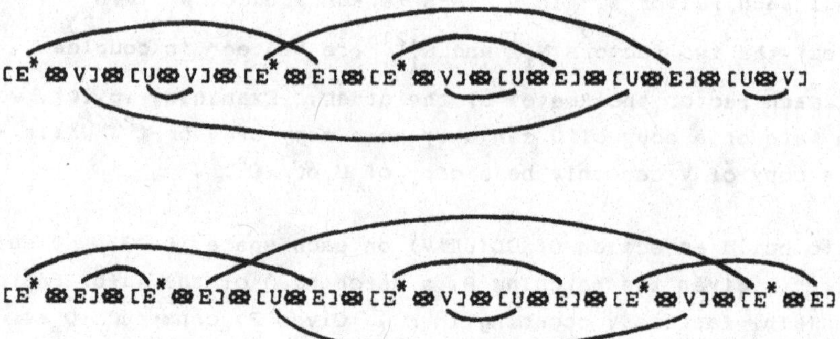

A key feature of our construction of Q is that Q is entirely independent of the ordering of the tensor factors $N_{i_1},\ldots,N_{i_p}$. Indeed, the construction just depends on the matching of copies of $(\mathbb{C}^p)^*$ with copies of $\mathbb{C}^p$, i.e., on the chosen GL(p)-invariant in N_u. Thus, summing over all (b+d)! matchings P, we get on the whole space $(N_u)^{GL(p)}$ an action of $GL(U\otimes V)$ which commutes with the action of $\mathcal{S}_p$. Hence we get such an action on $(L^{\otimes p})^{GL(p)}$, proving the theorem. []

<u>Remark 6.15.</u> If we replace x and y by finite truncations x⟨n⟩ and y⟨n⟩ in Lemma 6.1, then the Lemma still holds (with the same proof) for symmetric functions e and f of index n or less. In this finite case, another way to formulate the Lemma is that if U and V are n-dimensional complex vector spaces, then (finite-dimensional algebraic) representations of $GL(U\otimes V)$ of index n or less are completely determined by their restriction to $GL(U)\times GL(V)$. (This is a very special property for a group, a subgroup, and a class of representations of the big group to enjoy.) The original Lemma says in some rough manner that this holds true for <u>all</u> representations <u>stably</u>.

The Theorem is equivalent to

<u>Corollary 6.16.</u> For each symmetric function f and complete homogeneous symmetric function h_r, the skew-quotient $D_{x;y}(f(xy)/h_r(xy))$ is

fully symmetric in the product set xy.

Proof. Take homogeneous components. □

Remark 6.17. Lemma 6.1 and Corollary 6.16 generalize to the case of finitely many variable sets. See Corollary 7.6 for the full statement. The proof comes from generalizing the proof of the two variable set case, or, more simply, by arguing directly from the two variable set case.

6.17 establishes the existence of a "new" symmetric function $f//h_r$, which is completely characterized by the identity

$$(6.18)\qquad (f//h_r)(xy) = D_{x;y}(f(xy)/h_r(xy)).$$

In the next section we give a more general theory of this new operation //, which we call double skew-division, defining f//g for arbitrary pairs f,g of symmetric functions. Then in §8, we give some applications of the Hidden Symmetry Theorem to the computation of stable multiplicities. The reader may wish to turn ahead to §8 right away to see the power of the hidden symmetry.

Let us note already that the fact (same notations as Theorem 6.3) that mts $\underline{F}(M_\infty)$ is symmetric in xy can be rephrased as follows in terms of the stable multiplicities:

Corollary 6.19. Let $\underline{F}$ be a w-graded polynomial functor, and let α and β be partitions of common size. Then the value of $g_\infty^F(\alpha,\beta)$ depends only on the symmetric group representation $J_\alpha \otimes J_\beta$. More precisely, suppose $\alpha_i, \beta_i, \gamma_j, \delta_j$, i=1,...,s and j=1,...,t, are partitions of p. If the equality

$$(6.20)\qquad \sum_{i=1}^{s} J_{\alpha_i} \otimes J_{\beta_i} = \sum_{j=1}^{t} J_{\gamma_j} \otimes J_{\delta_j}$$

holds in the representation ring $\underline{R}(\mathcal{S}_p)$, then the equality

$$(6.21)\qquad \sum_{i=1}^{s} g_\infty^F(\alpha_i,\beta_i) = \sum_{j=1}^{t} g_\infty^F(\gamma_j,\delta_j).$$

holds in $(\Lambda(x;y))[[w]]$. In particular,

$$(6.22)\qquad g_\infty^F(\alpha,\beta) = g_\infty^F(\alpha',\beta').$$

Proof. The stable multiplicity $g^F_\infty(\alpha,\beta)$ just depends on the internal product $s_\alpha * s_\beta$ because:

$$(6.23)\qquad g^F_\infty(\alpha,\beta) = \langle s_\alpha(x)s_\beta(y),\ \mathrm{mts}\ \underline{F}(M_\infty)\rangle_{x;y}$$

$$= \langle s_\alpha(x)s_\beta(y),\ (F//H)(xy)\rangle_{x;y} = \langle s_\alpha * s_\beta,\ F//H\rangle .$$

But we know (see Theorem 2.24 and §2,IV) that under the natural chain of isomorphisms $\underline{R}(\mathcal{S}) \to \underline{R}_0(GL)) \to \Lambda$, tensor product of symmetric group representations goes over into internal product of symmetric functions. []

The Hidden symmetry and its consequences fail when n is small compared to the degree of $\underline{F}$.

Example 6.24. Let $\underline{F}$ be the q-graded symmetric algebra $\underline{\mathrm{Sym}}_q$. Then

$$g^F_n((2),(2)) = q^2(1-q^n)(1-q^{n-1})/[(1-q)(1-q^2)],$$

while

$$g^F_n((1,1),(1,1)) = q^2(1-q^n)(1-q^{n-3})/[(1-q)(1-q^2)].$$

(See [G],[B2].) On the other hand, these two expressions have the same limit:

$$g^F_\infty((2),(2)) = g^F_\infty((1,1),(1,1)) = q^2/[(1-q)(1-q^2)].$$

§7. DOUBLE SKEW-DIVISION OF SYMMETRIC FUNCTIONS.

The main result here, Theorem 7.1, is a generalization of Corollary 6.16. While our proof of Theorem 6.3 actually constructed an action of $GL(U\otimes V)$ on a very special sort of $GL(U)\times GL(V)$-representation, our proof below works by a calculation in the character ring. In particular, our proof below is independent of the work in §3-§6, (save for Lemma 6.1) and gives an alternative proof of 6.16 and hence of 6.3. As always, x and y are infinite variable sets.

General Hidden Symmetry Theorem 7.1. For any symmetric functions f and g, the skew-quotient $D_{x;y}(f(xy)/g(xy))$ is fully symmetric in the product variable set xy, i.e., the skew-quotient lies in $\Lambda(xy)$.

<u>Proof.</u> We will prove this by working in the basis of $\Lambda_{\mathbb{Q}}$ by power-sum symmetric functions $(p_\lambda)_\lambda$, given by $p_t(x):=x_1^t+x_2^t+\ldots$, and $p_\lambda=p_{\lambda_1}p_{\lambda_2}\cdots$. We have the multiplicativity property $p_\lambda(x)p_\lambda(y)=p_\lambda(xy)$, for all λ.

For each partition λ, set

$$z_\lambda := \prod_{i\geq 1} i^{m(i;\lambda)} m(i;\lambda)!,$$

where $m(i;\lambda)$ is the multiplicity of i as a part of λ. The internal product in Λ of power-sum symmetric functions is given by (see [M]):

(7.2) $$p_\lambda * p_\mu = \delta_{\lambda\mu} z_\lambda p_\lambda ,$$

and $(p_\lambda)_\lambda$ is an orthogonal basis of Λ (under the natural inner product) with $\langle p_\lambda, p_\lambda\rangle = z_\lambda$.

Thus, expanding f(xy) in the basis $(p_\lambda(x)p_\mu(y))_{\lambda\mu}$ of $\Lambda_{\mathbb{Q}}(x;y)$, we get using (2.2")

(7.3) $$f(xy) = \sum_{\lambda\mu} (z_\lambda z_\mu)^{-1}(f, p_\lambda * p_\mu)p_\lambda(x)p_\mu(y)$$
$$= \sum_\lambda (z_\lambda)^{-1}(f,p_\lambda)p_\lambda(x)p_\lambda(y) = \sum_\lambda (z_\lambda)^{-1}(f,p_\lambda)p_\lambda(xy).$$

So the subring $\Lambda_{\mathbb{Q}}(xy)$ (the image of the $\mathbb{Q}$-linear extension of (6.2)) of $\Lambda_{\mathbb{Q}}(x;y)$ has as basis the set $(p_\lambda(xy))_\lambda$.

Thus, in order to show that $\Lambda(xy)$ is closed under the skew-division operation $D_{x;y}(\ /\ )$, it suffices to prove that for any two partitions λ and μ, $D_{x;y}(p_\lambda(xy)/p_\mu(xy))$ lies in $\Lambda(xy)$.

Given partitions $\lambda=(1^{m(1,\lambda)}2^{m(2,\lambda)}\ldots)$ and $\mu=(1^{m(1,\mu)}2^{m(2,\mu)}\ldots)$, set $\lambda-\mu := (1^{m(1,\lambda)-m(1,\mu)}2^{m(2,\lambda)-m(2,\mu)}\ldots)$, if $m(i,\lambda)\geq m(i,\mu)$ for all $i\geq 1$, and set $\lambda-\mu := \emptyset$, otherwise. Then we have [see [M,I,5,Ex.3]) the skew-division formula

(7.4) $$D(p_\lambda/p_\mu) = q_{\lambda\mu}p_{\lambda-\mu} ,$$

where the $q_{\lambda\mu}$ are some non-zero rational coefficients.

Apply this by using multiplicativity to separate the variables:

(7.5) $$D_{x;y}(p_\lambda(xy)/p_\mu(xy)) = D_x(p_\lambda(x)/p_\mu(x))D_y(p_\lambda(y)/p_\mu(y))$$

$$= (q_{\lambda\mu})^2 p_{\lambda-\mu}(xy).$$

This completes the proof. []

Let us record

Corollary (of proof) 7.6. Let f and g be symmetric functions, and let $z^{(1)},\ldots,z^{(k)}$ be k infinite sets variables. Let z be the product variable set $z := z^{(1)}\cdots z^{(k)}$. Then $D_{z^{(1)};\ldots;z^{(k)}}(f(z)/g(z))$ is fully symmetric in the variable set z.

As in the last section, the General Hidden Symmetry Theorem and Lemma 6.1 enable us to make

Definition 7.7. Given symmetric functions f and g, f//g is the symmetric function uniquely determined by the equation

(7.8) $$(f//g)(xy) = D_{x;y}(f(xy)/g(xy))$$

We call the operation // on Λ double skew-division.

We may also write $D^{(2)}(f/g)$ for f//g. By means of the Corollary, we can similarly define the operator $D^{(k)}(\ /\ )$, skew-division of order k, for any positive integer k. There should also be a limiting operation $D^{(\infty)}(\ /\ )$ on symmetric functions.

The general problem of computing double skew-division (and also higher order skew-division) formulae seems interesting. Any such formula can, in priciple and often in practice, be verified by working with the basis of power-sum symmetric functions. On the other hand, proofs in the basis of Schur functions usually carry more information and make connections with the representation theory of symmetric functions.

The "first" double skew-division formula is

Proposition 7.9. Let f be any symmetric function. Then

$$(7.10) \qquad f//h_1 = D(f/h_1) + h_1 D(f/(h_1^2)).$$

Proof. We will prove (7.10) in the case f is a power-sum symmetric function p_λ by computing both sides. We retain the notations of the proof of the Theorem.

It follows from [M,I,5,Ex. 3] that $D(p_\lambda/p_1)=m(1,\lambda)p_{\lambda-(1)}$. Let $r=m(1,\lambda)$. Then by (7.5), $p_\lambda//p_1 = r^2 p_{\lambda-(1)}$. But also

$$D(p_\lambda/p_1) + p_1 D(p_\lambda/(p_1^2)) = rp_{\lambda-(1)} + p_1(r(r-1)p_{\lambda-(1,1)})$$

$$= rp_{\lambda-(1)} + (r^2-r)p_{\lambda-(1)} = r^2 p_{\lambda-(1)}. \qquad []$$

It is curious that this result compares double skew-division with iterated ordinary skew-division, in that $D(f/(h_1^2))= D(D(f/h_1)/h_1)$.

§8. APPLICATIONS OF HIDDEN SYMMETRIES.

In this section, we use the hidden symmetry to compute the stable mixed tensor symbol, the generating function of stable multiplicities, of the full symmetric algebra of M_n. Our method is to first make an explicit computation of stable first layer (see below) multiplicities, and to then obtain the stable mts through symmetrization, by force of the Hidden Symmetry.

There are two general approaches to developing stable formulae. The first is to take the limit of formulae valid for finite n. The second is to work directly in the stable environment, where many calculations become much easier (e.g., see the proof of Proposition 4.1). In this paper, we are following the second approach, while leaving the theory of the mixed tensor symbol as a function of n to [B2] (see also [G]). However, we do note that our original proof of Proposition 8.7 below (in 1981) came from the first method.

Let $\underline{F}$ be a w-graded functor with characteristic function F, over a variable set w. Let x and y be infinite variable sets as usual.

Definition 8.1. The layer number lay(V) of an irreducible PGL_n-representation V is the integer deg(V)/n.

So lay(V) is the smallest value of p such that V occurs in $(\mathbf{C}^n)^{\otimes pn}$. We saw in §3 that the degree of a non-zero $V^{(n)}_{\alpha\beta}$ is $n\beta_1$, so its layer number is just β_1.

First layer representations are then those that appear in $(\mathbf{C}^n)^{\otimes n}$; $V^{(n)}_{\alpha\beta}$ is first layer iff β is a column partition, i.e., iff $\beta = (1^r)$ for some r. First layer representations are the ones most closely related to symmetric group representations (cf. §2), and are generally the easiest to study via symmetric function methods.

All stable multiplicities of $(\underline{F}(M_n))_{n\geq 1}$ are deducible from the first layer stable multiplicities; let us set, for all α,

$$g^F_\infty(\alpha) := g^F_\infty(\alpha,(1^{|\alpha|}))$$

__Proposition 8.2.__ (i) We have the identity in $(\Lambda(x;y))[[w]]$ (in fact in $(\Lambda(xy))[[w]]$)

$$\text{(8.3)} \qquad \text{mts } \underline{F}(M_\infty) = \sum_\alpha g^F_\infty(\alpha')s_\alpha(xy).$$

__Proof.__ By Theorem 6.3, mts $\underline{F}(M_\infty) = (F//H)(xy)$. Expand $F//H$ in the orthonormal basis of Λ by Schur functions with some coefficients $n^F(\alpha)$; so $F//H = \sum_\alpha n^F(\alpha)s_\alpha$. Set $r := |\alpha|$, and compute using (6.23) to get

$$g^F_\infty(\alpha') = \langle s_{\alpha'} * a_r\ ,\ F//H\rangle = \langle s_\alpha\ ,\ F//H\rangle = n^F(\alpha),$$

where $r = |\alpha|$. To compute the internal product $s_{\alpha'} * a_r$, we evaluate:

$$J_{\alpha'} \otimes J_{(1^r)} = (J_\alpha \otimes \text{sgn}_r) \otimes \text{sgn}_r = J_\alpha. \qquad \square$$

A useful consequence is

__Corollary 8.4.__ If there exists some variable set z such that, for all α, $g^F_\infty(\alpha') = s_\alpha(z)$, then mts $\underline{F}(M_\infty) = H(xyz)$.

__Proof.__ Use the Cauchy expansion of H in (2.29). $\square$

Note there are many analogous statements.

__Remarks 8.5.__ It turns out that another very tractable family of multiplicities are those of the $V^{(n)}_{\alpha\beta}$ with β a __row__ partition. Clearly, we get the same sort of result: these stable multiplicities determine the stable symbol and

$$(8.6) \qquad \text{mts } \underline{E}(M_\infty) = \sum_\alpha g_\infty^F(\alpha,(|\alpha|))s_\alpha(xy).$$

<u>Proposition 8.7.</u> The first layer stable multiplicities in $(\underline{\mathrm{Sym}}_q(M_n))_{n\geq 1}$ are given by, for every r and partition α of r,

$$(8.8) \qquad g_\infty^{\mathrm{Sym}}(\alpha,(1^r)) = (\prod_{p\geq 1}(1-q^p)^{-1}) \cdot s_{\alpha'}(q,q^2,\ldots).$$

In "lambda-ring notation", the right side of (8.8) is just $H(q/(1-q))s_{\alpha'}(q/(1-q))$.

To prove Proposition 8.7, we first state and prove a finer calculation. Define the coefficients

$$(8.9) \qquad e(\alpha,\beta;\mu) := \langle s_\alpha(x)s_\beta(y),\ \text{mts } T_{\mu\mu}^{(\infty)}\rangle_{x;y} = \lim_{n\to\infty}\langle V_{\alpha\beta}^{(n)},\ T_{\mu\mu}^{(n)}\rangle,$$

on triples of partitions, with the first two of common magnitude. We know (4.1) that given α,β of size r, and μ of length m, the limit is attained for $n\geq r+m$.

Now for each pair α,β, form the stable generating function

$$(8.10) \qquad f_{\alpha\beta}(x) = \sum_\mu e(\alpha,\beta;\mu)x_\mu \in \mathbb{Z}[x],$$

where $x_\mu := x_{\mu_1}x_{\mu_2}\cdots$. Write f_α for $f_{\alpha\beta}$ when $\beta=(1^{|\alpha|})$.

<u>Proposition 8.11.</u> For each r and partition α of r, $f_\alpha(x)$ is a symmetric power series in x and we have the identity in $\Lambda((x))$,

$$(8.12) \qquad f_\alpha(x) = s_{\alpha'}(x)H(x) = s_{\alpha'}(x)(1+h_1(x)+h_2(x)+\ldots).$$

<u>Proof.</u> We prove (8.12) as an identity in $\mathbb{Z}[[x]]$, thereby demonstrating the symmetry of $f_\alpha(x)$.

Since $s_{\alpha'}(x)$ and $H(x)$ are, respectively, the generating functions $\sum_S x_S$ of row-strict tableaux S of shape α and $\sum_P x_P$ of all row tableaux P, the product $s_{\alpha'}(x)H(x)$ is the generating function $\sum_{S,P} x_Sx_P$ of the family of such pairs (S,P). But this last description also fits $f_{\alpha}(x)$, in view of Corollary 4.7. □

Remarks 8.13. (i) The function $f_{\alpha\beta}(x)$ will be symmetric in x iff the stable multiplicity (8.9) depends only on the *multiplicities* of μ, not on the sizes of its parts. This symmetry fails in general. For instance, if $\alpha=\beta=(2)$, then $V^{(n)}_{\alpha\beta}$ occurs in $T^{(n)}_{(p)(p)}$ for $n\geq 2$, but $V^{(n)}_{\alpha\beta}$ never occurs in $T^{(n)}_{(1)(1)}$, the direct sum of the adjoint and trivial representations. So the coefficient of x_1 in $f_{(2)(2)}(x)$ is different from that of x_p, $p\geq 2$. In fact,

$$f_{(2)(2)}(x) = (x_2+x_3+\ ..\ ) + \text{(higher order terms)}.$$

(ii) In view of (4.3), 8.12 is equivalent to the combinatorial identity, with α a fixed partition of r and $\gamma=(1^r)$,

$$\sum_{\mu} \langle s_{\mu/\alpha}, s_{\mu/\gamma}\rangle x_{\mu} = s_{\alpha'}(x)H(x). \tag{8.14}$$

If we restrict (8.14) to terms of degree p or less, then we get

$$\sum_{\ell(\mu)\leq p} \langle s_{\mu/\alpha}, s_{\mu/\gamma}\rangle x_{\mu} = s_{\alpha'}(x)(1+h_1(x)+..+h_{p-r}(x)). \tag{8.14'}$$

Proof of Proposition 8.7. The Cauchy decomposition says

$$\underline{S}^p(M_n) = \bigoplus_{\mu\vdash p} T^{(n)}_{\mu,\mu}. \tag{8.15}$$

So for all α and β, the stable multiplicity $g^{Sym}_{\infty}(\alpha,\beta)$ is equal to $f_{\alpha\beta}(q,q^2,...)$. Specializing (8.12), we get (8.8). □

Define the variable set $Q := (q,q^2,...)$. So $H(Q) = \prod_{p\geq 1} (1-q^p)^{-1}$.

Theorem 8.16 The stable mixed tensor symbol of the q-graded symmetric algebra of nXn matrices is

$$\text{mts}\ \underline{\text{Sym}}_q(M_{\infty}) = H(Q)H(xyQ) = H(Q)\prod_{p,i,j\geq 1} (1-q^p x_i y_j)^{-1}. \tag{8.17}$$

In "lambda-ring" notation: $Q = q/(1-q)$; $H(Q)H(xyQ)=H(q(1+xy)/(1-q))$.

Proof. Immediate from Proposition 8.7 and Corollary 8.4. □

The graded representations $\underline{Sym}(M_n)$ are especially interesting because they are so closely related to the <u>generalized exponents</u> of finite-dimensional irreducible PGL_n-representations V. These are fundamental invariants defined by Kostant in [K] for irreducible representations of any complex semi-simple Lie group of adjoint type. We will state the definitions and key properties here only in the PGL_n case, but the same statements go through in general (see [K]).

Let P_n be the cone of nilpotent matrices in M_n. Then the multiplicity of $p \geq 0$ as a generalized exponent of an irreducible V is defined to be the multiplicity of V in the space $R^p(P_n)$ of regular functions on P_n of degree p. It turns out that a given V has only finitely many generalized exponents; the number of them (counted with multiplicity, always) is the dimension of the "zero-weight" space of V, the space of invariants under the conjugation action of the diagonal matrices.

The ring of invariants I_n in the graded coordinate ring $R(M_n) = \underline{Sym}(M_n)^*$ of regular functions on M_n is a polynomial ring in n algebraically independent polynomials (e.g., the traces of the first n powers of a matrix). So the Hilbert series of I_n is

$$\text{(8.18)} \qquad \text{HS } I_n = \sum_{p \geq 0} \dim(I_n^p) q^p = \prod_{p=1}^{n} (1-q^p)^{-1} = H(q,\ldots,q^n).$$

In "lambda-ring notation", $H(q,\ldots,q^n) = H((q-q^{n+1})/(1-q))$.

Moreover, there is a tensor splitting $I_n \otimes \mathcal{H}_n = R(M_n)$, where $\mathcal{H}_n$ is the space of harmonic polynomial functions on M_n (i.e., functions annihilated by all positive degree PGL_n-invariant differential operators with constant coefficients). The defining ideal of P_n as a subvariety of M_n is generated by the positive degree elements of I_n. Thus the natural restriction map $R(M_n) \to R(P_n)$ gives a linear, graded, PGL_n-equivariant isomorphism from $\mathcal{H}_n$ to $R(P_n)$. Write $R_q(P_n)$ for the graded representation, so that $\text{ch } R_q(P_n) = \sum_{p \geq 0} (\text{ch } R^p(P_n)) q^p$. Then the characters satisfy

$$\text{(8.19)} \qquad H(q,\ldots,q^n)\, \text{ch}_{PGL_n} R_q(P_n) = \text{ch}_{PGL_n} \underline{Sym}_q(M_n).$$

In the expansion

(8.20) $$\text{mts } R_q(P_n) = \sum_{p\geq 0} (\text{mts } R^p(P_n))q^p = \sum_{\alpha\beta} j_n^{\alpha\beta}(q) s_\alpha(x) s_\beta(y),$$

the coefficients $j_n^{\alpha\beta}(q)$ are polynomials in q, and $j_n^{\alpha\beta}(q)$ is the natural generating function of generalized exponents of $V_{\alpha\beta}^{(n)}$. I.e., the coefficient of q^p in $j_n^{\alpha\beta}(q)$ is the multiplicity of p as a generalized exponent of $V_{\alpha\beta}^{(n)}$.

So a stable theory of the mixed tensor symbol of the family $(R(P_n))_{n\geq 1}$ of coordinate rings is the same thing as a stable theory of the generalized exponents of PGL_n.

__Theorem 8.21.__ The family of graded reprentations $(R(P_n))_{n\geq 1}$ is stable, with stable mixed tensor symbol

(8.22) $$\text{mts } R_q(P_\infty) = H(xyQ).$$

For each pair α,β of equal size partitions, the limit of $j_n^{\alpha\beta}(q)$ exists, and has value

(8.23) $$j_\infty^{\alpha\beta}(q) := \lim_{n\to\infty} j_n^{\alpha\beta}(q) = (s_\alpha * s_\beta)(Q).$$

__Proof.__ Equation (8.19) carries over to the mixed tensor symbol, so we have for all n

(8.24) $$\text{mts } R_q(P_n) = (1-q)\cdots(1-q^n)[\text{mts } \underline{\text{Sym}}_q(M_n)].$$

For each p, the terms on the right side of degree p or less are independent of n if $n\geq 2p$. So the limit of mts $R_q(P_n)$ as $n\to\infty$ exists in $(\Lambda(x;y)[[q]]$, and is given by

(8.25) $$\text{mts } R_q(P_\infty) = [(1-q)(1-q^2)\cdots][\text{mts } \underline{\text{Sym}}_q(M_\infty)].$$

Plugging (8.17) into (8.25) gives (8.22). []

The next step is to investigate the form of the stable mixed tensor symbol coefficients $j_\infty^{\alpha\beta}(q)$ as power series in q. Given a partition α of r, let $h_1(\alpha),\ldots,h_r(\alpha)$ be the combinatorial __hook-lengths__ (in any order) of α (see [M,I,1,Ex. 1). The __hook-length polynomial__ $H_\alpha(q)$ is

(8.26) $$H_\alpha(q) := \prod_{i=1}^{r} (1-q^{h_i(\alpha)}).$$

Also set $e(\alpha) := \sum_{i\geq 1} i\alpha_i$.

<u>Corollary 8.27.</u> If α is a partition of r, then

$$j_\infty^{\alpha,(r)}(q) = j_\infty^{\alpha',(1^r)}(q) = q^{e(\alpha)}/H_\alpha(q). \tag{8.28}$$

<u>Proof.</u> Both stable multiplicities are equal to $s_\alpha(Q)$, which in turn has the classical expansion $q^{e(\alpha)}/H_\alpha(q)$ (see [M,I,3,Ex. 2]). []

<u>Conjecture 8.29.</u> For every pair α,β of partitions of common size, there exists a polynomial $p_{\alpha\beta}(q)$ with non-negative integral coefficients such that

$$\sum_{p\geq 0} \lim_{n\to\infty} \langle V_{\alpha\beta}^{(n)}, R^p(P_n)\rangle q^p = p_{\alpha\beta}(q)/H_\alpha(q). \tag{8.30}$$

The left hand side is, by definition, $j_\infty^{\alpha\beta}(q)$.

We can deduce from Theorem 8.21 the weaker statement

<u>Corollary 8.31.</u> For every pair α,β of partitions of common size r, there exists a polynomial $t_{\alpha\beta}(q)$ with non-negative integral coefficients such that

$$\sum_{p\geq 0} \lim_{n\to\infty} \langle V_{\alpha\beta}^{(n)}, R^p(P_n)\rangle q^p = t_{\alpha\beta}(q)/[(1-q)\cdots(1-q^r)]. \tag{8.32}$$

<u>Proof.</u> By (8.23), $j_\infty^{\alpha\beta}(q)$ is equal to $\sum_\gamma c_{\alpha\beta\gamma} s_\gamma(Q)$. But each Schur function specialization $s_\gamma(Q)$ is known (see [S1]) to be the quotient of a polynomial with non-negative integral coefficients by the product $(1-q)\cdots(1-q^r)$. As all the coefficients $c_{\alpha\beta\gamma}$ are non-negative by their definition as symmetric group multiplicities, we get the Corollary. []

The author made Conjecture 8.29 in January 1982 on the basis of (i) the second equality in Corollary 8.27, which she had proven by calculating the multiplicities $j_n^{\alpha\beta}(q)$ as functions of n for column partitions β (see [G]) and then taking the limit and (ii) her conjectural table of the $j_\infty^{\alpha\beta}(q)$ for partitions pairs α,β of size 5 and less, which she made by constructing families of functions on nilpotent orbits. Between June 1981 and January 1982, the author constructed the mixed tensor parametrization and the families $(V_{\alpha\beta}^{(n)})$ where n varies as α and β are held fixed, and she had defined the notions of layer and depth. She proved Prop. 4.1 and the resulting stabilities of $(\underline{\mathrm{Sym}}_q(M_n))_{n\geq 1}$ and $(R_q(P_n))_{n\geq 1}$. However, there was no notion of "hidden symmetries".

Subsequent discussions with Richard Stanley in July 1982 led him and the author to conjecture the formula (8.23). In the light of this, it was natural to rewrite 8.29 as the purely combinatorial

Conjecture 8.33 (Gupta-Stanley) ([G-S]). For every pair α,β of partitions of common size, there exists a polynomial $p_{\alpha\beta}(q)$ with non-negative integral coefficients such that

$$s_\alpha * s_\beta(q/(1-q)) = p_{\alpha\beta}(q)/H_\alpha(q). \tag{8.34}$$

Shortly after, Stanley ([S2]) proved (8.23), and he also proved Conjecture 8.33, and hence Conjecture 8.29, save for the non-negativity of the coefficients of $p_{\alpha\beta}(q)$. The non-negativity question is still open at this writing!

In the spring of 1982, Phil Hanlon and the author together investigated the multiplicities of first layer representations in the graded exterior algebra $\underline{E}(M_n)$. They conjectured a formula ([G-H]) which was later proven by John Stembridge ([St1]) in his MIT thesis under Stanley.

Many developments in the first layer and stable theories have been made subsequently by Stanley ([S3]), Hanlon ([H]), Stembridge ([St2]), and the author. The author's work on these topics has opened some new lines - studying the mixed tensor symbol of $(R_q(P_n))_{n\geq 1}$ as a function of n ([G], [B2]), analyzing the stability of general multiplicative functors ([B3]), and interpreting the mixed tensor symbol by means of classical invariant theory and quotient variety constructions under GL_n ([B1],[B3]).

REFERENCES

[B1] R.K. Brylinski, Matrix concomitants with the mixed tensor model I, preprint (1987).

[B2] R.K. Brylinski, The mixed tensor character of the nullcone and Hall-Littlewood symmetric functions, in preparation.

[B3] R.K. Brylinski, Stable calculus of the mixed tensor character II, in preparation.

[G] R.K. Gupta, Generalized Exponents via Hall-Littlewood symmetric functions, Bull. Amer. Math. Soc. (new series) 16, no. 2, April (1987), 287-291.

[G-H] R.K. Gupta and P. Hanlon, Problem 5 in Problem Session, Combinatorics and Algebra, ed. by C. Greene, Contempory Mathematics 34, Amer Math. Soc. (1984), p. 305-307.

[G-S] R.K. Gupta and R. Stanley, Problem 7 in Problem Session, Combinatorics and Algebra, ed. by C. Greene, Contempory Mathematics 34, Amer Math. Soc. (1984), p. 308.

[H] P. Hanlon, On the decomposition of the tensor algebra of the classical Lie algebras, Adv. in Math. 56 (1985), 238-282.

[Ki] R.C. King, Generalized Young tableaux and the general linear group, Jour. of Math. Phys. 11, no. 1 (1970), p. 280-293.

[Ki-Pl] R.C. King, S.P.O. Plunkett, The evaluation of weight multiplicities using characters and S-functions, Jour. of Phys. A 9 (1976), p.863-887.

[Kk] K. Koike, On the decomposition of tensor products of the representations of the classical groups - by means of the universal characters-, preprint.

[K] B. Kostant, Lie group representations on polynomial rings, Amer. J. Math. 85 (1963), 327-404.

[L] D.E. Littlewood, On invariant theory under restricted groups, Phil. Trans. of. Royal Soc. 239 A (1944), p. 387-417.

[M] I.G. Macdonald, Symmetric Functions and Hall Polynomials, Oxford, Clarendon Press 1979.

[Mt] J. Matsuzawa, On the generalized exponents of classical Lie groups, preprint.

[S1] R. Stanley, Invariants of finite groups and their applications to combinatorics, Bull AMS (new series) 1,no 3 (1981), 475-511.

[S2] R. Stanley, The q-Dyson conjecture, generalized exponents, and the internal product of Schur functions, in AMS Contemp. Math. 34 (1984), 81-93.

[S3] R. Stanley, The stable behavior of some characters of SL(n,ℂ), Lin. and Mult. Alg. 16 (1984), 3-27.

[St1] J.R. Stembridge, Combinatorial decompositions of characters of SL(n,ℂ), Ph. D. Thesis, M.I.T. (1985).

[St2] J.R. Stembridge, First layer formulas for characters of SL(n,ℂ), Trans. Amer. Math. Soc. 229, No. 1 (1987), p. 319-350.

[W] H. Weyl, *The Classical Groups*, Princeton University Press 1939.

Filtrations of right ideals related to projectivity of left ideals

Vlastimil Dlab and Claus Michael Ringel

[This paper is in final form and no version of it will be submitted for publication else-where].

Let k be a field and A a finite-dimensional k-algebra. Since the endomorphism ring of the right A-module A_A is A itself, one must be able to describe all properties of A, for example properties of left ideals of A, in terms of the right A-module A_A. The aim of the present note is to show that the projectivity of certain left ideals can be characterized by the existence of suitable filtrations of right ideals.

As an application, we deal with quasi-hereditary rings. They have been defined by Scott **[S]** using heredity chains of ideals, thus using an inductive procedure of enlarging algebras. In this way one deals with a total ordering $e_1, \ldots, e_n$ of a complete set of primitive idempotents, with e_n being added last. But there is a reverse procedure based on investigations of Mirollo and Vilonen **[MV]**, and described in **[DR2]**: there we construct A from $\varepsilon_2 A \varepsilon_2$ where $\varepsilon_2 = e_2 + e_3 + \cdots + e_n$. We characterize quasi-hereditary algebras such that the class of modules with Weyl filtrations is closed under submodules in terms of the two recursive procedures. And we show that algebras which satisfy this and the opposite condition have global dimension at most 2. It follows that the deep algebras introduced in **[DR3]**, as well as the peaked ones defined in this paper have global dimension at most 2.

1. The main results

Unless otherwise stated, modules will be (finitely generated) *right* A-modules. Let $\mathcal{M}$ be a set of A-modules. Given a module X_A, an *$\mathcal{M}$-filtration* of X_A is a chain of submodules $0 = X_0 \subset X_1 \subset \cdots \subset X_t = X$ such that for all $1 \le i \le t$, the module X_i/X_{i-1} is isomorphic to a module in $\mathcal{M}$.

Let N be the (Jacobson) radical of A. Let $e_1, \ldots, e_n$ be a complete set of primitive (and orthogonal) idempotents. Let $E(i) = E(e_i)$ be the simple A-module not annihilated by e_i; thus $E_i \cong e_iA/e_iN$. Let $P(i) = P(e_i)$ be a projective cover of $E(i)$; thus $P(i) \cong e_iA$. Given a primitive idempotent e, we denote by $\hat{e}(i)$ the maximal quotient of $P(i)$ of Loewy length at most 2, whose radical is a direct sum of copies of $E(e)$. The set of modules $\hat{e}(i)$, with $1 \le i \le n$, is denoted by $\hat{e}$. The number of composition factors (in a composition series) of a module X which are isomorphic to $E(i)$ will be denoted by $\ell_i(X)$. We recall that a module is said to be *torsionless* provided it is isomorphic to a submodule of a projective module.

Theorem 1. *Let e be a primitive idempotent of A. The following statements are equivalent:*

(*i*) *The left ideal Ne is a projective left module.*

(*ii*) *A_A has an $\hat{e}$-filtration and $Ext^1_A(E(e), E(e)) = 0$.*

(*ii'*) *Every right ideal has an $\hat{e}$-filtration.*

(ii″) *Every torsionless module has an ê-filtration.*

Let $\varepsilon_i = e_i + \cdots + e_n$ for $1 \le i \le n$, and $\varepsilon_{n+1} = 0$. We denote by $\Delta(i)$ the largest factor module of $P(i)$ with all composition factors of the form $E(j)$, with $j \le i$; thus $\Delta(i) = e_iA/e_iA\varepsilon_{i+1}A$. The set of modules $\Delta(i)$, with $1 \le i \le n$, is denoted by Δ, note that these modules $\Delta(i)$ depend on the chosen ordering $e_1, \ldots, e_n$. Let $I_i = A\varepsilon_{n-i+1}A$, thus $0 = I_0 \subset I_1 \subset \cdots \subset I_n = A$ is a saturated chain of idempotent ideals of A. Note that $(I_i)_i$ is a heredity chain if and only if first, A_A has a Δ–filtration, and second, $\ell_i(\Delta(i)) = 1$, for all $1 \le i \le n$: in this case, A is said to be quasi-hereditary. (In case that A is quasi-hereditary, the Δ–filtrations of a module X are also called "Weyl filtrations" **[PS]**. Also, X has a Δ–filtration if and only if its filtration $0 = XI_0 \subseteq XI_1 \subseteq \cdots \subseteq XI_n = X$ is "good" in the sense of **[DR2]**; this follows from Lemma 1* in section 2.)

Theorem 2. *Assume that $(I_i)_i$ is a heredity chain, where $I_i = A\varepsilon_{n-1+1}A$, and let $C_i = \varepsilon_iA\varepsilon_i$. Then the following conditions are equivalent:*

(i) *ε_iNe_i is a projective left C_i–module, for $1 \le i \le n$,*

(i′) *$\varepsilon_{i+1}Ne_i$ is a projective left C_{i+1}–module, for $1 \le i \le n-1$,*

(ii) *rad $\Delta(i)$ has a Δ-filtration, for $1 \le i \le n$,*

(ii′) *every right ideal has a Δ–filtration,*

(ii″) *every torsionless module has a Δ–filtration,*

(ii‴) *submodules of modules with a Δ–filtration have a Δ–filtration.*

The left modules $\Delta^*(i)$ and $\Delta^* = \{\Delta^*(i) | 1 \le i \le n\}$ are defined similarly as $\Delta(i)$ and Δ, namely: $\Delta^*(i)$ is the largest factor module of $P^*(i)$ with all composition factors of the form $E^*(j)$ with $j \le i$, thus $\Delta^*(i) = Ae_i/A\varepsilon_{i+1}Ae_i$. The fact that $(I_i)_i$ is a heredity chain may be expressed in a similar way in terms of Δ^*. In the next theorem we deal with those algebras A such that both A and its opposite satisfy the equivalent conditions of Theorem 2.

Theorem 3. *Let $(I_i)_i$ be a heredity chain. Assume that any right ideal of A has a Δ-filtration and that any left ideal of A has a Δ^*-filtration. Then gl.dim.$A \le 2$.*

Corollary 1. *Deep quasi-hereditary algebras have global dimension at most 2.*

We recall that the quasi-hereditary algebra A is said to be *deep* **[DR3]** if, for every $1 \le i \le n$, both the right A–module rad $\Delta(i)$ and the left A–module rad $\Delta^*(i)$ are projective.

The proofs of these results will be given in section 2, 3, and 4 of the paper. Section 5 contains a construction of a class of quasi-hereditary algebras of global dimension 2 which we call the *peaked* algebras. These are examples of algebras A such that both A and A^{opp} satisfy the conditions of Theorem 2.

2. Preliminaries on filtrations of modules.

First, let $\mathcal{M}$ be an arbitrary set of modules. We consider modules which have an $\mathcal{M}$-filtration. It is sometimes necessary to arrange the various quotients occurring in a filtration. In order to be able to do so, we will use the following well-known lemma.

Lemma 1. *Assume that some $M \in \mathcal{M}$ satisfies $Ext^1_A(M', M) = 0$ for all $M' \in \mathcal{M}$. Let $\mathcal{M}' = \mathcal{M} \backslash \{M\}$. If a module X has an $\mathcal{M}$-filtration, then it has a submodule X' with an $\mathcal{M}'$-filtration such that X/X' is a direct sum of copies of M.*

Proof. Let X'' be a submodule of X with an $\mathcal{M}$-filtration such that X/X'' belongs to $\mathcal{M}$. By induction, there is a submodule X''' of X'' with an $\mathcal{M}'$-filtration such that X''/X''' is a direct sum of copies of M. Since $\mathrm{Ext}^1_A(X/X'', X''/X''') = 0$, there is a submodule Y of X with $Y \cap X'' = X'''$ and $Y + X'' = X$. If X/X'' belongs to $\mathcal{M}'$, let $X' = Y$; otherwise, let $X' = X'''$.

Lemma 1*. *Assume that some $M \in \mathcal{M}$ satisfies $Ext^1_A(M, M') = 0$ for all $M' \in \mathcal{M}$. Let $\mathcal{M}' = \mathcal{M} \backslash \{M\}$. If a module Y has an $\mathcal{M}$-filtration, then it has a submodule Y' which is a direct sum of copies of M such that Y/Y' has an $\mathcal{M}'$-filtration.*

Clearly, this is the dual assertion. Both results have been used by Cline–Parshall–Scott [**CPS**] for dealing with modules over quasi-hereditary rings, or, more generally, with objects in highest weight categories.

We will be interested to know whether submodules of modules with an $\mathcal{M}$-filtration again have $\mathcal{M}$-filtrations. The following is a useful criterion in this direction.

Lemma 2. *Assume that for any $M \in \mathcal{M}$, every maximal submodule of M has an $\mathcal{M}$-filtration. Then submodules of modules with an $\mathcal{M}$-filtration have an $\mathcal{M}$-filtration.*

Proof. Let $0 = X_0 \subset X_1 \subset \cdots \subset X_t = X$ be an $\mathcal{M}$-filtration of the module X, let Y be a submodule of X. We claim that Y has an $\mathcal{M}$-filtration. By induction on the length of X/Y, we may assume that Y is a maximal submodule of X. Choose i minimal with $X_i \not\subseteq Y$. Then $X_i \cap Y$ is a maximal submodule of X_i containing X_{i-1}. By assumption, $X_i \cap Y/X_{i-1}$ has an $\mathcal{M}$-filtration. Using it, we may refine the filtration $0 = X_0 \subset \ldots X_{i-1} \subseteq X_i \cap Y \subset \cdots \subset X_t \cap Y = Y$ in order to obtain an $\mathcal{M}$-filtration for Y.

We return to the complete set $e_1, \ldots, e_n$ of primitive idempotents of A, and we denote $e = e_1$. *We assume that* $\mathrm{Ext}^1_A(E(e), E(e)) = 0$. Let $\mathcal{M}(e) = \{\hat{e}(i) | 2 \leq i \leq n\}$, and let $\bar{\mathcal{M}}(e)$ be the set of non-zero quotient modules of modules in $\mathcal{M}(e)$.

Lemma 3. *A module X has an $\bar{\mathcal{M}}(e)$-filtration if and only if $Hom_A(X, E(e)) = 0$.*

Proof. If M is in $\bar{\mathcal{M}}(e)$, then $\mathrm{Hom}_A(M, E(1)) = 0$. Thus, if X has an $\bar{\mathcal{M}}(e)$-filtration, $\mathrm{Hom}_A(X, E(1)) = 0$. Conversely, assume $\mathrm{Hom}_A(X, E(1)) = 0$. We may assume $X \neq 0$, thus let X' be a maximal submodule of X. Then $X/X' \cong E(j)$ for some $2 \leq j \leq n$. Let $X'' = \mathrm{rad}\, X'$. There are (uniquely determined) submodules Y, Y' of X' containing

X'' such that $X'/X'' = Y/X'' \oplus Y'/X''$ with Y'/X'' a direct sum of copies of $E(1)$, and Y/X'' a direct sum of various $E(i)$, with $2 \leq i \leq n$. We claim that X/Y belongs to $\bar{\mathcal{M}}(e)$. For, the submodule X'/Y of X/Y is a direct sum of copies of $E(1)$, the quotient is $X/X' \cong E(j)$, and $\mathrm{Hom}_A(X/Y, E(1)) = 0$, thus $X'/Y = \mathrm{rad}(X/Y)$. On the other hand, $\mathrm{Hom}_A(Y, E(1)) = 0$, since otherwise $\mathrm{Ext}^1_A(E(1), E(1)) \neq 0$. By induction, Y has an $\bar{\mathcal{M}}(e)$–filtration and thus X has an $\bar{\mathcal{M}}(e)$–filtration.

The length of the module X will be denoted by $\ell(X)$; hence $\ell(X) = \sum_{i=1}^{n} \ell_i(X)$. Let $s_i = \ell(\hat{e}(i))$.

Lemma 4. *Assume that X has an $\bar{\mathcal{M}}(e)$–filtration. Then*

$$\ell(X) \leq \sum_{i=2}^{n} \ell_i(X) s_i\,;$$

moreover the following assertions are equivalent:

(*i*) $$\ell(X) = \sum_{i=2}^{n} \ell_i(X) s_i\,,$$

(*ii*) *the module X has an $\mathcal{M}(e)$–filtration,*

(*iii*) *any $\bar{\mathcal{M}}(e)$–filtration of X is an $\mathcal{M}(e)$–filtration.*

Proof. Let $0 = X_0 \subset X_1 \subset \cdots \subset X_t = X$ be an $\bar{\mathcal{M}}(e)$–filtration, with $X_j/X_{j-1} \cong \hat{e}(\sigma(j))/U_j$, where $U_j \subseteq \mathrm{rad}\, \hat{e}(\sigma(j))$, and $2 \leq \sigma(j) \leq n$. Clearly, for $2 \leq i \leq n$, the number $\ell_i(X)$ is just the number of j's with $\sigma(j) = i$. Thus

$$\begin{aligned}\ell(X) &= \sum_{j=1}^{t} \ell(X_j/X_{j-1}) = \sum_{j=1}^{t} \ell(\hat{e}(\sigma(j))) - \sum_{j=1}^{t} \ell(U_j)\\ &= \sum_{i=2}^{n} \ell_i(X)s_i - \sum_{j=1}^{t} \ell(U_j) \leq \sum_{i=2}^{n} \ell_i(X)s_i\,,\end{aligned}$$

and we have equality if and only if all $U_j = 0$, that is if and only if the given filtration is an $\mathcal{M}(e)$–filtration.

Lemma 5. *Assume that X has an $\mathcal{M}(e)$–filtration, and let e' be an idempotent of A with $eAe' \subseteq N$. Then also $X/Xe'A$ has an $\mathcal{M}(e)$–filtration.*

Proof. Since $\mathrm{Hom}_A(Xe'A, E(e)) = 0$, the module $Xe'A$ has an $\bar{\mathcal{M}}(e)$–filtration according to Lemma 3. Since X has an $\bar{\mathcal{M}}(e)$–filtration, also $X/Xe'A$ has one, and therefore X has an $\bar{\mathcal{M}}(e)$–filtration passing through $Xe'A$. But by Lemma 4, any $\bar{\mathcal{M}}(e)$–filtration is an $\mathcal{M}(e)$–filtration.

Lemma 6. *Assume X has an $\hat{e}$–filtration. Then there is a submodule X' of X with an $\mathcal{M}(e)$–filtration such that X/X' is a direct sum of copies of $E(e)$.*

Proof. Since $\mathrm{Ext}^1(E(e), E(e)) = 0$, we have $\mathrm{Ext}^1(\hat{e}(i), E(1)) = 0$ for all $1 \leq i \leq n$. Now we apply Lemma 1.

3. Proof of Theorem 1.

As before, we deal with a complete set $e = e_1, e_2, \ldots, e_n$ of primitive idempotents.

If the left ideal Ne is a projective left module, its indecomposable summands have to be of the form Ae_i, with $2 \leq i \leq n$. Since Ae cannot be embedded into Ne, but $\mathrm{Ext}^1_A(E(e), E(e)) = 0$.

We are going to establish the equivalence of assertions(i) and (ii) in Theorem 1, so we may assume from the beginning that $\mathrm{Ext}^1_A(E(e), E(e)) = 0$.

Recall that the species $\mathcal{S} = (D_i, {}_iM_j)_{i,j}$ of A is defined as follows: D_i is the division ring e_iAe_i/e_iNe_i, and ${}_iM_j$ is the D_i–D_j–bimodule $e_iNe_j/e_iN^2e_j$. Let $d_i = \dim_k D_i, d_{ij} = \dim({}_iD_j)_{D_j}, d'_{ij} = \dim_{D_i}({}_iM_j)$; thus $\dim_k({}_iM_j) = d_id'_{ij} = d_{ij}d_j$. We observe that $\mathrm{rad}\,\hat{e}(i) = d_{i1}E(1)$ (thus $s_i = d_{i1} + 1$).

The simple left A–modules will be denoted by $E^*(i) = Ae_i/Ne_i$, their projective covers by $P^*(i) = Ae_i$. The top of the left A–module Ne is isomorphic to $\bigoplus_{i=1}^{n} d'_{i1}E^*(i)$, and we consider the projective cover $p : {}_AP \longrightarrow {}_ANe$ of left A–modules: here, ${}_AP \cong \bigoplus_{i=1}^{n} d'_{i1}P^*(i)$. Actually, the assumption $\mathrm{Ext}^1_A(E(e), E(e)) = 0$ can be reformulated as ${}_1M_1 = 0$; thus $d_{11} = 0 = d'_{11}$. Let ${}_AY$ be the kernel of p.

We decompose $A_A = e'A \oplus e''A$, where $e'A$ is a direct sum of copies of eA, and $eAe'' \subseteq N$. Let $X_A = e'N \oplus e''A$, thus $Xe = Ne$, and $Xe_i = Ae_i = P^*(i)$ for $2 \leq i \leq n$. In particular, for $2 \leq i \leq n$, we have

$$\dim_k P^*(i) = \dim_k Xe_i = \ell_i(X)d_i\,;$$

therefore

$$\dim_k P = \sum_{i=2}^{n} d'_{i1}\dim_k P^*(i) = \sum_{i=2}^{n} \ell_i(X)d_id'_{i1} = \sum_{i=2}^{n} \ell_i(X)d_{i1}d_1\,.$$

Since $\mathrm{Ext}^1_A(E(e), E(e)) = 0$, we have $\mathrm{Hom}_A(X_A, E(e)) = 0$. Hence Lemma 3 asserts that X_A has an $\bar{\mathcal{M}}(e)$–filtration, say $0 = X_0 \subset X_1 \subset \cdots \subset X_t = X$ with $X_j/X_{j-1} \cong \hat{e}(\sigma(j))/U_j$

for some submodule U_j of $\hat{e}(\sigma(j))$ and $2 \leq \sigma(j) \leq n$. The number of j's with $\sigma(j) = i$ is $\ell_i(X)$. Since

$$\ell_1(X_j/X_{j-1}) = \ell_1(\hat{e}(\sigma(j))) - \ell_1(U_j) = d_{\sigma(j),1} - \ell(U_j),$$

we have

$$\begin{aligned}\dim_k Xe &= \sum_{j=1}^{t} \ell_1(X_j/X_{j-1})d_1 = \sum_{j=1}^{t}(d_{\sigma(j),1} - \ell(U_j))d_1 \\ &= \sum_{i=1}^{n} \ell_i(X)d_{i1}d_1 - \sum_{j=1}^{t} \ell(U_j)d_1 .\end{aligned}$$

Comparing the dimensions of P and $Ne = Xe$, we obtain the dimension for the kernel Y of p

$$\dim_k Y = \sum_{j=1}^{t} \ell(U_j)d_1 .$$

If we assume that ${}_ANe$ is a projective left A-module, then p is bijective, thus $Y = 0$. Therefore all $U_j = 0$, and our $\bar{\mathcal{M}}(e)$–filtration of X_A is an $\mathcal{M}(e)$–filtration. Since A_A/X_A is a direct sum of copies of $E(e) = \hat{e}(1)$, we conclude that A_A has an $\hat{e}$–filtration.

Conversely, assume that A_A has an $\hat{e}$–filtration. According to Lemma 6, we obtain a submodule $\bar{X}_A$ of A_A with an $\mathcal{M}(e)$–filtration such that $A_A/\bar{X}_A$ is a direct sum of copies of $E(e)$. Clearly, $\bar{X}_A = X_A$, so X_A has an $\mathcal{M}(e)$–filtration. It follows that $U_j = 0$ for all j, consequently p is bijective, and therefore ${}_ANe$ is a projective left A–module.

This shows the equivalence of assertions (i) and (ii). Every module $\hat{e}(i)$ in $\hat{e}$ has a unique maximal submodule, and this submodule is a direct sum of copies of $\hat{e}(1) = E(e)$. Hence, it has an $\hat{e}$–filtration. Lemma 1 asserts that submodules of modules with $\hat{e}$–filtrations have $\hat{e}$–filtrations. Under the assumption of (ii), any free module has an $\hat{e}$–filtration, thus any torsionless module has an $\hat{e}$–filtration. This shows (ii) $\Rightarrow$ (ii″), and trivially (ii″) $\Rightarrow$ (ii′).

Finally, we show the implication (ii′) $\Rightarrow$ (ii). Take a right ideal Y_A of minimal length having $E(e)$ as a composition factor.

Clearly, Y_A has a unique maximal submodule Y', and $Y/Y' \cong E(e)$, whereas Y' has no composition factor of the form $E(e)$. Take an $\hat{e}$–filtration $0 = Y_0 \subset Y_1 \subset \cdots \subset Y_t = Y$ of Y. Then $Y_{t-1} \subseteq Y'$, and $Y'/Y_{t-1} = \operatorname{rad}(Y_t/Y_{t-1})$. Since $Y/Y' \cong E(e)$, we see that $Y_t/Y_{t-1} \cong \hat{e}(1)$. Since Y'/Y_{t-1} has no composition factor $E(1)$, it follows that $Y'/Y_{t-1} = 0$. Thus $\hat{e}(1) = E(e)$, and therefore $\operatorname{Ext}^1_A(E(e), E(e)) = 0$.

4. Proof of theorem 2.

We assume that $(I_i)_i$ is a heredity chain, where $I_i = A\varepsilon_{n-i+1}A$, with $\varepsilon_i = e_i + \cdots + e_n$, for $1 \le i \le n$, and $\varepsilon_{n+1} = 0$, and we denote $C_i = \varepsilon_i A \varepsilon_i$.

Lemma 7. *The left ideal ${}_ANe_1$ is a projective left A-module if and only if $\varepsilon_2 Ne_1$ is a projective left C_2-module.*

Proof. First, assume that ${}_ANe_1$ is projective. Then ${}_ANe_1$ is isomorphic to a module of the form $\bigoplus_{i=2}^{n} m_i Ae_i$, for some $m_i \in \mathbb{N}_0$, since Ae_1 cannot be embedded into Ne_1. Thus $\varepsilon_2 Ne_1 \cong \bigoplus_{i=2}^{n} m_i(\varepsilon_2 Ae_i)$, as a left C_2-module. But $\varepsilon_2 Ae_i$ is a projective left C_2-module for $2 \le i \le n$, since $\varepsilon_2 = e_2 + \cdots + e_n$ with orthogonal idempotents $e_2, \ldots, e_n$.

Conversely, assume $\varepsilon_2 Ne_1$ is a projective left C_2-module. Since $A\varepsilon_2 A$ belongs to a heredity chain, we know that the multiplication map

$$A\varepsilon_2 \otimes_{C_2} \varepsilon_2 A \longrightarrow A\varepsilon_2 A$$

is bijective (Prop. 7 of [**DR2**]). Multiplying from the right by e_1, we obtain an isomorphism $A\varepsilon_2 \bigotimes_{C_2} \varepsilon_2 Ae_1 \cong A\varepsilon_2 Ae_1$ of left A-modules. Since $A\varepsilon_2$ is a projective left A-module, and $\varepsilon_2 Ae_1 = \varepsilon_2 Ne_1$ is a projective left C_2-module, it follows that $A\varepsilon_2 Ae_1$ is a projective left A-module. It remains to be shown that $A\varepsilon_2 Ae_1 = Ne_1$. First of all, $\varepsilon_2 Ae_1 \subseteq N$, thus $A\varepsilon_2 Ae_1 = A\varepsilon_2 Ne_1$. Second, $e_1 Ne_1 = e_1 N^2 e_1$, thus the left A-module Ne_1 is generated by $A\varepsilon_2$, consequently $A\varepsilon_2 Ne_1 = Ne_1$.

Note that the left A-module Ne_1 is projective if and only if the left C_1-module $\varepsilon_1 Ne_1$ is projective. This an immediate consequence of the Morita equivalence of A and C_1.

The equivalence of the assertions (i) and (i') in Theorem 1 is an immediate consequence of Lemma 7: we apply it to the rings C_i and their corresponding heredity chains ([**DR1**], statement 10). The implication (ii) $\Rightarrow$ (ii''') is asserted in Lemma 2. Since A_A has a Δ-filtration, the same is true for any free A-module, thus (ii''') $\Rightarrow$ (ii''). The implications (ii'') $\Rightarrow$ (ii') is trivial. In order to prove the implication (ii') $\Rightarrow$ (ii), we assume that the right ideals $e_i N$ have Δ-filtrations. Then there are Δ-filtrations of $e_i N$ passing through $e_i N\varepsilon_{i+1}A$, and therefore also $\operatorname{rad}\Delta(i) = e_i N / e_i N\varepsilon_{i+1}A$ has a Δ-filtration.

It remains to verify the equivalence of the conditions (i) and (ii). We will use induction on n. The algebra C_2 has the heredity chain $0 = \varepsilon_2 I_0 \varepsilon_2 \subset \varepsilon_2 I_1 \varepsilon_2 \subset \cdots \subset \varepsilon_2 I_{n-1}\varepsilon_2 = C_2$, and for C_2, we deal with the modules $\Delta_2(i) = e_i A\varepsilon_2 / e_i A\varepsilon_{i+1}A\varepsilon_2 = \Delta(i)\varepsilon_2$, where $2 \le i \le n$.

First, we assume that $\operatorname{rad}\Delta(i)$ has a Δ-filtration, for $1 \le i \le n$. Then $\operatorname{rad}\Delta_2(i)$ has a Δ_2-filtration, for $2 \le i \le n$, thus, by induction, $\varepsilon_i Ne_i$ is a projective left C_i-module, for $2 \le i \le n$. We want to show that Ne_1 is a projective left A-module. According to

Theorem 1, it suffices to show that A_A has an $\hat{e}$–filtration where $e = e_1$. Now A_A has a Δ–filtration, so we use the following lemma.

Lemma 8. *Assume that* rad $\Delta(i)$ *has a* Δ*–filtration, for all* $1 \le i \le n$. *Then any module with a* Δ*–filtration has an* $\hat{e}$*–filtration.*

Proof. Let X be a module with a Δ–filtration. We use induction on $\ell(X)$. We may assume $X = \Delta(i)$ for some i. If $\ell(\Delta(i)) = 1$, then $\mathrm{Ext}^1_A(E(i), E(j)) = 0$ for all $j \le i$; in particular, $\mathrm{Ext}^1_A(E(i), E(1)) = 0$. Hence $\hat{e}(i) = E(i) = \Delta(i)$. Now assume $\ell(\Delta(i)) > 1$. Let $X = \mathrm{rad}\,\Delta(i)$. By induction, X has an $\hat{e}$–filtration, thus there is a submodule X' with an $\mathcal{M}(e)$–filtration such that X/X' is a direct sum of copies of $E(1)$. It follows that $X' = e_i N \varepsilon_2 A$, thus $\Delta(i)/X' = \hat{e}(i)$. Since X' has an $\hat{e}$–filtration, we see that $\Delta(i)$ has an $\hat{e}$–filtration.

Finally, we verify the implication (i) $\Rightarrow$ (ii). For $1 \le i \le n$, let $\varepsilon_i N e_i$ be a projective left C_i–module. By induction we know that $\mathrm{rad}\,\Delta_2(i)$ has a Δ_2–filtration, for $2 \le i \le n$. Since Ne_1 is a projective left A–module, Theorem 1 asserts that A_A has an $\hat{e}$–filtration. We are going to show that $\mathrm{rad}\,\Delta(j)$, with $1 \le j \le n$, has a Δ–filtration. Since $\Delta(1) = E(1)$, we may assume $2 \le j \le n$. Consider $Z_{jr} = (\mathrm{rad}\ \Delta(j))\varepsilon_r A/(\mathrm{rad}\ \Delta(j))\varepsilon_{r+1} A$, with $1 \le r \le n$. We claim that Z_{jr} is a direct sum of copies of $\Delta(r)$. Again the case $r = 1$ is trivial, so assume $2 \le r \le n$. First of all, $\mathrm{top}\, Z_{jr}$ is clearly a direct sum of copies of $E(r)$, say $\mathrm{top}\, Z_{jr} = m_{jr} E(r)$. Since $\Delta(r)$ is the projective $A/A\varepsilon_{r+1}A$–cover of $E(r)$, and Z_{jr} is annihilated by $A\varepsilon_{r+1}A$, it follows that there is a surjective map $Y \longrightarrow Z_{jr}$ with $Y = m_{jr}\Delta(r)$. In order to show that this is an isomorphism, we are going to prove that $\ell(Y) = \ell(Z_{jr})$. First, we claim that both Y and Z_{jr} have $\mathcal{M}(e)$–filtrations. For, $e_r A$ has an $\hat{e}$–filtration, and $\mathrm{Hom}_A(e_r A, E(1)) = 0$, since $r \ge 2$; thus $e_r A$ has an $\mathcal{M}(e)$–filtration by Lemma 6. According to Lemma 5, $\Delta(r) = e_r A/e_r A\varepsilon_{e+1} A$ has an $\mathcal{M}(e)$–filtration, thus the same is true for Y. Since $\Delta(j)$ has an $\hat{e}$–filtration, also $\Delta(j)e_r A$ has one, according to Lemma 2. Using again $r \ge 2$, Lemma 5 and Lemma 6, we see that Z_{jr} has an $\mathcal{M}(e)$–filtration. Given any A–module X, and $i \ge 2$, the number $\ell_i(X)$ coincides with the number $\ell^{(2)}(X\varepsilon_2)$ of composition factors of the C_2–module $X\varepsilon_2$ which are of the form $E(i)\varepsilon_2 = e_i A\varepsilon_2/e_i N\varepsilon_2$. We use Lemma 4 in order to express $\ell(Y)$ and $\ell(Z_{jr})$ as follows:

$$\ell(Y) = \sum_{i=2}^{n} \ell_i(Y)s_i = \sum_{i=2}^{n} \ell_i^{(2)}(Y\varepsilon_2)s_i,$$

$$\ell(Z_{jr}) = \sum_{i=2}^{n} \ell_i(Z_{jr})s_i = \sum_{i=2}^{n} \ell_i^{(2)}(Z_{jr}\varepsilon_2)s_i.$$

On the other hand,

$$\begin{aligned} Z_{jr}\varepsilon_2 &= (\mathrm{rad}\ \Delta(j))\varepsilon_r A\varepsilon_2/(\mathrm{rad}\ \Delta(j))\varepsilon_{r+1} A\varepsilon_2 = \\ &= (\mathrm{rad}\ \Delta_2(j))\varepsilon_r C_2/(\mathrm{rad}\ \Delta_2(j))\varepsilon_{r+1} C_2 \end{aligned}$$

is a direct sum of copies of $\Delta_2(r)$, since $\Delta_2(j)$ has a Δ_2–filtration. It follows that $Z_{jr}\varepsilon_2 \cong m_{jr}\Delta_2(r) = Y\varepsilon_2$. As a consequence, $\ell(Y) = \ell(Z_{jr})$. This completes the proof of the implication (i) $\Rightarrow$ (ii).

5. Algebras of global dimension 2.

We are going to present the proof of Theorem 3 as well as some related examples. As before let $e_1, \ldots, e_n$ be a complete set of primitive and orthogonal idempotents, and let $\varepsilon_i = e_1 + \cdots + e_n$ for $1 \le i \le n$. Again, we assume that $(I_i)_i$ is a heredity chain, where $I_i = A\varepsilon_{n-i+1}A$.

Lemma 9. *Let $\varepsilon = \varepsilon_2$. Let $C = \varepsilon A \varepsilon$. Assume that $\varepsilon N e_1$ is a projective left C-module and that $e_1 N$ is a projective right A-module. Then proj.dim.$E(1)_A \le 1$, and proj.dim.$E(i)_A \le$ max $\{2, \text{proj.dim.}(E(i)\varepsilon)_C\}$ for $2 \le i \le n$.*

Proof. Since $E(1) = e_1A/e_1N$, it follows that proj.dim.$E(1)_A \le 1$. Consider now $E(i)$, where $2 \le i \le n$. We can assume that proj.dim. $(E(i)\varepsilon)_C$ is finite; let

$$0 \longrightarrow P^{(m)} \longrightarrow \ldots \longrightarrow P^{(1)} \longrightarrow P^{(0)} \longrightarrow E(i)\varepsilon \longrightarrow 0$$

be a projective resolution of the C-module $(E(i)\varepsilon)_C$. We tensor this sequence with ${}_C(\varepsilon A)$. Note that ${}_C(\varepsilon A)$ is a direct sum of copies of ${}_C(\varepsilon A e_j)$, with $1 \le j \le n$. For $2 \le j \le n$, the left C-module ${}_C(\varepsilon A e_j)$ is projective, since e_j is an idempotent of C, and ${}_C(\varepsilon A e_1) = {}_C(\varepsilon N e_1)$ is projective by assumption. Thus

$$0 \longrightarrow P^{(m)} \otimes_C \varepsilon A \longrightarrow \ldots \longrightarrow P^{(0)} \otimes_C \varepsilon A \longrightarrow E(i)\varepsilon \otimes_C \varepsilon A \longrightarrow 0$$

is exact. Since the A-modules $P^{(s)} \otimes {}_C(\varepsilon A)$ are projective, it follows that proj.dim. $E(i)\varepsilon \otimes {}_C(\varepsilon A)_A \le m$. The exact sequence $0 \longrightarrow e_iN \longrightarrow e_iA \longrightarrow E(i) \longrightarrow 0$ yields first by multiplying with ε and then tensoring with ${}_C(\varepsilon A)$, the exact sequence

$$0 \longrightarrow e_iN\varepsilon \otimes_C \varepsilon A \longrightarrow e_iA\varepsilon \otimes_C \varepsilon A \longrightarrow E(i)\varepsilon \otimes_C \varepsilon A \longrightarrow 0 .$$

Since $A\varepsilon A$ belongs to a heredity chain, we can identify $A\varepsilon \otimes_C \varepsilon A$ with $A\varepsilon A$ and therefore $e_iA\varepsilon \otimes_C \varepsilon A$ with $e_iA\varepsilon A = e_iA$. We see that $E(i)\varepsilon \otimes_C \varepsilon A \cong e_iA/e_iN\varepsilon A = \hat{e}(i)$. Thus proj.dim.$\hat{e}(i)_A \le m$. There is the exact sequence

$$0 \longrightarrow d_{i1}E(1) \longrightarrow \hat{e}(i) \longrightarrow E(i) \longrightarrow 0.$$

Since proj.dim.$E(1) \le 1$, it follows that

$$\text{proj.dim.}E(i) \le \max\{2, \text{proj.dim.}\hat{e}(i)_A\} = \max\{2, m\} .$$

Proof of Theorem 3. We use induction on n. Condition (i) of Theorem 2 applied to A and to its opposite shows that $C = C_2$ satisfies the corresponding assumptions (every right ideal of C_2 has a Δ_2-filtration, every left ideal of C_2 has a Δ_2^*-filtration). Thus gl.dim.$C \leq 2$. Also, $\varepsilon_2 N e_1$ is a projective left C_2-module by condition (i′) of Theorem 2. And $e_1 N \varepsilon_1$ is a projective right C_1-module by condition (i) of Theorem 2, applied to the opposite of A, thus $e_1 N$ is a projective A-module. We apply Lemma 9 and conclude that gl.dim$A \leq 2$.

Let us remark that not all algebras of global dimension 2 satisfy the conditions of Theorem 2: A simple example is provided by the path algebra of the graph

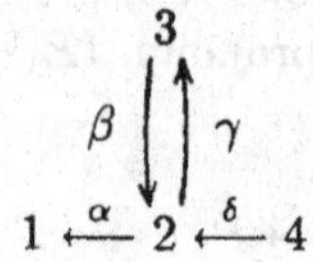

modulo the ideal $\langle \beta\alpha, \beta\gamma, \delta\gamma \rangle$:

$$A_A = 1 \oplus \begin{matrix} & 2 & \\ 1 & & 3 \\ & 2 & \end{matrix} \oplus \begin{matrix} 3 \\ 2 \end{matrix} \oplus \begin{matrix} 4 \\ 2 \\ 3 \\ 2 \end{matrix}.$$

Here,

$$\Delta(1) = 1, \quad \Delta(2) = \begin{matrix} 2 \\ 1 \end{matrix}, \quad \Delta(3) = \begin{matrix} 3 \\ 2 \end{matrix}, \quad \Delta(4) = \begin{matrix} 4 \\ 2 \\ 3 \\ 2 \end{matrix},$$

thus rad $\Delta(4)$ has no Δ-filtration. On the other hand, the path algebra of

$$1 \xrightarrow{\alpha_1} 2 \xrightarrow{\alpha_2} 3 \xrightarrow{\alpha_3} \cdots \xrightarrow{\alpha_{n-1}} n$$

modulo $\langle \alpha_{i-1}\alpha_i \,|\, 2 \leq i \leq n-1 \rangle$ satisfies the conditions of Theorem 2, but has global dimension $n-1$. Of course, for $n \geq 4$ this implies that its opposite algebra does not satisfy these conditions. Observe that, for $n = 3$ this is an example of an algebra of global dimension 2 whose dimension (namely 5) is less than the dimension of the corresponding peaked algebra (of dimension 6) as defined in the next section.

6. Peaked algebras

In this last section, we intend to give a construction of a new class of quasi-hereditary algebras of global dimension 2 which may be of further interest. Let $\mathcal{S} = (D_i, {}_iM_j)_{1\le i,j\le n}$ be a labelled species without loops [**DR3**]: thus ${}_iM_i = 0$ for all i, and the index set $\{1, 2, \ldots, n\}$ is considered with its natural ordering. As in [**DR3**], let

$$T = T(n) = \{(t_0, t_1, \ldots, t_m) \,|\, 0 \le t_i \le n \text{ are integers, } m \ge 1, \text{ and } t_{i-1} \ne t_i \text{ for all } 1 \le i \le m\};$$

for every $t = (t_0, t_1, \ldots, t_m) \in T$, let

$$M(t) = {}_{t_0}M_{t_1} \otimes_{D_{t_1}} {}_{t_1}M_{t_2} \otimes_{D_{t_2}} \cdots \otimes_{D_{t_{m-1}}} {}_{t_{m-1}}M_{t_m},$$

and for $T' \subseteq T$, let

$$M(T') = \bigoplus_{t \in T'} M(t).$$

We define the ideal $M(W^0)$ of the tensor algebra $\mathcal{T}(\mathcal{S})$ by specifying the subset W^0 of T as follows:

$$W^0 = W^0(n) = \{(t_0, t_1, \ldots, t_m) \in T \,|\, \text{there is } 0 < i < m \text{ such that } t_{i-1} > t_i < t_{i+1}\}.$$

Let W be the complement $T \backslash W^0$, thus

$$W = \{(t_0, t_1, \ldots, t_m) \in T \,|\, \text{there is } 0 \le i \le m \text{ such that } t_0 < t_1 < \cdots < t_i > \cdots > t_{m-1} > t_m\}.$$

Hence

$$[M(T)]^{2n-1} \subseteq M(W^0) \subseteq M(T)$$

and thus $M(W^0)$ is an admissible ideal. Let

$$\mathcal{P}(\mathcal{S}) = \mathcal{T}(\mathcal{S})/M(W^0).$$

Observe that the Loewy length of $\mathcal{P}(\mathcal{S})$ is at most $2n - 1$, and that, as an abelian group, $\mathcal{P}(\mathcal{S})$ can be identified with

$$\prod_{i=1}^{n} D_i \oplus M(W).$$

We call $\mathcal{P}(\mathcal{S})$ the *peaked* algebra with labelled species $\mathcal{S}$.

Proposition. *Let $\mathcal{P}(\mathcal{S})$ be the peaked algebra with labelled species $\mathcal{S}$. Then $\mathcal{P}(\mathcal{S})$ is quasi-hereditary, every right ideal of $\mathcal{P}(\mathcal{S})$ has a Δ-filtration, every left ideal of $\mathcal{P}(\mathcal{S})$ has a Δ^*-filtration. In particular, gl.dim.$\mathcal{P}(\mathcal{S}) \leq 2$.*

Proof. For any $1 \leq i \leq n$, we claim that $\operatorname{rad}\Delta(i)$ is a direct sum of various $\Delta(j)$. Since $\Delta(1)$ is simple, we can assume $2 \leq i \leq n$. Let

$$T_i = \{(i, t_1, \ldots, t_m) \in T \mid i > t_1 > \cdots > t_m\}.$$

Then $\Delta(i)$ may be identified with $D_i \oplus M(T_i)$, thus

$$\operatorname{rad}\Delta(i) = M(T_i) = \bigoplus_{(i,j,t_2,\ldots,t_m)\in T_i} d_{ij}\Delta(j),$$

where, as before, $d_{ij} = \dim({}_iM_j)_{D_j}$.

In comparison with the deep algebras over a given labelled species (whose global dimension is also at most 2), the dimensions of the peaked algebras are considerably smaller. For instance, for $\mathcal{S}_n = (D_i, {}_iM_j)_{1\leq i,j\leq n}$, where $D_i = k$ for all i and ${}_iM_j = {}_kk_k$ for all $i \neq j$ and ${}_iM_i = 0$ for all i, the dimensions $p(n)$ of $\mathcal{P}(\mathcal{S}_n)$ clearly satisfy

$$p(n+1) = p(n) + 4^n,$$

and thus, for all n,

$$p(n) = \frac{1}{3}(4^n - 1).$$

On the other hand, let $d(n)$ be the dimension of a deep algebra over $\mathcal{S}_n$. We have $d(5) = 3263441$ while $p(5) = 341$, and $d(10) \approx 2.7 \times 10^{208}$ (!) while $p(10) = 349525$. Even $p(20)$ is "only" 366503875925.

References

[CPS] Cline, E., Parshall, B., and Scott, L.: Finite dimensional algebras and highest weight categories. J.Reine.Ang.Math. **391**(1988), 85–99

[DR1] Dlab, V., and Ringel, C.M.: Quasi-hereditary algebras. Illinois J.Math. (to appear)

[DR2] Dlab, V., and Ringel, C.M.: A construction for quasi-hereditary algebras. Compositio Math. (to appear)

[DR3] Dlab, V., and Ringel, C.M.: The dimension of a quasi-hereditary algebra. Proceedings Banach Center Warszawa (to appear)

[MV] Mirollo, R., and Vilonen, K.: Bernstein–Gelfand–Gelfand reciprocity on perverse sheaves. Ann.Scient.Ec.Norm.Sup.4e série **20**(1987), 311–324

[PS] Parshall, B.J., and Scott, L.L.: Derived categories, quasi-hereditary algebras and algebraic groups. Proc. Ottawa–Moosonee Workshops in Algebra. Carleton Univ. Notes No.**3**(1988)

[S] Scott, L.L.: Simulating algebraic geometry with algebra I.: Derived categories and Morita theory. Proc. Symp.Pure Math., Amer.Math.Soc., Providence **47** (1987), part 1, 271-282.

V. Dlab
Department of Mathematics
Carleton University
Ottawa K1S 5B6
Canada

C.M. Ringel
Fakultät für Mathematik
Universität
D-4800 Bielefeld 1
West Germany

Hochschild cohomology of finite-dimensional algebras

Dieter Happel

[This paper is in final form, and no version of it will be submitted for publication elsewhere] .

The aim of this article is to report on some recent results on computing Hochschild cohomology groups of finite-dimensional algebras.
For this let k be an algebraically closed field and A a finite-dimensional k-algebra (associative, with unit). By mod$A$ we denote the category of finitely generated left A-modules.
Let ${}_AX_A$ be a finitely generated A-bimodule. The Hochschild cohomology groups $H^i(A,X)$ ($i \geq 0$) were introduced by Hochschild [**Ho**] (for a definition see section 1). The low-dimensional groups ($i \leq 2$) have a very concrete interpretation of classical algebraic structures such as derivations and extensions. It was observed by Gerstenhaber [**Ge**] that there are also connections to algebraic geometry. In fact, $H^2(A,A)$ controls the deformation theory of A. And it was shown that the algebras A which satisfy $H^2(A,A) = 0$ are rigid. For a similar approach we also mention an article of Gabriel [**Ga 1**].
Despite this very little was done in actual computations for particular classes of finite-dimensional algebras. In section 1 we briefly review the fundamental definitions of Hochschild cohomology and include an alternative description which one often uses for direct computations. In section 2 we present some computations. This includes a report on results due to Cibils [**C1**], [**C2**], [**C3**] and Gerstenhaber and Schack [**GS**]. For some of these results we have included proofs and some examples. In section 3 we deal with derivations.
In the remaining two sections we outline how recently emerged methods in the representation theory of finite-dimensional algebras yield information on the Hochschild cohomology.

1. Hochschild cohomology groups.

This section contains the basic definitions of the Hochschild cohomology groups. We will also give an alternative description which one often uses for direct computations. Moreover this section contains some elementary examples. The exposition follows closely the original approach in [**Ho**] and [**CE**]. We omit the proofs of these standard results.

1.1. Let k be an algebraically closed field and A a basic and connected finite-dimensional k-algebra. Let ${}_AX_A$ be an A-bimodule which is finite-dimensional over k. We define the Hochschild complex $C^{\bullet} = (C^i, d^i)_{i\in\mathbb{Z}}$ associated with this data as follows:
$C^i = 0, d^i = 0$ for $i < 0, C^o = {}_AX_A, C^i = \operatorname{Hom}_k(A^{\otimes i}, X)$ for $i > 0$ (where $A^{\otimes i}$ denotes the i-fold tensor product over k of A with itself), $d^o : X \to \operatorname{Hom}_k(A,X)$ with $(d^o x)(a) = ax - xa$ for $x \in X$ and $a \in A, d^i : C^i \to C^{i+1}$ with

$$\begin{aligned}(d^i f)(a_1 \otimes \cdots \otimes a_{i+1}) = a_1 f(a_2 \otimes \cdots \otimes a_{i+1}) \\ + \sum_{j=1}^{i}(-1)^j f(a_1 \otimes \cdots \otimes a_j a_{j+1} \otimes \cdots \otimes a_{i+1}) \\ + (-1)^{i+1} f(a_1 \otimes \cdots \otimes a_i)a_{i+1}\end{aligned}$$

for $f \in C^i$ and $a_1, \ldots, a_{i+1} \in A$.

It is a direct verification that this indeed gives a complex.
So we may define $H^i(A,X) = H^i(C^\bullet) = \ker d^i / \operatorname{im} d^{i-1}$ and we call it the i-th cohomology group of A with coefficients in the bimodule X.
Of particular interest to us is the example ${}_AX_A = {}_AA_A$. In this case $H^i(A,A)$ is simply denoted by $H^i(A)$.

1.2. Let us pause for a moment to recall the interpretation of the low-dimensional groups. Clearly $H^o(A,X) = X^A = \{x \in X \mid ax = xa \ \forall a \in A\}$. In particular, $H^o(A)$ coincides with the center of A.
Let $\operatorname{Der}(A,X) = \{\delta \in \operatorname{Hom}_k(A,X) \mid \delta(ab) = a\delta(b) + \delta(a)b\}$ be the k-vectorspace of derivations of A on X. By $\operatorname{Der}^o(A,X)$ we denote the subspace of inner derivations. Thus $\operatorname{Der}^o(A,X) = \{\delta_x : A \to X \mid \delta_x(a) = ax - xa,\ x \in X\}$. It follows immediately from the definition that $H^1(A,X) = \operatorname{Der}(A,X)/\operatorname{Der}^o(A,X)$. Let $\delta \in \operatorname{Der}(A,X)$. Then δ is called an outer derivation if the residue class of δ in $H^1(A,X)$ is different from zero.
For example let $A = k[x]/(x^2)$ and ${}_AX_A = {}_AA_A$. Then $\delta : A \to A$ given by $\delta(\lambda \cdot 1_A + \mu\bar{x}) = \mu\bar{x}$ is an outer derivation.

Let $f \in \operatorname{Hom}_k(A^{\otimes 2}, X)$, then we may form $A \underset{f}{\ltimes} X$ (the extension of A by X along f). As a k-vectorspace this is simply $A \oplus X$. The multiplication is defined by $(a,x)\cdot(a',x') = (aa', ax' + xa' + f(a \otimes a'))$. If $f \in \ker d^2$ then $A \underset{f}{\ltimes} X$ is an associative algebra with unit.
The following is an easy verification.

LEMMA. *Let $f, g \in \ker d^2$. Then $A \underset{f}{\ltimes} X \xrightarrow{\sim} A \underset{g}{\ltimes} X$ if $\bar{f} = \bar{g}$ in $H^2(A,X)$.*

1.3. A different way of approaching the cohomology groups is to consider the envelopping algebra $A^e = A \underset{k}{\otimes} A^*$, where A^* is the opposite algebra. For $a \in A$ we denote by a' the corresponding element in A^*. An A-bimodule ${}_AX_A$ can be considered as a left A^e-module by $(a \otimes b') \cdot x = axb$ for $a, b \in A$ and $x \in X$.
In particular A is an A-bimodule and we will therefore consider A always as a left A^e-module. We will now construct a projective resolution of A over A^e. For each integer $i \geq -1$, let $S_i(A)$ denote the $(i+2)$-fold tensor product over k of A with itself. Then $S_i(A)$ becomes a left A^e-module by setting $(a \otimes b') \cdot (a_o \otimes \cdots \otimes a_{i+1}) = a \cdot a_o \otimes a_1 \otimes \cdots \otimes a_i \otimes a_{i+1}b$ for $a, b \in A$ and $a_o \otimes \cdots \otimes a_{i+1} \in S_i(A)$. We define an A^e-homomorphism $\delta_i : S_i(A) \to S_{i-1}(A)$ (for $i \geq 0$) by

$$\delta_i(a_o \otimes \cdots \otimes a_{i+1}) = \sum_{j=o}^{i} (-1)^j a_o \otimes \cdots \otimes a_j a_{j+1} \otimes \cdots \otimes a_{i+1}.$$

Then $S_\bullet = (S_i(A), \delta_i)_{i \geq 0}$ is a projective resolution of A over A^e. Note that $S_{-1}(A) = A$ and $\delta_o : A \otimes A \to A$ is surjective. It is called the standard resolution [CE, p.174]. It is frequently convenient to write $S_i(A)$ in the form $S_i(A) = A \underset{k}{\otimes} \tilde{S}_i(A) \underset{k}{\otimes} A = A^e \underset{k}{\otimes} \tilde{S}_i(A)$, where $\tilde{S}_i(A) = A^{\otimes i}$.
In computing the cohomology groups we use

$$\operatorname{Hom}_{A^e}(S_i(A), X) = \operatorname{Hom}_{A^e}(A^e \underset{k}{\otimes} \tilde{S}_i(A), X) = \operatorname{Hom}_k(\tilde{S}_i(A), X) = \operatorname{Hom}_k(\tilde{S}_i(A), X) = C^i$$

It is easy to see that the two complexes $\mathrm{Hom}_k(\tilde{S}_\bullet, X)$ and $C^\bullet$ can be identified. In particular we see that $\mathrm{Ext}^i_{A^e}(A, X) = H^i(A, X)$.

1.4. The last observation can be used to define the cohomology algebra $H(A) = \bigoplus_{i \in \mathbb{Z}} H^i(A)$. The multiplication is induced by the Yoneda product. In this way $H(A)$ is a $\mathbb{Z}$-graded algebra. We will see in 1.6 below that $H(A)$ is a finite-dimensional k-algebra if A has finite global dimension ($gl.$ dim $A < \infty$).

The converse seems to be not known.

1.5. We want to construct a minimal projective resolution of A over A^e. For this let $e_1, \dots, e_n$ be a complete set of primitive orthogonal idempotents in A. By $P(i)$ we denote the indecomposable projective A-module Ae_i. Note that in this way we obtain a complete set of representatives from the isomorphism classes of indecomposable projective A-modules. We let $S(i) = top\, P(i)$ be the corresponding simple A-module.
Then $e_i \otimes e'_j$ $(1 \leq i, j \leq n)$ is a complete set of primitive orthogonal idempotents in A^e. We denote by $P(i, j')$ the indecomposable projective A^e-module $A^e(e_i \otimes e'_j)$. And we denote by $S(i, j') = top\, P(i, j')$ the corresponding simple A^e-module.
Observe that $S(i, j') \simeq \mathrm{Hom}_k(S(i), S(j))$.

Lemma. *Let*

$$\dots R_n \to R_{n-1} \to \dots \to R_1 \to R_o \to A \to 0$$

be a minimal projective resolution of A over A^e. Then

$$R_n = \bigoplus_{i,j} P(i, j')^{\dim \mathrm{Ext}^n_A(S(i), S(j))}.$$

Proof: Let $R_n = \bigoplus_{i,j} P(i, j')^{r_{ij}}$. Then by definition we have that

$$\begin{aligned} r_{ij} = \ & \dim \mathrm{Ext}^n_{A^e}(A, S(i, j')) = \dim \mathrm{Ext}^n_{A^e}(A, \mathrm{Hom}_k(S(i), S(j))) \\ & = \dim H^n(A, \mathrm{Hom}_k(S(i), S(j))) \\ & = \dim \mathrm{Ext}^n_A(S(i), S(j)). \end{aligned}$$

The last equality follows from corollary 4.4, p.170 of [CE].

In particular we see that $pd_{A^e} A = gl.$ dim A.
We will use this projective resolution in section 2 for some direct computations.

1.6. As an illustration we will compute the cohomology groups of finite-dimensional hereditary k-algebras. Our assumptions imply that there exists a finite connected quiver $\vec{\Delta}$ without oriented cycles such that $A \xrightarrow{\sim} k\vec{\Delta}$, where $k\vec{\Delta}$ denotes the path algebra of $\vec{\Delta}$. The set of vertices is denoted by Δ_o and the set of arrows is denoted by Δ_1. For an arrow α in $\vec{\Delta}$ we denote by $s(\alpha)$ the starting point and by $e(\alpha)$ the end point of α. The vertices of $\vec{\Delta}$ when considered as trivial paths form a complete set of primitive orthogonal idempotents in $k\vec{\Delta}$.

Let α be an arrow in $\vec{\Delta}$ and let $\nu(\alpha) = \dim_k s(\alpha)(k\vec{\Delta})e(\alpha)$. Moreover let n be the number of vertices in $\vec{\Delta}$.

PROPOSITION.

$$H^o(k\vec{\Delta}) = k, \ \dim H^1(k\vec{\Delta}) = 1 - n + \sum_{\alpha\in\Delta_1} \nu(\alpha), H^i(k\vec{\Delta}) = 0 \ for \ i \geq 2.$$

PROOF: Note that $\dim \mathrm{Ext}^1_{k\vec{\Delta}}(S(i), S(j))$ coincides with the number of arrows from i to j. Clearly $H^o(k\vec{\Delta}) = k$ and $H^i(k\vec{\Delta}) = 0$ for $i \geq 2$, for $gl.\dim k\vec{\Delta} \leq 1$.
Let $0 \to R_1 \to R_o \to k\vec{\Delta} \to 0$ be the minimal projective resolution constructed above. Then $R_o = \bigoplus_{i\in\Delta_o} P(i, i')$ and $R_1 = \bigoplus_{\alpha\in\Delta_1} P(s(\alpha), e(\alpha)')$. Applying $\mathrm{Hom}_{(k\vec{\Delta})^e}(-, k\vec{\Delta})$ to the above exact sequence yields:

$$0 \to \mathrm{Hom}_{(k\vec{\Delta})^e}(k\vec{\Delta}, k\vec{\Delta}) \to \mathrm{Hom}_{(k\vec{\Delta})^e}(R_o, k\vec{\Delta}) \to \mathrm{Hom}_{(k\vec{\Delta})^e}(R_1, k\vec{\Delta}) \to 0$$

Clearly $\mathrm{Hom}_{(k\vec{\Delta})^e}(k\vec{\Delta}, k\vec{\Delta}) \tilde{\to} k$ and $\mathrm{Hom}_{(k\vec{\Delta})^e)}(R_o, k\vec{\Delta}) \tilde{\to} k^n$ and $\mathrm{Hom}_{(k\vec{\Delta})^e}(R_1, k\vec{\Delta}) = \bigoplus_{\alpha\in\Delta_1} k^{\nu(\alpha)}$. Thus $\dim H^1(k\vec{\Delta}) = 1 - n + \sum_{\alpha\in\Delta_1} \nu(\alpha)$.

COROLLARY. *Let $\vec{\Delta}$ be a finite quiver without oriented cycle. Then $H^1(k\vec{\Delta}) = 0$ if and only if $\vec{\Delta}$ is a tree.*

1.7. There is the dual concept of Hochschild homology. Again let ${}_AX_A$ be a finitely-generated A-bimodule. Then the Hochschild homology groups $H_i(A, X)$ can be defined as $\mathrm{Tor}_i^{A^e}(X, A)$. We denote by D the standard duality on mod A. Then it is well-known that $H_i(A, X) \simeq H^i(A, D(X))$ [CE].
Suppose that A is a factor-algebra of a finite-dimensional hereditary k-algebra. Then it is an easy consequence from 1.5 that $H_i(A, A) = 0$ for $i > 0$. This was first shown in [C4].

2. Survey of results.

In this section we report on some results on computing the Hochschild cohomology groups for particular classes of algebras. Some proofs are given.

2.1 Incidence algebras [C3], [GS].

Let $P = (P, \leq)$ be a finite partially ordered set. We may assume that $P = 1, 2, \ldots, n$ is a labelling of the vertices. With P we may associate its incidence algebra $I(P)$. This is by definition the subalgebra of the algebra of $(n \times n)$-matrices over k with elements $(X_{ij}) \in M_n(k)$ satisfying $X_{ij} = 0$ if $i \not\leq j$. An equivalent definition is given as follows. Let $\vec{\Delta}$ be the following quiver attached to P. $\vec{\Delta}$ has n vertices and there is an arrow from s to t in $\vec{\Delta}$ if $t < s$ and there is no $u \in P$ with $t < u < s$. Let w, w' be two paths in $\vec{\Delta}$. We say that w, w' are parallel if w and w' have the same starting and endpoint. Let I be the two sided ideal in $k\vec{\Delta}$ generated by the differences of parallel paths. Then it is easy to see that $I(P) = k\vec{\Delta}/I$. With P we may associate a simplicial complex $\Sigma_P = (C_i, d_i)$, where

$$C_i = \{s_o > s_1 > \cdots > s_i \mid s_j \in P\}$$

and d_i is the obvious boundary map. The cohomology of Σ_P with coefficients in k is the cohomology of the following complex:

$$0 \to \operatorname{Hom}(kC_o, k) \xrightarrow{b_1} \operatorname{Hom}(kC_1, k) \xrightarrow{b_2} \cdots \longrightarrow \operatorname{Hom}(kC_i, k) \xrightarrow{b_{i+1}} \operatorname{Hom}(kC_{i+1}, k) \longrightarrow \cdots$$

where

$$(b_{i+1}f)(s_o, \cdots, s_{i+1}) = \sum_{j=o}^{i+1} (-1)^j f(s_o, \cdots, \hat{s}_j, \cdots, s_{i+1}).$$

THEOREM.

$$H^i(\Sigma_P, k) \xrightarrow{\sim} H^i(I(P))$$

Moreover, let Σ be an arbitrary finite simplicial complex and P_Σ the corresponding partially ordered set. So the elements of P_Σ are the simplices of Σ and the partial order is defined by inclusion. Then $H^i(\Sigma, k) \xrightarrow{\sim} H^i(I(P_\Sigma))$. This follows immediately from the theorem above and the observation that Σ_{P_Σ} is the barycentric subdivision of Σ.

Example: Consider the following partially ordered set P

P =

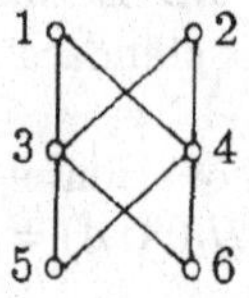

Then the incidence algebra $I(P)$ is given by:

$$I(P) = \begin{pmatrix} k & 0 & k & k & k & k \\ 0 & k & k & k & k & k \\ 0 & 0 & k & 0 & k & k \\ 0 & 0 & 0 & k & k & k \\ 0 & 0 & 0 & 0 & k & 0 \\ 0 & 0 & 0 & 0 & 0 & k \end{pmatrix}$$

Thus $\dim I(P) = 18$.

And Σ_P is the octahedron which is displayed below.

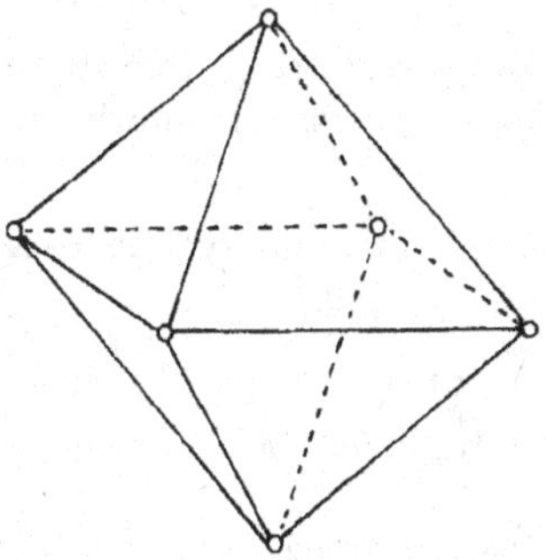

2.2 Narrow quivers [C1].

In this subsection we give a slight generalisation of a result of Cibils [C1]. Let $\vec{\Delta}$ be a finite connected quiver without oriented cycles. Let I be a two-sided ideal in $k\vec{\Delta}$ generated by paths of length at least two. Let $A = k\vec{\Delta}/I$. We will further assume that A is schurian (i.e. dim $\mathrm{Hom}_A(P, P') \leq 1$ for P, P' indecomposable projective A-modules) and that A is semicommutative (i.e. if w, w' are parallel paths in $\vec{\Delta}$ with $w \in I$ then $w' \in I$). In 1.6 we have introduced a valuation ν on the arrows of a quiver $\vec{\Delta}$. Note that our assumptions imply that $\nu(\alpha) = 1$ for $\alpha \in \Delta_1$. Let $|\Delta_o|$ (resp.$|\Delta_1|$) denote the number of vertices of $\vec{\Delta}$ (resp. the number of arrows of $\vec{\Delta}$). The Euler characteristic of $\vec{\Delta}$ is defined by $\chi(\vec{\Delta}) = 1 - |\Delta_1| + |\Delta_1|$.

THEOREM. *Let $A = k\vec{\Delta}/I$ be as above. Then $H^o(A) = k$, dim $H^1(A) = \chi(\vec{\Delta})$ and $H^i(A) = 0$ for $i \geq 2$.*

PROOF: Since $\vec{\Delta}$ does not contain an oriented cycle we infer that $H^o(A) = k$. Also note that $\dim_k \mathrm{Ext}^1_A(S,T) \leq 1$ for any pair S, T of simple A-modules and that $\mathrm{Ext}^1_A(S,T) \neq 0$ if and only if there is an arrow from the vertex of $\vec{\Delta}$ corresponding to S to the vertex of $\vec{\Delta}$ corresponding to T. Let

$$\cdots R_n \to R_{n-1} \to \cdots \to R_1 \to R_o \to A \to 0$$

be the minimal projective resolution of A over A^e constructed in 1.5.
Applying $(-, A) = \mathrm{Hom}_{A^e}(-, A)$ to this exact sequence yields:

$$0 \to (A, A) \to (R_o, A) \to (R_1, A) \to (R_2, A) \to \cdots$$

Clearly we have that $(A, A) = k$, $(R_o, A) = k^{|\Delta_o|}$ and by the considerations above we infer that $(R_1, A) = k^{|\Delta_1|}$ (note that A is supposed to be schurian). By the lemma below we infer that $(R_i, A) = 0$ for $i \geq 2$). So the assertion follows.

Lemma. *Let $A = k\vec{\Delta}/I$ be as above. Let e_i, e_j be primitive orthogonal idempotents corresponding to the vertices $i, j \in \Delta_o$. If $\mathrm{Ext}^t_A(S(i), S(j)) \neq 0$ for $t \geq 2$ then $e_i A e_j = 0$.*

Proof: If $\mathrm{Ext}^t_A(S(i), S(j)) \neq 0$ for $t \geq 2$ then $P(j)$ occurs in a minimal projective resolution of $S(i)$, yielding w from i to j in $\vec{\Delta}$. It is a straightforward verification that our assumptions imply that $w \in I$. Using again that A is semicommutative we infer that $e_i A e_j = 0$.

We say that a finite connected quiver $\vec{\Delta}$ is narrow if there is at most one path between any pair of vertices of $\vec{\Delta}$. For an arbitrary quiver $\vec{\Delta}$ we denote by $\vec{\Delta}_+$ the two-sided ideal in $k\vec{\Delta}$ generated by the arrows of $\vec{\Delta}$. Finally we recall that a two-sided ideal I in $k\vec{\Delta}$ is said to be admissible if $(\vec{\Delta}_+)^m \subset I \subset (\vec{\Delta}_+)^2$ for some $m \geq 2$.

Corollary [C1]. *Let $\vec{\Delta}$ be a narrow quiver and I be an amissible ideal in $k\vec{\Delta}$. Then $H^o(k\vec{\Delta}/I) = k$, $\dim H^1(k\vec{\Delta}/I) = \chi(\vec{\Delta})$ and $H^i(k\vec{\Delta}) = 0$ for $i \geq 2$.*

Remark: The class of narrow quivers clearly includes those quivers whose underlying graph is a tree. But it also contains for example the following quiver:

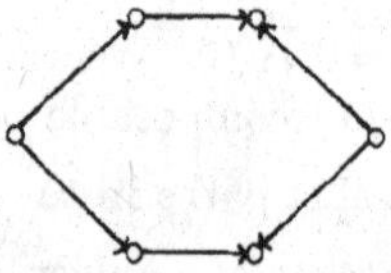

2.3 2-nilpotent algebras.

Let A be a finite-dimensional k-algebra. The radical of A is denoted by $radA$. We say that A is t-nilpotent if $(radA)^t = 0$ but $(radA)^{t-1} \neq 0$.
If A is a basic, connected 2-nilpotent algebra, then there exists a finite and connected quiver $\vec{\Delta}$ such that $A \xrightarrow{\sim} k(\vec{\Delta})/(\vec{\Delta}_+)^2$.

Proposition. *Let $A = k\vec{\Delta}/(\vec{\Delta}_+)^2$ with $\vec{\Delta}$ connected. The following statements are equivalent:*

(i) $H^i(A) = 0$ *for* $i > o$
(ii) $H^1(A) = 0$
(iii) $\vec{\Delta}$ *is a tree.*

Proof: (i) $\Rightarrow$ (ii) trivial and (iii) $\Rightarrow$ (i) follows from 2.2. So it remains to show (ii) $\Rightarrow$ (iii). Suppose $\vec{\Delta}$ is not a tree and let $\alpha \in \Delta_1$ be an arrow belonging to a cycle of Δ (the underlying graph of $\vec{\Delta}$). We will construct $\delta \in \mathrm{Der}(A)$ such that the residue class of δ in $H^1(A)$ is different from zero. Since A is 2-nilpotent there exists a k-basis of A induced by the vertices and arrows of $\vec{\Delta}$. We denote this basis by $e_1, \cdots, e_n$ (for the vertices) and $\alpha_1, \cdots, \alpha_r$ (for the arrows). We may assume that $\alpha = \alpha_1$. Define $\delta : A \to A$ by $\delta(e_i) = 0$ for $1 \leq i \leq n$, $\delta(\alpha_1) = \alpha_1$ and $\delta(\alpha_j) = 0$ for $2 \leq j \leq r$. An easy computation shows that $\delta \in \mathrm{Der}(A)$. Suppose that $\delta = \delta_x$ for some $x \in A$ where δ_x is the inner derivation associated with x. Then $x = \Sigma_{i=1}^{n}\lambda_i e_i + \Sigma_{j=1}^{r}\mu_j \alpha_j$ for some $\lambda_i, \mu_j \in k$. Since $\delta(\alpha_1) = \alpha_1 = \delta_x(\alpha_1) = x\alpha_1 - \alpha_1 x = \lambda_{s(\alpha_1)}\alpha_1 - \lambda_{e(\alpha_1)}\alpha_1$ it follows that $\lambda_{s(\alpha_1)} - \lambda_{e(\alpha_1)} = 1$. Since $\delta(\alpha_j) = 0 = \delta_x(\alpha_j) = x\alpha_j - \alpha_j x = \lambda_{s(\alpha_j)}\alpha_j - \lambda_{e(\alpha_j)}\alpha_j$ for $j \geq 2$ it follows that $\lambda_{s(\alpha_j)} = \lambda_{e(\alpha_j)}$. Since $\Delta \setminus \{\alpha_1\}$ is connected we infer that $\lambda_1 = \cdots = \lambda_n$. Thus the residue class of δ in $H^1(A)$ is different from zero. This finishes the proof.

THEOREM [C2]. *Let* $A = k\vec{\Delta}/(\vec{\Delta}_+)^2$ *with* $\vec{\Delta}$ *connected. The following statements are equivalent:*

(i) $H^2(A) = 0$

(ii) $\vec{\Delta}$ *does not contain a loop,* $\vec{\Delta}$ *does not contain a triangle, and* $\vec{\Delta}$ *is not* .

Note that a loop is an arrow α with $s(\alpha) = e(\alpha)$ and a triangle is a quiver of the following form

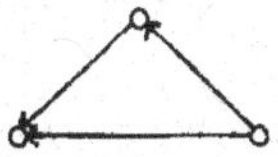

We indicate the easy direction ((i) $\Rightarrow$ (ii)) and refer for the converse to [C2].

Suppose $\vec{\Delta}$ contains a loop α. As above we may choose a basis of A induced by the vertices and arrows, say $x_1 = \alpha, x_2, \cdots, x_m$. We define $f \in \mathrm{Hom}_k(A \otimes A, A)$ by $f(x_i \otimes x_j) = x_1$ for $i = 1 = j$ and zero otherwise. It is an easy verification to show that $f \in \ker d^2$ (compare 1.1) and that the residue class of f in $H^2(A)$ is different from zero.

Suppose that $\vec{\Delta}$ contains a triangle

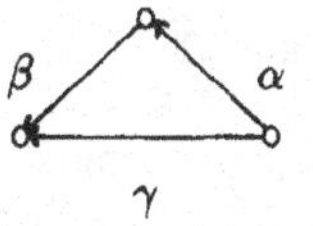

Choose a basis $x_1, \cdots, x_m$ of A as above with $x_1 = \alpha, x_2 = \beta, x_3 = \gamma$. So we may define $f \in \mathrm{Hom}_k(A \otimes A, A)$ by $f(x_i \otimes x_j) = x_3$ for $i = 1$ and $j = 2$ and equal to zero otherwise. Again $f \in \ker d^2$, but $f \notin im\, d^1$.

Finally suppose that $\vec{\Delta}$ is the following quiver

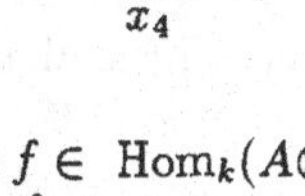

Choose a basis of A as indicated. Define $f \in \mathrm{Hom}_k(A \otimes A, A)$ by $f(x_2 \otimes x_4) = x_1$, $f(x_4 \otimes x_2) = x_3$ and zero otherwise. Again $f \in \ker d^2$, but $f \notin im\, d^1$. So in all three cases we see that $H^2(A)$ is different from zero.

3. Derivations.

In this section we will concentrate on computing derivations. Again let A be a finite-dimensional k-algebra. For expository reasons we will assume that the characteristic of k is zero. Let $e_1, \cdots, e_n$ be a complete set of primitive orthogonal idempotents in A. Let $A = \bigoplus_{i,j} e_i A e_j$ be the two-sided Pierce decomposition. Recall that we have denoted by $\mathrm{Der}(A, A) = \mathrm{Der}(A)$ the vectorspace of derivations of A. We denote by $\mathrm{Der}^o(A)$ the vectorspace of inner derivations.

3.1. LEMMA. *Let $\delta \in Der(A)$. Then there exists $\delta' \in Der^o(A)$ such that $\delta(e_i) - \delta'(e_i) = 0$ for $1 \leq i \leq n$.*

PROOF: Since $\delta(e_i) = \delta(e_i^2) = e_i\delta(e_i) + \delta(e_i)e_i$, we infer that $e_i\delta(e_i)e_i = 0$. Thus $e_i\delta(e_i) \in e_iA(1 - e_i)$ and $\delta(e_i)e_i \in (1 - e_i)Ae_i$. Moreover we have for $i \neq j$ that $0 = \delta(e_ie_j) = e_i\delta(e_j) + \delta(e_i)e_j$. Thus there exists $r_{ij} \in e_iAe_j, r_{ii} = 0$ such that $e_i\delta(e_i) = \Sigma_{j=1}^n r_{ij}$ and $\delta(e_i)e_i = -\Sigma_{j=1}^n r_{ji}$, thus $\delta(e_i) = \Sigma_{j=1}^n r_{ij} - r_{ji}$. Let $r = \Sigma_{i,j} r_{i,j}$ and let $\delta' = \delta_{-r}$ the inner derivation attached to $-r$. Thus $\delta_{-r} = -re_i + e_ir = -\Sigma_{l=1}^n r_{li} + \Sigma_{l=1}^n r_{il} = \delta(e_i)$.

We denote by $\mathrm{Der}^n(A) = \{\delta \in \mathrm{Der}(A) \mid \delta(e_i) = 0\}$ the subspace of normalized derivations. Let $\delta \in \mathrm{Der}^n(A)$ be a normalized derivation. Then δ respects the two-sided Pierce decomposition. In fact, let $\alpha \in e_iAe_j$, then $\delta(\alpha) = \delta(e_i\alpha e_j) = e_i\delta(\alpha e_j) = e_i\delta(\alpha)e_j \in e_iAe_j$. By $\mathrm{Der}^{n,o}$ we denote the subspace $\mathrm{Der}^n(A) \cap \mathrm{Der}^o(A)$. It is now an immediate consequence of the lemma above that $\mathrm{Der}^n(A)/\mathrm{Der}^{n,o}(A) \xrightarrow{\sim} H^1(A)$.

3.2. Let $\vec{\Delta}$ be a finite quiver and $k\vec{\Delta}$ the corresponding path algebra. We will consider $k\vec{\Delta}$ as a graded algebra, where the grading is induced by assigning to a path w in $\vec{\Delta}$ as degree the length of w.

THEOREM. *Let I be an admissible ideal in $k\vec{\Delta}$ which is homogeneous with respect to the grading. Let $A = k\vec{\Delta}/I$. If $H^1(A) = 0$ then $\vec{\Delta}$ is directed.*

PROOF: For $w \in k\vec{\Delta}$ we denote by $l(w)$ the length of w. We define $\delta : k\vec{\Delta} \to k\vec{\Delta}$ by $\delta(w) = l(w)w$. Then $\delta \in \mathrm{Der}^n(k\vec{\Delta})$. Since I is homogeneous, we infer that δ induces a normalized derivation on A again denoted by δ. By assumtion there exists $a \in A$ such that $\delta = \delta_a$. Since $\delta \in \mathrm{Der}^n(A)$ we must have that $a = \Sigma_{i=1}^n \mu_i e_i + y$ for $\mu_i \in k$ and $y \in \bigoplus_i e_i(rad\, A)e_i$. Let α be an arrow in $\vec{\Delta}$ and $\overline{\alpha}$ the residue class of α in A.
Then $\overline{\alpha} = \delta(\overline{\alpha}) = \delta_\alpha(\overline{\alpha}) = a\overline{\alpha} - \overline{\alpha}a = \mu_{s(\alpha)}\overline{\alpha} - \mu_{e(\alpha)}\overline{\alpha} + z$, where $z \in rad^2 A$.
Thus $\mu_{s(\alpha)} - \mu_{e(\alpha)} = 1$. Now suppose that

$$1 \xrightarrow{\alpha_1} 2 \xrightarrow{\alpha_2} 3 \cdots r-1 \xrightarrow{\alpha_{r-1}} r = 1$$

is an oriented cycle in $\vec{\Delta}$. Then we must have that $\mu_i - \mu_{i+1} = 1$ for $1 \leq i < n$ and $\mu_n - \mu_1 = 1$. It is easily seen that in $char\, k = 0$ this system of linear equations has no solution, a contradiction.

3.3. Recall that in 2.3 we have defined t–nilpotent algebras.

COROLLARY. *Let $A = k\vec{\Delta}/I$ be 3-nilpotent and suppose that $H^1(A) = 0$, then $\vec{\Delta}$ is directed.*

PROOF: It is well–known and easy to see that a 3–nilpotent algebra $k\vec{\Delta}/I$ is isomorphic to its associated graded algebra. With I we have denoted an admissible ideal in $k\vec{\Delta}$.

3.4. For the next corollary we need some additional terminology. Let A be a representation–finite algebra (i.e. up to isomorphism there exist only finitely many indecomposable A–modules.) Let $X_1, \cdots, X_r$ be a list of representatives from the isomorphism classes of indecomposable A–modules and let $X = \bigoplus_i X_i$. Then $\Lambda = \operatorname{End} X$ is called the Auslander–algebra of A [**Ga2**]. Recall that A is said to be standard if Λ is isomorphic to its associated graded algebra. For different characterizations we refer to [**BrG**]. Let $\Lambda = k\vec{\Delta}/I$, then A is called representation–directed if $\vec{\Delta}$ is directed. For an equivalent definition we refer to [**Ri**]. (Note that $\vec{\Delta}$ is the Auslander–Reiten quiver of A and I is the mesh–ideal.)

COROLLARY. *Let A be a representation–finite, standard k–algebra and Λ be the Auslander algebra of A. If $H^1(A) = 0$ then A is representation–directed.*

We will come back to the computation of the Hochschild cohomology of Auslander algebras in a subsequent publication.

We conclude this section by some examples. We leave out the rather elementary computations.

a) Let $A = k[x]/(x^n)$ and Λ be the Auslander algebra of A. Then $\dim H^o(\Lambda) = n$, $\dim H^1(\Lambda) = n-1$ and $\dim H^2(\Lambda) = n-1$.
(Note that $gl.\dim \Lambda \leq 2$, so $H^i(\Lambda) = 0$ for $i > 2$.)

b) Let $A = k\vec{\Delta}/I$ with $\vec{\Delta} = \alpha \circlearrowleft \underset{1}{\cdot} \xrightarrow{\beta} \underset{2}{\cdot} \circlearrowright \gamma$ and $I = <\alpha^2, \alpha\beta, \beta\gamma, \gamma^2>$.
Let Λ be the Auslander algebra of A. Then the Auslander–Reiten quiver of A is given by:

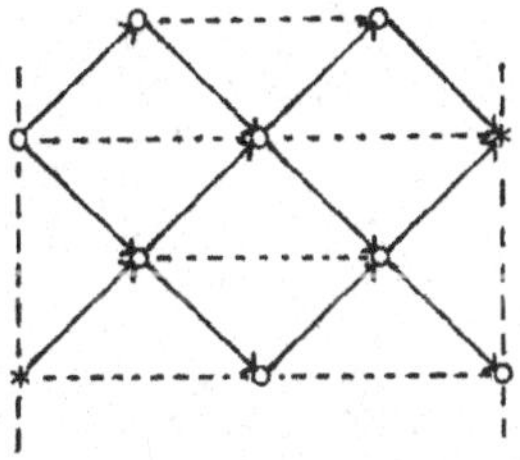

The identification is along the vertical dotted lines and the horizontal dotted lines indicate the Auslander–Reiten translation. Then $\dim H^o(\Lambda) = 2$, $\dim H^1(\Lambda) = 2$ and $H^2(\Lambda) = 0$.

4.Tilting invariance.

4.1. Let A be a finite-dimensional k-algebra. An A-module ${}_AM$ is called an r-tilting module if (i) $pd_A M \leq r$, (ii) $\mathrm{Ext}^i_A(M,M) = 0$ for $1 \leq i \leq r$, (iii) there exists an exact sequence

$$0 \to_A A \to M^o \to M^1 \to \cdots \to M^r \to 0$$

With $add\, M$ we have denoted the additive category generated by M.
For some information about tilting modules as well as for the proofs of the results we will use here we refer to **[CPS]**,**[Ha1]**,**[Ha2]**.
We say that a finite-dimensional k-algebra B is tiltable to A if there exists a family $(A_i, M^i, A_{i+1} = \mathrm{End}\, M^i)_{o \leq i < m}$ such that $A_o = B, A_m = A$ and M^i is an r-tilting module over A_i.
Let us give an easy example. Let $A = k\overrightarrow{\Delta}$, where

$$\overrightarrow{\Delta} = \underset{1}{\circ} \longleftarrow \underset{2}{\circ} \longleftarrow \underset{3}{\circ} \cdots \underset{n-1}{\circ} \longleftarrow \underset{n}{\circ}$$

Note that A is just the algebra of all $n \times n$-upper triangular matrices over k. Let $B = k\overrightarrow{\Gamma}/I$. Then B is tiltable to A if and only if $\overrightarrow{\Gamma}$ is a full and connected subquiver of the infinite quiver $\overrightarrow{\Lambda}$ defined below having n vertices containing the central vertex and I is generated by all possible paths $\alpha\beta$ and $\beta\alpha$. Below we have defined the infinite quiver $\overrightarrow{\Lambda}$. The arrows on the diagonals pointing down are labelled with α and those on the diagonals pointing upwards are labelled with β.

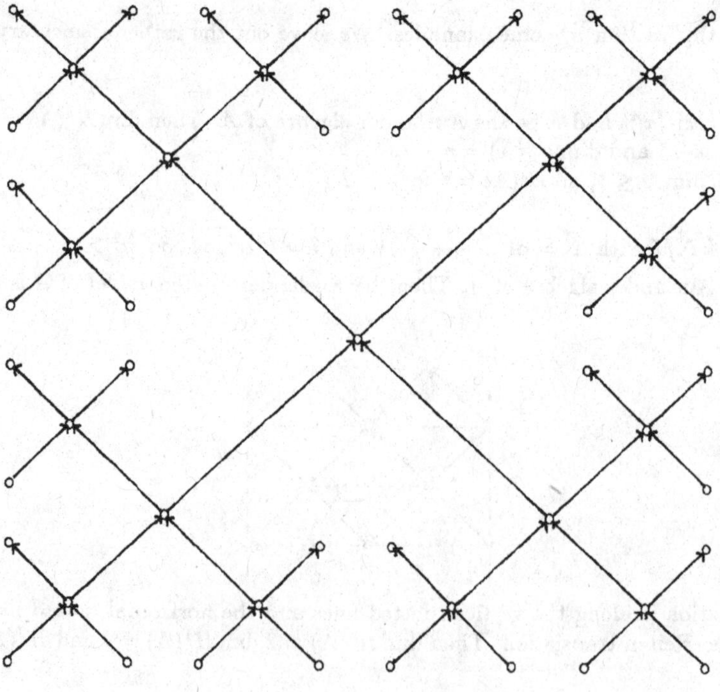

Let $D^b(A)$ be the derived category of bounded complexes over $\operatorname{mod} A$ [V]. Recall that $\operatorname{mod} A$ is fully embedded into $D^b(A)$ by sending a module ${}_AX$ to a complex which is concentrated in degree zero. We will identify $\operatorname{mod} A$ with its image in $D^b(A)$. Let T be the translation functor on $D^b(A)$. Then $\operatorname{Hom}_{D^b(A)}(X, T^iY) \simeq \operatorname{Ext}^i_A(X,Y)$ for $X, Y \in \operatorname{mod} A$ and $i \in \mathbb{Z}$.
Let ${}_BM$ be an r–tilting module and $A = \operatorname{End}_B M$. Then there exists a triangle–equivalence $F : D^b(A) \longrightarrow D^b(B)$ such that $F({}_BM) = {}_AA$. In fact F is the right derived functor of $\operatorname{Hom}_B(M, -)$.

4.2. In the following result we show that Hochschild cohomology is invariant under tilting.

THEOREM. *Let A and B be finite–dimensional k–algebras. If B is tiltable to A then $H(B)$ and $H(A)$ are isomorphic as $\mathbb{Z}$–graded algebras.*

PROOF: Let ${}_BM$ be an r–tilting module and $A = \operatorname{End}_B M$. Clearly it is enough to show that $H(B)$ and $H(A)$ are isomorphic as $\mathbb{Z}$–graded algebras. We will construct below a triangle–equivalence $\tilde{F} : D^b(B^e) \longrightarrow D^b(A^e)$ such that $\tilde{F}({}_{B^e}B) = {}_{A^e}A$. From this the assertion follows immediately. In fact, $H^i(A) = \operatorname{Ext}^i_{A^e}(A, A) = \operatorname{Hom}_{D^b(A^e)}(A, T^iA) \simeq \operatorname{Hom}_{D^b(B^e)}(B, T^iB) = H^i(B)$.
Since $\tilde{F}$ preserves the composition of morphisms, we see that $\tilde{F}$ induces an isomorphism of the $\mathbb{Z}$–graded algebras $H(A)$ and $H(B)$.
We now turn to the construction of $\tilde{F}$. Recall that $A^e = A \underset{k}{\otimes} A^*, B^e = B \underset{k}{\otimes} B^*$. Let $C = B \underset{k}{\otimes} A^*$. Note that ${}_BM_A$ is a left C–module. It is easily checked that $X = M \underset{k}{\otimes} A^*$ is an r–tilting module over C such that $\operatorname{End}_C X = A^e$. Thus we obtain a triangle–equivalence $F : D^b(C) \to D^b(A^e)$ such that $F({}_CX) = {}_{A^e}A^e$. It follows from the description above that $F({}_CM) = {}_{A^e}A$. Similarly, let $Y = B \underset{k}{\otimes} M$. Then Y is an r–tilting module over C such that $\operatorname{End}_C Y = B^e$. Thus we obtain a triangle–equivalence $G : D^b(C) \to D^b(B^e)$ such that $G({}_CY) = {}_{B^e}B^e$. Moreover it follows that $G({}_CM) = {}_{B^e}B$. Now let G' be a quasi–inverse to G. Then $\tilde{F} = FG'$ is the required triangle–equivalence from $D^b(B^e)$ to $D^b(A^e)$.

4.3. COROLLARY. *Let $\vec{\Delta}$ be a finite quiver without oriented cycle and let B be a finite–dimensional k–algebra which is tiltable to $k\vec{\Delta}$. Then $H^i(B) = 0$, for $i \geq 2$, $H^0(B) = k$ and $\dim H^1(B) = \dim H^1(k\vec{\Delta})$. In particular B is rigid. If Δ is not a tree and $B = k\vec{\Gamma}/I$, then Γ is not a tree.*

PROOF: The first assertions follow from 4.2 and 1.6. The second statement follows from this and 2.2.

4.4. Examples:

1) Let $B = k\vec{\Gamma}/I$ where $\vec{\Gamma}$ and I are given as follows:

$\vec{\Gamma}$ =

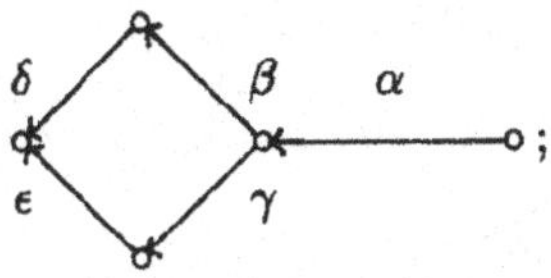

; $I = < \alpha\beta\delta - \alpha\delta\varepsilon >$

Then B is tiltable to $k\vec{\Delta}$ (in one step) where

$\vec{\Delta} =$ 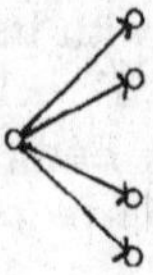

In particular $H^1(B) = 0$. Let

$\vec{\Delta'} =$ 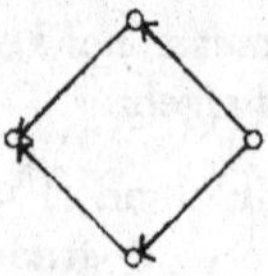

so $A' = k\vec{\Delta'}$ is a factor algebra of B generated by a primitive idempotent. By 1.6 we infer that $\dim H^1(A') = k$. This shows that even for well–behaved factors it is not always possible to extend non–trivial outer derivations.

2) Let $A = k\vec{\Delta}/I$ where

$\vec{\Delta} =$ 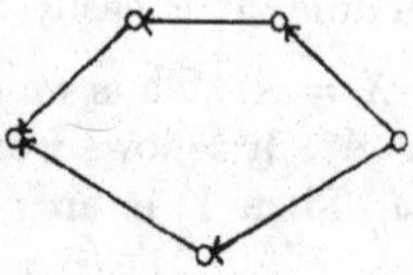 *and* $I = (\vec{\Delta}_+)^2$.

Then A satisfies the assumptions of 2.2, so $H^o(A) = k, H^1(A) = k$ and $H^i(A) = 0$ for $i \geq 2$. We choose as indecomposable summands of a 1–tilting module ${}_AM$ the indecomposable modules which correspond to the vertices marked with $*$ in the Auslander–Reiten quiver of A

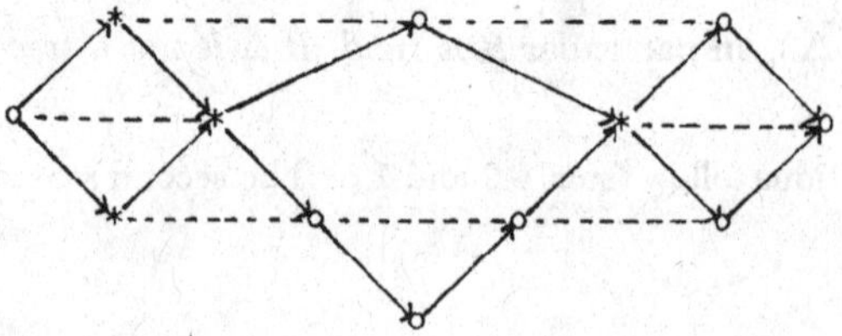

Then $B = \mathrm{End}_A M = k\vec{\Gamma}/I$ where

$\vec{\Gamma} =$ 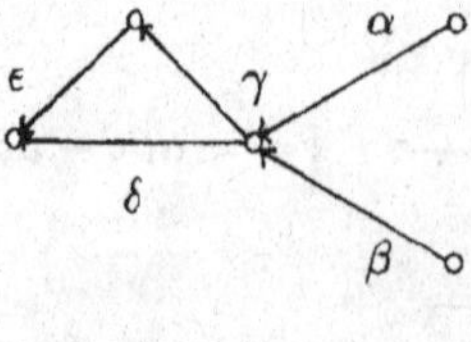

and $I = < \alpha\gamma, \beta\delta, \gamma\epsilon >$.

So in particular we see that $H^2(B) = 0$ despite the fact that B admits a factor algebra C by an ideal generated by primitive idempotents such that $H^2(C) \neq 0$.

4.5. In the final part of this section we want to report on some joint work with M. Schaps [HS] which is related to the theorem above. The proof of the theorem below will be omitted.

Let Alg_d be the affine variety of algebra structures (associative, with unit) on a vector space V of dimension d over the field k. This variety is the set of bilinear maps from $V \times V$ to V that are associative and admit a unit; for a formal definition and some of its properties the reader is referred to [**Ga1**]. To a point p of Alg_d we associate the finite-dimensional k-algebra A_p. Let $G = Gl(V)$. The canonical linear operation of G on V induces an operation of G on Alg_d. For $p \in Alg_d$ we denote by O_p the orbit of p under G. Let $p, q \in Alg_d$. We say that A_q deforms to A_p (or A_p degenerates to A_q) if $q \in \overline{O}_p$, the Zariski-closure of O_p. The algebra A_p is then called an algebra deformation of A_q.
For example $M_2(k)$ (the matrix algebra of (2×2)-matrices over k) is an algebra deformation of $k\vec{\Delta}/I$ where

$\vec{\Delta} =$ 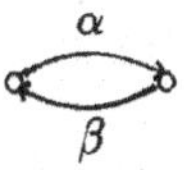

$and\ I = <\alpha\beta, \beta\alpha>$.

If A_p is a non-trivial algebra deformation of A_q (i.e. $A_p \not\cong A_q$), then $H^2(A_q) \neq 0$ [Ge].

4.6. THEOREM. *Let A be a finite-dimensional basic k-algebra and let $\tilde{A}$ be an algebra deformation of A. Let ${}_AM$ be a 1-tilting module with endomorphism algebra $\mathrm{End}_A M = B$. Then there exists a 1-tilting module ${}_{\tilde{A}}\tilde{M}$ such that $\tilde{B} = \mathrm{End}_{\tilde{A}}\tilde{M}$ is an algebra deformation of B.*

We point out that even in the case where ${}_AM$ is multiplicity-free (i.e. each indecomposable direct summand of ${}_AM$ occurs with multiplicity one) the tilting module ${}_{\tilde{A}}\tilde{M}$ will usually not be multiplicity-free. In other words, $\tilde{B}$ will usually not be a basic k-algebra. An example for this will be given below.

4.7. Examples:

1) Let $A = k\vec{\Delta}/I$, where

$\vec{\Delta} =$ 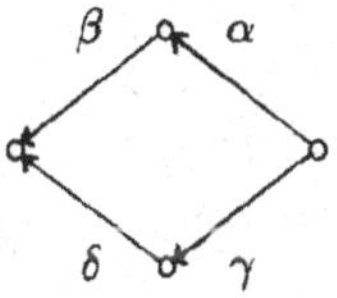

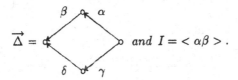

$and\ I = <\alpha\beta>$.

Then A admits a non-trivial deformation $\tilde{A} = k\vec{\Delta}/I'$ where $I' = <\alpha\beta - \gamma\delta>$. We choose as indecomposable summands of a 1-tilting module ${}_AM$ the indecomposable modules which correspond to the vertices marked with $*$ in the Auslander-Reiten quiver of A (the identification is along the vertical dotted lines).

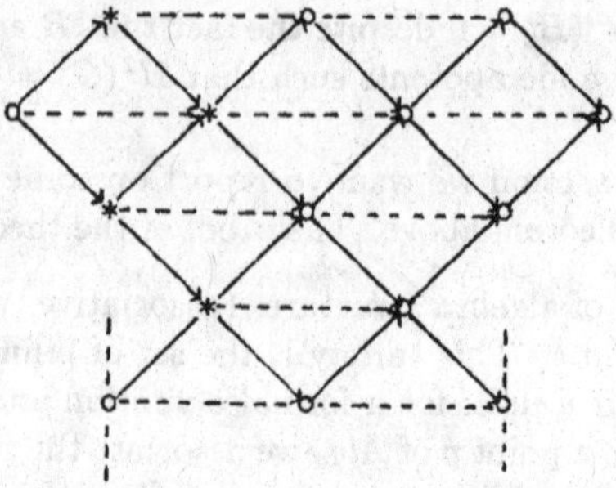

Then $\text{End}_A M = k\overrightarrow{\Gamma}/J$ where

$\overrightarrow{\Gamma} =$ 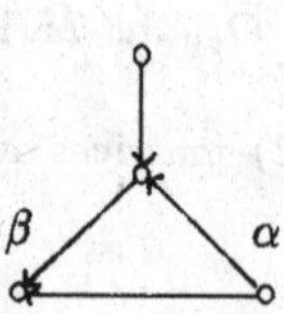 *and* $J = <\alpha\beta>$.

Then ${}_{\tilde{A}}M$ of the theorem above corresponds to the vertices marked with $*$ in the Auslander–Reiten quiver of $\tilde{A}$:

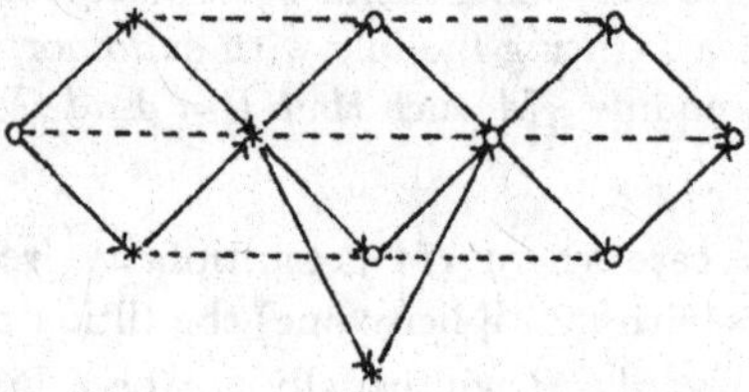

So $\text{End}_{\tilde{A}}\tilde{M} = k\overrightarrow{\Lambda}$ where

$\overrightarrow{\Lambda} =$ 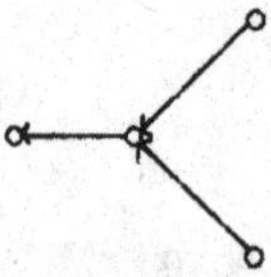

which clearly degenerates to $\text{End}_A M$.

2) Let $A = k\overrightarrow{\Delta}/I$, where $\overrightarrow{\Delta} =$ 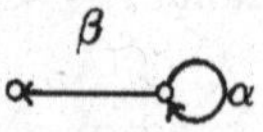 and $I = <\alpha^2, \alpha\beta>$. Then A admits a

non-trivial deformation $\tilde{A} = k\overrightarrow{\Delta'}$ where

$\Delta' =$ 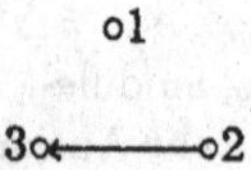

note that Δ' is not connected).
We choose as indecomposable summands of a 1-tilting module ${}_AM$ the indecomposable modules which correspond to the vertices marked with $*$ in the Auslander–Reiten quiver of A (the identification is along the vertical dotted lines).

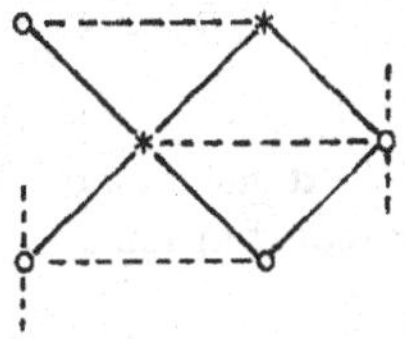

Then $\mathrm{End}_A M = k\vec{\Gamma}/J$ where

$\vec{\Gamma} =$ 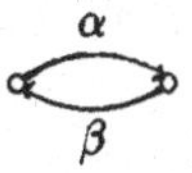 *and* $J = < \alpha\beta\alpha >$

and ${}_{\tilde{A}}\tilde{M}$ of the theorem above is $P(1) \oplus P(1) \oplus P(2) \oplus P(3)$. So $\mathrm{End}_{\tilde{A}}\tilde{M} = M_2(k) \times \begin{pmatrix} k & k \\ 0 & k \end{pmatrix}$.

5. One–point extensions.

5.1. In this section we present for a special class of algebras a way of computing the Hochschild cohomology of a given algebra by knowing the cohomology of a particular factor algebra. To be more precise we have to recall the definition of a one–point extension algebra. For more details and the representation–theoretic tools available in this context we refer to [Ri]. Let A be a finite–dimensional k–algebra and $M \in \mathrm{mod}\, A$. The one–point extension algebra $A[M]$ of A by M is by definition the finite–dimensional k–algebra

$$A[M] = \begin{pmatrix} A & M \\ 0 & k \end{pmatrix}$$

with multiplication

$$\begin{pmatrix} a & m \\ 0 & \lambda \end{pmatrix} \begin{pmatrix} a' & m' \\ 0 & \lambda' \end{pmatrix} = \begin{pmatrix} aa' & am' + m\lambda' \\ 0 & \lambda\lambda' \end{pmatrix}$$

where $a, a' \in A, m, m' \in M$ and $\lambda, \lambda' \in k$.
For example let $T_n(k)$ be the algebra of $n \times n$– upper triangular matrices over k. Then $T_n(k)$ operates on an n–dimensional k–vector space by left multiplication. We denote this $T_n(k)$–module by M. Then an easy verification shows that $T_n(k)[M] \simeq T_{n+1}(k)$.
From the definition of the one–point extension we see that a necessary condition for an algebra B to be of the form $A[M]$ for some algebra A and an A–module M is that there is a simple injective B–module. This clearly is sufficient. Indeed, if an algebra B admits a simple injective module S. Let $P(S)$ be a projective cover of S and let $e \in B$ be an idempotent such that $P(S) = Be$. Let $A = B/\langle e \rangle$ (where $\langle e \rangle$ denotes the two–sided ideal in B generated by

the idempotent e) and let $M = rad\,P(S)$. Then M is an A–module and it is easily checked that $B \simeq A[M]$.
There is the dual concept of a one–point coextension. We denote by M^* the k–dual of M. Then the one–point coextension $[M]A$ of A by M is by definition the finite–dimensional k–algebra

$$[M]A = \begin{pmatrix} A & 0 \\ M^* & k \end{pmatrix}$$

with the obvious multiplication.
We will state our results for one–point extensions and leave it to the reader to derive the corresponding statements for one–point coextensions.

5.2. Let A be a finite–dimensional k–algebra and let $M \in \operatorname{mod} A$. We denote by B the one–point extension of A by M. Let $e \in B$ be a primitive idempotent such that $M = rad\,Be$. Further let $I = \langle e \rangle$ be the two–sided ideal generated by e. Then we have that $A = B/I$.
In the following lemma we have collected some useful homological information about this data. As the proofs are easy we omit them.

LEMMA. *With the above notation we have:*

(i) ${}_{B^e}I \simeq P(e, e') \simeq Hom_k(S(e), P(e))$
(ii) $Ext^j_{A^e}(A, A) \simeq Ext^j_{B^e}(B, B)$
(iii) $Ext^i_B(S(e), P(e)) \simeq Ext^{i-1}_A(M, M)$ for $i \geq 2$
(iv) $Ext^1_B(S(e), P(e)) \simeq Hom_A(M, M)/k$
(v) $Hom_B(S(e), P(e)) = 0$

5.3 THEOREM. *Let $B = A[M]$ be as above. Then there exists the following long exact sequence connecting the Hochschild cohomology of A and B.*

$$0 \to H^o(B) \to H^o(A) \to Hom_A(M, M)/k \to H^1(B) \to H^1(A) \to Ext^1_A(M, M) \to \cdots$$
$$\cdots \to Ext^i_A(M, M) \to H^{i+1}(B) \to H^{i+1}(A) \to Ext^{i+1}_A(M, M) \to \cdots$$

PROOF:

$$(*)\ 0 \to I \to B \to A \to 0$$

By construction and 5.2(i) we infer that $Ext^i_{B^e}(I, A) = 0$ for $i \geq 0$. Applying $Hom_{B^e}(-, A)$ to $(*)$ then yields $Ext^i_{B^e}(A, A) \tilde{\to} Ext^i_{B^e}(B, A)$ for $i \geq 0$. Note that $Ext^i_{B^e}(B, I) = H^i(B, I) = H^i(B, Hom_k(S(e), P(e))) = Ext^i_B(S(e), P(e))$ which can be computed by 5.2 (iii),(iv) and (v). Apply $Hom_{B^e}(B, -)$ to $(*)$. This yields the long exact sequence

$$0 \to Hom_{B^e}(B, B) \to Hom_{B^e}(B, A) \to Ext^1_{B^e}(B, I) \to Ext^1_{B^e}(B, B) \to Ext^1_{B^e}(B, A) \to \ldots$$

which by the identifications above gives the assertion.

5.4. For the next corollaries we have to recall the definition of the S–condition **[BrG]**,**[BLS]**. Let $B = k\vec{\Delta}/J$ with $\vec{\Delta}$ directed. An indecomposable projective B–module $P(i)$ for $i \in \Delta_0$ is said to have a separated radical if distinct indecomposable summands of $rad\,P(i)$ have upport on different connected components of $\overrightarrow{\Delta}_i$, where $\overrightarrow{\Delta}_i$ is the full subquiver of $\overrightarrow{\Delta}$ with ertices j such that there is no path from j to i. Finally we say that B satisfies the S–condition f all indecomposable projective B–modules have separated radical.

 COROLLARY. *If B is representation–directed then $H^i(B) = 0$ for $i \geq 2$.*

PROOF: We proceed by induction on the number of distinct simple B–modules. We may ssume that $B = A[M]$ and $A = A_1 \times \cdots \times A_r$ with A_i being representation–directed for $1 \leq i \leq r$. Since B is representation–directed we infer that $M = \coprod_{i=1}^r M_i$, with $M_i \in \operatorname{mod} A_i$ atisfies $\operatorname{Ext}^j(M_i, M_i) = O$ for $j \geq 1$. By induction we have that $H^j(A_i) = 0$ for $j \geq 2$ and $1 \leq i \leq r$. The assertion now follows from 5.3.

.5. COROLLARY. *Let B be representation–directed. Then B satisfies the S–condition if nd only if $H^1(B) = 0$.*

PROOF: Suppose that $B = A[M]$ and $A = A_1 \times \cdots \times A_r$, $M = \coprod_{i=1}^r M_i$ with $M_i \in \operatorname{mod} A_i$. f B satisfies the S–condition so do $A_1, \cdots, A_r$. By induction we infer that $H^1(A) = 0$. Moreover we see that M_i is indecomposable. As A_i is representation–directed we have that End $M_i = k$. Clearly $H^o(B) = k$, $H^o(A) = k^r$ and $\dim \operatorname{Hom}_A(M, M)/k = r - 1$, so $H^1(B) = 0$.
Conversely, if $H^1(B) = 0$ and $B = A[M]$ we see as above that $\operatorname{Ext}_A^1(M, M) = 0$, for B is epresentation–directed. By induction we infer that A satisfies the S–condition. As above et $A = A_1 \times \cdots \times A_r$ and $M = \coprod_{i=1}^r M_i$. If $H^1(B) = 0$ then $\dim \operatorname{Hom}_A(M, M)/k = r - 1$. Thus $\dim \operatorname{End} M_i = 1$ and in particular we see that the M_i are indecomposable. Hence B atisfies the S–condition.

5.6. COROLLARY. *Let A be a connected finite–dimensional k–algebra with $H^o(A) = k$ and $H^i(A) = 0$ for $i > 0$. Let $M \in \operatorname{mod} A$ and $B = A[M]$ the corresponding one–point extension. Then $H^o(B) = k$, $H^1(B) \simeq \operatorname{Hom}_A(M, M)/k$ and $H^i(B) \simeq \operatorname{Ext}_A^{i-1}(M, M)$ for $i > 1$.*

We conclude by giving some examples.

Let $A = k\overrightarrow{\Delta}$ and Δ a Dynkin diagram of type $\mathbb{A}_n, \mathbb{D}_n, \mathbb{E}_6, \mathbb{E}_7, \mathbb{E}_8$ and let $M \in \operatorname{mod} A$ be ndecomposable. Then $H^i(A[M]) = 0$ for $i \geq 1$.

Let $A = k\overrightarrow{\Delta}$ and Δ an affine diagram of type $\tilde{\mathbb{D}}_n, \tilde{\mathbb{E}}_6, \tilde{\mathbb{E}}_7, \tilde{\mathbb{E}}_8$ and let $M \in \operatorname{mod} A$ be indecomposable such that $\operatorname{Ext}_A^1(M, M) \simeq k$. Then $H^1(A[M]) = 0$ and $H^2(A[M]) \simeq k$. Note hat such modules exist. This example includes the canonical tubular algebras of type $\tilde{\mathbb{D}}_{4,\lambda}$ of Ringel [**Ri**].

REFERENCES

[BLS] Bautista, R.; Larrion, F.; Salmeron, L., *On simply connected algebras,*, J. London Math. Soc. **27** (1983), 212-220.

[BrG] Bretscher, O.; Gabriel, P., *The standard form of a representation-finite algebra*, Bull. Soc. Math. France **111** (1983), 21-40.

[C1] Cibils, C., *Cohomologie de Hochschild d'algèbres de dimension finie*, preprint.

[C2] Cibils, C., *2-nilpotent and rigid finite-dimensional algebras*, J. London Math. Soc. **36** (1987), 211-218.

[C3] Cibils, C., *Cohomology of incidence algebras and simplicial complexes*, preprint.

[C4] Cibils, C., *Hochschild homology of an algebra whose quiver has no oriented cycles*, Springer Lecture Notes, Heidelberg **1177** (1986), 55-59.

[CE] Cartan, H.; Eilenberg, S., *Homological Algebra*, Princeton University Press (1956).

[CPS] Cline, E.; Parshall, B.; Scott, L., *Derived categories and Morita theory*, J. Algebra **104** (1986), 397-409.

[Ga1] Gabriel, P., *Finite representation type is open*, Springer Lecture Notes, Heidelberg **488** (1975), 132-155.

[Ga2] Gabriel, P., *Auslander-Reiten sequences and representation-finite algebras*, Springer Lecture Notes, Heidelberg **831** (1980), 1-71.

[Ge] Gerstenhaber, M., *On the deformations of rings and algebras*, Ann. of Math. **79** (1964), 59–103.

[GS] Gerstenhaber, M.; Schack, S.P., *Simplicial homology is Hochschild cohomology*, J. Pure and Appl. algebra **30** (1983), 143-156.

[Ha1] Happel, D., *On the derived category of a finite-dimensional algebra*, Comment. Math. Helv. **62** (1987), 339-389.

[Ha2] Happel, D., LMS, Cambridge University Press **119** (1988), "Triangulated categories in the representation theory of finite-dimensional algebras,".

[HS] Happel, D.; Schaps, M., *Deformations of tilting modules*, preprint.

[Ho] Hochschild, G., *On the cohomology groups of an associative algebra*, Ann. of Math. **46** (1946), 58-67.

[Ri] Ringel, C.M., *Tame algebras and integral quadratic forms*, Springer Lecture Notes, Heidelberg **1099** (1984).

[V] Verdier, J.L., *Catégories dérivées, état 0*, Springer Lecture Notes, Heidelberg **569** (1977), 262-311.

Fakultät für Mathematik, Universität Bielefeld, Postfach 8640, 4800 Bielefeld 1, FRG

SIMULTANEOUS EQUIVALENCE
OF SQUARE MATRICES

Lieven Le Bruyn
Department of Mathematics U.I.A.
Research Associate N.F.W.O.

February 1987 87-01

[*This paper is in final form and no version of it will be submitted for publication else-where*].

Abstract :

Recent results of M. Maruyama on vector bundles over the projective plane give new information on the problem of classifying m-tuples of n by n matrices upto simultaneous equivalence.
As an application we also give a linearization procedure for partial differential equations of the form $\sum_{i+j+k=n} a_{ijk} \frac{\partial^n \psi}{\partial x^i \partial y^j \partial z^k} = c^n \psi$.

AMS-classification : (primary) 15 A 21, (secondary) 16 A 64, 14 F 05.

Key Words : Simultaneous equivalence, vector bundles, representation theory, partial differential equations.

1. Introduction.

In this paper we aim to show how certain recent results of M. Maruyama [14], [15] can be applied to obtain some grip on the following "hopeless" problem.

Question 1 : Parametrize m-tuples of n by n matrices under simultaneous equivalence.

That is, we aim to study the orbits of the group $GL_n(\mathbb{C}) \times GL_n(\mathbb{C})$ acting on $M_n(\mathbb{C})^{\oplus m}$ by

$$(\alpha, \beta).(A_1, \ldots, A_m) = (\alpha A_1 \beta, \ldots, \alpha A_m \beta)$$

If $m = 1$, different orbits correspond to different ranks. If $m = 2$ and A_1 is invertible, then the orbit of (A_1, A_2) is completely determined by the Jordan normal form of $A_1^{-1}.A_2$.
If $m \geq 3$, however, this problem is known to be wild since it corresponds to classifying representations of the wild quiver

$$0 \; \begin{matrix} \xrightarrow{\varphi_1} \\ \vdots \\ \xrightarrow[\varphi_m]{} \end{matrix} \; 0$$

upto equivalence. In this paper we will restrict attention to representations which are generic, i.e. in sufficiently general position. To be more precise, we want to find an open subvariety $U_{m,n}$ of $M_n(\mathbb{C})^{\oplus m}$, and a morphism $\pi : U_{m,n} \longrightarrow V_{m,n}$ having the property that for each $\xi \in V_{m,n}$ the fiber $\pi^{-1}(\xi)$ consists of precisely one orbit. The main open problem concerning these parametrizing varieties $V_{m,n}$ is :

Question 2 :

Is $V_{m,n}$ (stably) rational ?

In the next two sections we will see that a positive solution to this question would be important also for the study of Brauer groups of functionfields and for the study of vectorbundles over the projective plane. We mention that question 2 would follow immediately from a positive solution to a rather daring conjecture of V. Kač [11]. He conjectured that, for any quiver Q and any dimension vector α, the variety parametrizing isoclasses of indecomposable α-representations would allow a cellular decomposition into locally closed subvarieties each isomorphic to some affine space. Since generic equidimensional representations of the two point quivers are indecomposable thin would imply rationality of $V_{m,n}$.

2. Connection with Brauer groups of functionfields.

For a long time, one of the main open problems on Brauer groups has been.

Question 3 : If X is a variety over $\mathbb{C}$, is the Brauer group $Br\mathbb{C}(X)$ of the functionfield $\mathbb{C}(X)$ generated by cyclic algebras ?

Recall that a cyclic algebra is an n^2-dimensional central simple $\mathbb{C}(X)$-algebra generated by two elements x and y satisfying the relations :

$$x^n = a, y^n = b, xy = \omega yx$$

where $a, b \in \mathbb{C}(X)^*$ and ω is a primitive n-th root of unity.
Question 3 was recently solved (in a more general setting) by Merkurjev and Suslin using heavy tools from algebraic K-theory, see for example [24].
Ringtheorists have tried to solve this problem by using generic methods. Let us briefly scetch their approach. For any $m, n \in \mathbb{N}$, consider the polynomial ring

$$P_{m,n} = \mathbb{C}[x_{ij}(k) : 1 \leq i, j \leq n; 1 \leq k \leq m]$$

The ring of m generic n by n matrices, $\mathbb{G}_{m,n}$, is the subring of $M_n(P_{m,n})$ generated by the generic matrices

$$X_k = (X_{ij}(k))_{i,j} \in M_n(P_{m,n})$$

It is well-known, see for example [17], that $\mathbb{G}_{m,n}$ is a left and right Õre-domain so we can consider its ring of fractions $\Delta_{m,n}$ which is a division algebra of dimension n^2 over its center $K_{m,n}$.
$\Delta_{m,n}$ is called the generic division algebra. Procesi [19] has shown that if all $\Delta_{m,n}$ are Brauer equivalent to a product of cyclic algebras, then question 3 has a positive solution.
Moreover, S. Bloch [4] has shown that if L is a field containing $\mathbb{C}$ then $Br(L)$ is generated by cyclic algebras if and only if $Br(L(x_1, \ldots, x_r))$ is generated by cyclic algebras.

So, by applying Bloch's result (twice), (stable) rationality of $K_{m,n}$ would imply the Merkurjev-Suslin result for fields containing $\mathbb{C}$.

What is known about the rationality of $K_{m,n}$? Procesi [17,] has proved that $K_{m,n}, m \geq 3$, is rational over $K_{2,n}$ thereby reducing the problem to two generic matrices. Moreover, he proved that $K_{2,2} = \mathbb{C}\,(Tr(X_1), Tr(X_2), \mathrm{Det}(X_1), \mathrm{Det}(X_2), Tr(X_1, X_2))$ setling the problem for 2 by 2 matrices.
Formanek, [8] and [9], has proved rationality of $K_{m,3}$ and $K_{m,4}$.

Apart from these results, D. Saltman [20], [21], has obtained some partial results : he proved that the unramified Brauer group of $K_{m,n}$ is trivial and that $K_{m,n}$ is retract rational for n prime.
Further, we note that the obvious approach (i.e. trying to prove that $K_{2,n}$ is rational over $K_{1,n}$) fails for $n = 4$. This was shown by R. Snider and mentioned in [9] and [19]. A proof of this fact is contained in the paper [7] by Colliot-Thélène and Sansuc. D. Saltman has communicated to us that $K_{2,n}$ cannot be stably rational over $K_{1,n}$ for any non-squarefree n.

What does all this have to do with our problem ? The connection is given by the (easy) proof of the wildness of the quivers for $m \geq 3$. Consider the open subvariety

$$U_{m,n} = GL_n(\mathbb{C}) \oplus M_n(\mathbb{C})^{\oplus m-1}$$

of $M_n(\mathbb{C})^{\oplus m}$, then a representant for the orbit under action of the first component of $GL_n(\mathbb{C}) \times GL_n(\mathbb{C})$ is of the form

$$(I_n; A_2; \ldots; A_m)$$

Now, we have to calculate the action of the second component on these representants

$$\begin{aligned}(I_n, \beta).(I_n, A_2, \ldots, A_m) \\ = (\beta, A_2\beta, \ldots, A_m\beta) \\ = (I_n.\beta^{-1}A_2\beta, \ldots, \beta^{-1}A_m\beta)\end{aligned}$$

That is, the orbit structure of $GL_n(\mathbb{C}) \times GL_n(\mathbb{C})$ acting on $U_{m,n}$ is the same as that of $GL_n(\mathbb{C})$ acting on $(m-1)$-tuples of n by n matrices by componentswise conjugation.
Procesi [18] has shown that the quotient variety $M_n(\mathbb{C})^{\oplus m-1}/GL_n(\mathbb{C})$ has as its function field the field $K_{m-1,n}$. Therefore,

Proposition 1. : $V_{m,n}(m \geq 3)$ is birational to the quotient variety

$$M_n(\mathbb{C})^{\oplus m-1}/GL_n(\mathbb{C})$$

Therefore, rationality of $V_{m,n}$ would imply a positive answer to question 3. Moreover, from Procesi's observation that $K_{m,n}$ is rational over $K_{2,n}$ we also obtain the following (perhaps surprising) result :

Proposition 2 : $V_{m,n}(m \geq 3)$ is birational to $V_{3,n} \times \mathbb{A}^{(m-3)n^2}$ which basically reduces question 1 to the special case of triples of n by n matrices. That is, from now on we will restrict attention to the isomorphism problem of equidimensional generic representations of

$$0 \;\overset{\longrightarrow}{\underset{\longrightarrow}{\longrightarrow}}\; 0$$

3. Connection with vectorbundles over $\mathbb{P}_2$

A very coarse classification of all vectorbundles over the projective plane $\mathbb{P}_2$ is given by topological invariants such as the rank and the chern classes, which can be interpreted as integers in the case of projective spaces.
Given the numbers r, c_1, c_2 one wants to study how sufficiently general bundles with these invariants look like. Such bundles will turn out to be stable, i.e. for all coherent subsheaves $\mathcal{F} \in \mathcal{E}$ we have $\frac{c_1 \mathcal{F}}{rk(\mathcal{F})} < \frac{c_1}{r}$.
Precisely as parametrizing problems in representation theory one therefore wants to study a variety $M(r, c_1, c_2)$ whose points correspond to isomorphism classes of stable bundles $\mathcal{E}$ of rank r and with Chern numbers $c_1(\mathcal{E}) = c_1$ and $c_2(\mathcal{E}) = c_2$. Again, a major open problem is

Question 4 : which of these moduli spaces $M(r, c_1, c_2)$ are (stably) rational ?

The motivation is that in case they are rational we can find additional algebraic invariants of bundles such that they classify freely and completely sufficiently general bundles with the invariants r, c, and c_2. What is known about this problem ? Barth has shown in [1] that $M(2,0,2)$ is rational. Moreover, $M(2,0,2)$ is just the variety of nonsingular plane conics. Unfortunately, his proof of the rationality of $M(2,0,n)$ contains a gap. The only other moduli space $M(2,0,n)$ which is known to be rational is $n = 4$ and will by published by Le Potier. Apart from this, nothing seems to be known about the rationality of $M(r,0,n)$. What has this to do with our problem ?
Hulek [10] gave the following elegant description of $M(r,0,n)$ which can be rephrased in terms of represention. He calls an equidimensional representation

$$\mathcal{A} = (A_0, A_1, A_2) \text{ of } 0 \begin{array}{c}\longrightarrow\\ \longrightarrow\\ \longrightarrow\end{array} 0 \text{ prestable if and only if for every } v \in \mathbb{C}^n :$$

$$\dim \langle A_0 v, A_1 v, A_2 v\rangle \geq 2$$
$$\dim \langle A_0^\tau v, A_1^\tau, v, A_2^\tau v\rangle \geq 2$$

Remark that this is an open condition on $M_n(\mathbb{C})^{\oplus 3}$. With such a representation one can construct a bundle over $\mathbb{P}_2$ in the following way.
Let $V = \Gamma(\mathcal{O}_{\mathbb{P}_2}(1))^*$ with basis u, v, w dual to the usual x, y, z basis of $\Gamma(\mathcal{O}_{\mathbb{P}_2}(1))$, then we can define a linear map

$$\varphi_{\mathcal{A}} : \mathbb{C}^n \otimes V \longrightarrow \mathbb{C}^n \otimes V^*$$

given by the matrix

$$\begin{bmatrix} 0 & A_2 & -A_1 \\ -A_2 & 0 & A_0 \\ A_1 & -A_0 & 0 \end{bmatrix} = \psi_{\mathcal{A}}$$

and denote $U = Im\varphi_{\mathcal{A}}$. Further, $s : \Gamma(\mathcal{O}_{\mathbb{P}2}(1)) \otimes (\mathcal{O}_{\mathbb{P}_2} \rightarrow \mathcal{O}_{\mathbb{P}_2}(1)$ is the natural multiplication map and s^* is its dual. Then we have a complex of vectorbundles :

$$\begin{array}{ccccc}
\mathbb{C}^n \otimes \mathcal{O}_{\mathbb{P}_2}(-1) & \xrightarrow{a} & U \otimes \mathcal{O}_{\mathbb{P}_2} & \xrightarrow{b} & \mathbb{C}^n \otimes \mathcal{O}_{\mathbb{P}_2}(1) \\
\downarrow {\scriptstyle 1 \otimes s^*} & \nearrow {\scriptstyle \varphi_{\mathcal{A}} \otimes 1} & \downarrow & \nearrow {\scriptstyle 1 \otimes s} & \\
\mathbb{C}^n \otimes V \otimes \mathcal{O}_{\mathbb{P}_2} & & \mathbb{C}^n \otimes V^* \otimes \mathcal{O}_{\mathbb{P}_2} & &
\end{array}$$

where a is a mono and b is epi. The cohomology of this complex is a bundle $\mathcal{E}_A$ of rank dim $U - 2n$ and with Chern-numbers $c_1(\mathcal{E}_A) = 0, c_2(\mathcal{E}_A) = n$. Moreover. Hulek has shown that isoclasses of prestable representations and isoclasses of the corresponding bundles coincide. Clearly, for sufficiently general representations, the corresponding bundle will be of rank n and stable. Therefore, we obtain :

Proposition 3 : $V_{3,n}$ is birational to the moduli space of stable rank n bundles over $\mathbb{P}_2$ with Chern-numbers $c_1 = 0, c_2 = n$; $M(n,0,n)$

In particular, rationality of $V_{3,n}$ would imply rationality of the moduli spaces $M(n,0,n)$.

4. Some consequences.

In the foregoring two sections we have reduced our original problem to that of two existing varieties :

$$\begin{array}{ccccc} & & V_{3,n} & & \\ & \sim & & \sim & \\ M_n(\mathbb{C})^{\oplus z}/GL_n(\mathbb{C}) & & & & M(n,0,n) \end{array}$$

where $\sim$ denotes birationality. Note that we did not have to use anything but the wildness of the quiver $0 \begin{array}{c}\longrightarrow\\ \longrightarrow\\ \longrightarrow\end{array} 0$ to prove these results. Nevertheless, we get using Formanek's result on the rationality of the quotient varieties $M_n(\mathbb{C})^{\oplus 2}/GL_n(\mathbb{C})$ the following rather surprising result :

Proposition 4 : The moduli spaces $M(3,0,3)$ and $M(4,0,4)$ are rational.

As far as I know these results were not known. In [16] M. Maruyama claims stable rationality of $M(n,0,n)$ for all $n \in \mathbb{N}$. As we have mentioned before such a result would immediately imply the Merkurjev-Suslin result for function fields of varieties. Unfortunately, in view of Sniders's remark mentioned before the method of proof of [16] cannot be correct. In fact, D. Saltman, K. Hulek and Le Potier have communicated to us a gap in [16]. For more details we refer the reader to [13].

5. Plane Curves and their Jacobians.

Although we have reduced reduced our original problem on classifying simultaneous equivalence classes of m-tuples of square matrices to some existing varieties it is by no means clear that these varieties are easier to handle than $V_{3,n}$.

In this section, we will outline an elegant approach originally due to Maruyama [15] and rediscovered in [13] and [23].

Let us start by considering the open subvariety

$$U_{3,n} = \{A = (A_0, A_1, A_2) : \det(A_0x + A_1y + A_2z) \neq 0\}$$

of $M_n(\mathbb{C})^{\oplus 3}$. With such an A we can associate a monomorphism of vectorbundles

$$\lambda_A = A_0x + A_1y + A_2z : \mathcal{O}_{\mathbb{P}_2}(-1)^{\oplus n} \longrightarrow \mathcal{O}_{\mathbb{P}_2}^{\oplus n}$$

and we can consider its cokernel :

$$0 \to \mathcal{O}_{\mathbb{P}_2}(-1)^{\oplus n} \xrightarrow{\lambda} \mathcal{O}_{\mathbb{P}_2}^{\oplus n} \longrightarrow \mathcal{L}(1) \longrightarrow 0$$

then it follows that $\mathcal{L}$ is a torsion sheaf satisfying $H^0(\mathbb{P}_2, \mathcal{L}) = H^1(\mathbb{P}_2, \mathcal{L}) = 0$ and $c_1(\mathcal{L}) = n$.

Conversely, suppose we start off with a torsion coherent sheaf $\mathcal{L}$ on $\mathbb{P}_2$ with $H^0(\mathbb{P}_2, \mathcal{L}) = H^1(\mathbb{P}_2, \mathcal{L}) = 0$ and $c_1(\mathcal{L}) = n$, then it follows that $\mathcal{L}$ is 1-regular in Mumford's terminology [12]. That is, $H^q(\mathcal{L}(1-q))$ has to vanish for all $q \geq 1$. A pleasant consequence of this is that $\mathcal{L}(n)$ is generated by its sections $H^0(\mathcal{L}(n))$ for all $n \geq 1$ [12]. That is, we get an exact sequence :

$$0 \longrightarrow \mathcal{F} \longrightarrow \mathcal{O}_{\mathbb{P}_2}^{\oplus n} \longrightarrow \mathcal{L}(1) \longrightarrow 0$$

Because $\mathcal{L}$ is torsion, $\mathcal{F}$ is a vector bundle and using Horrock's classification of vector bundles which are sums of line bundles, see for example [3], one can show that $\mathcal{F}$ is $\bigoplus_{i=1}^{n} \mathcal{O}(a_i)$. Finally, using $c_1(\mathcal{L}) = n$ one gets $\mathcal{F} = \mathcal{O}_{\mathbb{P}_2}(-1)^{\oplus n}$. That is, we have an exact sequence

$$0 \longrightarrow \mathcal{O}_{\mathbb{P}_2}(-1)^{\oplus n} \xrightarrow{\lambda} \mathcal{O}_{\mathbb{P}_2}^{\oplus n} \longrightarrow \mathcal{L}(1) \longrightarrow 0$$

and hence a representation $\mathcal{A}_1 = (A_0, A_1, A_2)$ of $0 \;\substack{\longrightarrow \\ \longrightarrow \\ \longrightarrow}\; 0$. Moreover, isoclasses of representations correspond to isoclasses of torsions sheaves. Therefore we have :

Proposition 5 : Orbits in $U_{3,n}$ correspond bijectively to isomorphism classes of torsion coherent sheaves $\mathcal{L}$ on $\mathcal{L}$ on $\mathbb{P}_2$ satisfying $H^0(\mathbb{P}_2, \mathcal{L}) = H^1(\mathbb{P}_2, \mathcal{L}) = 0$ and $c_1(\mathcal{L}) = n$.

At first sight we have not gained much. We reduced our problem again to some moduli problem.

However, it is fairly easy to show that the torsion coherent sheaf $\mathcal{L}$ associated to a representation $\mathcal{A} = (A_0, A_1, A_2)$ lives on the curve $C_{\mathcal{A}} \hookrightarrow \mathbb{P}_2$ of degree n determined by

$$\det(A_0 x + Ay + A_2 z) = 0$$

Moreover, Barth [2] has shown that if $C_{\mathcal{A}}$ is reduced and if x is a smooth point of $C_{\mathcal{A}}$, then $\mathcal{L}_x$ is invertible. Therefore, if $C_{\mathcal{A}}$ is a nonsingular curve of degree n, and hence of genus $g = \frac{(n-1)(n-2)}{2}$, then $\mathcal{L}$ is a divisor over it. Because $H^0(c\mathcal{L}) = H^1(c\mathcal{L}) = 0$ we obtain from the Riemann-Roch theorem for C :

$$\chi(\mathcal{L}) = \deg \mathcal{L} + 1 - g$$

that the degree of $\mathcal{L}$ is $d = g - 1 = \frac{n(n-3)}{2}$.

Conversely, let $C \in \mathbb{P}_2$ be a smooth plane curve of degree n, then the degree determines an exact sequence

$$0 \longrightarrow \mathrm{Jac}(C) \longrightarrow \mathrm{Pic}(C) \xrightarrow{\deg} \mathbb{Z} \longrightarrow 0$$

where Jac (C) is the Jacobian variety of C. Restricting to the set of all divisors of degree d we get an homogeneous space $\mathrm{Pic}_d(C)$ over $\mathrm{Jac}(C)$.

Moreover, there exists an open subvariety $\mathrm{Pic}'_{g-1}(C) \subset \mathrm{Pic}_{g-1}(C)$ consisting of those divisors $\mathcal{L}$ s.t. $H^0(C,\mathcal{L}) = H^1(C,\mathcal{L}) = 0$. One can show, see for example [6], that $\mathrm{Pic}'_{g-1}(C)$ is precisely the complement in $\mathrm{Pic}_{g-1}(C)$ of the image of the natural map

$$\pi_{g-1} : \underbrace{C \times \ldots \times C}_{g-1 \text{ copies}} \longrightarrow \mathrm{Pic}_{g-1}(C)$$

associating to a $(g-1)$-tuple of points of C the divisor $\theta(p_1 + \ldots + p_{g-1})$.
In view of the vast amount of theory on Jacobian varieties and the explicit description of π_{g-1} one can consider $\mathrm{Pic}'_{g-1}(C)$ as a tractable variety.

Now, let us restrict attention to the following open subvariety of $U_{3,n}$:

$$U'_{3,n} = \{A = (A_0, A_1, A_2) | C_A \text{ is smooth}\}$$

then we have the following answer to our problem 1 :

Theorem 1 : Orbits of $GL_n(\mathbb{C}) \times GL_n(\mathbb{C})$ acting on $U'_{3,n}$ by simultaneous equivalence can be parametrized by couples $(C, \mathcal{L})$ where

(1) C is a smooth plane curve of degree n

(2) $\mathcal{L} \in \mathrm{Pic}'_{g-1}(C)$

Let us check the dimension of this variety. Smooth plane curves of degree n form an open subvariety of $\mathbb{P}_{\frac{(n+1)(n+2)}{2}-1}$, i.e. is of dimension $\frac{(n+1)(n+2)}{2} - 1$ and the dimension of $\mathrm{Pic}'_{g-1}(C)$ is equal to that of $\mathrm{Jac}(C)$ which is known to be equal to the genus $= \frac{(n-1)(n-2)}{2}$. Therefore, the dimension of the parametrizing variety $V_{3,n}$ is

$$\frac{(n+1)(n+2)}{2} + \frac{(n-1)(n-2)}{2} - 1 = n^2 + 1$$

as expected.

Purists who like to know the structure of this variety rather than a description of its points may consult [15], [13] or [23]. In short, the variety is a relative Picard scheme of a smooth family of curves.

6. Linearization of partial differential equations.

The Schrödinger equation of a free particle in relativistic free quantum mechanics is

$$i\hbar \frac{\partial \psi(\bar{r},t)}{\partial t} = H(\bar{r}, \frac{\hbar}{i}\nabla)\psi(\bar{r},t) \tag{1}$$

where $|\psi(\bar{r},t)|^2$ is the probability that the particle is in place $\bar{r}$ at time t and H is the Hamiltonian, i.e.

$$H = \frac{p^2}{2m} + V(\bar{r},t)$$

One is primarily interested in solutions of the form

$$\psi(\bar{r},t) = \psi(\bar{r})e^{-iEt/\hbar}$$

where E denotes the energy. Substituting this form in equation (1) we obtain

$$E\psi(\bar{r},t) = H\psi(\bar{r},t)$$

where the energy E is an eigenvalue of the Hamiltonian operator. For a free particle ($V(\bar{r},t)=0$) we obtain

$$H=\frac{p^2}{2m}=\frac{1}{2m}\left(\frac{\partial^2}{\partial x^2}+\frac{\partial^2}{\partial y^2}+\frac{\partial^2}{\partial z^2}\right)$$

and we therefore obtain the equation

$$\frac{\partial^2\psi}{\partial x^2}+\frac{\partial^2\psi}{\partial y^2}+\frac{\partial^2\psi}{\partial z^2}=2mE\psi=(\sqrt{2}mc)^2\psi \tag{2}$$

In order to solve this equation (2) one tries to replace it by a system of first order partial differential equations. In this case, the Pauli matrices :

$$A=\begin{pmatrix}0&i\\-i&0\end{pmatrix}B=\begin{pmatrix}-1&0\\0&1\end{pmatrix}C=\begin{pmatrix}0&-1\\-1&0\end{pmatrix}$$

do this so called linearization trick. For,

$$\left(A\frac{\partial}{\partial x}+B\frac{\partial}{\partial y}+C\frac{\partial}{\partial z}\right)^2=\left(\frac{\partial^2}{\partial x^2}+\frac{\partial^2}{\partial y^2}+\frac{\partial^2}{\partial z^2}\right)I_2$$

and therefore any solution to the system of first order partial differential equations :

$$A\frac{\partial\psi}{\partial x}+B\frac{\partial\psi}{\partial y}+C\frac{\partial\psi}{\partial z}=(\sqrt{2}mc)\psi.S$$

where $S\in M_2(\mathbb{C})$ s.t. $S^2=I_2$ is a solution to (2).

Of course, one can try to generalize this linearization procedure to more variables and/or higher order equations. The present knowledge of this problem is as follows, see for example [5] :
In two variables partial differential equations of any order can be linearized and for any number of variables second order equations can be linearized. The first open case, that of three variables and a third order partial differential equation was (under some extra assumptions) recently solved by M. Van den Bergh [22].

In this section we will use the results of the foregoing section to give a linearization procedure for partial differential equations of the form :

$$\sum_{i+j+k=n}a_{ijk}\frac{\partial^n\psi}{\partial x^i\partial y^i\partial z^k}=c^n\psi \tag{3}$$

Proposition 6 : Partial differential equations of type (3) can always be linearized.

Proof. Suppose that we can find n by n matrices (A_0,A_1,A_2) such that

$$\det\left(A_0\frac{\partial}{\partial x}+A_1\frac{\partial}{\partial y}+A_2\frac{\partial}{\partial z}\right)=\sum a_{ijk}\frac{\partial^n}{\partial x^i\partial y^j\partial z^k}$$

then one can show using the Cayley-Hamilton polynomial that any solution to the system of first order equations

$$A_0\frac{\partial\psi}{\partial x}+A_1\frac{\partial\psi}{\partial y}+A_2\frac{\partial\psi}{\partial z}=c\psi S$$

where $S\in SL_n(\mathbb{C})$ is a solution to (3).

For notational reasons let us write $X=\frac{\partial}{\partial x}, Y=\frac{\partial}{\partial y}, Z=\frac{\partial}{\partial z}$. If $\sum a_{ijk}X^iY^jZ^k$ determines a smooth curve such matrices exist by §5.
If $\sum a_{ijk}X^iY^jZ^k$ determines an irreducible curve C having singularities one can look at the desingularization

$$f:\tilde{C}\to C\hookrightarrow \mathbb{P}_2$$

and take an element $\mathcal{L}\in \mathrm{Pic}'_{g-1}(\tilde{C})$, then $f_*(\mathcal{L})$ is a torsion coherent sheaf on $\mathbb{P}_2$ having the required properties and hence it determines the wanted triple of matrices.

In case $\sum a_{ijk}X^iY^jZ^k=\prod_{l=1}^{s} f_l(X,Y,Z)$ is a factorization, one can construct torsion sheafs on each of the $f_l(X,Y,Z)$, say $\mathcal{L}_l$, and then $\oplus\mathcal{L}_l$ is a torsion sheaf of the required type. ∎

References.

[1] W. Barth : *Moduli of vector bundles on the projective plane*, Invent. Math. 42 (1977) 63-91.

[2] W. Barth : *Some properties of stable rank 2 vector bundles on $\mathbb{P}_n$*, Math. Ann. 226 (1966) 125-150.

[3] W. Barth, K. Hulek : *Monads and moduli of vector bundles*, Manuscripta Math. 25 (1978) 323-347.

[4] S. Bloch : *Torsion algebraic cycles, K_2 and the Brauer group of functionfields*, Bull. AMS 80 (1974).

[5] L. Childs : *Linearization of n-ic forms and generalized Clifford algebras.*, Lin. Mult. Alg. 5 (1978) 267-278.

[6] Clemens : *Scrapbook on complex curve theory.*

[7] J.L Colliot-Thélène, J. Sansuc : *Principal homogeneous spaces under flasque tori with applications to various problems*, J. of Algebra, to appear.

[8] E. Formanek : *The center of the ring of 3 by 3 generic matrices*, Lin. Mult. Alg. 7 (1979) 203-212.

[9] E. Formanek : *The center of the ring of 4 by 4 generic matrices*, J. Algebra 62 (1980) 304-319.

10] K. Hulek : *On the classification of stable rank r vectorbundles over the projective plane*, Birkhaüser PM 7 (1980) 113-144.

[11] V. Kač : *Root systems, representations of quivers and invariant theory*, LNM 996, 74-108.

[12] S. Kleiman : *Les théorèmes de finitude pour le foncteur de Picard*, LNM 225, 616-666.

[13] L. Le Bruyn : *Some remarks on rational matrix invariants*, to appear.

[14] M. Maruyama : *Vector bundles on $\mathbb{P}_2$ and torsion sheaves on the dual plane*, Proceedings Tata 1984.

[15] M. Maruyama : *The equations of plane curves and the moduli spaces of vector bundles on $\mathbb{P}_2$*

[16] M. Maruyama : *Stable rationality of some moduli spaces of vector bundles on $\mathbb{P}_2$*, LNM 1194, 80-89.

[17] C. Procesi : *Rings with polynomial identities*, Marcel (1973).

[18] C. Procesi : *Invariant theory of n by n matrices*, Adv. Math. 19 (1976) 306-381.

[19] C. Procesi : *Relazioni tra geometrica algebrica ed algebra non commutative*, Boll. Un. Math. Ital. (5) 18-A (1981) 1-10.

[20] D. Saltman : *Retract rational fields and cyclic Galois extensions*, Israel J. Math. 46 (1983).

[21] D. Saltman : *The Brauer group and the center of generic matrices*, J. Alg. (1986).

[22] M. Van den Bergh : *Linearization of binary and ternary forms*, to appear.

[23] M. Van den Bergh : *Center of generic division algebras*, to appear.

[24] W. Van der Kallen : *The Merkurjev-Suslin Theorem*, in : Integral Representations and applications; Springer LNM (1986).

The Auslander condition on noetherian rings

by Jan-Erik Björk
Department of Mathematics
University of Stockholm
Box 6701
11385 Stockholm (Sweden)

[This paper is in final form and no version of it will be submitted for publication elsewhere].

Introduction.

In the late sixties *M. Auslander* introduced various homological conditions on noetherian, or more generally coherent rings which then were investigated both for commutative and non-commutative rings. Auslander's conditions were expressed in terms of the vanishing of certain Ext-group. In this work we shall only consider noetherian rings and study the condition below on categories of finitely generated left or right modules.

Auslander's condition.

Let A be a left and a right noetherian ring. A finitely generated left or right A-module M is said to satisfy the Auslander condition if the following holds : for every integer v and every submodule N of $\mathrm{Ext}_A^v(M,A)=0$ we have $\mathrm{Ext}_A^i(N,A)=0$ for every $i<v$.

If A is a left and a right noetherian ring with a *finite injective dimension*, i.e. if both the left and the right A-module A has a bounded injective resolution, then we say that the ring A is *Auslander-Gorenstein* if every finitely generated left or right A-module satisfies Auslander condition.

We can also impose the stronger condition that the ring A has a *finite global homological dimension*. If gl.dim(A) is finite and every finitely generated left or right A-module satisfies Auslander's condition, then we say that the ring A is *Auslander regular*.

Concerning the terminoly introduced above, we remark that in the work [FGR], Auslander-Gorenstein rings were called *μ-Gorenstein rings* with μ=inj.dim(A). Auslander's condition was studied in a more general context in [FGR], i.e. both for certain additive categories as well as for categories of modules over rings. Let us also remark that

Auslander's condition was inspired by previous results for *commutative* rings. In his work [Ba], *H. Bass* proved that if A is a commutative noetherian ring with a finite injective dimension, then Auslander's condition holds for finitely generated A-modules, i.e. with our present terminology this means that the ring A is Auslander-Gorenstein. For commutative noetherian rings with finite global homological dimension, the Auslander condition was established in an earlier work by *J-E. Roos* in [Ro:1] who also made a complete investigation of the so called ***Bidualizing complex***. Let us also mention [F-F] contains a study of the bidualizing complex for commutative Auslander-Gorenstein rings, inspired by [Ba] and [Ro:1].

For *non-commutative* rings Auslander's condition is not a direct consequence of finite injective dimension, and not even of finite global homological dimension. An instructive example occurs in the thesis by *I. Reiten* [Re] and we present it here.

Reiten's example (See [Re, example 2.4.6]). Let k be a commutative field and Λ denotes the matrix algebra $\begin{pmatrix} k & 0 \\ V & k \end{pmatrix}$ where V is a 2-dimensional vector space. Then $\mathrm{gl.dim}(\Lambda)=2$ but Λ is not Auslander regular for if the two-sided ideal $\begin{pmatrix} 0 & 0 \\ V & 0 \end{pmatrix}$ is denoted by J and if J for example is treated as a left Λ-module, then it is easy to show that $\mathrm{Ext}^0_\Lambda(\mathrm{Ext}^1(\Lambda/J,\Lambda),\Lambda)\neq 0$, so Λ/J does not satisfy Auslander's condition.

Work related to Auslander's condition which appeared before 1972 was mainly devoted to study matrix algebras and so called semi-trivial extensions. Then a new area started when *filtered* noetherian rings were studied and homological conditions on these were found for some important cases. For example, *J-E Roos* used homological methods, based upon Auslander's condition to estimate Krull dimensions of $U(\zeta)$-modules of the form $U(\zeta)/\mathbb{P}$, where ζ is a finite dimensional Lie algebra and $\mathbb{P}$ a primitive ideal. His work in [Ro:2], combined with his proof that the global homological dimension of the Weyl algebra $A_n(k)$ over a commutative field of characteristic zero is equal to n, suggested that $A_n(k)$ is an Auslander regular ring. This is indeed true and the Auslander-regularity of $A_n(k)$ is obtained as a special case of a general study rings with *positive filtrations* whose associated graded rings are commutative noetherian rings with finite global homological dimension. The fact that any such filtered ring is Auslander regular is proved in Chapter 2 of [Bj:1].

The theory of $\mathcal{D}$ and $\mathcal{E}$-modules, to a large extent created by *M. Sato, M. Kashiwara and T. Kawai* also established Auslander's condition in important cases. For example, the fact that the stalks of $\mathcal{D}_X$, (resp. of $\mathcal{E}_X$) are Auslander regular rings for any complex manifold X was already asserted in [SKK] and complete proofs appear in Kshiwara's article [Ka:1].

A general study of filtered rings, where the filtration no longer has to be positive was done in [Bj:2], based upon lectures at a meeting in Oberwolfach in January 1983. However, this work was still restricted to the case when the rings have finite global homological dimension. A systematic study of filtered noetherian rings with ***finite injective dimension*** was done by ***T. Levasseur***. See [Lev:1]. An elegant example of a class of Auslander-Gorenstein rings occur in [Lev:1]. For example, in his thesis Levasseur proves that if $n \geq 1$ and G is a finite subgroup of $GL(n,\mathbb{C})$ without any pseudo-reflexion, then the natural G-action on $A_n(\mathbb{C})$ gives rise to the G-invariant subring and ***it is proved that this ring always is Auslander-Gorenstein.***

After this discussion about the background in the subject we begin to expose the material in this work. To some extent it is expositary since many results already are wellknown. However, it appears to be important to offer some detailed proofs in connection with Auslander-Gorenstein rings where properties of the so called ***Bidualizing complex*** are less understood, as compared with the case of Auslander regular rings. For this matter we refer to the ***open problem,*** announced as "Question" after Proposition 1.11.

Another reason for the present work is that we include a study of filtrations which need not satisfy the separation axiom on finitely generated modules. That is, using the terminology in the theory of filtered rings we study cases where the filtration ***fails to be Zariskian.*** We remark that this is quite relevant since such important cases as ***V-filtrations on modules over rings of differential operators are non-Zariskian.*** See also the ***Remark*** after Theorem 3.6 for this matter.

Let us finish the introduction by briefly describe the contents of the subsequent sections. In §1 we review an important result due to L. Levasseur in [Lev1] who proves the regularity of the non-trivial spectral sequence which appears in the bidualizing complex for finitely generated modules over Auslander-Gorenstein rings. Thanks to this result we are then able to establish results for modules over Auslander-Gorenstein rings, with particular emphasis on so called ***pure modules***. The main result is Theorem 1.14, referred to as ***Gabber's maximality principle***, since it was O. Gabber who established this result in his lectures at Université Paris VI in 1982.

Theorem 1.14 will later on be used in §4 to describe the ***characteristic ideal***, and more generally the ***characteristic cycle*** of certain quotients of pure modules over filtered rings. We refer to §4 for this and mention only that Theorem 4.7 is an important result whose proof requires the homological study in §1 as well as in §3, where we establish Auslander conditions on filtered rings. In order to be able to offer a reasonable self-contained presentation we have included a section about filtrations. The various conditions which are used to study filtered noetherian rings and the interplay between these are investigated. In section 2 we also define and study associated Rees rings of filtered rings.

In this connection we refer to the article by *E-K. Ekström* in these lecture notes for a study of Auslander's condition on graded rings and on Rees rings.

1. Auslander-Gorenstein rings

Let A be a ring such that every left or right ideal is finitely generated. Then we simply say that the ring is noetherian. If there exists an integer μ such that $\mathrm{Ext}_A^{\nu}(M,A)=0$ for every $\nu>\mu$ and every finitely generated left or right A-module M, then we say that A has a ***finite injective dimension***. Of course, this is equivalent to the condition that both the left and the right A-module A has a bounded injective dimension. We refer to [Z] for the proof that the left and the right injective dimensions are the same.

From now on we assume that A is a noetherian ring with a finite injective dimension. Let us now briefly recall the construction of the ***bidualizing complex*** of a finitely generated left A-module M. (A similar bidualizing complex is obtained for right A-modules). First we take a projective resolution $0 \longleftarrow M \longleftarrow P_0 \longleftarrow P_{-1} \longleftarrow$ where every $P_{-\nu}$ is a finitely generated projective A-module. Then, by standard methods from [C-E] we can construct a projective resolution $Q_{\bullet\bullet}$ of the complex of right A-modules $Hom_A(P_\bullet,A)$. Then the double complex $\check{Q}_{\bullet\bullet}=Hom_A(Q_{\bullet\bullet},A)$ represents the bidualizing complex of A. We notice that $\check{Q}_{\bullet\bullet}$ is placed in the second quadrant.

$$\begin{array}{ccccc} & \uparrow & & \uparrow & \\ \to & \check{Q}_{-1,1} & \to & \check{Q}_{01} & \\ & \uparrow & & \uparrow & \\ \to & \check{Q}_{-1,0} & \to & \check{Q}_{00} & \end{array}$$

As usual consider the *diagonal complex* $\dot{\Delta}$, where we have $\Delta^m=\oplus_v \check{Q}_{-v,v+m}$. The so-called *second filtration* on $\check{Q}_{\bullet\bullet}$ gives a spectral sequence where the first term E_1 is reduced to a single row complex $\longrightarrow \check{\check{P}}_{-1} \longrightarrow \check{\check{P}}_0 \longrightarrow 0$. Here $\check{\check{P}}=Hom_A(Hom_A(P,A),A)$ and since finitely generated projective A-modules are reflexive, it follows that $E_2^{p,q}=0$ if $(p,q)\neq 0$, while $E_2^{0,0}=M$. From this one easily deduces that the total cohomology groups $H^m(\Delta^\bullet)$ are zero if $m\neq 0$, while $H^0(\Delta^\bullet)=M$.

Concerning the *first filtration* it is easily verified that the second term $E_2^{\bullet\bullet}$ is a double complex where

$$E_2^{-p,q}=Ext_A^q(Ext_A^p(M,A),A)$$

In particular, if μ=inj.dim(A), then $E_2^{-p,q}=0$ when $p>\mu$ or $q>\mu$. So $E_2^{\bullet\bullet}$ is a **bounded double complex** and hence it is trivial that there exists an integer r such that $E_s^{\bullet\bullet}=E_r^{\bullet\bullet}$ for every $s>r$. However, this does not ensure that the spectral sequence associated with the first filtration is regular because the double complex $Q_{\bullet\bullet}$ is not bounded.

By a quite subtle study, ***T. Levasseur has proved that the spectral sequence indeed is regular***. His proof in [Lev.1] is very detailed and we mention that Levasseur uses the so called ***Ischebeck complex*** as a tool for the proof of the regularity of the spectral sequence above.

We are going to use this result by Levasseur for the special class of so called ***Auslander-Gorenstein rings*** which is based upon the following

1.1. Definition. Let A be a noetherian ring with a finite injective dimension. A finitely generated left or right A-module M satisfies the ***Auslander condition*** if the following hold : for every integer v and every submodule N of $Ext_A^v(M,A)$, it follows that $Ext_A^i(N,A)=0$ when $i<v$.

1.2. Definition. An ***Auslander-Gorenstein ring*** is a noetherian ring A with a finite injective dimension and such that every finitely generated A-module satisfies the Auslander condition.

The following result was proved by Levasseur in [Lev.1: Corollaire 4.3].

1.3. Theorem. *Let* A *be Auslander-Gorenstein. Then, if* M *is a finitely generated* A*-module, there exists a sequence of* A*-submodules* $M=B_0(M) \supset B_{-1}(M) \supset \ldots \supset B_{-\mu}(M) \supset 0$, *where* μ *is the injective dimension of* A. *Moreover, if* $0 \le p \le \mu$ *there exists an exact sequence of* A*-modules* :

$$0 \longrightarrow B_{-p}(M)/B_{-p-1}(M) \longrightarrow \mathrm{Ext}_A^p(\mathrm{Ext}_A^p(M,A),A) \longrightarrow Q(p) \longrightarrow 0$$

where the A*-module* Q(p) *is a subquotient of the* A*-module given as* :

$$\bigoplus_{i \ge 1} \mathrm{Ext}_A^{p+i+1}(\mathrm{Ext}_A^{p+i}(M,A),A)$$

Remark. In the case when A has a finite *global homological dimension* there exist ***bounded*** projective resolutions. So the bidualizing complex of a finitely generated A-module is bounded and there is no problem to ensure the regularity of the spectral sequence associated with the first filtration. In this case the result in Theorem 1.3 occurs in [Bj.1:Chapter 2]. Let us also give the following :

1.4. Definition. An *Auslander regular ring* is a noetherian ring A with a finite global homological dimension and such that every finitely generated A-module satisfies the Auslander condition.

Of course, it follows that every Auslander regular ring is Auslander Gorenstein. In the sequel we mostly announce results for Auslander Gorenstein rings and then similar results follow for Auslander regular rings.

We are going to investigate the B-filtration on finitely generated modules over Auslander-Gorenstein rings. For this purpose we give

1.5. Definition. Let A be Auslander-Gorenstein. If M is a non-zero and finitely generated A-module, then the smallest non-negative integer k such that $\mathrm{Ext}_A^k(M,A) \ne 0$ is called the *grade number of* M and denoted by $j_A(M)$.

1.6. Proposition. *Let* A *be Auslander-Gorenstein. Then, if* M *is a finitely generated* A*-module, we have* :

(1) $j(B_{-v}(M)) \ge v$ *for every* v ;

(2) $j(\mathrm{Ext}_A^{j(M)}(M,A))=j(M)$.

Proof. The Auslander condition and the exact sequence in Theorem 1.4 yield $j(B_{-v}(M)/B_{-v-1}(M))\geq v$ for every v. Next we apply Lemma 1.6 and by an induction which starts with $v=\mu+1$ where $B_{-\mu-1}=0$ we obtain (1). To prove (2) we notice that $j(Ext^{j(M)}(M,A))\geq j(M)$ follows from the Auslander condition. If the inequality is strict, then $M=B_{-j(M)-1}(M)$ and this gives $j(M)\geq j(M)+1$ by (1). This cannot occur and hence (2) holds.

1.7. Lemma. *Let* A *be a Noetherian ring with a finite injective dimension. Then, if* $0 \longrightarrow M' \longrightarrow M \longrightarrow M'' \longrightarrow 0$ *is an exact sequence of finitely generated* A*-modules, it follows that :*

$$j_A(M)\geq\inf\{j_A(M'),j_A(M'')\}$$

Proof. Follows from the long exact Ext-sequence which contains

$$\longrightarrow Ext_A^v(M'',A) \longrightarrow Ext_A^v(M,A) \longrightarrow Ext_A^v(M',A) \longrightarrow$$

1.8. Proposition. *Let* A *be Auslander-Gorenstein. Then, if* $0 \longrightarrow M' \longrightarrow M \longrightarrow M'' \longrightarrow 0$ *is an exact sequence of finitely generated* A*-modules, we have :* $j_A(M)=\inf\{j_A(M'),j_A(M'')\}$

Proof. We notice first that the long exact Ext-sequence gives $j_A(M'')\geq\inf\{j_A(M'),j_A(M)+1\}$. So by Lemma 1.6 there remains only to show that $j_A(M')\geq j_A(M)$. Assume the contrary, i.e. let $k=j_A(M')<j_A(M)$. The long exact Ext-sequence contains

$$0 \longrightarrow Ext^k(M',A) \longrightarrow Ext^{k+1}(M'',A) \longrightarrow Ext^{k+1}(M,A) \longrightarrow$$

Then we obtain $j_A(Ext^k(M',A))\geq k+1$ from the Auslander condition. But this contradicts (2) in Proposition 1.7, applied with the A-module M'.

Purity. Let A be Auslander-Gorenstein. A finitely generated A-module M is called *pure* if $j(N)=j(M)$ holds for every non-zero submodule N.

1.9. Proposition. *A finitely generated* A*-module* M *is pure if and only if* $Ext_A^v(Ext_A^v(M,A),A)=0$ *for every* $v\neq j(M)$.

Proof. Assume first that M is pure. Proposition 1.7 shows that $B_{-v}(M)=0$ for every $v>j(M)$. So if $v>j(M)$, then Theorem 1.4 gives $j(Ext^v(Ext^v(M,A),A))\geq v+2$. Now, if $Ext^v(Ext^v(M,A),A)$ is non-zero we notice that $v=j(Ext^v(M,A))$ and then (2) in Proposition 1.7 gives $j(Ext^v(Ext^v(M,A),A))=v$. So we conclude that the double Ext-group $Ext^v(Ext^v(M,A),A)=0$ for every $v>j(M)$.

Conversely, assume that $Ext^{v}(Ext^{v}(M,A)A)=0$, $v>j(M)$. Let N be a submodule of M and assume that $k=j(N)>j(M)$. The long exact Ext-sequence contains

$$Ext^{k}(M,A) \longrightarrow Ext^{k}(N,A) \longrightarrow Ext^{k+1}(M/N,A)$$

Here $j(Ext^{k}(M,A))>k$ by the hypothesis. Then Lemma 1.6 and the inequality $j(Ext^{k+1}(M/N,A))\geq k+1$ yield $j(Ext^{k}(N,A))\geq k+1$. But this contradicts (2) in Proposition 1.7 and hence $j(N)>j(M)$ cannot hold. So the A-module M is pure.

Remark. In the case when A is Auslander regular we use bounded projective resolutions and then the bidualizing complex is of bounded amplitude and hence regular. Also, by projective resolutions the differentials of the associated spectral sequence are A-linear. Then, since the total cohomology is concentrated in degree zero, a straight forward argument yields :

(1.10) $$j(Ext^{v}(Ext^{j(M)}(M,A),A))\geq v+2$$

holds for every $v>j(M)$ and every finitely generated A-module M. See also [Bj.2 : Proposition 1.18 and Remark on p. 66].

1.11. Proposition. *Let* A *be Auslander regular. Then* $Ext^{j(M)}(M,A)$ *is a pure* A*-module for every f.g.* A*-module* M.

Proof. Put $N=Ext^{j(M)}(M,A)$. Then (1.10) implies that

$$Ext^{v}(Ext^{v}(N,A),A)=0 \text{ for every } v>j(M).$$

Since $j(M)=j(N)$ by Proposition 1.7 we get the purity of N from Proposition 1.9.

A Question. Let A be Auslander-Gorenstein. Is it true that $Ext_{A}^{j(M)}(M,A)$ is a pure A-module for every finitely generated A-module ?

Let us remark that this holds if A is *commutative*. For then the higher differentials of the spectral sequence associated with the first filtration of the *Ischebeck complex are* A*-linear* and by the regularity of this spectral sequence we can prove (1.10) in that case and then establish the purity of $Ext_{A}^{j(M)}(M,A)$. But in the general case I do not know the answer to the Question. Let us remark that if we put $N=Ext^{j(M)}(M,A)$, then it is easy to show that the A-module N is isomorphic with $Ext^{j(N)}(Ext^{j(N)}(N,A),A)$. So if $Ext^{j(M)}(M,A)$ is not pure, then the double Ext-group above fails to be pure. Again, this cannot occur for Auslander regular rings. We shall describe this below.

Assume for a while that A is *Auslander regular*. Hence Proposition 1.11 holds. We notice that this implies that if M is a finitely generated A-module, then $Ext^{j(M)}(Ext^{j(M)}(M,A),A)$ is pure. Concerning this pure A-module we can establish more. First we give :

1.12. Définition. Let M be a pure A-module. A *tame and pure extension of* M is a finitely generated A-module M′ such that $M \subset M'$, M′ is pure and $j(M'/M) \geq j(M)+2$.

1.13. Proposition. *Let* M *be a pure* A-*module and let* $\tilde{M}$ *denote* $Ext^{j(M)}(Ext^{j(M)}(M,A),A)$. *Then there does not exist a tame and pure extension* M′ *of* $\tilde{M}$ *with* $M'/\tilde{M} \neq 0$.

Proof. Suppose that $0 \longrightarrow \tilde{M} \longrightarrow M' \longrightarrow M'/\tilde{M} \longrightarrow 0$ exists where $M'/M \neq 0$. Put $k=j(M'/\tilde{M})$ so that $k \geq j(\tilde{M})+2$. We obtain

$$Ext^{k-1}(\tilde{M},A) \longrightarrow Ext^{k}(M'/\tilde{M},A) \longrightarrow Ext^{k}(M',A)$$

Now (1.10) gives $j(Ext^{k-1}(\tilde{M},A)) \geq k+1$ and then it would follow that $j(Ext^{k}(M'/\tilde{M},A) \geq k+1$. Since $k=j(M'/\tilde{M})$ this contradicts (2) in Proposition 1.7 and hence $M'/\tilde{M}=0$.

At this stage we announce a result which in the case of Auslander regular rings is due to *O. Gabber* who used this result as a tool for investigating characteristic varieties of certain modules over enveloping algebras. Gabber's results occur in notes by *T. Levasseur*, taken from Gabber's lectures at Université Paris VI in 1982. Since Levasseur's notes no longer are easy to obtain we review results from these. In addition we are going to establish Gabber's results for Auslander-Gorenstein rings, using a new method based upon certain *"multiplicities"* which are obtained by Theorem 1.17 below.

1.14. Theorem. *Let* A *be Auslander-Gorenstein. Let* M *be a pure* A-*module and let* $\mathcal{M}$ *be some* A-*module which contains* M *such that every finitely generated submodule of* $\mathcal{M}$ *is pure. Then* $\mathcal{M}$ *contains a unique largest tame and pure extension of* M.

Remark. We may refer to Theorem 1.14 as *Gabber's maximality principle*. Notice that we do not assume that $\mathcal{M}$ is a finitely generated A-module.

Following Gabber's original methods, kindly communicated to me by *T. Levasseur*, we first offer a proof a Theorem 1.14 when A is *Auslander regular*. In order to do this we begin with some preliminary remarks. In general, let M be a pure A-module. Then, if $\tilde{M}=Ext^{j(M)}(Ext^{j(M)}(M,A),A)$, it follows that $\tilde{M}$ is a tame and pure extension of M. The injective map $M \longrightarrow \tilde{M}$ is also *canonical* when we use projective resolutions to construct this map. More precisely, we take a ***bounded*** projective resolution $P_{\bullet}$ of M and a ***bounded*** projective resolution $Q_{\bullet\bullet}$ of $Hom_A(P_{\bullet},A)$. Then the bidualizing complex of M is $Hom_A(Q_{\bullet\bullet},A)$. This complex is unique up to homotopy so the associated spectral sequence with respect to the first filtration is also unique and yields the *canonical* imbedding of M into $\tilde{M}$. The canonical imbedding is functorial with respect to M, i.e this follows from the existence of simultanous projective resolutions. In particular we have the following :

1.15. Lemma. *Let* M *be a pure* A*-module and* M' *a tame and pure extension of* M. *Then* $\tilde{M}=\tilde{M}'$ *and the following diagram commutes :*

$$\begin{array}{ccc} M & \xrightarrow{\alpha} & M' \\ \downarrow & & \downarrow \\ \tilde{M} & \xrightarrow[\sim]{\tilde{\alpha}} & \tilde{M}' \end{array}$$

where the vertical arrows are the canonical maps and $\alpha:M \longrightarrow M'$ *the given injective map from* M *into its tame and pure extension* M', *while* $\tilde{\alpha}:\tilde{M} \longrightarrow \tilde{M}'$ *is an isomorphism.*

Proof of Theorem 1.14 ***when*** A ***is Auslander regular***. The existence of a unique maximal tame and pure extension in $\mathcal{M}$ follows if we can show that every increasing sequence $M \subset M_1 \subset \ldots$ of tame and pure extensions of M is stationary. To prove this we apply Lemma 1.15 and conclude that we can construct injective maps $\alpha_\nu : M_\nu \longrightarrow \tilde{M}$ such that if $i_\nu : M \longrightarrow M_\nu$ is the inclusion map in $\mathcal{M}$, then $\alpha_\nu o i_\nu$ is equal to the canonical imbedding of M into $\tilde{M}$. Now $\alpha_\nu(M_\nu)$ increase in the Noetherian A-module $\tilde{M}$ and hence this sequence is stationary. Then we conclude that $\{M_\nu\}$ is stationary in $\mathcal{M}$.

In the case when A is ***Auslander-Gorenstein*** we do not know if a similar proof as above is available. One obstacle is that we cannot be sure to obtain A-linear maps α_ν as in the previous proof. However, we are going to prove Theorem 1.14 by another method. But first we are going to construct a multiplicity of every finitely generated A-module.

1.16. Definition. Let M be a finitely generated A-module. A *fast chain* in M consists of an increasing sequence of submodules $0 \subset M_1 \subset M_2 \subset \ldots$ such that $j(M_1)=j(M_2/M_1)=\ldots=j(M)$.

1.17. Theorem. *Let* A *be Auslander-Gorenstein. Then, if* M *is a finitely generated* A*-module there exists a positive integer* $\varepsilon(M)$ *such that every fast chain in* M *has length* $\varepsilon(M)$ *at most.*

Before we prove Theorem 1.17 we need some preliminaries. Let us say that a finitely generated A-module M is ***chain simple*** if there does not exist a submodule N such that $j(N)=j(M/N)=j(M)$. We notice that if $k=j(M)$ and $M'=M/B_{-k-1}(M)$ for a finitely generated A-module, then M is chain simple if and only if M' is chain simple. The proof follows easily if we use Proposition 1.8.

1.18. Lemma. *Let* M *be a finitely generated* A*-module and put* $k=j(M)$. *Then there exists a submodule* N *of* M *such that* $j(N)=k$ *and* N *is chain simple.*

Proof. By the observation above it suffices to prove Lemma 1.18 with M replaced by $M/B_{-k-1}(M)$ so then M is pure and M is a submodule of $\tilde{M}=Ext^k(Ext^k(M,A),A)$. Let us now consider the A-module $Ext^k(M,A)$. Since this A-module is Noetherian we easily obtain a submodule S such that $j(Ext^k(M,A)/S)=j(M)$ and $Ext^k(M,A)/S$ is chain simple.

Using Theorem 1.4 and Proposition 1.8 it is easily seen that $Ext^k(Ext^k(M,A)/S,A)$ is chain simple too. Now this A-module is a submodule of $\tilde{M}$, denoted by W. Then $M \cap W$ is a chain simple submodule of M with $j(M \cap W)=j(M)$.

Proof of Theorem 1.17. Let M be a finitely generated A-module and put $k=j(M)$. Since the A-module M is Noetherian, it follows from Lemma 1.18 that there exists a bounded sequence of submodules

$$0 \subset M_1 \subset \ldots \subset M_s=M$$

such that $M_1, M_2/M_1, \ldots, M/M_{s-1}$ all have grade number k and are chain simple. Using this "maximal" chain we prove that $\varepsilon(M)=s$. Namely, if $0 \subset N_1 \subset \ldots \subset N_t=M$ is some fast chain in M, then we first choose the unique smallest integer v_0 such that $j((N_{v_0}+M_{s-1})/M_{s-1})=k$. From Proposition 1.8 it follows easily that $\{N_v \cap M_{s-1}: 1 \le v \le v_0-1\}$ is a fast chain in M_{s-1}. Similarly we show that $\{N_v \cap M_{s-1}: v_0<v\le t\}$ is a fast chain. We also have $j((N_{v_0+1} \cap M_{s-1})/N_{v_0-1} \cap M_{s-1}))=k$. By an induction over s we may assume that every fast chain inside M_{s-1} has length s-1 at most. Then we see that $t-1 \le s-1$ so the induction step is proved and then $t \le s$ follows. This proves that $\varepsilon(M)$ exists and we notice that we also have proved that every fast chain in M with chain simple quotients is of length $\varepsilon(M)$.

Proof of Theorem 1.14. Let us consider some integer $k\geq j(M)+2$. The purity of M gives $j(Ext^{k-1}(M,A))\geq k$. Next, let M' be a tame and pure extension and suppose that $j(M'/M)=k$. The long exact Ext-sequence contains :

$$Ext^{k-1}(M,A) \longrightarrow Ext^{k}(M'/M,A) \longrightarrow Ext^{k}(M',A)$$

The purity of M' gives $j(Ext^{k}(M',A)\geq k+1$. By proposition 1.7 we have $j(Ext^{k}(M'/M,A)=k$ and hence $j(Ext^{k-1}(M,A)=k$ must hold if M'/M is non-zero with $j(M'/M)=k$. Also, the exact sequence above show that $e(M'/M)\leq e(Ext^{k-1}(M,A))$.

Now we conclude that if $M=M_0 \subset M_1 \subset \ldots$, where each M_v is a tame and pure extension of M and $j(M_v/M_{v-1})=k$ for some $k\geq j(M)+2$ holds for every v, then at most $e(Ext^{k-1}(M,A))$ many terms can occur. Indeed, we have the obvious additivty $e(M_v/M)=e(M_{v-1}/M)+e(M_v/M_{v-1})$.

At this stage we easily finish the proof of Theorem 1.14. Namely, let $\mathcal{F}$ denote the family of tame and pure extensions of M inside $\mathcal{M}$. With $k=j(M)+2$ the previous argument shows that there exists some $M' \in \mathcal{F}$ such that $j(M''/M')=k$ cannot occur for any M" in $\mathcal{F}$ which contains M'. Then we start with the pure A-module M' and use $k=j(M)+3$ and obtain some $M'' \supset M'$ such that we cannot find any N in $\mathcal{F}$ which contains M" and has $j(N/M'')=j(M)+3$. We continue the process and after a finite number of steps we obtain some M^* in $\mathcal{F}$ which is relatively maximal, i.e. M^* is not contained in any other M' from the class $\mathcal{F}$. But then the hypothesis on $\mathcal{M}$ gives that M^* is maximal in $\mathcal{F}$, i.e. if $M' \in \mathcal{F}$ then M^*+M' is a pure and tame extension of M^* and hence contained in M^*, so $M' \subset M^*$.

Theorem 1.14 will be applied to study quotient modules $M/\varphi(M)$, where $\varphi: M \longrightarrow M$ is an injective A-linear map and M is a pure A-module.

1.19. Theorem. *Assume that* A *is Auslander-Gorenstein. Let* M *be a pure* A-*module and* $\varphi: M \longrightarrow M$ *an injective* A-*linear map such that* $M/\varphi(M)$ *is non-zero. Then* $j(M/\varphi(M)=j(M)+1$ *and moreover, there exists a submodule* M' *of* M *satisfying the following :*

(1) $\varphi(M') \subset M'$ *and* $\varphi^{k}(M) \subset M'$ *for some* $k>1$;

(2) $M'/\varphi(M')$ *is pure with grade number* $j(M)+1$.

Proof. Let m be the universal φ-inverting module. Thus, m is the A-module $M[T,T^{-1}]/\mathbb{R}$, where T is an indeterminate and $\mathbb{R}$ is a A-submodule generated by $\{\varphi(m)-Tm \ ; \ m \in M\}$. Since φ is not surjective we see that the A-module m is not finitely generated. In fact, we have $m=\bigcup_{k\geq 1}\varphi^{-k}M$, where $M \subset \varphi^{-1}M \subset \varphi^{-2}M \subset \ldots$ increases strictly. The A-modules M and $\varphi^{-k}M$ are isomorphic for each k. Also, if M′ is a finitely generated submodule of m, then $M' \subset \varphi^{-k}M$ for some k. This implies that M′ is pure so the conditions in Theorem 1.14 hold. Now we find the ***unique largest tame and pure extension*** of M inside m. Let us denote it by M^*. Since $M+\varphi(M^*)$ also is a tame and pure extension of M, it follows that $\varphi(M^*) \subset M^*$. We also have $M^* \subset \varphi^{-k}M$ for some integer k. If $M^*=\varphi^{-1}M^*$ it would follow that $m=\varphi^{-k}M$. This cannot happen since we have noticed that the A-module m is not finitely generated. The A-modules $\varphi^{-1}M^*/M^*$ and $M^*/\varphi(M^*)$ are isomorphic so $M^*/\varphi(M^*)$ is non-zero.

Next, if N is a non-zero submodule of $M^*/\varphi(M^*)$, then $\varphi^{-1}N$ is a strictly larger than M^*. So by the maximality of M^*, it follows that $j(\varphi^{-1}N/M^*) \leq j(M)+1$. We conclude that $j(N) \leq j(M)+1$ holds for every non-zero submodule of $M^*/\varphi(M^*)$. In particular $j(M^*/\varphi(M^*)) \leq j(M)+1$.

Next, we choose k so large that $\varphi^kM^* \subset M$. So if $M'=\varphi^kM^*$, then M′ is a submodule of M and $j(N) \leq j(M)+1$ for every non-zero submodule of $M'/\varphi(M')$. At this stage we conclude that if we have shown that $j(M/\varphi(M))=j(M'/\varphi(M'))=j(M)+1$, then $M'/\varphi(M')$ is pure and Theorem 1.19 follows. To prove that $j(M/\varphi(M))=j(M'/\varphi(M'))=j(M)+1$ we first notice that $M'/\varphi(M')$ is a subquotient of $M/\varphi^{k+1}(M)$. Also, since the A-modules $\varphi^v(M)/\varphi^{v+1}(M)$ are isomorphic for every v, it follows from Proposition 1.8 that $j(M'/\varphi(M')) \geq j(M/\varphi^{k+1}(M))=j(M/\varphi(M))$.

We have already shown that $j(M'/\varphi(M')) \leq j(M)+1$. So the proof is finished if we can prove that $j(M/\varphi(M)) \geq j(M)+1$. To prove this inequality we use the exact sequence $0 \longrightarrow M \xrightarrow{\varphi} M \longrightarrow M/\varphi(M) \longrightarrow 0$. Then, if we have $j(M/\varphi(M))=j(M)$, it follows by additivity of multiplicties that $e(M)=e(M)+e(M/\varphi(M))$, so $e(M/\varphi(M))$ would be zero. This contradiction shows that $j(M/\varphi(M))$ must be strictly larger than $j(M)$.

2. Filtrations

We are going to study various conditions which can be imposed for filtrations on rings. By a filtration on a ring A we mean an increasing sequence of additive subgroupe $\{A_v\}$, indexed by integers and satisfying $\cup A_v=A$, $\cap A_v=0$ and $A_vA_k \subset A_{v+k}$. We also assume that A has a multiplicative identity which belongs to A_0. Notice that A_0 is a subring of A. If A is a filtered ring we can construct the ***associated graded ring***, defined by $\oplus A_v/A_{v-1}$ and denoted by G(A).

<u>The filtered topology</u>. The filtration on the ring A enables us to construct a topology on A, using the distance function $d(x,y)=2^{v}$ for a pair x,y in A such that x−y has *order* v. That is, x−y belongs to $A_v \setminus A_{v-1}$. Since $\cap A_v=0$ we have $d(x,y)>0$ if $x\neq y$. The distance function gives the filtered topology on the set A. We may then refer to closed subsets of A with respect to the filtered topology. This leads to the following :

2.1. Definition. The filtration on A satisfies the *closure condition* if every finitely generated left or right ideal is a closed subset of A.

2.2. Definition. The filtration satisfies the *strong closure condition* if the following holds : for every finite subset $x_1,\dots,x_s$ of A and integers $v_1,\dots v_s$, it follows that the subsets of A defined by $A_{v_1}x_1+\dots+A_{v_s}x_s$, respectively $x_1A_{v_1}+\dots+x_sA_{v_s}$ are closed.

Next, let $L=Ax_1+\dots Ax_s$ be a finitely generated left ideal. The filtered topology on A induces a topology on the subset L which we refer to as the *induced topology* on L. We also have the surjective A-linear map $\pi:A^s \longrightarrow L$ which sends the basis elements ε_i of A^s into x_i. The product of the filtered topology on A gives a topology on A^s and then the *quotient topology* on L is obtained, using the distance function δ defined by $\delta(x,y)=\pi^{v}$ if $x-y\in L_v\setminus L_{v-1}$, where $L_v=A_vx_1+\dots+A_vx_s$. Notice that if k is an integer such that $x_i\in A_k$ for every i, then L_v is contained in $L\cap A_{v+k}$. It follows that convergence in the quotient topology implies convergence in the induced topology, i.e. the induced topology is weaker than the quotient topology.

2.3. Definition. The filtration on A satisfies *the weak comparison condition* if the quotient topology is equal to the induced topology for every finitely generated left or right ideal.

2.4. Definition. The filtration on A satisfies *the comparison condition* if the following holds : for every finite set $x_1,\dots,x_s$ in A there exists an integer w such that

$$A_v\cap Ax_1+\dots+Ax_s\subset A_{v+w}x_1+\dots+A_{v+w}x_s$$

$$\text{resp. } x_1A+\dots+x_sA\cap A_v\subset x_1A_{v+w}+\dots+x_sA_{v+w}$$

<u>Remark</u>. Notice that the comparison condition implies the weak comparison condition, i.e. the comparison condition imposes a *uniform* homeomorphism with respect to induced and quotient topologies on

finitely generated ideals.

Principal symbols. If $0\neq x \in A$ the unique integer v for which $x \in A_v \setminus A_{v-1}$ is the *order* of x. The image of x in $gr_v(A)=A_v/A_{v-1}$ is called its *principal symbol* and is denoted by $\sigma(x)$. If L is a left ideal of A, then $\sigma(L)$ denotes the graded left ideal of gr(A) generated by principal symbols of elements in L.

At this stage we begin to establish some results which relate the previous conditions. It turns out that the hypothesis we shall need is that G(A) is a noetherian ring, i.e. left and right ideals of G(A) are finitely generated.

2.5. Proposition. *If* G(A) *is Noetherian and the filtration on* A *satisfies the closure condition, then* A *is a Noetherian ring.*

Proof. Let L be a left ideal. We can choose a finite set $x_1,\dots x_s$ such that $\sigma(L)$ is generated by their principal symbols. A trivial recursion shows that

(i) $$L \cap A_v \subset A_{v-k_1}x_1+\dots+A_{v-k_s}x_s+A_{v-w}$$

hold for every v and each $w\geq 1$, where k_i is the order of x_i.

Keeping v fixed we see that as $w>>0$, it follows that $L \cap A_v$ belongs to the closure of the left ideal generated by $x_1,\dots,x_s$. So the closure condition gives $L \cap A_v \subset Ax_1+\dots+Ax_s$. Since v was arbitrary we get $L=Ax_1+\dots+Ax_s$.

2.6. Proposition. *If* G(A) *is Noetherian and the strong closure condition holds, then the closure condition and the comparison condition hold.*

Proof. First we prove the closure condition. Let $L=Ax_1+\dots+Ax_s$ be a finitely generated left ideal. Enlarging the set of generators if necessary we may assume that $\sigma(x_1),\dots\sigma(x_s)$ generate the ideal $\sigma(L)$. Then we obtain

$$L \cap A_v \subset A_{v-k_1}x_1+\dots+A_{v-k_s}x_s+A_{v-w}$$

Keeping v fixed while $w>>0$, it follows from the strong closure condition applied to $A_{v-k_1}x_1+\dots+A_{v-k_s}x_s$, that $L \cap A_v$ is equal to this set. In particular, $L \cap A_v$ is a closed subset of A.

Next, let x belong to the closure of L. We have $x \in A_v$ for some v and for any $w\geq 1$ we can find $x' \in A_{v-w}$ such that x-x' belongs to L. Then

$x-x' \in L \cap A_v$ and since $L \cap A_v$ is closed and w can be made arbitrary large we obtain $x \in L \cap A_v$. This proves the closure condition. There remains to prove the comparison condition. So let $x_1,\dots,x_s$ be a finite set and let L be the left ideal generated by $x_1,\dots,x_s$. Choose $y_1,\dots y_t$ in L so that $\sigma(y_1),\dots,\sigma(y_t)$ generate $\sigma(L)$. Then the closedeness of $L \cap A_v$ and the trivial recursion during the proof of Proposition 2.5 yield :

$$L \cap A_v = A_{v-k_1}y_1+\dots+A_{v-k_t}y_t \text{ for every } v$$

Now we find an integer w so that $y_i \in A_{w+k}x_1+\dots+A_{w+k}x_s$ for every i. This gives $L \cap A_v \subset A_{v+w}x_1+\dots+A_{v+w}x_s$.

2.7. Proposition. *The weak comparison condition and the closure condition imply together the strong closure condition.*

Proof. Let x belong to the closure of $A_{v_1}x_1+\dots+A_{v_s}x_s$. In particular x belongs to the closure of the left ideal generated by $x_1,\dots,x_s$. So by the closure condition we have x in $Ax_1+\dots+Ax_s=L$. Next, if $w\geq1$ we find x' in A_{-w} so that $x-x' \in A_{v_1}x_1\dots+A_{v_s}x_s$. The weak comparison condition shows that if w is sufficiently large, then the fact that $x' \in A_{-w} \cap L$ gives x' in $A_{v_1}x_1+\dots+A_{v_s}x_s$ and then $x \in A_{v_1}x_1+\dots+A_{v_s}x_s$. This proves the strong closure condition.

2.8. Corollary. *Assume that* G(A) *is Noetherian. Then the following are equivalent :*

(1) Strong closure condition holds ;

(2) Closure and comparison conditions hold ;

(3) Closure and weak comparison conditions hold.

2.9. Definition. A filtration is called *Zariskian* if G(A) is Noetherian and (1)-(3) in Corollary 2.8 hold.

Remark. By Corollary 2.8 we see that we can impose the Zariskian condition in several equivalent ways.

2.10. Positive filtrations. If $A_v=0$ when $v<0$ we say that *the filtration is positive*. In this case the filtered topology is discrete and hence every subset of A is closed. In particular the strong closure condition holds. We conclude that if the filtration is discrete and G(A) Noetherian, then the filtration is Zariskian. Let us also observe the following.

2.11. Proposition. *Let* A *be a positively filtered ring. Then, if* L *is a left ideal of* A *such that* $\sigma(L)$ *is a finitely generated ideal in* G(A), *it follows that* L *is finitely generated.*

Proof. Choose $x_1,\dots,x_s$ in L so that $\sigma(x_1),\dots,\sigma(x_s)$ generate $\sigma(L)$. The trivial recursion from the proof of Proposition 2.5 gives $L=Ax_1+\dots+Ax_s$.

The Rees ring. Let A be a filtered ring. Then we construct a graded ring R as follows. The additive group $R=\oplus A_v$. If v is an integer, then j_v denotes the map from the subset A_v of A into the homogeneous component $gr_v(R)=A_v$. The ring structure on R is defined by $j_v(x)j_k(y)=j_{v+k}(xy)$ when $x\in A_v$ and $y\in A_k$ for some pair of integers. The graded ring R is called the *associated Rees ring*.

The identity in the ring A belongs to A_1 and we put $T=j_1(1)$. We notice that T is a *central element* in the graded ring R. Also, T is *homogeneous of degree one* and since $Tj_v(x)=j_{v+1}(x)$ holds if $x\in A_v$, it follows that the T-kernel on R cannot contain a non-zero homogeneous element. From this we deduce that T is a non-zero divisor. We notice that $j_v(R)/Tj_{v-1}(R)=A_v/A_{v-1}=gr_v(A)$. This gives easily :

2.12. Proposition. *Denote by* (T) *the two-sided ideal of* R *generated by* T. *Then the rings* R/(T) *and* gr(A) *are isomorphic.*

Next, every element in R is a finite sum $\Sigma j_v(x_v)$ and then we can choose an integer K so that $x_v\in A_K$ for every v. It follows that $\Sigma T^{K-v}j_v(x_v)=j_K(\Sigma x_v)$. We notice that $\Sigma j_v(x_v)-\Sigma T^{K-v}j_v(x_v)$ belongs to the two-sided ideal of R generated by 1-T. This observation easily gives :

2.13. Proposition. *Let* $\pi:R\longrightarrow A$ *be the map defined by* $\pi(\Sigma j_v(x_v))=\Sigma x_v$. *Then* π *is a surjective ring homomorphism whose kernel is the two-sided ideal of* R *generated by* 1-T. *Hence the rings* R/(1-T) *and* A *are isomorphic.*

Next, let X be an indeterminate and consider the ring $A[X,X^{-1}]$. We notice that there exists an injective ring homomorphism θ from R into $A[X,X^{-1}]$, defined by $\theta(\Sigma j_v(x_v))=\Sigma x_vX^v$. The image $\theta(R)$ is in general only a subring of $A[X,X^{-1}]$. But if we consider the localisation R_T, i.e. given as the so called universal T-inverting ring obtained from the multiplicative subset of R given by powers of T, then θ extends to a ring isomorphism between R_T and $A[X,X^{-1}]$. So we have proved

2.14. Proposition. *The rings* R_T *and* $A[X,X^{-1}]$ *are isomorphic.*

Let us now consider a left ideal L of A. Then we construct a ***graded*** left ideal $\tilde{L}$ in the ring R as follows :

$$\tilde{L} = \oplus j_v(L \cap A_v) \qquad (2.15.)$$

We refer to $\tilde{L}$ as the extended Rees ideal of L. Let us now try to reverse the construction, i.e. let $\mathcal{L}$ be a ***graded*** left ideal of R. The ring homomorphism $\theta:R \longrightarrow A$ gives the left ideal $\theta(\mathcal{L})$ in A.

2.16. Proposition. *The graded Rees-ideal* $\widetilde{\theta(\mathcal{L})}$ *contains* $\mathcal{L}$ *and the* R-*module* $\widetilde{\theta(\mathcal{L})}/\mathcal{L}$ *has* T-*torsion, i.e every element in this* R-*module is annihilated by some power of* T.

Proof. First we notice that $\theta(gr_v(R))=A_v$. Decompose the graded ideal $\mathcal{L}$ into homogeneous components, i.e. $\mathcal{L}=\oplus\mathcal{L}_v$. Since $\mathcal{L}_v \subset gr_v(R)$ we obtain $\theta(\mathcal{L}_v) \subset A_v \cap L$. This means that if $x \in \mathcal{L}_v$, then there exists some $\alpha \in A_v \cap L$ so that $\theta(x-j_v(\alpha))=0$ and hence $x-j_v(\alpha)$ belongs to (1-T). But then $x=j_v(\alpha)$ follows in the ring R since it is obvious that the ideal (1-T) cannot contain a non-zero homogeneous element. This gives $\mathcal{L} \subset \widetilde{\theta(\mathcal{L})}$. We obtain an exact sequence of graded R-modules given by

$$0 \longrightarrow \mathcal{L} \longrightarrow \widetilde{\theta(\mathcal{L})} \longrightarrow \widetilde{\theta(\mathcal{L})}/\mathcal{L} \longrightarrow 0$$

Let us notice that $\theta(\widetilde{\theta(\mathcal{L})})=\theta(\mathcal{L})$ follows from the construction of the map θ. This equality shows that if we put $M=\widetilde{\theta(\mathcal{L})}/\mathcal{L}$, then the R-module $M/(1-T)M=0$. Now we denote the T-torsion part of M by $M_o=\{m:T^v m=0 \text{ for some } v\}$. Since the R-element T is homogeneous, it follows that M_o is a graded submodule of M and we want to show that $M=M_o$. To prove this we put $N=M/M_o$. Since $M/(1-T)M=0$ we obtain $N/(1-T)N=0$. But this cannot happen unless N is zero. Indeed, N is a graded R-module and T is a non-zero divisor of N. Then we observe that a non-zero ***homogeneous*** element of N cannot belong to (1-T)N. Hence N=0 and Proposition 2.16 is proved.

Remark. We refer to $\widetilde{\theta(\mathcal{L})}$ as the ***saturation*** of the graded ideal $\mathcal{L}$. Notice that if we put $L=\theta(\mathcal{L})$, then the A-module L has a filtration defined by $L_v=\theta(\mathcal{L}_v)$. Thus, $\{L_v\}$ is an increasing sequence of additive subgroups such that $\cup L_v=L$ and $A_k L_v \subset L_{k+v}$. We have seen that $L_v \subset L \cap A_v$, i.e. the filtration on $L=\theta(\mathcal{L})$, derived from the graded ideal $\mathcal{L}$ ***increases slower than the induced filtration*** on the left ideal L of the ring A. We shall later see that this gives an interesting interplay between filtrations on left ideals in the ring A and graded left ideals of the Rees ring.

So far Noetherian assumptions have been made. The next result clarifies when R is a Noetherian ring.

2.17. Proposition. *Let* A *be a filtered ring. Then the following are equivalent :*

(1) R *is a Noetherian ring ;*

(2) A *and* G(A) *are Noetherian rings and the filtration satisfies the comparison condition.*

Before we prove Proposition 2.17 we shall need the result below which is due to *J.R. Matijevic* (see [Ma]).

2.18. Theorem. *Let* $A=\oplus A_v$ *be a* $\mathbb{Z}$*-graded ring. Then, if every graded left or right ideal of* A *is finitely generated, it follows that the ring* A *is Noetherian.*

Proof. Let A[t] be the polynomial ring over A. If $x \in A$ we decompose x into homogenous elements and write $x=\Sigma x_v$, $x_v \in A_v$. Let v_o be the largest number such that $x_{v_o} \neq 0$. Then we put

$$x^*=\Sigma t^{v_o-v} x_v$$

and refer to x^* as the ***external homogenization*** of x. Now we take a left ideal L in the ring A. Denote by L^* the ideal of A[t] generated by $\{x^* : x \in L\}$. Next, consider the ***positive filtration*** on A[t] defined by $\Sigma_v=A+\ldots+At^v$. Then we construct the ideal $\sigma(L^*)$ in $gr_\Sigma(A[t])$. Of course, the rings $gr_\Sigma(A[t])$ and A[t] are, the same and $\sigma(L^*)=\oplus(L^* \cap \Sigma_v)/L^* \cap \Sigma_{v-1})$. The construction of external homogenizations easily show that the left ideal $\sigma(L^*)$ in the ring A[t] is of the form $\sigma(L^*)=\oplus J(v)t^v$, where each J(v) is a ***graded*** left ideal of A. Now the hypothesis means that every increasing sequence of graded left ideals of the ring A is stationary. ***By Hilbert's classical argument***, it follows that $\sigma(L^*)$ is a finitely generated left ideal. Then, since the Σ-filtration is positive, it follows from Proposition 2.11 that L^* is finitely generated.

Next, let $\varphi: A[t] \longrightarrow A$ be defined by $\varphi(\Sigma x_v t^v)=\Sigma x_v$. We see that φ is a surjective ring homomorphism and obviously $\varphi(L^*)=L$. We conclude that the left ideal L is a finitely generated A-module and hence A is a left Noetherian ring. The same proof work of course with left replaced by right and Theorem 2.18 follows.

Proof of Proposition 2.17. Assume first that R is Noetherian. Since gr(A)=R/(T) and A=R/(1-T), it follows that A and gr(A) are Noetherian. To prove (1)⇒(2) there remains to verify the comparison condition. So let L be a left ideal in A. Then $\tilde{L}$ is a left ideal in R and hence finitely generated. Thus we can find a finite set $x_1,\ldots,x_s$ in L and integers $v_1,\ldots v_s$ so that $\tilde{L}$ is generated by $j_{v_1}(x_1),\ldots,j_{v_s}(x_s)$. The construction of $\tilde{L}$ gives

$$A_v \cap L = A_{v-v_1}x_1+\ldots+A_{v-v_s}x_s \text{ for every } v.$$

From this we see that the induced topology on the left ideal L is uniformly homeomorphic with its quotient topology and hence the comparison condition holds.

Now we prove (2)⇒(1). By Theorem 2.18 it suffices to prove that every graded ideal $\mathcal{L}$ of R is finitely generated. To prove this we put $L=\theta(\mathcal{L})$. We choose a finite set $x_1,\ldots,x_s$ in L which generate L as a left A-module and in addition their principal symbols generate $\sigma(L)$.

We obtain $A_v \cap L \subset A_{v-v_1}x_1+\ldots+A_{v-v_s}x_s+A_{v-w} \cap L$. Here w is any positive integer. Keeping v fixed we use the comparison condition and find w so that $A_{v-w} \cap L \subset A_{v-v_1}x_1+\ldots+A_{v-v_s}x_s$. We conclude that $A_v \cap L=\Sigma A_{v-v_i}x_i$ holds for every v. Then the Rees ideal $\tilde{L}$ is generated by $j_{v_1}(x_1),\ldots,j_{v_s}(x_s)$.

Next, Proposition 2.16 gives $\mathcal{L} \subset \tilde{L}$ and the R-module $\tilde{L}/\mathcal{L}$ has T-torsion. Since $\tilde{L}$ is a finitely generated R-module, it follows that there exists a positive integer w such that T^w annihilates $\tilde{L}/\mathcal{L}$. This means that $T^w\tilde{L} \subset \mathcal{L} \subset \tilde{L}$. Now we notice that the hypothesis that gr(A) is Noetherian implies that $R/(T)^w$ is a Noetherian ring and then $\tilde{L}/T^w\tilde{L}$ is a Noetherian $R/(T)^w$-module. It follows that the R-module $\mathcal{L}/T^w\tilde{L}$ is finitely generated. Since the R-module $T^w\tilde{L}$ also is finitely generated we conclude that $\mathcal{L}$ is a finitely generated R-module. This finishes the proof of Proposition 2.17.

Gooo filtrations. From now on we assume that A is a filtered ring whose associated Rees ring R is Noetherian. In particular both A and G(A) are Noetherian. A filtration on a left A-module M consists of an increasing sequence of additive subgroups $\{M_v\}$ satisfying $\cup M_v=M$ and $A_kM_v \subset M_{v+k}$. If $\cap M_v=0$ we say that $\{M_v\}$ is a ***separated*** filtration. If $\{M_v\}$ is a filtration on some left A-module, then we construct the ***Rees module*** $\tilde{M}=\oplus M_v$. This is a graded R-module with $gr_v(\tilde{M})=M_v$. The map from the subset M_v of M into $gr_v(M)$ is denoted by j_v.

2.19. Definition. A filtration $\{M_v\}$ on a left A-module is called a *good filtration* if the Rees module $\oplus M_v$ is a finitely generated R-module.

2.20. Proposition. *If a left* A*-module* M *has a good filtration* $\{M_v\}$, *then the* A*-module* M *is finitely generated and there exists a finite set* $x_1,\ldots,x_s$ *in* M *and integers* $v_1,\ldots,v_s$ *such that :*

$$M_v = A_{v-v_1}x_1 + \ldots + A_{v-v_s}x_s \quad \text{for every } v.$$

Proof. We can find a finite set $\{x_i \in M_{v_i}\}$ such that $\{j_{v_i}(x_i)\}$ generate the Rees module. Then Proposition 2.20 follows.

Obviously every finitely generated left (or right) A-module has a good filtration. If $\{M_v\}$ is a good filtration on some finitely generated A-module M and $N \subset M$ is a submodule, then the *induced* filtration $\{N \cap M_v\}$ is good. To see this we observe that the Rees module $\oplus N \cap M_v$ is a submodule of $\oplus M_v$ and therefore finitely generated. Thus, good filtrations induce good filtrations on submodules. Of course, they also induce good filtrations on quotient modules.

2.21. Proposition. *Let* $\{M_v\}$ *and* $\{M'_v\}$ *be a pair of good filtrations on some finitely generated* A*-module* M. *Then there exists an integer* w *such that* $M_{v-w} \subset M'_v \subset M_{v+w}$ *hold for every* v.

Proof. Follows easily from the description of good filtrations in Proposition 2.20.

2.22. Proposition. *Let* $\{M_v\}$ *be a filtration on a finitely generated* A*-module* M. *Then, if there exists a good filtration* $\{\Gamma_v\}$ *on* M *such that* $M_v \subset \Gamma_v$ *for every* v, *it follows that* $\{M_v\}$ *is good.*

Proof. The Rees module $\oplus M_v$ is a submodule of the noetherian R-module $\oplus \Gamma_v$. Hence $\oplus M_v$ is a finitely generated R-module so $\{M_v\}$ is good.

Remark. In the case when A is a *positively* filtered ring and G(A) is Noetherian, then we know that the filtration is Zariskian and the Rees ring is Noetherian. If M is a finitely generated A-module it is easy to show that a filtration $\{M_v\}$ is good if and only if $\oplus M_v/M_{v-1}$ is a finitely generated G(A)-module.

However, for non-positive filtrations it is unclear if we can conclude that a filtration $\{M_v\}$ is good from the sole hypothesis that

$\oplus M_v/M_{v-1}$ is a finitely generated G(A)-module. In fact, we ask the following :

Question. Let A be a ring with a Zariskian filtration. Let M be a finitely generated A-module and $\{M_v\}$ is a *separated* filtration such that $\oplus M_v/M_{v-1}$ is a *finitely generated* G(A)-module. *Does it follow that* $\{M_v\}$ *is good ?* I believe there exists a counterexample but is unable to give one. To find an example might be cumbersome. If we *assume* that the topology on M defined by the filtration $\{M_v\}$ is the same as the topology on M defined by a good filtration, then it is easy to show that $\{M_v\}$ is good. So an eventual counter example requires the construction of a rather ackward filtration.

A special case concerning the question above occurs if A is the stalk $\mathcal{E}_{X,p}$ of the sheaf $\mathcal{E}_X$ of micro-differential operators and p is outside the zero-section of the cotangent bundle. For example, take $X=\mathbb{C}^{n+1}$ with coordinates $x_1,\dots,x_n$, t and p=(0,0,0,dt). Put $A=\mathcal{E}_{X,p}$ and A is filtered by $A_m=\mathcal{E}_{X,p}(m)$, where $\mathcal{E}_X(m)$ is the sheaf of micro-differential operators of order m at most. Then D_t is an invertible element in the ring A which has order 1. It follows that $A_m=D_t^m A_o=A_o D_t^m$ for every m. Also, if $\{M_v\}$ is an exhaustive filtration on a left A-module M, it follows that $M_v=D_t^v M_o$ for every v and here M_o is an A_o-submodule of M such that $M=AM_o$. Then we see that the associated graded module $\oplus M_v/M_{v-1}$ is a finitely generated G(A)-module if and only if the A_o/A_{-1}-module $M_o/D_t^{-1}M_o$ is finitely generated. Also, the filtration is separated if and only if $\cap D_t^{-v}M_o=0$. So we aks the following :

Question. Let $A=\mathcal{E}_{X,p}$ and M is a finitely generated A-module. Does there exist an A_o-submodule M_o of M such that M_o is not finitely generated as a left A_o-module while $\cap D_t^{-v}M_o=0$ and $M_o/D_t^{-1}M_o$ is a finitely generated $A_o/D_t^{-1}A_o$-module ?

Final Remarks. The study of the filtered sheaf of micro-differential operators, where the stalks have non-positive filtrations was originally the reason for studying properties of filtrations as exposed in this section. The proof that the stalks of $\mathcal{E}_X$ are Noetherian rings with Zariskian filtrations rely on a quite involved analysis and makes use of special analytic properties of $\mathcal{E}_X$. The original proof was carried out by M. Kashiwara in his Master's Thesis from Tokyo in 1971. A detailed presentation of Kashiwara's proof is given in [Bj.1 : Theorem in Chapter 4] where we remark that the so called *strong closure condition* is established. At that time the present terminology was not yet made, i.e. there was no specific reference to *Zariskian filtrations*. Of course, we have seen that various conditions coincide, i.e. Corollary 2.8 and Proposition 2.17 show that a filtration satisfies the strong closure condition if and only if the Rees ring is Noetherian and left or right ideals are closed.

The use of Rees rings occured in Kashiwara's lectures at Paris-Nord in 1975 [Ka:2]. See also [Sch]. The more systematic use of Rees rings has been carried out by F. Oystaeyen and we refer to [AO] for a study of filtered rings which has inspired the presentation in this section.

3. The Auslander condition on filtered rings

Let A be a filtered noetherian ring. We assume that the Rees ring is noetherian. Hence the so called *comparison condition* holds and we recall that every finitely generated A-module M can be equipped with a good filtration. Let M be a left A-module and choose a good filtration Γ on M. Then we construct a *filt-free* resolution $F_\bullet$ of M. To be precise, we have $\longrightarrow F_1 \longrightarrow F_0 \longrightarrow M \longrightarrow 0$, where each F_ν is a *filtered free* A-module and the differentials are filter preserving and strict in the sense that $d(\Gamma_\nu(F_k))=d(F_k)\cap\Gamma_\nu(F_{k-1})$ for every ν and $k\geq1$. Also $d(\Gamma_\nu(F_0))=d(F_0)\cap\Gamma_\nu(M)$.

Remark. Recall that a filtered free A-module is a finite direct sum of A-modules of the form A[w], where $A[w]_\nu=A_{\nu+w}$ defines the filtration on the free A-module A[w].

The construction of filtered free resolutions is standard. Moreover, if $F_\bullet$ and $G_\bullet$ is a pair of filt-free resolutions of the filtered R-module M, then $F_\bullet$ and $G_\bullet$ are *homotopic* as complexes of A-modules where the homotopy maps from F_ν into $G_{\nu+1}$ are filter preserving.

If $F_\bullet$ is a filt-free resolution of M we consider the complex $\mathrm{Hom}_A(F_\bullet,A)$ which becomes a filtered complex of right A-modules. To be precise, $\Gamma_\nu(\mathrm{Hom}_A(F_k,A))=\{\varphi\in\mathrm{Hom}_A(F_k,A):\varphi(\Gamma_j(F_k))\subset A_{j+\nu}\}$.

As usual we construct the spectral sequence of the filtered

complex $Hom_A(F_\bullet,A)$. The associated graded complex is given by $Hom_{G(A)}(gr(F_\bullet),G(A))$. We notice that $G(F_\bullet)$ is a free resolution of the $G(A)$-module $gr_\Gamma(M)$.

So $Ext^k_{G(A)}(gr_\Gamma(M),G(A))$ appear as cohomology groups of the graded complex. Next, the spectral sequence converges (or is regular). This follows because the right A-modules $B_k=d(Hom_A(F_k,A))$ are finitely generated and the good filtration on B_k induced from $Hom_A(F_{k-1},A))$ is *comparable* with the good filtration on B_k induced from the filtration on $Hom_A(F_k,A)$ under the surjective and A-linear map $Hom_A(F_k,A) \longrightarrow B_k$. To be precise, we apply Proposition 2.21 to this pair of good filtrations. This comparison between the induced, respectively the quotient filtration on every B_k yields the convergence of the spectral sequence in each given degree. In particular we can conclude :

3.1. Proposition. *Let* Γ *be a good filtration on some finitely generated A-module* M. *Then if* $F_\bullet$ *is a filtered free resolution we obtain induced good filtrations on the cohomology groups* $\{Ext^k_A(M,A)\}$ *of the filtered complex* $Hom_A(F_\bullet,A)$. *Moreover,* $gr(Ext^k_A(M,A))$ *is a subquotient of the right* $G(A)$*-module* $Ext^k_{G(A)}(gr_\Gamma(M),G(A))$ *for every* k.

Remark. By the homotopy relation between a pair of filtered free resolutions of the given filtered A-module M, it follows that the good filtration on every right A-module $Ext^k_A(M,A)$ is intrinsic, i.e. independent of the particular filtered free resolution. But of course, the good filtration on $Ext^k_A(M,A)$ depends on the good filtration on M. So we may refer to the *canonical* Γ*-filtration* on $Ext^k_A(M,A)$ obtained by a given good filtration Γ on M.

The case when A has a positive filtration. If the filtration on A is positive we can draw more conclusions. Namely, let N be another left A-module with some good filtration Ω. Choose a filtered free resolution $F_\bullet$ of the Γ-filtered A-module M as before. Now $Hom_A(F_\bullet,N)$ is a complex of filtered abelian groups. Since the filtration on A is positive we notice that if k is an integer, then there exists some v_0 such that $v<v_0$ gives $\Gamma_v(Hom_A(F_\bullet,N))=0$. Now we recall the elementary result which says that if $G^\bullet$ is a complex of filtered abelian groups such that the filtration $\Gamma_v(G^k)$ is discrete in the sense that $\Gamma_v(G^k)=0$ if $v<<0$ for every k, then if m is an integer and the cohomology group of degree m of the graded complex is zero, it follows that $H^m(G^\bullet)=0$. Using this we conclude :

3.2. Proposition. *Let* A *be a Noetherian ring with a positive filtration. Let* (M,Γ) *and* (N,Ω) *be a pair of finitely generated A-modules with good filtrations. Then, if* k *is some integer such that* $\mathrm{Ext}^k_{G(A)}(gr_\Gamma(M),gr_\Omega(N))=0$, *it follows that* $\mathrm{Ext}^k_A(M,N)=0$.

Remark. If we only assume that the filtration on A is *Zariskian* then we ask if Proposition 3.2 still holds. I do not know of any counter example when A has a non-positive but Zariskian filtration. In [Bj.2 : Corollary 3.12] it was claimed that Proposition 3.2 holds for Zariskian filtrations. But the proof is incomplete. This was pointed out to me by *Li Huishi* and has led to the open problem above.

The case when G(A) is Auslander-Gorenstein. We assume here that A is a filtered ring whose associated Rees ring is Noetherian and G(A) is Auslander-Gorenstein. Since we do not assume that the filtration is Zariskian, it may occur that $gr_\Gamma(M)=0$ when Γ is a good filtration on some finitely generated A-module M. Notice that $gr_\Gamma(M)=0$ means that $\Gamma_v=M$ for every v. So by Proposition 2.21, the vanishing of $gr_\Gamma(M)$ is intrinsic, i.e. independent of the good filtration on M. If $\mathcal{C}$ is the family of finitely generated left A-modules M for which $gr_\Gamma(M)=0$ we notice that $\mathcal{C}$ is a subcategory of $\mathrm{Mod}_f(A)$. In the sequel we are only interested in modules outside the class $\mathcal{C}$. Let us first show :

3.3. Proposition. *Let* M *be a finitely generated A-module outside the class* $\mathcal{C}$. *Then* $j_{G(A)}(gr_\Gamma(M)=j_{G(A)}(gr_\Omega(M))$ *for each pair of good filtrations* Γ *and* Ω.

Proof. Set $\Gamma[w]_v=\Gamma_{v+w}$. Then $\Gamma[w]$, is a good filtration for each w and Proposition 2.21 gives an integer w such that

$$\Gamma[-w]_v \subset \Omega_v \subset \Gamma[w]_v \text{ for every } v.$$

If $-w\leq i\leq w$ we put $\Sigma_v(i)=\Gamma_v[i]\cap\Omega_v$. Then $\Sigma_\bullet(i)$ is a good filtration for every i. In fact this follows from Proposition 2.22. Notice that Proposition 3.3 follows if $j_{G(A)}(gr_{\Sigma(i)}(M))$ is independent of i since $\Sigma(w)=\Omega$ and $\Sigma(-w)=\Gamma[-w]$.

Therefore there only remains to show that :

(1) $$j_{G(A)}(gr_{\Sigma(i)}(M))=j_{G(A)}(gr_{\Sigma(i+1)}(M)) \text{ for every } i.$$

To prove (1) we notice that $\Sigma_v(i)\subset\Sigma_v(i+1)\subset\Sigma_{v+1}(i)$ holds for every v. Then, if $k=\oplus\Sigma_v(i)/\Sigma_{v-1}(i+1)$ and $S=\oplus\Sigma_{v-1}(i+1)/\Sigma_v(i)$, i.e. here $gr_v(S)=\Sigma_{v-1}(i)/\Sigma_v(i)$, then S and K are graded G(A)-modules and we have two exact sequences :

$$0 \longrightarrow S \longrightarrow gr_{\Sigma(i)}(M) \longrightarrow K \text{ and } 0 \longrightarrow K \longrightarrow gr_{\Sigma(i+1)}(M) \longrightarrow S[1] \longrightarrow 0$$

Then Proposition 1.8 gives $j_{G(A)}(gr_{\Sigma(i)}(M))=j_{G(A)}(gr_{\Sigma(i+1)}(M))$.

Remark. The method for proving Proposition 3.3 was used by Kashiwara in [Ka] to clarify the existence of so called *characteristic cycles* of coherent $\mathcal{E}$-or $\mathcal{D}$-modules. So the proof above is important in the special case when G(A) is commutative. We return to this in §4.

The next result which originally was proved in [Bj.2 : Theorem 4.4] is essential for the further study which relates grade numbers between filtered A-modules and their assocated graded G(A)-modules.

3.4. Proposition. *Let* M *be a finitely generated* A-*module outside the class* $\mathcal{C}$. *Let* Γ *be a good filtration on* M *and put* $k=j_{G(A)}(gr_\Gamma(M))$. *Then, if we take the canonical good filtration on* $Ext^k_A(M,A)$, *it follows that* $j_{G(A)}(gr(Ext^k_A(M,A))=k$.

Proof. Choose a filtered free resolution $F_\bullet$ of the Γ-filtered A-module M. Then, if $E^\bullet_r$ denotes the r:th term of the spectral sequence associated with the filtered complex $Hom_A(F_\bullet,A)$, we have $E^\nu_1=Ext^\nu_{G(A)}(gr_\Gamma(M),G(A))$. So $\nu<k$ gives $E^\nu_1=0$. So if $r\geq 1$, then no coboundaries occur in degree k of the complex $E^\bullet_r$. Also, for every $r\geq 1$, the spectral sequence gives an exact sequence of G(A)-modules :

$$0 \longrightarrow E^k_{r+1} \longrightarrow E^k_r \longrightarrow Q_r(k) \longrightarrow 0$$

Here $Q_r(k)$ is a subquotient of $E^{k+1}_1=Ext^{k+1}_{G(A)}(gr_\Gamma(M),G(A))$. Hence the Auslander condition gives $j_{G(A)}(Q_r(k))\geq k+1$ for every r. Then, by Proposition 1.8 and an induction over r, we get $j_{G(A)}(E^k_r))=k$ for every r. If r is sufficiently large we have $E^k_r=gr(Ext^k_A(M,A))$ and Proposition 3.4 follows.

The case when both A and G(A) are Auslander-Gorenstein. Let us assume this and also that the filtration on A satisfies the comparison condition. Then it follows that the Rees ring is noetherian by Proposition 2.17.

If M is a finitely generated A-module and Γ a good filtration we notice that Proposition 3.4 gives

(3.5) $$j_A(M)\leq j_{G(A)}(gr_\Gamma(M)$$

In general the inequality is strict. For example, we may imagine the case when $M=M'\oplus M''$, where M'' belongs to the class $\mathscr{C}$ and $j_A(M')>j_A(M'')$. But if we avoid such cases we do have the following :

3.6. Theorem. *Let* M *be a pure* A-*module outside the class* $\mathscr{C}$. *Then*

$$j_A(M)=j_{G(A)}(gr_\Gamma(M))$$

Proof. Put $k=j_{G(A)}(gr_\Gamma(M))$. If $k>j_A(M)$, then the purity of the A-module M gives $j_A(Ext_A^k(M,A))>k$. So then (3.5) shows that if we use the canonical good filtration on this Ext-module, then we have $j_{G(A)}(gr(Ext_A^k(M,A)>k$. This contradicts Proposition 3.5.

Remark. The inequality in (3.5) can be used to prove holonomicity of certain modules over rings, or sheaves of differential operators. For example, let $A=A_n(\mathbb{C})$ be the *Weyl algebra*. Consider the so called Fuchsian filtration along the hyperplane $x_n=0$. This means that $F_\cdot$ is a filtration on A such that $x_n\in F_{-1}, \partial_n\in F_1$ while x_ν and ∂_ν belong to F_0 if $\nu\neq n$. Then $gr_F(A_n)=A_n$ and $F_\cdot$ satisfies the comparison condition. If we take a *holonomic* A_n-module M,i.e. M is a finitely generated A_n-module such that $j_A(M)=n$, where we recall that n is the global dimension of A_n, then we can show that if Γ is a good filtration on M and $gr_\Gamma(M)$ is non-zero, it follows that $gr_\Gamma(M)$ is holonomic. In fact, (3.5) gives the inequality $j_{A_n}(gr_\Gamma(M))\geq j_{A_n}(M)=n$ so the A_n-module $gr_\Gamma(M)$ is holonomic. This result is then used to establish ***existence of certain*** b-*functions*. We remark that the argument also work for holonomic sheaves of $\mathscr{E}$-modules, using Fuchsian filtrations on with respect to non-singular hyperplanes, or more generally submanifolds. We refer to [La-Sch] for further details about this case and remark only that (3.5) offers algebraic proofs of some results in [La-Sch].

Let us again consider a pure A-module M outside the class $\mathscr{C}$. So then $j_A(M)=j_{G(A)}(gr_\Gamma(M))$, where Γ is some good filtration on M. With $k=j_A(M)$ we put $N=Ext_A^k(M,A)$. We know that $j_A(N)=k$ and notice that if N belongs to the class $\mathscr{C}$, then Proposition 3.1 implies that $Ext_A^k(N,A)\in\mathscr{C}$. But this cannot occur since the pure A-module M is a submodule of $Ext_A^k(N,A)$. So then $k=j_{G(A)}(gr(N))$ by Theorem 3.6.

Now the proof of Proposition 3.4 shows that $Ext_A^k(N,A)$ has a good filtration such that $gr(Ext_A^k(N,A))\subset Ext_{G(A)}^k(gr(N),G(A))$.

Also, $M\subset Ext_A^k(N,A)$ so the good filtration on $Ext_A^k(N,A)$ induces a good filtration on M and then we have

(3.7) $$gr(M) \subset Ext^k_{G(A)}(gr(N), G(A))$$

where gr(N) is a graded G(A)-module whose grade number is k.

Now, if G(A) is Auslander regular or if the ring G(A) is commutative, we know from the discussion after Proposition 1.11 that $Ext^k_{G(A)}(gr(N),G(A))$ is a *pure* G(A)-module. So we have proved :

3.8. Theorem. *Let* A *be a filtered ring such that the comparison condition holds. If* A *is Auslander-Gorenstein and* G(A) *is Auslander regular or a commutative Auslander-Gorenstein ring, it follows that if* M *is a pure* A*-module outside the class* $\mathcal{C}$, *then there exists a good filtration* Γ *on* M *such that* $gr_\Gamma(M)$ *is a pure* G(A)*-module and* $j_A(M)=j_{G(A)}(gr_\Gamma(M))$.

The case when A is Zariskian. Let A be a noetherian ring with a Zariskian filtration. Then good filtrations on finitely generated A-modules are separated. So the subquotient relations in Proposition 3.1 show that if Γ is a good filtration on some A-module M and R some integer such that $Ext^k_{G(A)}(gr_\Gamma(M),G(A))=0$, then $Ext^k_A(M,A)=0$. Now we can easily prove :

3.9. Theorem. *Assume that the filtration is Zariskian and that* G(A) *is Auslander-Gorenstein. Then* A *is Auslander-Gorenstein.*

Proof. Put μ=inj.dim(G(A)). Proposition 3.1 implies that $Ext^\nu_A(M,A)=0$ for every $\nu>\mu$ and every finitely generated A-module M. So inj.dim(A)$\leq\mu$. Next, let Γ be a good filtration on a finitely generated A-module M. Then (3.5) combined with the observations above Theorem 3.9 yield

(1) $$j_A(M)=j_{G(A)}(gr_\Gamma(M))$$

Now we verify the Auslander condition. So let $N \subset Ext^\nu_A(M,A)$. By Proposition 3.1 $gr(Ext^\nu_A(M,A))$ is a subquotient of $Ext^\nu_{G(A)}(gr_\Gamma(M),G(A))$ and using the induced good filtration on the submodule N, it follows that gr(N) is a subquotient of $Ext^\nu_{G(A)}(gr_\Gamma(M),G(A))$. Then we have $j_{G(A)}(gr(N))\geq\nu$ and (1) gives $j_A(N)\geq\nu$.

Remark. If G(A) is Auslander regular, then the same proof shows that A is Auslander regular since it is a wellknown fact that gl.dim(A)≤gl.dim(G(A)) holds when A has a Zariskian filtration. In the case when G(A) is Auslander regular and A has a *positive* filtration, Theorem 3.9 was established in Chapter 2 of [Bj.1]. We also mention that Theorem 3.9 was proved for certain classes of rings of differential operators, in particular the Auslander regularity was proved for stalks of the sheaf $\mathcal{E}_X$ of micro-differential operators in Kashiwara's work [Ka:2].

In the Zariskian case, keeping the hypothesis that G(A) is Auslander regular, the Auslander regularity of A was proved in [Bj.2 : Theorem 4.1]. The case of Auslander-Gorenstein rings was settled by T. Levasseur in [Lev:2].

4. The case when G(A) is commutative

In this section we consider a filtered noetherian ring A such that G(A) is commutative. We assume that the filtration satisfies the comparison condition. So by Proposition 2.17, it follows that the Rees ring is noetherian and then we can consider good filtrations on finitely generated A-modules. Since G(A) is commutative we can construct certain invariants of finitely generated A-modules. Let us briefly describe this. First we recall some constructions in commutative algebra.

4.1. Some commutative algebra. Let R be a commutative noetherian ring. If N is a finitely generated R-module, then its *characteristic ideal* is the radical of the annihilating ideal (0:N). It is denoted by J(N). Let $\mathbb{P}_1,\ldots,\mathbb{P}_s$ be the *minimal prime divisors* of J(N). So we have $J(N)=\mathbb{P}_1\cap\ldots\cap\mathbb{P}_s$. Then, the localisation $N_{\mathbb{P}_i}$ is a finitely generated module over the local ring $R_{\mathbb{P}_i}$ which is annihilated by some power of the maximal ideal $\mathbb{P}_i R_{\mathbb{P}_i}$ and hence the $R_{\mathbb{P}_i}$-module $N_{\mathbb{P}_i}$ has a finite length. It is denoted by $e_{\mathbb{P}_i}(N)$ and called the *multiplicity* of N along the minimal prime divisor $\mathbb{P}_i$ of J(N).

4.2. Definition. The *characteristic cycle* of a finitely generated R-module N is given by $\Sigma e_{\mathbb{P}_i}(N)\cdot\mathbb{P}_i$; $J(N)=\mathbb{P}_1\cap\ldots\cap\mathbb{P}_s$. The characteristic cycle is denoted by Ch(N).

After this digression we return to the filtered ring A. Let M be a finitely generated A-module. If Γ and Ω is a pair of good filtrations on M, then the existence of an integer w such that $\Gamma_{v-w} \subset \Omega_v \subset \Gamma_{v+w}$ easily gives that $J(gr_\Gamma(M))=J(gr_\Omega(M))$. So we define the *characteristic ideal of the A-module* M to be $J(gr_\Gamma(M))$ where Γ is any good filtration on M. We denote the characteristic ideal of the A-module M by J(M).

Moreover, using the proof of Proposition 3.3 it can also be shown that $Ch(gr_\Gamma(M))=Ch(gr_\Omega(M))$ for every pair of good filtrations.

We conclude that we can define the *characteristic cycle* of any finitely generated A-module M. By definition we have :

(4.3) $Ch(M)=Ch(gr_\Gamma(M))$: Γ is a good filtration on M.

Now we are going to establish an important result which has been used to compute characteristic cycles of various modules over rings of differential operators or enveloping algebras. Theorem 4.6 below is essentially due to *O. Gabber* and *M. Kashiwara* who treated special, but important cases. The more general content was pointed out in the article by *Ginsburg* [Gi]. The subsequent proof is also inspired by Ginsburg's article.

Before we announce Theorem 4.6 we need some notations. If $\mathbb{P}$ is a prime ideal of the ring G(A) and a an element of G(A), then $(\mathbb{P},a)$ denotes the ideal generated by $\mathbb{P}$ and a. If ϑ is a prime ideal in G(A) which contains $(\mathbb{P},a)$, where a does not belong to $\mathbb{P}$ and if there does not exist any prime ϑ' strictly in between $\mathbb{P}$ and ϑ, then the R_ϑ-module $(\mathbb{P},a)R_\vartheta$ is zero-dimensional. Hence we can define the *multiplicity* $e_\vartheta(R/(\mathbb{P},a))$ which by definition is the length of the $R_\vartheta/(\mathbb{P},a)_\vartheta$.

Next, return to the ring A. If L is a left ideal and M=A/L, then every A-linear map φ:M $\longrightarrow$ M is induced by the right multiplication in the ring A by some element a satisfying $La \subset L$. We see that $M/\varphi(M)$ is equal to A/(L+Ra).

Recall that if L is a left ideal of A, then $\sigma(L)$ denotes the ideal of G(A) generated by principal symbols of elements in L. We notice that $J(A/L)=\sqrt{\sigma(L)}$ for every left ideal L. In particular we have $J(M/\varphi(M))=\sqrt{\sigma(L+Aa)}$. Since it is obvious that $\sigma(L) \subset \sigma(L+Aa)$ and $\sigma(a) \in \sigma(L+Aa)$, it follows that

(4.4) $$\sqrt{(\sigma(L),\sigma(a)} \subset \sqrt{\sigma(L+Aa)} = J(M/\varphi(M))$$

In general the inclusion in (4.4) is strict. We are going to show that equality holds under certain assuptions. First we shall impose the following condition on the filtered ring A :

4.5. Condition. Both A and G(A) are Auslander-Gorenstein and the filtration satifies the comparison condition.

Fron now on we assume that Condition 4.5 holds. Then we have

4.6. Theorem. *Let* M=A/L *be a cyclic* A-*module and* φ:M ⟶ M *an* A-*linear map defined by the right multiplication of some* a *in* A. *Then, if the* A-*module* M *is pure,* φ *is injective but not surjective and* σ(a) *does not belong to any minimal prime divisor of* J(M), *it follows that the equality in* (4.4) *holds.*

The proof of Theorem 4.6 requires several steps. Actually the subsequent proof enables us to express the characteristic cycle of M/φ(M) with the aid of Ch(M) and the G(A)-element σ(a). This results, which of course is a refinement of Theorem 4.6 is announced in Theorem 4.7 below.

A multiplicity formula. Let M, φ and a be as in Theorem 4.6. Let $\vartheta_1,\dots,\vartheta_t$ be the minimal prime divisors of $\sqrt{(\sigma(L),\sigma(a))}$ and $\mathbb{P}_1,\dots,\mathbb{P}_s$ are the minimal prime divisors of $J(M)=\sqrt{\sigma(L)}$.

Then $J(M/\varphi(M))=\vartheta_1 \cap\dots\cap \vartheta_t$ by Theorem 4.6 and the characteristic cycle of M/φ(M) is given as follows :

4.7. Theorem. $Ch(M/\varphi(M))= \sum_{i=1}^{t} e_{\vartheta_i}(M/\vartheta(M))\cdot\vartheta_i$, *where*

$$e_{\vartheta_i}(M/\varphi(M))= \sum_{v=1}^{v=s} e_{\vartheta_i}(G(A)/(\mathbb{P}_v,\sigma(a))\cdot e_{\mathbb{P}_v}(M)$$

holds for each $1\geq i\geq t$.

The proof of Theorem 4.7 is based upon some wellknown constructions in commutative algebra which we begin to explain. Denote by $\mathcal{K}$ the additive *Grothendieck group* generated by finitely generated G(A)-modules K such that the characteristic ideal J(K) *contains* J(M). Let K be some finitely generated G(A)-module with $J(K) \supset J(M)$.

Then, if ϑ is one of the minimal prime divisors of $\sqrt{(\sigma(L),\sigma(a))}$, it follows that the $G(A)_\vartheta$-modules $G(A)_\vartheta \otimes_{G(A)} (0:\sigma(a))_K$ and $G(A)_\vartheta \otimes_{G(A)} K/\sigma(a)K$ both are zero-dimensional, where we have put $(0:\sigma(a))_K=\{x \in K:\sigma(a)x=0\}$. This leads to :

4.8. Definition. The integer $e_\vartheta(K/\sigma(a)K)-e_\vartheta((0:\sigma(a))_K)$ is denoted by $I_\vartheta(K)$ and called the *ϑ-index of* K.

Now $k \longrightarrow I_\vartheta(K)$ extends to an additive function on the Grothendieck group $\mathcal{K}$, i.e. this follows by "counting multiplicities" and the flatness of the G(A)-module $G(A)_\vartheta$.

In particular we take K=gr(M), where the A-module M=A/L has the good filtration defined by $M_v=(A_v+L)/L$. So $K=G(A)/\sigma(L)$ and then we have :

4.9. Lemma. $I_\vartheta(grM)=\Sigma e_{\mathbb{P}_i}(M)e\vartheta(G(A)/(\mathbb{P}_i,\sigma(a))$ *with* Σ *extented over the minimal prime divisors* $\{\mathbb{P}_i\}$ *of* J(M).

Proof. The G(A)-module gr(M) has a bounded *prime chain*, i.e. we have $gr(M)=K_0 \supset K_1 \supset\ldots\supset K_w \supset 0$, where $K_v/K_{v+1}=G(A)/P_v$ for some prime ideal P_v in G(A). In this prime chain each $\mathbb{P}_i$ is repeated $e_{\mathbb{P}_i}(M)$ times. It is easily seen that the hypothesis that $\sigma(a)$ does not belong to any $\mathbb{P}_i$, and the observation that if a prime P_v enters via $K_v/K_{v+1}=G(A)/P_v$, then P_v must contain some $\mathbb{P}_i$, implies that we either have $P_v=\mathbb{P}_i$ for some i, or else $I_\vartheta(G(A)/P_v)=0$. Then Lemma 4.9 follows by the additivity of the ϑ-index.

Concerning the sum which appears in Lemma 4.9 we notice that if i is given, then there exists at least some $\mathbb{P}_i$ such that $P_i \subset \vartheta$. It follows that $e_\vartheta(G(A)/(\mathbb{P}_i,\sigma(a))>0$ and hence Lemma 4.9 gives :

4.10. Lemma. $I_\vartheta(gr(M))>0$ *for every minimal prime divisor* ϑ *of* $\sqrt{(\sigma(L),\sigma(a))}$.

Now we return to the A-module M. Let k be the order of a, i.e. the principal symbol $\sigma(a)$ is homogenous of order k. By M[k] we denote the same A-module as M, but with the filtration defined by $M[k]_v=M_{v+k}$. Then ϑ:M $\longrightarrow$ M[k] is a filter preserving map. On $M/\varphi(M)$ we can use the good filtration induced by the surjective A-linear map from M[k]. Let us now consider the complex :

(4.11) $$0 \longrightarrow gr(M) \xrightarrow{\sigma(a)} gr(M[k]) \longrightarrow 0$$

We notice that (4.11) is the associated graded complex of the filtered complex of A-modules, defined by $0 \longrightarrow M \xrightarrow{\varphi} M[k] \longrightarrow 0$. Consider the spectral sequence associated to this filtered complex. Then we obtain a sequence of complexes $E_r^{\bullet}$ of G(A)-modules, where $E_0^{\bullet}$ is the complex in (4.11) and $E_{r+1}^{v}=H^{v}(E_r^{\bullet})$ if $r\geq 0$. Of course, here only $H^0(E_r^{\bullet})$ and $H^1(E_r^{\bullet})$ can be non-zero. Since good filtrations on finitely generated A-modules are comparable, it follows that the spectral sequence converges and there exists some integer r such that $E_r^0=0$ while $E_r^1=gr(M/\varphi(M))$. Of course, here $E_r^0=0$ follows from the hypothesis that the A-linear map φ is injective.

Using this spectral sequence and the *additivity of the* θ*-index* for every minimal prime divisor θ of $\sqrt{(\sigma(L),\sigma(a))}$ we obtain :

4.12. Lemma. $e_{\theta}(M/\varphi(M))=I_{\theta}(gr(M))$ *holds for every minimal prime divisor* θ *of* $\sqrt{(\sigma(L),\sigma(a))}$.

At this stage we see that Lemma 4.9 gives Theorem 4.7 if we have $J(M/\varphi(M))=\sqrt{(\sigma(L),\sigma(a))}$. So there remains to prove this equality, i.e. to prove the equality asserted in Theorem 4.6. Notice that we already have obtained Lemma 4.12 so we will use this to obtain the following :

4.13. Lemma. *Let* $J(M/\varphi(M))=P_1\cap\ldots\cap P_w$ *and assume that for every* P_j, *there exists some* $\mathbb{P}_i$ *such that* $\mathbb{P}_i\subset P_j$ *and there does not exist any prime strictly in between* $\mathbb{P}_i$ *and* P_j. *Then* $J(M/\varphi(M))=\sqrt{(\sigma(L),\sigma(a))}$.

Proof. Take some P_j and notice that the *inclusion* in (4.4) implies that P_j contains $\theta_1\cap\ldots\cap\theta_t$ and then we find some θ_v so that $P_j\supset\theta_v$. We also have $\mathbb{P}_1\cap\ldots\cap\mathbb{P}_s\subset\theta_v$, so we find some $\mathbb{P}_i$ with $\mathbb{P}_i\subset\theta_v$. The last inclusion is strict since $\sigma(a)\in\theta_v$ while $\sigma(a)$ does not belong to $\mathbb{P}_i$. Then $P_j=\theta_v$ by the hypothesis in Lemma 4.13. After a suitable arrangement of the indices 1,...,t we conclude that there exists some $1\leq t'\leq t$ such that $J(M/\varphi(M))=\theta_1\cap\ldots\cap\theta_{t'}$. But finally, we have proved that $e_{\theta}(M/\varphi(M))>0$ for every minimal prime divisor θ of $\sqrt{(\sigma(L),\sigma(a))}$. So then we must have t=t' and Lemma 4.13 is proved.

At last there remains to show that the hypothesis made in Lemma 4.13 holds. To prove this we will use the purity of the A-module M and begin with a result which has an interest in its own.

4.14. Theorem. *Let* N *be a pure* A*-module and assume that* N *does not belong to the class* $\mathcal{E}$, i.e. $gr_\Gamma(N)\neq 0$ *for any good filtration* Γ. *Then, if* $k=j_A(N)$ *it follows that the injective dimension of the local ring* $G(A)_{\mathbb{P}}$ *is equal to* k *for every minimal prime divisor* $\mathbb{P}$ *of* J(N).

Proof. By Theorem 3.8 we can choose a good filtration Γ on N such that $gr_\Gamma(N)$ is a pure G(A)-module. We also have the equality $j_A(N)=j_{G(A)}(gr_\Gamma(N))$ and by definition $J(N)=J(gr_\Gamma(N))$. Let $\mathbb{P}$ be a minimal prime divisor of J(N) and let v be the injective dimension of the local ring $G(A)_{\mathbb{P}}$. Now $gr_\Gamma(N)_{\mathbb{P}}$ is a non-zero and zero-dimensional $G(A)_{\mathcal{P}}$-module. Since the ring G(A) is Gorenstein, it is wellknown that this implies

(1) $$Ext^v_{G(A)_{\mathcal{P}}}(Ext^v_{G(A)_{\mathcal{P}}}(gr_\Gamma(N)_{\mathcal{P}}),G(A)_{\mathcal{P}})\neq 0$$

Next, localisation commutes with Ext and hence (1) implies that $Ext^v_{G(A)}(Ext^v_{G(A)}(gr_\Gamma(N),G(A)),G(A))\neq 0$. The purity of the G(A)-module $gr_\Gamma(N)$ gives $v=j_{G(A)}(gr_\Gamma(N))=j_A(N)$.

Remark. If a Nullstellen-Satz is available which identifies radical ideals of G(A) with algebraic or complex subsets of manifolds, then Theorem 4.14 means that the "characteristic variety of a pure A-module is equi-dimensional". This applies to the case when A is an enveloping algebra or one of the standard rings of differential operators. We remark that Theorem 4.14 was obtained by Kashiwara in [Ka:1] for coherent $\mathcal{E}_X$-modules. The more general context was clarified more or less explicit in the lectures by O. Gabber at Paris VI in 1982.

Let us return to the pure A-module M and the injective A-linear map φ. Theorem 1.19 gives a submodule M′ of M such that $\varphi^k M \subset \varphi M' \subset M' \subset M$ and $M'/\varphi(M')$ is a pure A-module whose grade number is $j_A(M)+1$. It is easily seen that $J(M/\varphi(M))$ is equal to $J(M'/\varphi(M'))$. So if $k=j(M)$, then Theorem 4.14 applied to the pure A-module $M'/\varphi(M')$ gives

4.15. $inj.dim(G(A)_{\mathcal{P}})=k+1$ holds for every minimal prime divisor $\mathcal{P}$ of $J(M/\varphi(M))$.

Next, we have also inj.dim$(G(A)_{\mathbb{P}})=k$ for every minimal prime $\mathbb{P}$ divisor of J(M). Then we see that the hypothesis in Lemma 4.13 holds and the proof of Theorem 4.6, and hence also of Theorem 4.7, is complete.

REFERENCES

[Ao] Awami, A., Oystaeyen F., On filtered rings with noetherian associated graded rings, Ring Theory, (J.L. Buesco, P. Jara, B. Torrecillas, ed.). Lecture notes in Math. ***1328***, p. 8-27, Springer Verlag (1988).

[Ba] Bass, H., On the ubiquity of Gorenstein rings. Math. Z. ***82***, p. 8-28 (1963).

[Bj:1] Björk, J.-E., Rings of differential operators. North Holland. Math. Lib. Ser. ***21*** (1979).

[Bj:2] Björk, J.-E., Noetherian rings and their applications. Math. Surveys and monographs. N° ***24***. Amer. Math. Soc., p. 59-98 (1987).

[Ek] Ekström, E.-K., The Auslander condition on graded and filtered Noetherian rings. (This proceedings).

[C-E] Cartan, H., Eilenberg, S., Homological algebra. Princeton University Press (1956).

[F-F] Fossum, R., Foxby, H.-B., The category of graded modules, Math. Scand (1974) vol. 35, 288-300.

[FGR] Fossum, R., Griffith, P., Reiten, I., Trivial extensions of Abelian categories. Lecture notes in math. ***456***, Springer Verlag (1975).

[Gi] Ginsburg, V., Characteristic varieties and vanishing cycles. Inv. Math. (1986) 84 p. 327-402.

[Ka:1] Kashiwara, M., B-functions and holonomic systems. Inventiones Math. *38* (1976).

[Ka:2] Kashiwara, M., Systems of microdifferential equations. Progress in Math. *34*, Birkhäuser (1983).

[La-Sch] Laurent, V., Schapira, P., Images inverses des modules différentiels. Compositio Math., vol. 61, (1987), 229-251.

[Lev:1] Levasseur, T., Grade des modules sur certains anneaux filtrés. Communications in Alg. *9*, p. 1519-1532 (1981).

[Lev:2] Levasseur, T., Complexe bidualisant en algèbre non commutative. Sém. Dubreil-Malliavin. Lecture Notes in Math. *1146*, p. 270-287. Springer Verlag (1983-84).

[Lev:3] Levasseur, T., Anneau d'opérateurs différentiels. Sém. Dubreil-Malliavin. Lecture Notes in Math. *867*, p. 157-173 (1980).

[Ma] Matijevic, J.R., Thesis. Chicago Univ. (1973).

[M-R] McConnell, J.C., Robson, J.C., Noncommutative noetherian rings. Wiley series in pure and appl. math. (1987).

[N-O] Năstăsescu, C., Van Oystaeyen, F., Graded ring theory. North Holland Math. Library *28* (1982).

[Re] Reiten, I., Trivial extensions and Gorenstein rings. Thesis. Chicago Univ. (1971).

[Ro:1] Roos, J.-E., Bidualité et structure des foncteurs dérivés de $\varprojlim$ dans la catégorie des modules sur un anneau régulier. C.R. Acad. Sci. Paris *254* (1962).

[Ro:2] Roos, J.-E., Compléments à l'étude des quotients primitifs des algèbres enveloppantes des algèbres de Lie semi-simples. C.R. Acad. Sci. Paris *276* p. 447-450 (1973).

[Sch] Schapira, P., Microdifferential systems in the complex domain. Springer Verlag (1985).

[SKK] Sato, M., Kashiwara, M., Kawai, T. Microfunctions and pseudo-differential operators. Lecture Notes in Math. *287* Springer Verlag (1973).

[Z] Zaks, A., Injective dimension of semi-primary rings. J. of algebra *13* p.73-86 (1969).

GROUPE DES CLASSES DE DIVISEURS DES ALGEBRES GRADUEES NORMALES

P. CARBONNE
U.E.R. de Mathématiques, Université Toulouse-Le-Mirail
5, Allées Antonio Machado
31081 - TOULOUSE CEDEX

[*Cet article est dans sa forme définitive et il ne fera pas l'objet d'une autre publication*].

Les travaux de I.V. DOLGACHEV [7] et H. PINKHAM [13] ont montré l'intérêt des algèbres $\mathbb{N}$-graduées normales. Une importante publication de M. DEMAZURE [6] a été à l'origine de nombreux travaux sur le sujet. Les fibrations de Seifert des surfaces donnent de telles algèbres [1]. Dans [3] (voir aussi [5]) j'ai montré que ce résultat se généralise à la dimension n+1 moyennant une hypothèse que l'on peut qualifier de "Cohen-Macaulay homogène". Dans [5] je donne aussi une généralisation pour les algèbres $\mathbb{N}^n$-graduées normales du principal théorème de M. DEMAZURE (Th. 3.5 [6]). Cette généralisation fait appel à deux hypothèses :

- celle (dans une version $(\tilde{H})$ légèrement plus forte) que nous notons (H) et que nous rappellerons plus loin , qui est la traduction généralisée de la propriété : "Cohen-Macaulay homogène" [5].

- l'existence dans le corps des fractions de l'anneau d'éléments de degrés respectifs $\varepsilon_1,\ldots,\varepsilon_n$, où les ε_j $(1 \leq j \leq n)$ sont les générateurs canoniques du monoïde $\mathbb{N}^n$.

Nous allons étudier ici le groupe des classes de diviseurs des algèbres $\mathbb{N}^n$-graduées normales. Nous verrons que chacune de ces deux hypothèses séparément suffit pour ramener le calcul de ce groupe à un calcul sur les diviseurs homogènes. Nous démontrerons ensuite qu'il y a bijection entre les idéaux premiers homogènes de hauteur 1 de l'algèbre et les "pseudo-idéaux premiers" d'un monoïde. Nous illustrerons enfin ce travail par des exemples. Dans le cas où n = 1 on se reportera aussi à [11] et [16].

I - Résultats sur les idéaux des anneaux $\mathbb{N}^n$-gradués

I.1 - On peut établir sans trop de difficultés le résultat suivant -bien connu dans le cas des anneaux $\mathbb{N}$-gradués :

Proposition 1 :

Soit R un anneau $\mathbb{N}^n$-gradué et $\mathfrak{a}$ un idéal homogène de R tel que chaque fois que le produit xy de deux éléments homogènes x et y est dans $\mathfrak{a}$, l'un des facteurs x ou y est dans $\mathfrak{a}$, alors $\mathfrak{a}$ est un idéal premier de R.

Pour un idéal $\mathfrak{a}$ de R nous noterons $\mathfrak{a}^h$ l'idéal homogène engendré par les éléments homogènes appartenant à $\mathfrak{a}$. L'idéal $\mathfrak{a}^h$ est alors le plus grand idéal homogène inclus dans $\mathfrak{a}$. Nous avons le corollaire suivant :

Corollaire :

Si p est un idéal premier de R, alors p^h est aussi un idéal premier.

(Pour n = 1 le résultat est connu. Voir [17]).

I.2 - Nous allons démontrer maintenant un certain nombre de propriétés des idéaux des anneaux $\mathbb{N}^n$-gradués. Ces propriétés sont classiques dans le cas des $\mathbb{N}$-graduations mais nous ne les avons jamais rencontrées, démontrées, pour les $\mathbb{N}^n$-graduations.

Le lemme qui suit permet de ramener les $\mathbb{N}$-graduations aux $\mathbb{N}$-graduations quand c'est techniquement utile.

Lemme 1 :

Soient Ω et Ω' deux parties non vides de $\mathbb{N}^n$, alors il existe $\alpha = (\alpha_1,\dots,\alpha_n)$ appartenant à $\mathbb{N}^n$ tel que si l'on pose, pour $I = (i_1,\dots,i_n)$ appartenant à $\mathbb{N}^n$:

$$l(I) = \sum_{j=1}^{n} \alpha_j i_j,$$ il existe un unique couple (I_0,J_0) dans $\Omega\times\Omega'$ tel que I_0 réalise strictement le minimum de $l(\Omega)$ et J_0 celui de $l(\Omega')$.

Soit R un anneau commutatif unitaire $\mathbb{N}^n$-gradué par : $R = \bigoplus_{I\in\mathbb{N}^n} R_I$. Pour tout p de 1 à n et pour $(\alpha_1,\dots,\alpha_p)$ fixé dans $\mathbb{N}^p$ nous posons :

$$R_{\alpha_1,\dots,\alpha_p} = \bigoplus_{I=(\alpha_1,\dots,\alpha_p,i_{p+1},\dots,i_n)} R_I .$$

La composante de x dans $R_{\alpha_1,\ldots,\alpha_p}$ est notée $x_{\alpha_1,\ldots,\alpha_p}$. Nous avons bien sûr pour p = n : $R_{\alpha_1,\ldots,\alpha_n} = R_{(\alpha_1,\ldots,\alpha_n)}$ et $x_{\alpha_1,\ldots,\alpha_n} = x_{(\alpha_1,\ldots,\alpha_n)}$.

Remarque préliminaire :

Si $\mathfrak{a}$ est un idéal homogène de R (homogène pour la $\mathbb{N}^n$-graduation) et si x appartient à $\mathfrak{a}$, alors tous les $x_{\alpha_1,\ldots,\alpha_p}$ sont dans $\mathfrak{a}$. Ceci résulte aussitôt du fait que tous les x_I sont dans $\mathfrak{a}$.

Proposition 2 :

Si $\mathfrak{a}$ est un idéal homogène de R, $\sqrt{\mathfrak{a}}$ est aussi un idéal homogène de R.

Soit x appartenant à $\sqrt{\mathfrak{a}}$ et x_{a_1} sa première composante non nulle dans la décomposition : $R = \bigoplus_{\alpha_1 \in \mathbb{N}} R_{\alpha_1}$. Il existe $r \geq 1$ tel que $x^r \in \mathfrak{a}$. La première composante non nulle de x^r est $x_{a_1}^r$. D'après la remarque préliminaire $x_{a_1}^r \in \mathfrak{a}$ et donc x_{a_1} est dans $\sqrt{\mathfrak{a}}$. On applique le même raisonnement à $x - x_{a_1}$ et, par itération, on voit que toutes les composantes de x dans la décomposition $R = \bigoplus_{\alpha_1 \in \mathbb{N}} R_{\alpha_1}$ sont dans $\sqrt{\mathfrak{a}}$. Pour chacune des composantes non nulles x_{α_1} de cette première décomposition on procède de même avec la décomposition : $R_{\alpha_1} = \bigoplus_{\alpha_2 \in \mathbb{N}} R_{\alpha_1\alpha_2}$. On prouve ainsi que les $x_{\alpha_1\alpha_2}$ sont dans $\sqrt{\mathfrak{a}}$. Par itération on arrive au fait que tous les $x_{(\alpha_1,\ldots,\alpha_n)} = x_{\alpha_1,\ldots,\alpha_n}$ sont dans $\sqrt{\mathfrak{a}}$. L'idéal $\sqrt{\mathfrak{a}}$ est donc homogène pour la $\mathbb{N}^n$-graduation.

Corollaire :

Pour tout idéal de R nous avons : $(\sqrt{\mathfrak{a}})^h = \sqrt{\mathfrak{a}^h}$.

Se déduit de la proposition comme dans le cas des $\mathbb{N}$-graduations.

Proposition 3 :

Soit p un idéal premier homogène de R et $\mathfrak{q}$ un idéal p-primaire. alors $\mathfrak{q}^h$ est aussi p-primaire.

1) Nous avons : $\sqrt{\mathfrak{q}^h} = (\sqrt{\mathfrak{q}})^h = p^h = p$.

2) Soient maintenant x et y tels que $xy \in \mathfrak{q}^h$ et $x \notin \mathfrak{q}^h$. Notons x' la somme des composantes homogènes de x qui sont dans $\mathfrak{q}^h$ et x" celle des composantes qui ne sont pas dans $\mathfrak{q}^h$. Nous avons $x'' \neq 0$ et aucune de ses composantes n'est dans $\mathfrak{q}^h$ et donc n'est pas non plus dans $\mathfrak{q}$.

L'élément $x''y = xy-x'y$ appartient à $\mathfrak{q}^h$. Nous posons :

$\Omega = \{I \mid x''_I \neq 0\}$ et $\Omega' = \{J \mid Y_J \neq 0\}$ et nous choisissons 1 comme dans le lemme 1. Le produit $x''_{I_0}\ Y_{J_0}$ est une composante homogène de x"y dans la $\mathbb{N}$-graduation de R obtenue par :

$$R = \bigoplus_{m \in \mathbb{N}} \left(\bigoplus_{l(I)=m} R_I \right)$$

L'idéal $\mathfrak{q}^h$, homogène pour la $\mathbb{N}^n$-graduation, reste homogène pour la $\mathbb{N}$-graduation. Il en résulte que $x''_{I_0}\ y_{J_0}$ est dans $\mathfrak{q}^h$ et donc dans $\mathfrak{q}$. Comme x''_{I_0} n'est pas dans $\mathfrak{q}$ et que $\mathfrak{q}$ est p-primaire, nous en déduisons que y_{J_0} appartient à $\sqrt{\mathfrak{q}} = p$. On recommence le même raisonnement avec $y-y_{J_0}$. On montre ainsi, par itération, que y est dans p.

Corollaire :

Soit p un idéal premier homogène de R. Alors $p^{(n)}$ est un idéal homogène pour tout entier $n \geq 1$.

Posons : $\mathfrak{q} = p^{(n)}$. Alors $\mathfrak{q}$ est p-primaire et donc aussi $\mathfrak{q}^h$. L'idéal p étant homogène, p^n l'est aussi. Comme par ailleurs : $p^n \subset p^{(n)} = \mathfrak{q}$, nous avons donc $p^n \subset \mathfrak{q}^h$. Par définition $\mathfrak{q} = p^{(n)}$ est le saturé de p^n par rapport à p. L'idéal $\mathfrak{q}^h$ qui est p-primaire est saturé par rapport à p. Il vient donc $\mathfrak{q} = \mathfrak{q}^h$ et donc $p^{(n)}$ $(=\mathfrak{q})$ est homogène.

I.3 - L'anneau R est maintenant intègre.

Un idéal fractionnaire homogène $\mathfrak{a}'$ de R est, par définition, un idéal fractionnaire tel qu'il existe v homogène dans R tel que $v\mathfrak{a}'$ soit un idéal homogène de R.

Lemme 2 :

Soit $\mathfrak{a}'$ un idéal (fractionnaire) homogène de R tel que $(R:\mathfrak{a}') \subset R$. Alors $(R:\mathfrak{a}')$ est un idéal homogène.

Soit x appartenant à $(R:\mathfrak{a}')$. Alors x appartient à R et $x\mathfrak{a}' \subset R$. D'où : $xv\mathfrak{a}' \subset vR$. Par suite ceci revient à dire que x appartient à $(vR:v\mathfrak{a}')$ dans R. Les idéaux de A, $v\mathfrak{a}'$ et vR sont homogènes (le second parce que v est homogène). Or on sait que pour deux idéaux homogènes de R, l'idéal quotient (dans R) est homogène. Donc : $(R:\mathfrak{a}') = (vR:v\mathfrak{a}')$ est homogène.

Corollaire :

Soit $\mathfrak{a}$ un idéal homogène de R ; alors $(R:\mathfrak{a})$ est un idéal fractionnaire homogène.

Si x appartient à $(R:\mathfrak{a})$, nous avons pour tout u dans $\mathfrak{a}$: $ux \in R$. Ceci s'écrit : $u(R:\mathfrak{a}) \subset R$. On en déduit que : $(R : \frac{1}{u}\mathfrak{a}) = u(R:\mathfrak{a}) \subset R$. Prenons u homogène dans $\mathfrak{a}$ (ceci est possible car $\mathfrak{a}$ est homogène) et posons $\mathfrak{a}' = \frac{1}{u}\mathfrak{a}$. Comme $u\mathfrak{a}' = \mathfrak{a}$, $\mathfrak{a}'$ est un idéal u fractionnaire homogène. De plus nous venons de voir que $(R:\mathfrak{a}') \subset R$. Le lemme 2 nous dit que $(R:\mathfrak{a}')$ est un idéal homogène de A. Vu que u est homogène et que $u(R:\mathfrak{a}) = (R:\mathfrak{a}')$, l'idéal $(R:\mathfrak{a}')$ est fractionnaire homogène.

Prouvons encore trois lemmes .

Lemme 3 :

Soient $\mathfrak{a}$ un idéal homogène de R et $\mathfrak{b}$ un idéal fractionnaire homogène de R. Alors $\mathfrak{a}\mathfrak{b}$ est un idéal fractionnaire homogène de R.

1) Le cas où $\mathfrak{b}$ est un idéal de R est presque évident et se traite comme pour les $\mathbb{N}$-graduations.

2) Le cas où $\mathfrak{b}$ est fractionnaire s'en déduit grâce à l'élément β homogène qui fait de $\beta\mathfrak{b}$ un idéal homogène de R.

Lemme 4 :

Soient α et β dans R tels que αβ soit homogène. Alors α et β sont homogènes.

L'utilisation du lemme 1 ramène à une $\mathbb{N}$-graduation. La propriété est alors évidente.

Lemme 5 :

Si $\mathfrak{a}$ est un idéal fractionnaire principal homogène son générateur est homogène.

1) Si $\mathfrak{a} = aR$ est un idéal homogène de R et si a_I est une composante homogène de a, a_I appartient à $\mathfrak{a}$. Il existe donc b dans R tel que $a_I = ba$. Le lemme 4 permet de conclure.

2) Le cas fractionnaire se ramène au précédent et le générateur de l'idéal est quotient de deux éléments homogènes de R, donc homogène dans son corps des fractions.

II - Réduction du calcul des classes de diviseurs aux diviseurs homogènes

II.1 - Soit A un anneau noethérien, normal, $\mathbb{N}^n$-gradué. A est un anneau de Krull [17]. Nous renvoyons à [8] ou [14] pour la définition du groupe des classes de diviseurs Cl(A) de A. Nous utiliserons les notations classiques [8] :

$X^1(A)$ est l'ensemble des idéaux premiers de hauteur 1 de A et $X^1_h(A)$ l'ensemble de ceux qui sont de plus homogènes.

Div(A) désigne le groupe des diviseurs de A et $\mathrm{Div}_h(A)$ le sous-groupe engendré par les diviseurs homogènes. On dit qu'un diviseur $D = \sum_{p\in X^1(A)} n_p\, p$ de A est homogène si tous les p tels que $n_p \neq 0$ sont homogènes (i.e. sont dans $X^1_h(A)$).

Prin(A) note le sous-groupe des diviseurs principaux et nous posons :

$\mathrm{Prin}_h(A) = \mathrm{Prin}(A) \cap \mathrm{Div}_h(A)$.

Proposition 4 :

Si D est un diviseur homogène de A, l'idéal divisoriel correspondant IdD est un idéal homogène.

1) Supposons d'abord $D \geq 0$. Si $D = \sum_{i=1}^{s} n_i p_i$, alors $\mathrm{I}dD = \bigcap_{i=1}^{s} p_i^{(n_i)}$. Par hypothèse les p_i sont homogènes, donc aussi les $p_i^{(n_i)}$ (voir cor. prop. 3) et enfin leur intersection $\mathrm{I}dD$.

2) Dans le cas général on écrit : $D = D_1 - D_2$ où D_1 et D_2 sont positifs et homogènes. Nous venons de voir que $\mathrm{I}dD_1$ et $\mathrm{I}dD_2$ sont homogènes. Le corollaire du lemme 2 dit que $(A : \mathrm{I}dD_2)$ est homogène et le lemme 3 dit que $\mathrm{I}dD = (\mathrm{I}dD_1)\,(A : \mathrm{I}dD_2)$ est aussi homogène.

II.2 - Soit $A = \bigoplus_{I \in \mathbb{N}^n} A_I$. Nous notons : $A_+ = \bigoplus_{I \neq 0} A_I$ et pour $I = i_1\varepsilon_1 + \ldots + i_n\varepsilon_n \neq 0$ nous posons : $A_{(I)} = \prod_{j=1}^{n} A_{i_j\varepsilon_j} \subset A_I$. Nous dirons que A vérifie l'hypothèse (H) si :

(H) $\left\{\right.$ Tout x de A_+ se décompose en une somme finie d'éléments x_1 homogènes tels que pour chacun d'eux il existe un entier $m_1 > 0$ et tel que : $x_1^{m_1}$ soit égal à une somme finie d'éléments de $A_{(m_1 d^o x_1)}$.

Pour plus de précisions sur (H) et ses rapports avec d'autres hypothèses on se rapportera à [5].

Théorème 1 :

Soit A un anneau noethérien, normal, $\mathbb{N}^n$-gradué. On suppose en outre que A vérifie (H) ou que A est tel que, pour tout j de 1 à n, il existe T_j dans le corps des fractions Fr(A) de degré ε_j. Alors :

$$Cl(A) = {}^{Div_h(A)}/_{Prin_h(A)}.$$

1) <u>Preuve dans le cas où A vérifie (H)</u> :

a) Soit $\bar{A}$ le sous-anneau de A :

$$\bar{A} = \bigoplus_{I \in \mathbb{N}^n} A_{(I)} \qquad \text{(par convention : } A_{(0)} = A_0\text{).}$$

et S la réunion pour tous les I de $\mathbb{N}^n$ des $A_{(I)}$ privés de 0. Comme l'anneau A est intègre, que $A_{(I)}.A_{(J)} \subset A_{(I+J)}$ et que $1 \in A_0 = A_{(0)}$, S est une partie multiplicative de $\bar{A}$. L'anneau de fractions $S^{-1}\bar{A}$ est $\mathbb{Z}^n$-gradué par : $d^o(\frac{a}{s}) = d^o a - d^o s$ (pour a dans $\bar{A}$ et homogène).

Notons D l'ensemble des éléments de degré nul de $S^{-1}\bar{A}$. Nous avons :

$$D = \{\tfrac{a}{s} \mid a \text{ et } s \text{ appartiennent à un } A_{(I)} \text{ et } s \neq 0\}.$$

Il est clair que D est un corps.

b) Si $A = A_0$ il n'y a rien à démontrer. On peut donc supposer $A \neq A_0$. il existe alors $I \neq 0$ tel que $A_I \neq \{0\}$. L'hypothèse (H) permet alors de dire qu'il existe $m > 0$ tel que $A_{(mI)} \neq \{0\}$. Il en résulte que pour tout j de 1 à n : $A_{mi_j\epsilon_j} \neq \{0\}$. Il en résulte que le sous-anneau $A^j = \bigoplus_{\lambda \in \mathbb{N}} A_{\lambda\epsilon_j}$ de $\bar{A}$ est différent de A_0. ($A^j \subset \bar{A}$ car : $A_{(\lambda\epsilon_j)} = A_{\lambda\epsilon_j}$).

Comme $\bigoplus_{\lambda \in \mathbb{N}^*} (S^{-1}\bar{A})_{\lambda\epsilon_j}$ contient les éléments de A^j qui ne sont pas dans A_0, on en déduit que cette somme est $\neq \{0\}$. On peut donc introduire T_j un élément de cette somme de degré $\alpha_j\epsilon_j$ avec $\alpha_j > 0$ et minimum, T_j est alors transcendant sur D. En effet une relation de dépendance algébrique : $b_0 T_j^p + \ldots + b_p = 0$ (avec $b_0, \ldots, b_p$ dans D), compte tenu du fait que tous les termes ont des degrés distincts, donne $b_0 = \ldots = b_p = 0$. Nous avons donc :

$$D[T_1, \ldots, T_n] \subset S^{-1}\bar{A} \subset S^{-1}A \qquad \text{où} \qquad D[T_1, \ldots, T_n] \text{ est un anneau de}$$
polynômes.

Comme $T_j \neq 0$, son numérateur est dans S et donc $T_j^{-1} \subset S^{-1}\bar{A} \subset S^{-1}A$. Nous avons donc une inclusion d'anneaux gradués :

$$D[T_1, T_1^{-1}, \ldots, T_n, T_n^{-1}] \subset S^{-1}A.$$

c) Nous allons prouver qu'il y a égalité. Soit u dans $\bar{S}^1A$, homogène de degré I. Alors : $u = \frac{x}{s}$ où x et s sont homogènes et $d^o x = d^o s + I$. L'hypothèse (H) permet d'écrire : $x = x_1 + \ldots + x_\omega$ avec, pour l de 1 à ω, $d^o x_l = d^o s + I$ et pour chaque l un entier $m_l > 0$ tel que : $x_l^{m_l} = y_{1,1} \ldots y_{n,1} + \ldots + y_{1,\Theta} \ldots y_{n,\Theta}$ (avec $d^o y_{j,h} \in \mathbb{N}\varepsilon_j$). Par ailleurs nous avons : $s = s_1 \ldots s_n$ avec $d^o s_j \in \mathbb{N}\varepsilon_j$. Nous avons alors :

$$v_{j,k} = \frac{y_{j,k}}{s_j^{m_l}} \in \bigoplus_{\lambda \in \mathbb{N}} (\bar{S}^1\bar{A})_{\lambda\varepsilon_j} \quad . \quad (k = 1 \text{ à } \Theta).$$

Si nous faisons la division euclidienne :

$$d^o(v_{j,k}) = q_{j,k}\,\alpha_j + r_{j,k} \qquad 0 \leq r_{j,k} < \alpha_j,$$

nous avons :

$$T_j^{-q_{j,k}}\, v_{j,k} \in (\bar{S}^1\bar{A})_{r_{j,k}\varepsilon_j}$$

Comme le degré α_j de T_j est minimum nous en déduisons que $r_{j,k} = 0$. Il en résulte que

$$v_{j,k} \in D[T_1, T_1^{-1}, \ldots, T_n, T_n^{-1}].$$

Il en est de même pour :

$$\left(\frac{x_l}{s}\right)^{m_l} = \sum_{k=1}^{\Theta} \prod_{j=1}^{n} v_{j,k} \quad .$$

Comme D est un corps, $D[T_1, T_1^{-1}, \ldots, T_n, T_n^{-1}]$ est normal ; donc :

$\frac{x_l}{s} \in D[T_1, T_1^{-1}, \ldots, T_n, T_n^{-1}]$ et aussi $u = \frac{x}{s} = \sum_{l=1}^{\omega} \frac{x_l}{s}$.

Nous avons donc bien :

$$\bar{S}^1A = D[T_1, T_1^{-1}, \ldots, T_n, T_n^{-1}].$$

$\bar{S}^1A$ est donc un anneau factoriel.

d) Un corollaire du théorème de Nagata [8] nous dit que nous avons une suite exacte :

$$0 \to \ker \varphi \to Cl(A) \overset{\varphi}{\to} Cl(\bar{S}^1A) \to 0$$

où ker φ est engendré par les classes des idéaux premiers p de hauteur 1 tels que $p \cap S \neq \emptyset$. Comme S est formée d'éléments homogènes $p \cap S \neq \emptyset$ donne $p^h \cap S \neq \emptyset$ et comme $0 \notin S$ ceci donne $p^h \neq (0)$. p^h étant premier, inclus dans p qui est de hauteur 1, on en déduit que $p = p^h$. Ceci signifie que p est homogène.

Comme nous avons $Cl(S^{-1}A) = 0$ et donc $ClA \approx \ker\varphi$ nous avons bien le résultat cherché :

$$Cl(A) = {}^{Div_h(A)}/_{Prin_h A} \, .$$

2) <u>Preuve dans le cas où il existe des fractions T_j $(1 \leq j \leq n)$ de degrés respectifs ε_j :</u>

Cette fois ci on prend pour partie multiplicative S de A l'ensemble de tous les éléments homogènes non nuls de A. L'anneau $S^{-1}A$ est $\mathbb{Z}^n$-gradué et on note $D = (S^{-1}A)_0$ le sous-anneau des éléments de degré nul de $S^{-1}A$. Ici aussi D est un corps.

Par hypothèse, pour $j = 1,\dots,n$, il existe T_j de degré ε_j dans $S^{-1}A$. Le même raisonnement que dans le cas précédent montre que les T_j sont transcendants sur D et que donc l'anneau de polynômes $D[T_1,\dots,T_n]$ est inclus dans $S^{-1}A$. Ici aussi nous avons $T_j^{-1} \in S^{-1}A$ et donc : $D[T_1,T_1^{-1},\dots,T_n,T_n^{-1}]$ est un sous-anneau gradué de $S^{-1}A$. Soit maintenant u homogène appartenant à $S^{-1}A$ et $q_1\varepsilon_1+\dots+q_n\varepsilon_n \in \mathbb{Z}^n$ son degré. Nous avons alors : $d^o(T_1^{q_1}\dots T_n^{q_n}u) = 0$. Donc : $T_1^{q_1}\dots T_n^{q_n}u \in D$ et $u \in D[T_1,T_1^{-1},\dots,T_n,T_n^{-1}]$. Nous avons donc encore l'égalité :

$$S^{-1}A = D[T_1,T_1^{-1},\dots,T_n,T_n^{-1}]$$

qui montre que $S^{-1}A$ est factoriel.

On termine comme tout à l'heure.

<u>Remarque 1</u> :

Si $n = 1$ l'hypothèse (H) est automatiquement vérifiée et on retombe sur un résultat connu [8] [14].

Remarque 2 :

Les deux hypothèses envisagées sont indépendantes. Le contre-exemple donné dans [3] (I.7.4) vérifie la seconde et pas la première [5].

Dans [5] nous donnons (I.3.3 de la partie VI) un exemple d'anneau noethérien intègre qui vérifie (H) sans vérifier l'autre hypothèse. On vérifie sans mal que cet anneau est normal.

II.3 - On appelle fonction rationnelle homogène sur Spec A tout élément du corps des fractions Fr(A) de A quotient de deux éléments homogènes de A.

Proposition 5 :

Un élément f du corps des fractions Fr(A) de A est une fonction rationnelle homogène sur Spec A exactement lorsque le diviseur principal (f) associé à f est homogène.

1) Supposons que $f = \frac{x}{y}$ $(x \in A_I,\ y \in A_J)$ est une fonction rationnelle homogène sur Spec A. Par additivité on peut supposer que I-J appartient à $\mathbb{N}^n$. Il vient alors :

$$(f) = \sum_{p \in X^1(A)} \mathrm{ord}_p(f)\, p\ .$$

Dire que $\mathrm{ord}_p(f) \neq 0$ revient à dire que f appartient à p. Comme f est homogène, f est aussi dans p^h et donc $p^h \neq (0)$. Un raisonnement déjà fait dans la démonstration du théorème 1 donne alors $p \in X_h^1(A)$. Le diviseur (f) est homogène.

2) Supposons réciproquement le diviseur (f) homogène. Par définition : (f) = div(fA). L'idéal principal fA est divisoriel et donc :

$$fA = (A : (A:fA)) = \cap\, p_i^{(n_i)} = \mathrm{Id}(f) \qquad ([8]\ \text{p. } 26).$$

Le diviseur (f) étant homogène la proposition 4 nous dit que l'idéal fA = Id(f) est lui-même homogène. Le lemme 5 permet alors de dire que f est homogène.

Corollaire :

Sous les hypothèses du théorème 1 nous avons :

$$Cl(A) = {}^{Div_h(A)}/_{\{\text{diviseurs des fonctions rationnelles homogènes sur Spec } A\}}$$

Ces résultats sont à rapprocher d'un résultat de P. Samuel [8], [14], [15].

III - Etude des idéaux premiers homogènes de hauteur 1

III.1 - Un sous-monoïde Γ de $\mathbb{N}^n$ est une partie telle que $\Gamma+\Gamma \subset \Gamma$ et $0 \in \Gamma$. On appelle idéal de Γ toute partie H telle que $H+\Gamma \subset H$. H est un idéal permier de Γ si $H \neq \Gamma$ et si : "$I+J \in H$ est équivalent à I (ou J) $\in H$" (I et J étant dans Γ). Un idéal premier est de hauteur 1 s'il est minimal pour l'inclusion parmi les idéaux premiers non vides. Le bord ∂H de l'idéal H est :

$$\partial H = \{I \in H \mid I-K \notin H \ \forall K \in \Gamma \text{ et } K \neq 0\}.$$

Dire que H est un idéal de Γ est dire que : $H = \bigcup_{I \in \partial H} (I+\Gamma)$.

Si H est premier les éléments de ∂H sont indécomposables dans Γ et appartiennent donc nécessairement à tout système minimal de générateurs de Γ.

Soit A un anneau unitaire, intègre, $\mathbb{N}^n$-gradué par : $A = \bigoplus_{I \in \mathbb{N}^n} A_I$.

Notons : $\Gamma = \{I \mid A_I \neq (0)\}$. L'intégrité donne $\Gamma+\Gamma \subset \Gamma$ et le fait que l'unité est dans A_0 donne $0 \in \Gamma$. Γ est donc un sous-monoïde de $\mathbb{N}^n$. Quitte à modifier la graduation de A on peut supposer que :

$$\Gamma \subset \mathbb{N}d_1\varepsilon_1+\ldots+\mathbb{N}d_n\varepsilon_n \quad \text{entraine } d_1 = \ldots = d_n = 1.$$

Si H est un idéal de Γ, on voit facilement que $\alpha(H) = \bigoplus_{I \in H} A_I$ est un idéal homogène de A. Si H est premier, $\alpha(H)$ est aussi un idéal premier.

III.2 - Dans toute la suite on fait l'hypothèse que pour tout I dans $\mathbb{N}^n$, A_I est nul ou est un A_0 -module libre de rang 1. Pour I dans Γ on note ξ^I une A_0-base de A_I. On peut prendre $\xi^0 = 1$. Nous avons :

$$A = \bigoplus_{I \in \Gamma} A_0 \xi^I .$$

Pour I et J dans Γ il existe $a_{I,J} \in A_0$ tel que : $\xi^I \xi^J = a_{I,J} \xi^{I+J}$. La commutativité donne : $a_{J,I} = a_{I,J}$. L'associativité se traduit par : $a_{I,J}\, a_{I+J,K} = a_{I,J+K} a_{J,K}$. Du fait que A est intègre nous avons $a_{I,J} \neq 0$. Enfin nous avons : $a_{0,I} = 1$ pour tout I.

Nous posons une définition qui permet de généraliser ce qui a été dit au III.1.

Définition :

Soit p_0 un idéal premier de A_0. On dit qu'une partie Ω de Γ est un <u>p_0-pseudo-idéal premier de Γ</u> si :

a) $\Omega \neq \Gamma$

et

b) Il y a équivalence entre les deux propriétés :

(1) I ou J appartient à Ω

(2) I+J appartient à Ω ou $a_{I,J} \in p_0$.

Remarque 3 :

$\bar{\Omega} = C_\Gamma^\Omega$ est alors un sous-monoïde de Γ.

D'une part $a_{0,I} = 1 \notin p_0$. D'autre part il existe I tel que $I \notin \Omega$. Donc $I+0 \notin \Omega$. Par suite $0 \notin \Omega$, i.e. : $0 \in \bar{\Omega}$. Par ailleurs si I et J sont dans $\bar{\Omega}$ alors I+J est dans $\bar{\Omega}$.

<u>Exemples</u> :

1) Pour tout p_0, <u>Γ^* (Γ privé de 0) est un p_0-pseudo-idéal premier</u>. En effet, si $a_{I,J} \in p_0$, alors $a_{I,J} \neq 1$ et donc I et J sont $\neq 0$.

2) Pour tout p_0, <u>$\Omega_0 = \{I \mid \text{il existe K tel que } a_{I,K} \in p_0\}$ est un p_0-pseudo-idéal premier</u>.

a) De $a_{0;K} = 1 \notin p_0$ on tire $0 \notin \Omega_0$ et donc $\Omega_0 \neq \Gamma$.

b) Si $I \in \Omega_0$, alors il existe K tel que $a_{I;K} \in p_0$. Mais alors pour tout J nous avons : $a_{I,J}\, a_{I+J,K} = a_{I,K}\, a_{I+K,J} \in p_0$. Il en résulte que $a_{I,J} \in p_0$ ou sinon $a_{I+J,K} \in p_0$ et donc $I+J \in \Omega_0$.

c) Si $a_{I,J} \in p_0$ nous avons I et J dans Ω_0.

d) Si $I+J \in \Omega_0$, il existe K tel que $a_{I+J,K} \in p_0$. Il vient alors : $a_{I,K}\, a_{I+K,J} = a_{I,J}\, a_{I+J,K} \in p_0$. Par suite $a_{I,K} \in p_0$ et donc $I \in \Omega_0$ ou $a_{I+K,J} \in p_0$ et donc $J \in \Omega_0$.

Plus généralement, pour L fixé dans Γ,

$\Omega_L = \{I \mid$ il existe K tel que $a_{I,L+K} \in p_0\}$ est un p_0-pseudo-idéal premier.

La démonstration est la même.

3) Si aucun des $a_{I,J}$ n'appartient à p_0, les p_0-pseudo-idéaux premiers de Γ coïncident avec les idéaux premiers de Γ et $\Omega_0 = \emptyset$. Nous sommes en particulier dans cette situation lorsque $p_0 = (0)$.

Ceci justifie la terminologie adoptée.

Il est clair que réciproquement $\emptyset$ n'est un p_0-pseudo-idéal premier de Γ que si aucun des $a_{I,J}$ n'appartient à p_0.

Notations :

1) Soient p_0 un idéal premier de A_0 et Ω un p_0-pseudo-idéal premier de Γ.

On pose alors :

$$\boxed{J(\Omega) = \bigoplus_{I \in \bar{\Omega}} p_0\, \xi^I \oplus \bigoplus_{I \in \Omega} A_0 \xi^I .}$$

Si $p_0 = (0)$ nous avons : $J(\Omega) = \alpha(\Omega) = \bigoplus_{I \in \Omega} A_0 \xi^I$.

2) Soit p un idéal premier de A. On pose alors :

$$\boxed{\omega(p) = \{I \mid \xi^I \in p\}} \qquad \text{et} \qquad \boxed{p_0(p) = p \cap A_0} \quad .$$

Si p est tel que $p_0(p) = (0)$ alors $\omega(p) = \{I \mid p \cap A_I \neq (0)\}$ et $p^h = \bigoplus_{I\in\omega(p)} A_0\,\xi^I$.

En effet si $a \in A_0$ et $a \neq 0$ alors $a \notin p_0(p)$ donc $a \notin p$. Si $I \notin \omega(p)$, alors $\xi^I \notin p$ et donc $a\xi^I \in p$. D'où $p \cap A_I = (0)$. Si x est homogène non nul et appartient à p, nous avons I tel que $x \in A_I \cap p$. Donc $I \in \omega(p)$ et $x \in A_0\xi^I$. D'où :

$$p^h = \bigoplus_{I\in\omega(p)} A_0\xi^I .$$

On considère maintenant les couples $C = (p_0,\Omega)$ formés d'un idéal premier p_0 de A_0 et d'un p_0-pseudo-idéal premier Ω de Γ. On dit que $C' \subset C$ si $p_0' \subset p_0$ et $\Omega' \subset \Omega$. Le couple C est de hauteur 1 s'il est minimal pour cette relation d'ordre parmi les couples différents du couple $((0),\emptyset)$. Ce dernier est dit de hauteur nulle.

Théorème 2 :

1) Soient p_0 un idéal premier de A_0 et Ω un p_0-pseudo-idéal premier de Γ.
   a) Alors $J(\Omega)$ est un idéal premier homogène de A tel que :
   $p_0(J(\Omega)) = p_0$ et $\omega(J(\Omega)) = \Omega$.
   b) Si (p_0,Ω) est de hauteur 1, $J(\Omega)$ est minimal parmi les idéaux premiers homogènes non nuls.

2) Soit p un idéal premier de A.
   a) Alors $p_0(p)$ est un idéal premier de A_0 et $\omega(p)$ est un $p_0(p)$-pseudo-idéal premier de Γ tel que :
   $J(\omega(p)) = p^h$.
   b) Si p est homogène de hauteur 1 alors $(p_0(p),\omega(p))$ est de hauteur 1. Si p n'est pas homogène et est de hauteur 1 alors le même couple est de hauteur 0.

Preuve de 1.a :

1) Il est clair que $\mathcal{J}(\Omega)$ est homogène et stable pour l'addition. Soient $x = a\xi^I$ et $y = b\xi^J$ deux éléments homogènes, x étant dans $\mathcal{J}(\Omega)$. Nous avons :

$$xy = a\ b\ \xi^I\ \xi^J = ab\ a_{I,J}\xi^{I+J}.$$

Si $I \in \bar{\Omega}$ alors $a \in p_0$ donc aussi $ab\ a_{I,J}$ et $xy \in \mathcal{J}(\Omega)$. Si $I \in \Omega$ alors $I+J \in \Omega$ et donc $xy \in \mathcal{J}(\Omega)$ ou bien $a_{I,J} \in p_0$ donc aussi $ab\ a_{I,J}$ et $xy \in \mathcal{J}(\Omega)$.

$\mathcal{J}(\Omega)$ est donc un idéal homogène de A.

2) Soient x et y homogènes dans A et tels que : $xy = ab\ a_{I,J}\xi^{I+J} \in \mathcal{J}(\Omega)$. Si $I+J \in \Omega$ alors I (ou J) est dans Ω et donc x (ou y) est dans $\mathcal{J}(\Omega)$. Si $I+J \in \bar{\Omega}$ alors $ab\ a_{I,J} \in p_0$. Il en résulte que :

$a_{I,J} \in p_0$, donc I (ou J) est dans Ω et x (ou y) est dans $\mathcal{J}(\Omega)$, sinon :

a (ou b) $\in p_0$ et donc x (ou y) $\in \mathcal{J}(\Omega)$.

L'idéal $\mathcal{J}(\Omega)$ est donc premier.

3) Comme $0 \in \bar{\Omega}$ nous avons bien $p_0(\mathcal{J}(\Omega)) = p_0$.

4) Il est clair que par construction $\omega(\mathcal{J}(\Omega)) = \Omega$.

Preuve de 2.a :

1) On sait que $p_0(p) = p \cap A_0$ est un idéal premier de A_0.

2) Dire que I (ou J) $\in \omega(p)$ équivaut à ξ^I (ou ξ^J) $\in p$ et donc à : $\xi^I.\xi^J \in p$, c'est-à-dire à : $a_{I,J}\xi^{I+J} \in p$. Ceci équivaut à $\xi^{I+J} \in p$ ou $a_{I,J} \in p$ ce qui se traduit par $I+J \in \omega(p)$ ou $a_{I,J} \in p_0(p)$.

$\omega(p)$ est bien un $p_0(p)$-pseudo-idéal premier de Γ.

3) La relation $\mathcal{J}(\omega(p)) \subset p$ est évidente. Il en résulte que $\mathcal{J}(\omega(p)) \subset p^h$. Réciproquement si $x = a\xi^I \in p^h \subset p$, nous avons $I \in \omega(p)$ ou $a \in p \cap A_0 = p_0(p)$, donc dans tous les cas $x \in \mathcal{J}(\omega(p))$.

Preuve de 1.b :

Si p est homogène et non nul, nous avons : $\mathcal{J}(\omega(p)) = p^h = p \neq (0)$ et donc : $(p_0(p),\omega(p)) \neq ((0),\emptyset)$. Si maintenant $p \subset \mathcal{J}(\Omega)$ nous avons : $p_0(p) \subset p_0(\mathcal{J}(\Omega)) = p_0$ et $\omega(p) \subset \omega(\mathcal{J}(\Omega)) = \Omega$. Comme (p_0,Ω) est de hauteur 1, il vient : $(p_0(p),\omega(p)) = (p_0,\Omega)$. On en déduit que :

$$\mathcal{J}(\Omega) = \mathcal{J}(\omega(p)) = p^h = p.$$

Preuve de 2.b :

Si p est de hauteur 1 et n'est pas homogène nous avons : $p^h = (0)$. Donc $\mathcal{J}(\omega(p)) = (0)$ et par suite $(p_0(p),\omega(p)) = ((0),\emptyset)$. Si p est homogène et de hauteur 1, le même raisonnement donne $(p_0(p),\omega(p)) \neq ((0),\emptyset)$. Prenons alors $(p_0,\Omega) \subset (p_0(p),\omega(p))$ et tel que $(p_0,\Omega) \neq ((0),\emptyset)$. Nous avons.

$$(0) \neq \mathcal{J}(\Omega) = \bigoplus_{I\in\overline{\Omega}} p_0\xi^I \oplus \bigoplus_{I\in\Omega} A_0\xi^I \subset \bigoplus_{I\in\overline{\omega(p)}} p_0\xi^I \oplus \bigoplus_{I\in\omega(p)} A_0\xi^I \subset \bigoplus_{I\in\overline{\omega(p)}} p_0(p)\xi^I \oplus \bigoplus_{I\in\omega(p)} A_0\xi^I$$

$$= \mathcal{J}(\omega(p))$$

Donc :

$$(0) \neq \mathcal{J}(\Omega) \subset \mathcal{J}(\omega(p)) = p^h = p.$$

Comme p est de hauteur 1 nous avons $\mathcal{J}(\Omega) = p$. Il en résulte que $p_0 = p_0(p)$ et $\Omega = \omega(p)$. La hauteur de $(p_0(p),\omega(p))$ est bien 1.

Le calcul de Cl(A) est donc possible par la détermination des p_0-pseudo-idéaux premiers Ω de Γ tels que le couple (p_0,Ω) soit de hauteur 1 (pour $p_0 \in \operatorname{Spec} A_0$). Cette détermination étant faite on calcule les $v_\Omega(f)$ pour f fonction rationnelle homogène sur Spec A (v_Ω étant la valuation dans $A_{\mathcal{J}(\Omega)}$). Le calcul de $v_\Omega(f)$ se ramène à celui des $v_\Omega(\xi^I)$ et des $v_\Omega(\bar{a})$ (pour $a \neq 0$ et appartenant à A_0).

IV - Exemples et applications

IV.1 - Nous allons appliquer ce qui précède à l'anneau $A = k_0[\Sigma]$ associé à un semi-groupe de congruence Σ. Les notations sont les suivantes : k_0 est un corps ; $a_1,\ldots,a_n,b$ sont des entiers tels que : p.g.c.d. $(a_1,\ldots,a_n,b) = 1$ et $0 \leq a_k < b$ (pour k de 1 à n) ;

$$\Sigma = \{R = (i,j_1,\ldots,j_n) \in \mathbb{N}^{n+1} \mid i+a_1j_1+\ldots+a_nj_n \equiv 0(b)\} ;$$

$$A = k_0[\Sigma] = \bigoplus_{R\in\Sigma} k_0\, X^R \qquad (\text{où} : X^R = X_0^i X_1^{j_1}\ldots X_n^{j_n}).$$

On sait que A est une k_0-algèbre, $\mathbb{N}^{n+1}$-graduée et normale [2]. De plus A vérifie l'hypothèse (H) ; en effet si $x \in A_R$ (pour $R \neq 0$) alors $x^b \in A_{(bR)}$.

Le seul idéal premier de $A_0 = k_0$ est l'idéal (0). Les seuls pseudo-idéaux premiers de Σ sont donc les idéaux premiers de Σ. Ceux de hauteur 1 sont $g_0+\Sigma,\ldots,g_n+\Sigma$ où les g_k sont les éléments du système minimal de générateurs $G(\Sigma)$ de Σ (qui est unique) qui sont sur les axes. Plus précisément nous avons : $g_k = b'_k\,\varepsilon_k$ où $b'_0 = b$ et $b'_k = \dfrac{b}{\text{p.g.c.d}\,(a_k,b)}$ (pour k = 1 à n).

Les idéaux premiers de hauteur 1 de A sont les $J(g_k+\Sigma) = X^{b'_k\varepsilon_k}A = X_k^{b'_k}A$.

Nous avons :

$$\mathrm{Div}_h(A) = \langle X_h^1(A)\rangle \simeq \mathbb{Z}^{n+1} .$$

D'autre part :

$$\mathrm{Prin}_h(A) \simeq \langle\Sigma\rangle \simeq \mathbb{Z}^{n+1}.$$

La suite exacte :

$$0 \to \mathrm{Prin}_h(A) \to \mathrm{Div}_h(A) \to \mathrm{Cl}(A) \to 0$$

est donc isomorphe à une suite exacte :

$$0 \to \mathbb{Z}^{n+1} \overset{\Pi}{\to} \mathbb{Z}^{n+1} \to \mathrm{Coker}\,\Pi \to 0$$

On voit sans trop de mal que Π est donnée par :

$$\Pi(1,j_1,\ldots,j_n) = (bl-a_1j_1-\ldots-a_nj_n,j_1,\ldots,j_n).$$

Si l'on définit une application Θ de $\mathbb{Z}^{n+1}$ dans $\mathbb{Z}/_{b\mathbb{Z}}$ par :

$$\Theta(x,y_1,\ldots,y_n) = \text{classe mod. } b \text{ de } (x+a_1y_1+\ldots+a_ny_n)$$

il est clair que Θ est surjective et que $\operatorname{Im}\Pi = \ker\Theta$. Donc : $\operatorname{Coker}\Pi \simeq \mathbb{Z}/_{b\mathbb{Z}}$.

<u>Le groupe des classes de diviseurs de l'algèbre $k_0[\Sigma]$ du semi-groupe de congruence $\Sigma = \Sigma_{b,a_1,\ldots,a_n}$ est $\mathbb{Z}/_{b\mathbb{Z}}$.</u>

L'anneau $k_0[\Sigma]$ est factoriel lorsque $b = 1$, c'est-à-dire lorsqu'il est régulier. C'est alors l'anneau de polynômes $k_0[X_0,\ldots,X_n]$.

<u>Remarque 4</u> :

On peut retrouver ce résultat directement par le théorème de descente galoisienne [8] [14]. En effet nous avons [2] :

$$A = k_0[\Sigma] = (k_0[X_0,\ldots,X_n])^{\mathbb{Z}/_{b\mathbb{Z}}}$$

L'anneau $k_0[X_0,\ldots,X_n]$ est factoriel et tout idéal premier divisoriel de cet anneau est non-ramifié sur A. L'action de $\mathbb{Z}/_{b\mathbb{Z}}$ sur les unités de cet anneau $k_0[X_0,\ldots,X_n]$ (i.e. sur k_0^*) est $\gamma * x = x$. Nous avons donc :

$$\operatorname{Cl}(A) = H^1(\mathbb{Z}/_{b\mathbb{Z}}, k_0^*) = \mathbb{Z}/_{b\mathbb{Z}} .$$

IV.2 - Nous allons maintenant faire la même chose pour les anneaux $A = \bigoplus_{I\in\mathbb{N}^n} \mathcal{O}\xi^I$ qui apparaissent dans les fibrations de Seifert singulières. Nous avons étudié ces anneaux dans [3], auquel nous renvoyons pour les notations, et aussi dans [5]. Ils vérifient les hypothèses du théorème 1 [3] [5].

Ici l'anneau de valuation $A_0 = \mathcal{O}$ admet deux idéaux premiers : (0) et $t\mathcal{O}$. Pour (0) les pseudo-idéaux premiers de hauteur 1 sont les idéaux premiers de hauteur 1 de $\Gamma = \mathbb{N}^n$. Ce sont les $\varepsilon_j + \mathbb{N}^n$ (pour $j = 1$ à n).

Les idéaux premiers de hauteur 1 de A associés sont les $\mathcal{J}(\Omega_j) = \bigoplus_{I\in\mathbb{N}^n} \mathcal{O}\xi^{\varepsilon_j+I}$.

Nous avons vu que (exemple 2 de III.2) :

$$\Omega_0 = \{I \mid \text{il existe } K \text{ tel que } a_{I,K} \in t\mathcal{O}\}$$

c'est-à-dire :

$$\Omega_0 = \{I \mid \text{il existe } K \text{ tel que } \delta(I+K) > \delta(I) + \delta(K)\}$$

est un $t\mathcal{O}$-pseudo-idéal premier de $\mathbb{N}^n$.

Pour tout j de 1 à n nous avons : $b\varepsilon_j \in \Omega_j$ et $b\varepsilon_j \notin \Omega_0$. Il en résulte que $((0),\Omega_j) \not\subset (t\mathcal{O},\Omega_0)$ et donc que pour $\Omega \neq \emptyset$, $((0),\Omega) \not\subset (t\mathcal{O},\Omega_0)$.

Soit $I \in \Omega_0$, nous avons $2I \in \Omega_0$ ou $a_{I,I} \in t\mathcal{O}$. Dans le premier cas nous avons $3I \in \Omega_0$ ou $a_{2I,I} \in t\mathcal{O}$, etc ... Compte tenu de $bI \notin \Omega_0$ nous pouvons affirmer qu'il existe λ $(1 \leq \lambda \leq b-1)$ tel que $a_{\lambda I,I} \in t\mathcal{O}$.

Soit Ω un $t\mathcal{O}$-pseudo-idéal premier. Nous avons donc soit $\lambda I \in \Omega$ soit $I \in \Omega$. Dans le premier cas nous avons $(\lambda-1)I+I \in \Omega$ et donc si $\lambda \geq 2$ $(\lambda-1)I \in \Omega$ ou $I \in \Omega$. On voit ainsi que $I \in \Omega$ et donc que $\Omega_0 \subset \Omega$. Ceci prouve que $(t\mathcal{O},\Omega_0)$ est de hauteur 1.

Les fonctions rationnelles homogènes sur Spec A sont les $f = t^{\alpha}\dfrac{\xi^{I'}}{\xi^{I}}$ où

où $\alpha \in \mathbb{Z}$. Nous avons donc un isomorphisme φ de $\mathrm{Prin}_h(A)$ sur $\mathbb{Z}^{n+1}$ défini par :

$$\varphi(f) = (\alpha+\delta(I)-\delta(I')),\ i'_1-i_1,\dots,i'_n-i_n) .$$

Partant de $I = i_j\varepsilon_j+K$ (avec $K \notin \varepsilon_j+\mathbb{N}^n$) on obtient :

$$\xi^I\ t^{\delta(I)-\delta(K)} = \xi^K\ (\xi^{\varepsilon_j})^{i_j} .$$

Nous avons pour les valuations $A_{\mathfrak{J}(\Omega_j)}$:

$$v_j(\bar{t}) = 0,\ v_j(\overline{\xi^K}) = 0 \text{ et } v_j(\overline{\xi^{\varepsilon_j}}) = 1.\ \text{D'où : } v_j(\overline{\xi^I}) = i_j.$$

Le coefficient de $\mathfrak{J}(\Omega_j)$ dans le diviseur principal (f) est donc : i'_j-i_j.
Les relations :

$$(\xi^{\varepsilon'_j})^b = t^{a_j}\ \xi^{b\varepsilon_j},\ \xi^I\ t^{\delta(I)} = \prod_{j=1}^{n} (\xi^{\varepsilon_j})^{i_j} \text{ et } b\varepsilon_j \in \Omega_0$$

permettent de voir que dans $A_{\mathfrak{J}(\Omega_0)}$ nous avons :

$$v_0(\bar{t}) = b,\ v_0(\overline{\xi^{\varepsilon_j}}) = a_j \text{ et } v(\overline{\xi^I}) = \sum_{j=1}^{n} a_j i_j - b\delta(I).$$

Nous avons donc :

$$(f) = (b\alpha+b\delta(I)-b\delta(I')) + \sum_{j=1}^{n} a_j(i'_j-i_j))\mathcal{J}(\Omega_0) + \sum_{j=1}^{n} (i'_j-i_j)\mathcal{J}(\Omega_j).$$

L'isomorphisme φ entre $\mathrm{Prin}_h(A)$ et $\mathbb{Z}^{n+1}$ permet de dire que la suite exacte

$$0 \to \mathrm{Prin}_h(A) \to \mathrm{Div}_h(A) \to \mathrm{Cl}(A) \to 0$$

est isomorphe à une suite exacte :

$$0 \to \mathbb{Z}^{n+1} \overset{\Pi}{\to} \mathbb{Z}^{n+1} \to \mathrm{Coker}\,\Pi \to 0$$

où :

$$\Pi(l,j_1,\ldots,j_n) = (bl + \sum_{k=1}^{n} a_k j_k,\ j_1,\ldots,j_n).$$

En composant avec l'application Θ de $\mathbb{Z}^{n+1}$ dans $\mathbb{Z}/_{b\mathbb{Z}}$ définie par :

$$\Theta(x,y_1,\ldots,y_n) = \text{classe mod } b \text{ de } (x-a_1y_1-\ldots-a_ny_n)$$

On obtient : $\mathrm{Coker}\,\Pi \simeq \mathbb{Z}/_{b\mathbb{Z}}$. D'où :

$$\boxed{\mathrm{Cl}(A) = \mathbb{Z}/_{b\mathbb{Z}}}$$

Remarque 5 :

Une explication de ce résultat doit être la suivante :

Soient $\mathfrak{m}$ l'idéal maximal de A : $\mathfrak{m} = t\mathcal{O} \oplus \underset{I\neq 0}{A_I}$ et $\mathfrak{m}_0$ l'idéal maximal de $k_0[\Sigma]$: $\mathfrak{m}_0 = \bigoplus_{R\neq 0} k_0X^R$. Les complétés $\hat{A}$ et $\widehat{k_0[\Sigma]}$ respectifs de A et $k_0[\Sigma]$ pour les topologies $\mathfrak{m}$-adique et $\mathfrak{m}_0$-adique sont tous deux égaux à $k_0[[\Sigma]]$, l'algèbre des séries entières en X^R pour $R \in \Sigma$ et donc égaux entre eux [3].

D'autre part il n'y a que des singularités rationnelles [10]. Je pense donc que nous avons : $Cl(A) = Cl(\hat{A})$ et $Cl(k_0[\Sigma]) = Cl(\widehat{k_0[\Sigma]})$. D'où :

$$Cl(A) = Cl(k_0[\Sigma]) = \mathbb{Z}/_{b\mathbb{Z}} .$$

IV.3 - On reprend les hypothèses du paragraphe III.2, c'est-à-dire que A est un anneau noethérien, normal $\mathbb{N}^n$-gradué tel que pour tout I A_I est nul ou un A_0-module libre de rang 1.

Soit $\mathfrak{m}$ un idéal maximal de A_0. Alors $S_{\mathfrak{m}} = C_{A_0}\mathfrak{m}$ est une partie multiplicative de A formée d'éléments homogènes (de degré 0). L'anneau $S_{\mathfrak{m}}^{-1} A$ est noethérien, normal et $\mathbb{N}^n$-gradué par : $d^o(\frac{x}{s}) = d^o x$ (pour x homogène dans A). Le sous-anneau $(S_{\mathfrak{m}}^{-1} A)_0$ est égal à $A_{0_{\mathfrak{m}}}$ et est donc local. Si : $A = \bigoplus_{I \in \Gamma} A_0 \xi^I$

(voir III.2) nous avons : $S_{\mathfrak{m}}^{-1} A = \bigoplus_{I\in\Gamma} (S_{\mathfrak{m}}^{-1} A)_0 \zeta^I$, où $\zeta^I = \dfrac{\xi^I}{1}$.

Si A vérifie l'hypothèse (H) on vérifie aussitôt que $S_{\mathfrak{m}}^{-1}A$ vérifie aussi l'hypothèse (H).

Si A est tel que pour chaque j de 1 à n il existe T_j dans le corps des fractions Fr(A) de A de degré ε_j alors $T_j \in Fr(S_{\mathfrak{m}}^{-1}A) = Fr(A)$ et son degré est toujours ε_j. Donc si A vérifie les hypothèses du théorème 1, $S_{\mathfrak{m}}^{-1}A$ les vérifie aussi.

Un théorème de Nagata [8] [14] nous permet de dire qu'il existe une surjection φ :

$$Cl(A) \overset{\varphi}{\to} \prod_{\mathfrak{m}\in \mathrm{Max}\, A_0} Cl(S_{\mathfrak{m}}^{-1}A) \to 0$$

dont le noyau Ker φ est engendré par les classes des idéaux premiers p de hauteur 1 de A tels que : $p \cap S_{\mathfrak{m}} \neq \emptyset$ pour tout $\mathfrak{m} \in \mathrm{Max}\, A_0$. Mais alors $p \cap A_0$ est un idéal premier de A_0 donc il existe $\mathfrak{m}$ tel que $p \cap A_0 \subset \mathfrak{m}$ et par suite $p \cap S_{\mathfrak{m}} = \emptyset$. Il en résulte que Ker $\varphi = 0$ et

$$\boxed{Cl(A) \simeq \prod_{\mathfrak{m} \in \mathrm{Max}\, A_0} Cl(S_{\mathfrak{m}}^{-1}A)}$$

Ce résultat est à rapprocher d'un résultat de P. Samuel [8] cor. 10.3, ou [14] prop. 7.4.

Remarque 6 :

Si l'on pose $\mathfrak{m}' = \mathfrak{m} \oplus A_+ = \mathfrak{m} \oplus \bigoplus_{I \neq 0} A_I$ alors $\mathfrak{m}'$ est un idéal maximal de A

et le théorème de Nagata donne une surjection ψ :

$$Cl(A) \xrightarrow{\psi} \prod_{\mathfrak{m} \in \mathrm{Max}\, A_0} Cl(A_{\mathfrak{m}'}) \to 0.$$

Mais ici le même raisonnement donne $\mathrm{Ker}\,\psi \cap X_h^1(A) = \emptyset$ et il ne semble pas facile de voir que $\mathrm{Ker}\,\psi = 0$.

Remarque 7 :

Dans le cas où A_0 est un corps ou dans le cas où A_0 est un anneau de valuation il y a un seul idéal maximal et le résultat précédent perd tout intérêt. On obtient $Cl(A) \simeq Cl(A)$.

Remarque 8 :

Dans le cas où $A_0 = k_0[t]$ (où k_0 est un corps) et où A est une A_0-algèbre de type fini normale qui vérifie (H) nous avons :

$$\mathrm{Max}\, A_0 = \{(t-\alpha)A_0 \mid \alpha \in k_0\}$$

et

$$A_\alpha = S^1_{(t-\alpha)A_0} A = \bigoplus_{I \in \mathbb{N}^n} \mathcal{O}_\alpha \zeta^I$$

où $\mathcal{O}_\alpha = k_0[t]_{(t-\alpha)}$ est un anneau de valuation et $\zeta^I = \frac{\xi^I}{1}$. Alors A est une $\mathcal{O}_\alpha$-algèbre de type fini, normale et qui vérifie (H). Elle est donc du type étudié au IV.2. [5]. Donc il existe un certain b_α [3] tel que :

$$Cl(A_\alpha) = \mathbb{Z}/_{b_\alpha}\mathbb{Z}$$

et :

$$\boxed{Cl(A) = \prod_{\alpha \in k_0} \mathbb{Z}/_{b_\alpha}\mathbb{Z}}$$

IV.4 - On peut utiliser les résultats de IV.2 et IV.3 pour calculer le groupe des classes de diviseurs des anneaux B construits dans [3] (§ III) au moyen de n $\mathbb{Q}$-diviseurs de Weil pris sur une droite affine. Nous reprenons les notations de [3]. Ici Max B_0 = Max $\mathbb{C}[t] = \{(t-\alpha)\mathbb{C}[t] \mid \alpha \in \mathbb{C}\}$. On vérifie sans trop de difficulté que :

$$S^{-1}_{(t-\alpha)\mathbb{C}[t]}B = B_{\mathfrak{m}_\alpha} \qquad \text{où } \mathfrak{m}_\alpha = (t-\alpha)\mathbb{C}[t] \bigoplus_{I\neq 0} B_I .$$

Si $\alpha \notin \Delta$, $B_{\mathfrak{m}_\alpha}$ est un anneau de polynômes et donc $Cl(B_{\mathfrak{m}_\alpha}) = 0$ (en fait $b_\alpha = 1$ dans ce cas). Nous avons donc :

$$Cl(B) = \prod_{\alpha\in\mathbb{C}} Cl(B_{\mathfrak{m}_\alpha}) = \prod_{\alpha\in\Delta} Cl(B_{\mathfrak{m}_\alpha})$$

soit :

$$\boxed{Cl(B) = \prod_{\alpha\in\Delta} \mathbb{Z}/b_\alpha\mathbb{Z}}$$

IV.5 - Nous allons maintenant calculer Cl(A) pour $A = \bigoplus_{I\in\mathbb{N}^2} \mathcal{O}\,\xi^I$ où $\mathcal{O}$ est un anneau de valuation discrète et où la multiplication de ξ^I est donnée par :

$$\xi^I.\xi^J = t^{\delta(I+J)-\delta(I)-\delta(J)}\xi^{I+J} \quad \text{avec } \delta(I) = \inf(i_1,i_2).$$ t notant une uniformisante de $\mathcal{O}$. (Voir I.7.4. dans [3] et V de la partie III dans [5]).

Un tel anneau ne vérifie pas (H) mais vérifie l'autre version des hypothèses du théorème 1.

L'anneau $A_0 = \mathcal{O}$ a deux idéaux premiers : (0) et $t\mathcal{O}$. Les (0)-pseudo-idéaux premiers Ω de $\mathbb{N}^2$ tels que $((0),\Omega)$ soit de hauteur 1 sont les idéaux premiers de hauteur 1 de $\mathbb{N}^2$. Ils sont au nombre de deux : $\Omega_1 = \varepsilon_1 + \mathbb{N}^2$ et $\Omega_2 = \varepsilon_2 + \mathbb{N}^2$.

Les idéaux premiers de hauteur 1 de A correspondants sont :

$$\mathcal{J}(\Omega_1) = \bigoplus_{I\in\mathbb{N}^2} \mathcal{O}\xi^{\varepsilon_1+I} \quad \text{et} \quad \mathcal{J}(\Omega_2) = \bigoplus_{I\in\mathbb{N}^2} \mathcal{O}\xi^{\varepsilon_2+I} .$$

Il est facile de voir que :

$$\Omega_1' = \{I \mid i_1 > i_2\} \quad \text{et} \quad \Omega_2' = \{I \mid i_1 < i_2\}$$

sont des $t\mathcal{O}$-pseudo-idéaux premiers de $\mathbb{N}^n$. Les idéaux premiers correspondants sont :

$$J(\Omega_1') = \bigoplus_{I|i_1<i_2} t\mathcal{O}\xi^I \quad \bigoplus_{I|i_1>i_2} \mathcal{O}\xi^I = t\,A + \xi^{\varepsilon_1} A$$

et

$$J(\Omega_2') = \bigoplus_{I|i_1>i_2} t\mathcal{O}\xi^I \quad \bigoplus_{I|i_1<i_2} \mathcal{O}\xi^I = t\,A + \xi^{\varepsilon_2} A.$$

Nous avons : $\Omega_1 \not\subset \Omega_1'$ et $\Omega_2 \not\subset \Omega_1'$. Donc aucun couple du type $((0),\Omega)$ n'est inclus dans $(t\mathcal{O},\Omega_1')$ (pour $\Omega \neq \emptyset$) et, de même, n'est inclus dans $(t\mathcal{O},\Omega_2')$.

Si maintenant on prend un couple $(t\mathcal{O},\Omega)$, nous avons :

$\xi^{\varepsilon_1}.\xi^{\varepsilon_2} = t\xi^{\varepsilon_1+\varepsilon_2} \in J(\Omega)$. Donc ξ^{ε_1} (ou ξ^{ε_2}) $\in J(\Omega)$. Par suite $J(\Omega_1')$ (ou $J(\Omega_2')$) $\subset J(\Omega)$. Il en résulte que Ω_1' (ou Ω_2') est inclus dans Ω. Les couples $(t\mathcal{O},\Omega_1')$ et $(t\mathcal{O},\Omega_2')$ sont donc de hauteur 1.

On voit facilement que pour les valuations v_1, v_2, v_1', v_2' respectivement attachées à $A_{J(\Omega_1)}$, $A_{J(\Omega_2)}$, $A_{J(\Omega_1')}$ et $A_{J(\Omega_2')}$ nous avons :

$$v_j(\bar{t}) = 0,\ v_j(\overline{\xi^I}) = i_j,\ v_j'(\bar{t}) = 1 \text{ et } v_j'(\overline{\xi^I}) = i_j - \delta(I).$$

On en déduit que pour la fraction rationnelle homogène $f = t^\alpha \dfrac{\xi^{I'}}{\xi^I}$ sur Spec A, le diviseur associé est :

$$(f) = (i_1'-i_1)J(\Omega_1)+(i_2'-i_2)J(\Omega_2)+[\alpha+i_1'-i_1+\delta(I)-\delta(I')]J(\Omega_1')$$
$$+[(\alpha+i_2'-i_2+\delta(I)-\delta(I')]J(\Omega_2').$$

Ces diviseurs sont, parmi les diviseurs $m_1J(\Omega_1) + m_2J(\Omega_2) + m_1'J(\Omega_1') + m_2'J(\Omega_2')$, **ceux tels que :** $m_1' - m_1' + m_2 = m_2' = 0$. Il en résulte que :

$$\boxed{Cl(A) = \mathbb{Z}}$$

V - QUELQUES QUESTIONS

V.1. Le théorême 1 reste-t-il vrai si l'on suppose uniquement que A est noethérien, normal et $\mathbb{N}^n$- gradué ?

V.2. Que peut-on dire sur ht $\mathfrak{p}_0$ lorsque ht $(\mathfrak{p}_0, \Omega) = 1$?

V.3. Dans le point 1,b du théorême 2, que peut-on dire sur ht $(\mathfrak{I}(\Omega))$ lorsque ht $(\mathfrak{p}_0,\Omega) = 1$?

V.4. A la fin du III, peut-on trouver une description de la structure de Ω qui donne explicitement les valeurs de $v_\Omega(\xi^I)$ et $v_\Omega(a)$?

Post Scriptum (juin 1988) :

Le théorême 5.9 (p. 565) de l'article de D.D. ANDERSON et D.F. ANDERSON, "Divisorial Ideals and Invertible Ideals in a Graded Integral Domain" (J. of Algebra 76-2- 1982. p. 549-569) redonne notre théorême 1. Il permet de plus de répondre positivement à la question V.1.

Je remercie Monsieur le Professeur F. VAN OYSTAEYEN, de l'Université d'Antwerpen (Anvers) qui m'a donné ce renseignement. Il est l'auteur de nombreux articles sur les anneaux gradués et de deux ouvrages écrits en collaboration sur ce même thème : "Graded Ring Theory" (North Holland 1982) et "Graded Orders" (Birkhauser 1988).

BIBLIOGRAPHIE

[1] J. BERTIN. Automorphismes des surfaces non complètes et surfaces affines réglées. Thèse. Université Paul Sabatier, Toulouse (1981).

[2] P. CARBONNE. Sur les singularités cycliques quotient. Colloque d'Algèbre. Université de Rennes I (1980) p. 187-200.

[3] P. CARBONNE. Algèbres graduées normales. J. Of Algebra, Vol. 102, n° 2, Sept. 1986, p. 332-352.

[4] P. CARBONNE. Algèbres graduées normales II. Colloque d'Algèbre. Université de Rennes I (1985), p. 71-95.

[5] P. CARBONNE. Algèbres graduées normales III. Preprint (1987).

[6] M. DEMAZURE. Anneaux gradués normaux. Séminaire Demazure-Guiraud-Teissier sur les singularités de surfaces. Ecole Polytechnique (1979).

[7] I.V. DOLGACHEV. Automorphic forms and quasihomogeneous singularities. Funet. Anal. and Applic. (1976), p. 149-151.
Traduit de :
Funkt Anal. i Ego Prilo., Vol. 9, n° 2, (1975), p. 67-68.

[8] R.M. FOSSUM. The divisor-class group of a Krull domain. Ergebnisse der Mathematik n° 74, Springer Verlag (1973).

[9] S. GOTO et K. WATANABE. On graded rings II ($\mathbb{Z}^n$-graded rings). Tokyo J. Math., Vol. 1, n° 2, (1978), p. 237-261.

[10] G. KEMPF, F. KNUDSEN, D. MUMFORD et B. SAINT-DONAT. Toroidal Embeddings I. Lecture Notes n° 339, Springer Verlag (1973).

[11] S. MORI. Graded factorial domains. Japan. J. Math. Vol. 3, n° 2, (1977), p. 223-238.

[12] P. ORLIK et P. WAGREICH. Singularities of algebraic surfaces with $\mathbb{C}^*$-action. Math. Ann. 193 (1971) p. 121-135.

[12 bis] P. ORLIK et P. WAGREICH. Algebraic Surfaces with k*-action. Acta Math. 138 (1977) p. 43-81.

[13] H. PINKHAM. Normal singularities with $\mathbb{C}^*$-action. Math. Ann. 227 (1977), p. 183-193.

[14] P. SAMUEL. Lectures on Unique Factorization Domains. Tata Institute n° 30. Bombay (1964).

[15] P. SAMUEL. Sur les anneaux factoriels. Bull. S.M.F. 89, (1961), p. 155-173.

[16] K. WATANABE. Some remarks concerning Demazure's construction of normal graded rings. Nagoya Math. J. 83 (1981), p. 203-211.

[17] O. ZARISKI et P. SAMUEL. Commutative Algebra I et II. G.T.M. n° 28 et 29 (2ème ed.), Springer Verlag (1975).

Differential operators on smooth varieties

S.C. Coutinho
Departamento de Matemática,
Universidade Federal de Pernambuco,
50 730 Recife, Pernambuco, Brazil.

M.P. Holland
Department of Pure Mathematics,
University of Leeds,
Leeds LS2 9JT, England.

1. Introduction.

In recent years there has been a growing interest in rings of differential operators and their modules, the so-called $\mathcal{D}$-modules. Applications of these rings have been found in areas as diverse as partial differential equations, algebraic geometry and Lie algebras. This has led a number of algebraists to study these rings from the point of view of non-commutative ring theory. In this context, rings of differential operators occur naturally as generalisations of the Weyl algebra—if the underlying variety is affine—and as quotients of enveloping algebras—if the underlying variety is projective. The case of an affine variety has been the most extensively studied from this point of view: see the references in [McConnell and Robson 1987], for example. In this paper we study the smooth projective case in some detail.

As one would expect, the geometry of the underlying variety is responsible for many of the algebraic properties of its ring of differential operators. If the variety is smooth and affine, it is easy to keep the use of algebraic geometry to a minimum, whilst studying the ring of differential operators. In particular one can do without any sheaf theory at all. The same cannnot be said when the underlying variety is projective. In this case the ring of differential operators itself is most naturally defined as the global sections of a certain sheaf. Since ring theorists are the main intended audience for this paper we have tried to keep the use of sheaf theory and cohomology to a minimum. In particular, we have made an effort to define the sheaf of rings of differential operators as concretely as possible.

We should make it clear that the main results of this paper are not new. We only try to give a more concrete approach to the Beilinson-Bernstein theory of rings of differential operators over $\mathcal{D}$-affine varieties, with emphasis on the projective space, $\mathbf{P}^n$. We believe that $\mathcal{D}(\mathbf{P}^n)$ deserves to take its place alongside the Weyl algebra, $\mathcal{D}(\mathbf{A}^n)$, in the ring theorist's armoury of examples.

We now review, briefly, the contents of the sections. In section 2 we collect definitions, examples and some basic results. The reader's attention is drawn to the definition of a $\mathcal{D}$-variety. In essence, this is a variety for which the sheaf of differential operators is generated by its global sections. Affine varieties and complete homogeneous G-spaces for G a connected semi-simple algebraic group are examples of $\mathcal{D}$-varieties. The main result in this section is the Beilinson-Bernstein Localisation Theorem (Theorem 2.6). For rings of differential operators on a $\mathcal{D}$-variety, this theorem establishes an equivalence between sheaves of modules and their global

This paper is in final form and no version of it will be submitted for publication elsewhere.

sections. We do not give a proof of this result; an elementary proof, in the special case of projective space, can be found in [Hodges and Smith 1985c].

In section 3 we introduce locally free sheaves. As is well known, in the commutative case these sheaves correspond to vector bundles (and hence to projective modules, if the underlying variety is affine). Using these sheaves we define some rings of twisted differential operators. We also consider sheaves which are locally free over the ring of differential operators. In particular it is easy to achieve a classification of sheaves of locally free $\mathcal{D}$-modules of rank one, exactly as in the commutative case. This will be central to our work in section 4.

The special case of the (twisted) ring of differential operators on projective space is studied in detail in the last section. We present this ring as a subring of the Weyl algebra and study its two-sided ideal structure. At the end of the section we compare the twisted rings of diferential operators, for different twistings, and decide when they are Morita equivalent. These results depend on a detailed description of locally free modules of rank one over the ring of differential operators on $\mathbf{P}^n$.

Two comments are in order. Firstly, holonomic modules—and most of the ideas in that theory—do not arise in this context. In a sense they are "small" modules (torsion modules) whereas we deal only with "large" modules (locally free modules). For a survey on the theory of holonomic modules see [Borel 1987]. Secondly, the approach to (twisted) rings of differential operators in section 4 is an offshoot of the authors' work on locally free modules over rings of differential operators. A more detailed study of these modules will be found in the forthcoming [Coutinho and Holland 1988].

The first named author would like to thank the University of Leeds for its hospitality during a visit, when this paper was conceived. During this visit he received financial support from CNPq (Brazil). The second named author is grateful to the SERC (UK) for their financial support.

2. Rings of differential operators.

In this section we define the ring of differential operators, its associated sheaf, and establish some basic results. We assume that the reader is familiar with sheaf theory at the level of [Iitaka 1982, Chapter 1]. A rather painless introduction to the subject can be found in [Seebach et al 1970]. For the sake of briefness it is useful to fix the following nomenclature for the rest of the paper. A *variety* $\mathcal{X}$ will always be an algebraic irreducible smooth variety over $\mathbf{C}$ and all *topological* notions refer to the Zariski topology.

Let A be a commutative regular $\mathbf{C}$-algebra, which is a domain. Then *the ring of differential operators on* A, denoted by $\mathcal{D}(A)$, is the subring of $\mathrm{End}_{\mathbf{C}}A$ generated by $\mathrm{Der}_{\mathbf{C}}A$, the A-module of $\mathbf{C}$-linear derivations of A, and A (acting as multiplication). This ring has a natural filtration over A based on $\mathrm{Der}_{\mathbf{C}}A$ as a generating set. In other words, $\mathcal{D}^i(A)$, the i-th component of the filtration, is the A-module generated by all products of at most i derivations from $\mathrm{Der}_{\mathbf{C}}A$. One readily checks that if $\alpha \in \mathcal{D}^i(A)$ and $\beta \in \mathcal{D}^j(A)$ then $[\alpha, \beta] = \alpha\beta - \beta\alpha \in \mathcal{D}^{i+j-1}(A)$. Thus the associated graded ring

$$\operatorname{gr}\mathcal{D}(A) = \bigoplus_{i=0}^{\infty} \mathcal{D}^i(A)/\mathcal{D}^{i-1}(A)$$

is a commutative domain.

The most important example of this construction occurs when $A = \mathcal{O}(\mathcal{X})$, the coordinate ring of an affine variety $\mathcal{X}$. In this case we write simply $\mathcal{D}(\mathcal{X})$ instead of $\mathcal{D}(\mathcal{O}(\mathcal{X}))$. A few properties of $\mathcal{D}(\mathcal{X})$ are listed below.

PROPOSITION 2.1. *Let $\mathcal{X}$ be an affine variety and $A = \mathcal{O}(\mathcal{X})$, its coordinate ring.*

(a) *If $S \subseteq A \setminus \{0\}$ is a multiplicatively closed set then S is a left and right Ore set of $\mathcal{D}(\mathcal{X})$ and $\mathcal{D}(A_S) = \mathcal{D}(\mathcal{X})_S$.*

(b) *$\mathcal{D}(\mathcal{X})$ is a simple left and right noetherian domain.*

(c) *The Krull and global dimensions of $\mathcal{D}(\mathcal{X})$ are both equal to $\dim\mathcal{X}$.*

(d) *If $\mathfrak{m}$ is a maximal ideal of A then $\mathcal{D}(A_\mathfrak{m}) = A_\mathfrak{m}[x_1; \partial/\partial y_1] \cdots [x_n; \partial/\partial y_n]$ is an iterated Ore extension of $A_\mathfrak{m}$ by n commuting derivations, where $n = \dim\mathcal{X}$ and $\{y_1, \ldots, y_n\}$ generate $\mathfrak{m}A_\mathfrak{m}$.*

PROOF: [McConnell and Robson 1987, Chapter 15].

The reader should observe that, for a non-affine variety $\mathcal{X}$, our previous definition of $\mathcal{D}(\mathcal{X})$ may give a trivial ring. As an example consider the n-th complex projective space $\mathbb{P}^n_\mathbb{C}$. Its ring of global regular functions, $\mathcal{O}(\mathbb{P}^n)$, is $\mathbb{C}$. Thus the definition which we gave above would give $\mathcal{D}(\mathbb{P}^n) = \mathcal{D}(\mathcal{O}(\mathbb{P}^n)) = \mathbb{C}$. We avoid this by sheafifying our definition.

Consider first the case when $\mathcal{X}$ is affine. Our aim is to construct a sheaf of non-commutative rings which has $\mathcal{D}(\mathcal{X})$ as its ring of global sections. Recall the definition of the structure sheaf of $\mathcal{X}$, denoted by $\mathcal{O}_\mathcal{X}$. Let $U \subseteq \mathcal{X}$ be an open subset. The ring of sections of $\mathcal{O}_\mathcal{X}$ over U is defined by

$$\mathcal{O}(U) = \bigcap_{x \in U} \mathcal{O}_x \subset \mathbb{C}(\mathcal{X})$$

where $\mathcal{O}_x = \mathcal{O}(\mathcal{X})_x$, identifying every $x \in \mathcal{X}$ with its corresponding maximal ideal under the *Nullstellensatz*. As usual $\mathbb{C}(\mathcal{X})$ denotes the field of rational functions of $\mathcal{X}$, or equivalently, the quotient field of $\mathcal{O}(\mathcal{X})$. Clearly $\mathcal{O}(\mathcal{X}) = \Gamma(\mathcal{X}, \mathcal{O}_\mathcal{X})$, the ring of global sections of $\mathcal{O}_\mathcal{X}$.

To define the sheaf, $\mathcal{D}_\mathcal{X}$, of rings of differential operators on $\mathcal{X}$ it is enough to mimic the construction above. First a technical lemma.

LEMMA 2.2. *Let $\mathcal{X}$ be an affine variety and $U \subseteq \mathcal{X}$ an open set. Then*

$$\mathcal{D}(\mathcal{O}(U)) = \bigcap_{x \in U} \mathcal{D}(\mathcal{O}_x).$$

PROOF: Let $\theta \in \bigcap_{x\in U} \mathcal{D}(\mathcal{O}_x)$. Then $\theta(\mathcal{O}_x) \subseteq \mathcal{O}_x$ for every $x \in U$. Thus

$$\theta(\mathcal{O}(U)) = \theta(\bigcap_{x\in U} \mathcal{O}_x) \subseteq \bigcap_{x\in U} \theta(\mathcal{O}_x) \subseteq \mathcal{O}(U).$$

Hence $\mathcal{D}(\mathcal{O}(U)) \supseteq \bigcap_{x\in U} \mathcal{D}(\mathcal{O}_x)$. The other inclusion follows from a slightly more general version of Proposition 2.1(a) and is omitted.

If U is an open set of $\mathcal{X}$ the section $\mathcal{D}(U)$ of $\mathcal{D}_{\mathcal{X}}$ over U is now defined by

$$\mathcal{D}(U) = \mathcal{D}(\mathcal{O}(U)) = \bigcap_{x\in U} \mathcal{D}(\mathcal{O}_x).$$

If $U \subseteq V$ are open subsets of $\mathcal{X}$, then $\mathcal{O}(V) \subseteq \mathcal{O}(U)$. It follows that $\mathcal{D}(V) \subseteq \mathcal{D}(U)$ which immediately gives us the restriction maps for $\mathcal{D}_{\mathcal{X}}$. It is now easy to see that this defines a sheaf of rings over $\mathcal{X}$ and that $\mathcal{D}(\mathcal{X}) = \Gamma(\mathcal{X}, \mathcal{D}_{\mathcal{X}})$, is its ring of global sections.

We now consider a general variety $\mathcal{X}$. Once again let us start by defining the structure sheaf $\mathcal{O}_{\mathcal{X}}$. If $x \in \mathcal{X}$ it is easy to define the stalk $\mathcal{O}_x$. Choose an open affine subset V of $\mathcal{X}$ with $x \in V$. Now define $\mathcal{O}_x = \mathcal{O}(V)_x$. One sees immediately that this does not depend on the choice of the affine neighbourhood. The ring of sections of $\mathcal{O}_{\mathcal{X}}$ over an open set $U \subseteq \mathcal{X}$ is defined by $\mathcal{O}(U) = \bigcap_{x\in U} \mathcal{O}_x$, just as before. Analogously, the sheaf $\mathcal{D}_{\mathcal{X}}$ of rings of differential operators on $\mathcal{X}$ is defined by

$$\mathcal{D}(U) = \bigcap_{x\in U} \mathcal{D}(\mathcal{O}_x) \tag{2.3}$$

for each open subset U of $\mathcal{X}$. If $x \in \mathcal{X}$, the stalk is $\mathcal{D}(\mathcal{O}_x)$, which will be denoted by $\mathcal{D}_x$.

Let $\{V_1, \dots, V_n\}$ be an open affine cover of $\mathcal{X}$. It follows immediately from (2.3) that

$$\mathcal{D}(\mathcal{X}) = \bigcap_{i=1}^{n} \mathcal{D}(V_i). \tag{2.4}$$

This is the ring of global sections of $\mathcal{D}_{\mathcal{X}}$. From now on $\mathcal{D}(\mathcal{X})$ will be referred to as the *ring of differential operators on* $\mathcal{X}$. As we have seen, this is compatible with the previous definition when $\mathcal{X}$ is affine. The next example shows that in general $\mathcal{D}(\mathcal{X}) \neq \mathcal{D}(\mathcal{O}(\mathcal{X}))$.

EXAMPLE 2.5. Let $\mathbf{P}^1$ be projective 1-space over $\mathbf{C}$. In homogeneous coordinates the points of $\mathbf{P}^1$ can be written as $(x_0 : x_1)$, where $x_0, x_1 \in \mathbf{C}$. Let $V_i = \{(x_0 : x_1) : x_i \neq 0\}$ for $i = 0, 1$. Then $\{V_0, V_1\}$ is an open affine cover of $\mathbf{P}^1$. If $t = x_1/x_0$ then $\mathcal{O}(V_0) = \mathbf{C}[t]$ and $\mathcal{O}(V_1) = \mathbf{C}[t^{-1}]$. It is also easy to check that $\mathcal{D}(V_0) = \mathbf{C}[t, \partial]$ and $\mathcal{D}(V_1) = \mathbf{C}[t^{-1}, t^2\partial]$, where $\partial = \partial/\partial t$. Notice that $\mathcal{D}(V_i) \simeq A_1(\mathbf{C})$. By (2.4) it follows that:

$$\mathcal{D}(\mathbf{P}^1) = \mathcal{D}(V_0) \cap \mathcal{D}(V_1) \subset \mathcal{D}(V_0).$$

It is clear that $\partial, t\partial, t^2\partial \in \mathcal{D}(\mathbf{P}^1)$. An elementary calculation shows that these elements generate $\mathcal{D}(\mathbf{P}^1)$ over $\mathbf{C}$ as a subalgebra of $\mathcal{D}(V_0)$, and satisfy the following relations

$$[\partial, t\partial] = \partial \quad [\partial, t^2\partial] = 2t\partial \quad [t\partial, t^2\partial] = t^2\partial.$$

This can be used to show that $\mathcal{D}(\mathbf{P}^1)$ is a quotient of the enveloping algebra $U(\mathfrak{sl}_2)$. Recall that $\mathfrak{sl}_2$ is the Lie algebra generated by e, f, h satisfying

$$[h, e] = 2e \quad [e, f] = h \quad [h, f] = -2f.$$

Therefore there is a map:

$$\varphi : U(\mathfrak{sl}_2) \twoheadrightarrow \mathcal{D}(\mathbf{P}^1)$$

defined by $\varphi(e) = \partial$, $\varphi(f) = -t^2\partial$ and $\varphi(h) = -2t\partial$. It is possible to calculate the kernel of φ explicitly: $\ker\varphi = \Omega U(\mathfrak{sl}_2)$, where $\Omega = 4fe + h^2 + 2h$ is the *Casimir* element. This is a primitive ideal of $U(\mathfrak{sl}_2)$. It can also be shown that $\mathcal{D}(\mathbf{P}^1)$ has exactly one non-zero two-sided ideal, the image of the augmentation ideal of $U(\mathfrak{sl}_2)$ under φ.

Finally notice that if U is an open affine subset of $\mathbf{P}^1$ then $\mathcal{D}(U) = \mathcal{O}(U)\mathcal{D}(\mathbf{P}^1)$. This means that, similarly to the affine case, $\mathcal{D}(U)$ can be regarded as a *localisation* of the ring of global sections $\mathcal{D}(\mathbf{P}^1)$.

This last example is by no means unique. Many of its properties are shared by other rings of differential operators. Especially important is the *localisation* property that $\mathbf{P}^1$ has in common with affine varieties. In the next definition we isolate this and other properties that a variety will be expected to satisfy.

A variety $\mathcal{X}$ will be called a $\mathcal{D}$*-variety* if the following conditions are satisfied:

(a) $\mathcal{D}(\mathcal{X})$ is a noetherian domain with quotient division ring $Q(\mathcal{X})$. If U is an open affine subset of $\mathcal{X}$ then $\mathcal{D}(\mathcal{X}) \subseteq \mathcal{D}(U) \subset Q(\mathcal{X})$ and $\mathcal{D}(U) = \mathcal{O}(U)\mathcal{D}(\mathcal{X}) = \mathcal{D}(\mathcal{X})\mathcal{O}(U)$.

(b) $\mathcal{X}$ has a finite open affine cover $\{V_1, \ldots, V_m\}$ such that $\bigoplus_{j=1}^{m} \mathcal{D}(V_j)$ is a faithfully flat right $\mathcal{D}(\mathcal{X})$-module. The diagonal embedding $\mathcal{D}(\mathcal{X}) \hookrightarrow \bigoplus_{j=1}^{m} \mathcal{D}(V_j)$ obtained from the restriction maps determines the right module structure.

As is shown in [Hodges and Smith 1985a] if $\mathcal{X}$ is a $\mathcal{D}$-variety then (b) holds for every finite open affine cover $\{V_1, \ldots V_m\}$.

Note that (a) simply says that the sheaf $\mathcal{D}_{\mathcal{X}}$ is defined by *localising* its ring of global sections $\mathcal{D}(\mathcal{X})$. Condition (b) is more technical. The main reason for its introduction is that it allows one to prove the following localisation theorem, as shown in [Hodges and Smith 1985a]. First, however, some more notation. Let $\mathcal{D}(\mathcal{X})$-mod be the category of left $\mathcal{D}(\mathcal{X})$-modules and $\mathcal{D}_{\mathcal{X}}$-mod the category of sheaves of left $\mathcal{D}_{\mathcal{X}}$-modules which are quasi-coherent as $\mathcal{O}_{\mathcal{X}}$-modules. Let $\Gamma(\mathcal{X}, -)$ denote the global sections functor from $\mathcal{D}_{\mathcal{X}}$-mod to $\mathcal{D}(\mathcal{X})$-mod.

THEOREM 2.6. *If $\mathcal{X}$ is a $\mathcal{D}$-variety there is an equivalence of categories between $\mathcal{D}(\mathcal{X})$-mod and $\mathcal{D}_{\mathcal{X}}$-mod given by the mutually inverse functors:*

$$\mathcal{M} \longmapsto \Gamma(\mathcal{X}, \mathcal{M}) \qquad \text{and} \qquad M \longmapsto \mathcal{D}_{\mathcal{X}} \otimes_{\mathcal{D}(\mathcal{X})} M,$$

for $\mathcal{M} \in \mathcal{D}_{\mathcal{X}}$-mod and $M \in \mathcal{D}(\mathcal{X})$-mod. Under this equivalence finitely generated $\mathcal{D}(\mathcal{X})$-modules correspond to coherent sheaves in $\mathcal{D}_{\mathcal{X}}$-mod.

Next we list some basic facts about the ring $\mathcal{D}(\mathcal{X})$ when $\mathcal{X}$ is a $\mathcal{D}$-variety.

PROPOSITION 2.7. *Let $\mathcal{X}$ be a $\mathcal{D}$-variety. Then:*

(a) *$\mathcal{D}(\mathcal{X})$ has left Krull dimension equal to $\dim \mathcal{X}$.*

(b) *The global dimension of $\mathcal{D}(\mathcal{X})$ is greater than or equal to $\dim \mathcal{X}$ and less than than or equal to $2 \dim \mathcal{X}$. Moreover both bounds can be attained.*

PROOF: Let $\{V_j\}$ be the finite open affine cover in the definition of $\mathcal{D}$-variety. To do (a) and (b) first recall from Proposition 2.1 that each $\mathcal{D}(V_j)$ has Krull and global dimensions equal to $\dim \mathcal{X}$. Now, as noted in [Hodges and Smith 1985a], each $\mathcal{D}(V_j)$ is a *perfect left localisation* of $\mathcal{D}(\mathcal{X})$ (in the sense of [Stenström 1975]). Thus by [Gabriel 1962, Proposition IV.1] and [Stenström 1975, Proposition XI.3.10] the Krull and global dimensions of $\mathcal{D}(\mathcal{X})$ are greater than or equal to $\dim \mathcal{X}$. The reverse inequality for Krull dimension follows from [McConnell and Robson 1987, Lemma 6.5.3]. That the global dimension of $\mathcal{D}(\mathcal{X})$ is less than or equal to $2 \dim \mathcal{X}$ is a consequence of [Borel 1987, Theorem VI.1.10] and the equivalence of categories in Theorem 2.6. The lower bound for global dimension is always obtained for affine varieties by Proposition 2.1. The upper bound is obtained when $\mathcal{X} = \mathbf{P}^1$ as we shall see in 2.11.

We now turn to examples. The reader is advised to study example 2.11 carefully. In it we return to $\mathbf{P}^1$ and show, for example, that it does not satisfy (b) in the definition of $\mathcal{D}$-variety if the right module condition is replaced by its left module analogue.

EXAMPLE 2.8. An affine variety $\mathcal{X}$ is a $\mathcal{D}$-variety. We have already seen that (a) is verified for affine varieties. Notice that (b) is trivial; the open affine cover is $\{\mathcal{X}\}$. In this case the equivalence of categories described in Theorem 2.6 is a special case of the equivalence between sheaves of quasi-coherent $\mathcal{O}_{\mathcal{X}}$-modules and $\mathcal{O}(\mathcal{X})$-modules provided by the inverse functors

$$M \longmapsto \widetilde{M} \qquad \text{and} \qquad \mathcal{M} \longmapsto \Gamma(\mathcal{X}, \mathcal{M}),$$

for M an $\mathcal{O}(\mathcal{X})$-module and $\mathcal{M}$ a quasi-coherent $\mathcal{O}_{\mathcal{X}}$-module. Clearly $\mathcal{D}_{\mathcal{X}}$-mod is a subcategory of the category of sheaves of quasi-coherent $\mathcal{O}_{\mathcal{X}}$-modules and $\mathcal{D}(\mathcal{X})$-mod is a subcategory of the category of $\mathcal{O}(\mathcal{X})$-modules. We must check that if $M \in \mathcal{D}(\mathcal{X})$-mod then $\widetilde{M}$ is a sheaf of $\mathcal{D}_{\mathcal{X}}$-modules.

Recall the definition of the $\mathcal{O}_{\mathcal{X}}$-module $\widetilde{M}$. Let $f \in \mathcal{O}(\mathcal{X})$ and let $\mathcal{X}_f = \{x \in \mathcal{X} \ : \ f(x) \neq 0\}$ be the corresponding principal open set. Then $\Gamma(\mathcal{X}_f, \widetilde{M}) = M_f = \mathcal{O}(\mathcal{X})_f \otimes_{\mathcal{O}(\mathcal{X})} M$. If U is an open set of $\mathcal{X}$ put

$$\Gamma(U,\widetilde{M}) = \varprojlim_{\mathcal{X}_f \subset U} M_f.$$

But this is a $\mathcal{D}(U)$-module, as one easily checks. Therefore $\widetilde{M}$ is a $\mathcal{D}_{\mathcal{X}}$-module as claimed.

EXAMPLE 2.9. Let G be a connected semisimple algebraic group over $\mathbf{C}$ and P a parabolic subgroup. Let $\mathcal{X} = G/P$ denote the corresponding complete (projective) homogeneous space. Then $\mathcal{X}$ is a $\mathcal{D}$-variety. The proof of this statement is beyond the scope of this paper. It follows from [Beilinson and Bernstein 1981] and [Borho and Brylinski 1985, Theorem 1:9]. Let $\mathfrak{g}$ be the Lie algebra of G. Then $\mathcal{D}(\mathcal{X}) \simeq U(\mathfrak{g})/J$, where J is a completely prime primitive ideal of $U(\mathfrak{g})$. Some important examples of parabolic subgroups are the *Borel* subgroups of G. A subgroup of G is a Borel subgroup if it is maximal amongst all closed connected solvable subgroups of G.

Next we consider a more specific example.

EXAMPLE 2.10. Let $G = SL_{n+1}$. This is a connected semisimple algebraic group over $\mathbf{C}$. It is generated by the matrices $I + \alpha E_{ij}$, $i \neq j$, where E_{ij} is the $n+1 \times n+1$ matrix with 1 in the i,j position and zeroes elsewhere. There is an action of SL_{n+1} on $\mathbf{P}^n$ as follows:

$$(I + \alpha E_{ij})(x_1 : \ \dots \ : x_{n+1}) = (x_1 : \ \dots \ : x_i + \alpha x_j : \ \dots \ x_{n+1}).$$

It follows that:

$$\mathbf{P}^n = SL_{n+1}(1:0: \ \dots \ :0) \simeq SL_{n+1}/\mathrm{Stab}_G(1:0: \ \dots \ :0).$$

Therefore

$$\mathrm{Stab}_G(1:0: \ \dots \ :0) = \left\{ \begin{pmatrix} \alpha & * \\ 0 & M \end{pmatrix} : 0 \neq \alpha \in \mathbf{C} \quad \text{and} \quad \det M = 1/\alpha \right\}$$

is a closed parabolic subgroup of SL_{n+1}.

It follows from the previous example that $\mathcal{D}(\mathbf{P}^n) \simeq U(\mathfrak{sl}_{n+1})/J$, for some primitive ideal, J, of $U(\mathfrak{sl}_{n+1})$. This map will be studied in more detail in section 4.

EXAMPLE 2.11. In this final example we return to $\mathcal{D}(\mathbf{P}^1)$. We want to show that condition (b), in the definition of $\mathcal{D}$-variety, is not left-right symmetric. Denote by $\mathbf{m}$ the unique non-zero two-sided ideal of $\mathcal{D}(\mathbf{P}^1)$. It has generators $\partial, t\partial$ and $t^2\partial$ (as a left or right module). It is easy to check that $\mathbf{m}\mathcal{D}(V_i) = \mathcal{D}(V_i)$ for $i = 0,1$. Thus $\oplus_0^1 \mathcal{D}(V_i)$ is not a faithfully flat left $\mathcal{D}(\mathbf{P}^1)$-module. In fact it is not even flat, for then, by [Hodges and Smith 1985b], $\mathcal{D}(\mathbf{P}^1)$ would be hereditary. But, as is shown in [Stafford 1982], $\mathcal{D}(\mathbf{P}^1)$ has global dimension 2.

Section 3: Locally free sheaves.

Let $\mathcal{X}$ be an affine variety and P a finitely generated projective $\mathcal{O}(\mathcal{X})$-module. It is well known [Kunz 1985, IV.3.6] that this is equivalent to saying $P_{\mathfrak{m}}$ is a free $\mathcal{O}(\mathcal{X})_{\mathfrak{m}}$-module for every maximal ideal $\mathfrak{m}$ of $\mathcal{O}(\mathcal{X})$. This last condition can be restated in terms of sheaves. Let $\widetilde{P}$ be the $\mathcal{O}_{\mathcal{X}}$-module defined in Example 2.8. Then P is projective if and only if for every $x \in \mathcal{X}$ the stalk $\widetilde{P}_x = P_x$ is a free $\mathcal{O}_x$-module. We will isolate this last property in a definition.

Let $\mathcal{X}$ be a variety, and $\mathcal{A}$ a sheaf of rings (not necessarily commutative) over $\mathcal{X}$. A sheaf $\mathcal{L}$ is *locally free* if $\mathcal{L}_x$ is a free $\mathcal{A}_x$-module for every $x \in \mathcal{X}$. Equivalently, $\mathcal{L}$ is locally free if there exists a finite open affine cover $\{V_1, \ldots, V_m\}$ of $\mathcal{X}$ such that $\Gamma(V_j, \mathcal{L})$ is a free $\Gamma(V_j, \mathcal{A})$-module for $1 \leq j \leq m$. Locally projective sheaves are defined analogously.

Returning to the case of an affine variety, recall the following classical result of [Serre 1955].

THEOREM 3.1. *Let $\mathcal{X}$ be an affine variety. The category of finitely generated projective $\mathcal{O}(\mathcal{X})$-modules is equivalent to the category of coherent locally free sheaves of $\mathcal{O}_{\mathcal{X}}$-modules.*

To obtain this equivalence it is enough to restrict the functors of Example 2.8 to these subcategories. In this section we want to investigate a similar correspondence for rings of differential operators on $\mathcal{D}$-varieties. Once again we start with the affine case. Let $\mathcal{X}$ be an affine variety and P a finitely generated projective $\mathcal{D}(\mathcal{X})$-module. If $\mathfrak{m}$ is a maximal ideal of $\mathcal{O}(\mathcal{X})$ then $P_{\mathfrak{m}}$ is a projective $\mathcal{D}(\mathcal{X})_{\mathfrak{m}}$-module but it is not necessarily free. In fact, as will be shown later (in Corollary 3.6), if $\mathcal{X} = \mathbb{A}^n(\mathbb{C})$ then given a non-cyclic uniform, finitely generated projective $\mathcal{D}(\mathbb{A}^n)$-module, P, there exists a maximal ideal $\mathfrak{m}$ of $\mathcal{O}(\mathbb{A}^n)$ such that $P_{\mathfrak{m}}$ is *not* free. However the next best result holds.

THEOREM 3.2. *Let $\mathcal{X}$ be an affine variety. The category of finitely generated projective $\mathcal{D}(\mathcal{X})$-modules is equivalent to the category of coherent locally projective sheaves of $\mathcal{D}_{\mathcal{X}}$-modules.*

PROOF: The equivalence is established by the functors defined in Theorem 2.6. It is clear that if M is a finitely generated projective $\mathcal{D}(\mathcal{X})$-module then $\mathcal{D}_{\mathcal{X}} \otimes_{\mathcal{D}(\mathcal{X})} M$ is a locally projective sheaf of $\mathcal{D}_{\mathcal{X}}$-modules. We must show that if $\mathcal{M}$ is a locally projective coherent sheaf of $\mathcal{D}_{\mathcal{X}}$-modules then $\Gamma(\mathcal{X}, \mathcal{M})$ is projective. Let $M = \Gamma(\mathcal{X}, \mathcal{M})$. By [Schapira 1985, Lemma B.2.1] it is enough to show that each $\operatorname{Ext}^i(M, \mathcal{D}(\mathcal{X})) = 0$ for $i > 0$. But $\operatorname{Ext}^i(M, \mathcal{D}(\mathcal{X}))$ is naturally a right $\mathcal{D}(\mathcal{X})$-module and so if x is a maximal ideal of $\mathcal{O}(\mathcal{X})$ then $\operatorname{Ext}^i(M, \mathcal{D}(\mathcal{X}))_x = \operatorname{Ext}^i(M_x, \mathcal{D}(\mathcal{X})_x) = 0$, since M is locally projective. Therefore, for $i > 0$, $\operatorname{Ext}^i(M, \mathcal{D}(\mathcal{X})) = 0$, as required.

The idea in the above proof can also be used to prove:

PROPOSITION 3.3. *Let $\mathcal{X}$ be an affine variety and M a finitely generated $\mathcal{D}(\mathcal{X})$-module. Then*

$$\operatorname{pd} M = \sup\{\operatorname{pd}(M_x) : x \in \mathcal{X}\}$$

where pd *stands for projective dimension.*

The equivalence of Theorem 3.1 fails if $\mathcal{X}$ is not affine, and the same can be said of Theorem 3.2. However, unlike the commutative case, the correspondence between sheaves of $\mathcal{D}_{\mathcal{X}}$-modules and $\mathcal{D}(\mathcal{X})$-modules holds also for many non-affine varieties. This suggests that it might be interesting to study the $\mathcal{D}(\mathcal{X})$-modules, M, for which $\mathcal{D}_{\mathcal{X}} \otimes_{\mathcal{D}(\mathcal{X})} M$ is locally projective, at least if $\mathcal{X}$ is a $\mathcal{D}$-variety.

We have a more modest aim in mind, namely to study modules which are global sections of locally free sheaves of $\mathcal{D}_{\mathcal{X}}$-modules. For a more extensive study of these modules see [Coutinho and Holland 1988].

Using *cohomology* it is possible to classify all locally free sheaves of $\mathcal{O}$-modules of rank one. As we will see, the same is true over $\mathcal{D}_{\mathcal{X}}$. First we must say what should be understood by the rank of a locally free coherent sheaf $\mathcal{L}$ of $\mathcal{D}_{\mathcal{X}}$-modules. The function

$$\begin{aligned} \rho_0 : \mathcal{X} &\longrightarrow \mathbb{Z} \\ x &\longmapsto \rho(\mathcal{L}_x, 0) \end{aligned}$$

(where $\rho(\mathcal{L}_x, 0)$ is the rank of $\mathcal{L}_x$) is locally constant. Since $\mathcal{X}$ is irreducible, ρ_0 is in fact constant. The *rank* of $\mathcal{L}$ can now be defined as $\rho(\mathcal{L}, 0) = \rho_0(x)$, where $x \in \mathcal{X}$.

PROPOSITION 3.4. *Let $\mathcal{X}$ be a variety. The isomorphism classes of rank one locally free sheaves over $\mathcal{D}_{\mathcal{X}}$ are in bijective correspondence to $H^1(\mathcal{X}, \mathcal{O}^*_{\mathcal{X}})$, where $\mathcal{O}^*_{\mathcal{X}}$ is the sheaf of units over $\mathcal{O}_{\mathcal{X}}$.*

PROOF: The proof is an application of Čech cohomology, exactly as in the commutative case. We only indicate the crux of the argument. If $x \in \mathcal{X}$ then $\mathcal{D}_x$ is an iterated Ore extension of $\mathcal{O}_x$, by Proposition 2.1. Thus the group of units, $\mathcal{D}^*_x$ equals $\mathcal{O}^*_x$. Now if U is an open set of $\mathcal{X}$, we have

$$D(U)^* = \bigcap_{x \in U} \mathcal{D}^*_x = \bigcap_{x \in U} \mathcal{O}^*_x = \mathcal{O}(U)^*.$$

Now let $\mathcal{L}$ be a locally free $\mathcal{D}_{\mathcal{X}}$-module of rank one. Let U, V be open sets of $\mathcal{X}$ over which $\mathcal{L}$ is cyclic. Denoting by $\mathcal{L}|_U$ the restriction of the sheaf $\mathcal{L}$ to U, we have $(\mathcal{L}|_U)|_V \simeq (\mathcal{L}|_V)|_U$. This determines an automorphism φ, of $\mathcal{D}(U \cap V)$, which is itself completely defined by $\varphi(1)$. However $\varphi(1) \in \mathcal{D}(U \cap V)^* = \mathcal{O}(U \cap V)^*$, by the first paragraph. Hence, as in the commutative case [Hartshorne, Exercise III.4.5], we can associate to $\mathcal{L}$ an element of $H^1(\mathcal{X}, \mathcal{O}^*_{\mathcal{X}})$.

COROLLARY 3.5.

(a) *If $\mathcal{F}$ is a rank one locally free $\mathcal{D}_{\mathcal{X}}$-module then $\mathcal{F} \simeq \mathcal{D} \otimes_{\mathcal{O}} \mathcal{L}$ for some (rank one) locally free $\mathcal{O}_{\mathcal{X}}$-module, $\mathcal{L}$.*

(b) *If $\mathcal{L}$ and $\mathcal{M}$ are rank one locally free $\mathcal{O}_{\mathcal{X}}$-modules then $\mathcal{D} \otimes_{\mathcal{O}} \mathcal{L} \simeq \mathcal{D} \otimes_{\mathcal{O}} \mathcal{M}$ if and only if $\mathcal{L} \simeq \mathcal{M}$.*

Proposition 3.4 has interesting applications even for the Weyl algebra.

COROLLARY 3.6. *Let P be a projective non-free left ideal of A_n. Then there exists a maximal ideal $\mathfrak{m}$ of $\mathbb{C}[x_1, \ldots, x_n]$ such that $P_{\mathfrak{m}}$ is not a free left ideal of $(A_n)_{\mathfrak{m}}$.*

PROOF: Since $H^1(\mathbb{A}^n, \mathcal{O}^*_{\mathbb{A}^n}) = 0$ then $\widetilde{P}$ is locally free if and only if it is free.

Of course even for an affine curve $\mathcal{C}$ the group $H^1(\mathcal{C}, \mathcal{O}^*_{\mathcal{C}}) \simeq \mathrm{Pic}(\mathcal{C})$ may be non-trivial. It can even be infinitely generated as shown in [Hartshorne 1977, page 447]. The reader should be warned that for a variety which is *not* affine the module of global sections of a locally free sheaf need not be projective. Examples are given in the next section.

Suppose now that $\mathcal{X}$ is a $\mathcal{D}$-variety. By Theorem 2.6 every locally free sheaf of $\mathcal{D}_{\mathcal{X}}$-modules is determined by its module of global sections. The endomorphism rings of these modules can be calculated explicitly. They are the global sections of certain sheaves of twisted rings of differential operators. We proceed to define these sheaves.

Let $\mathcal{X}$ be a variety and $\mathcal{L}$ a locally free sheaf of $\mathcal{O}_{\mathcal{X}}$-modules of rank one. Then every stalk of $\mathcal{L}$ is cyclic; say $\mathcal{L}_x = s(x)\mathcal{O}_x$, for $x \in \mathcal{X}$. In fact it can be assumed that $s(x) \in \mathbb{C}(\mathcal{X})$, for every $x \in \mathcal{X}$. The sheaf $\mathcal{D}_{\mathcal{L}}$ of twisted differential operators with coefficients in $\mathcal{L}$ can now be defined. It is locally isomorphic to $\mathcal{D}_{\mathcal{X}}$, in fact

$$(\mathcal{D}_{\mathcal{L}})_x = s(x)\mathcal{D}_x s(x)^{-1} \subseteq \mathcal{D}(\mathbb{C}(\mathcal{X})),$$

for each $x \in \mathcal{X}$. If U is an open subset of $\mathcal{X}$ set

$$\mathcal{D}_{\mathcal{L}}(U) = \bigcap_{x \in U} (\mathcal{D}_{\mathcal{L}})_x = \bigcap_{x \in U} s(x)\mathcal{D}_x s(x)^{-1}.$$

It is easy to verify that $\mathcal{D}_{\mathcal{L}}$ is a sheaf of algebras with restriction maps $\mathcal{D}_{\mathcal{L}}(V) \subseteq \mathcal{D}_{\mathcal{L}}(U)$, for open sets $U \subseteq V \subseteq \mathcal{X}$.

For a more general definition of twisted differential operators see [Beilinson and Bernstein 1983]. The definition above is equivalent to saying that

$$\mathcal{D}_{\mathcal{L}} = \mathcal{L} \otimes_{\mathcal{O}} \mathcal{D}_{\mathcal{X}} \otimes_{\mathcal{O}} \mathcal{L}^{-1}.$$

Clearly $\mathcal{D}_{\mathcal{O}} = \mathcal{D}_{\mathcal{X}}$. Several examples of this construction can be found in the next section. We end this section with a proof that $\mathrm{End}_{\mathcal{D}(\mathcal{X})}\Gamma(\mathcal{X}, \mathcal{D}_{\mathcal{X}} \otimes \mathcal{L}^{-1}) \simeq \mathcal{D}_{\mathcal{L}}(\mathcal{X}) = \Gamma(\mathcal{X}, \mathcal{D}_{\mathcal{L}})$, where $\mathcal{X}$ is a $\mathcal{D}$-variety. But first a technical lemma.

LEMMA 3.7. *Let $\mathcal{X}$ be a $\mathcal{D}$-variety and $I \subseteq Q(\mathcal{X})$ a finitely generated left $\mathcal{D}(\mathcal{X})$-module. Then:*

(a) $I = \bigcap_{x \in U} \mathcal{D}_x I$.

(b) $\mathrm{End}_{\mathcal{D}(\mathcal{X})} I \hookrightarrow \mathrm{End}_{\mathcal{D}_x} \mathcal{D}_x I$, *for every $x \in \mathcal{X}$. Under this embedding, $\theta \in \mathrm{End}_{\mathcal{D}(\mathcal{X})} I$ is taken to $\widetilde{\theta} \in \mathrm{End}_{\mathcal{D}_x} \mathcal{D}_x I$, where $\widetilde{\theta}(\Sigma d_k y_k) = \Sigma d_k \theta(y_k)$, for $d_k \in \mathcal{D}_x$ and $y_k \in I$.*

(c) *Under the identification in (b)*

$$\mathrm{End}_{\mathcal{D}(\mathcal{X})} I = \bigcap_{x \in U} \mathrm{End}_{\mathcal{D}_x} \mathcal{D}_x I.$$

PROOF: Recall that for every $x \in \mathcal{X}$ and every neighbourhood U of x one has

$$\mathcal{D}(\mathcal{X}) \subseteq \mathcal{D}(U) \subseteq \mathcal{D}_x \subseteq \mathcal{D}(\mathrm{C}(\mathcal{X})) \subseteq Q(\mathcal{X}).$$

Let $J = \bigcap_{x \in \mathcal{X}} \mathcal{D}_x I$. Clearly $I \subseteq J$. To prove (a) we must show that this is in fact equality. Let $\{V_1, \ \dots \ , V_m\}$ be an open affine cover of $\mathcal{X}$. Since $\mathcal{X}$ is a $\mathcal{D}$-variety we have that $\oplus_1^m \mathcal{D}(V_i)$ is a faithfully flat right $\mathcal{D}(\mathcal{X})$-module. In particular each $\mathcal{D}(V_i)$ is flat and so

$$\mathcal{D}(V_i) \otimes_{\mathcal{D}(\mathcal{X})} J/I \simeq \mathcal{D}(V_i) J / \mathcal{D}(V_i) I.$$

Since $(\mathcal{D}(V_i)J)_x = (\mathcal{D}(V_i)I)_x$ for every $x \in V_i$ it follows that

$$\mathcal{D}(V_i) \otimes_{\mathcal{D}(\mathcal{X})} J/I = 0 \qquad \text{for } 1 \le i \le m.$$

Together with the faithful flatness of $\oplus_1^m \mathcal{D}(V_i)$ this implies that $J = I$ as required.

Let $x \in \mathcal{X}$. To prove (b) we must first show that $\widetilde{\theta} : \mathcal{D}_x I \longrightarrow \mathcal{D}_x I$ is well defined. Suppose that $\sum_1^k d_j y_j = 0$ for some $d_j \in \mathcal{D}_x, y_j \in I$. Since $\mathcal{D}_x \subseteq Q(\mathcal{X})$, there exist $c, r_j \in \mathcal{D}(\mathcal{X})$ with $c \neq 0$ and $d_j = c^{-1} r_j$. Then $\sum_1^k r_j y_j = 0$ and

$$\widetilde{\theta}(\sum_1^k d_j y_j) = \sum_1^k c^{-1} r_j \theta(y_j) = c^{-1} \theta(\sum_1^k r_j y_j) = 0,$$

thus showing that $\widetilde{\theta}$ is well-defined. It is now easy to check that the correspondence $\theta \longmapsto \widetilde{\theta}$ defines an injective homomorphism, as claimed in (b).

We now prove (c). The inclusion

$$\mathrm{End}_{\mathcal{D}(\mathcal{X})} I \subseteq \bigcap_{x \in \mathcal{X}} \mathrm{End}_{\mathcal{D}_x} \mathcal{D}_x I$$

follows from (b). If $\theta \in \bigcap_{x \in \mathcal{X}} \mathrm{End}_{\mathcal{D}_x} \mathcal{D}_x I$ then $\theta(I) \subseteq \bigcap_{x \in \mathcal{X}} \mathcal{D}_x I = I$, by (a), which proves the opposite inclusion.

PROPOSITION 3.8. *Let $\mathcal{X}$ be a $\mathcal{D}$-variety and $\mathcal{L}$ a rank one locally free sheaf of $\mathcal{O}_{\mathcal{X}}$-modules. Then*

$$\mathcal{D}_{\mathcal{L}}(\mathcal{X}) = \Gamma(\mathcal{X}, \mathcal{D}_{\mathcal{L}}) \simeq \mathrm{End}_{\mathcal{D}(\mathcal{X})} \Gamma(\mathcal{X}, \mathcal{D}_{\mathcal{X}} \otimes_{\mathcal{O}_{\mathcal{X}}} \mathcal{L}^{-1}),$$

where $\mathcal{L}^{-1} = \mathcal{H}om_{\mathcal{O}_{\mathcal{X}}}(\mathcal{L}, \mathcal{O}_{\mathcal{X}})$ is the dual sheaf of $\mathcal{L}$.

PROOF: Let $\mathcal{L}_x = \mathcal{O}_x s(x) \subseteq \mathrm{C}(\mathcal{X})$, for every $x \in \mathcal{X}$. Then $(\mathcal{L}^{-1})_x = \mathcal{O}_x s(x)^{-1}$. Thus

$$(\mathcal{D}_{\mathcal{X}} \otimes_{\mathcal{O}_{\mathcal{X}}} \mathcal{L}^{-1})_x = \mathcal{D}_x \otimes_{\mathcal{O}_x} \mathcal{O}_x s(x)^{-1} \simeq \mathcal{D}_x s(x)^{-1} \subseteq Q(\mathcal{X})$$

for every $x \in \mathcal{X}$. Without loss of generality it can be assumed that the isomorphism above is an equality. Let $L = \Gamma(\mathcal{X}, \mathcal{D}_\mathcal{X} \otimes_{\mathcal{O}_\mathcal{X}} \mathcal{L}^{-1})$. Then, under this identification, $L = \bigcap_{x\in\mathcal{X}} \mathcal{D}_x s(x)^{-1}$ is a finitely generated left $\mathcal{D}(\mathcal{X})$-submodule of $Q(\mathcal{X})$. Applying Lemma 3.7 one obtains

$$\operatorname{End}_{\mathcal{D}(\mathcal{X})} L = \bigcap_{x\in\mathcal{X}} \operatorname{End}_{\mathcal{D}_x}(\mathcal{D}_x L).$$

By Theorem 2.6 it follows that $\mathcal{D}_x L = \mathcal{D}_x s(x)^{-1}$ for every $x \in \mathcal{X}$. Hence $\operatorname{End}_{\mathcal{D}_x} \mathcal{D}_x L = s(x)\mathcal{D}_x s(x)^{-1}$ for every $x \in \mathcal{X}$. Combining these equalities one obtains

$$\operatorname{End}_{\mathcal{D}(\mathcal{X})} L = \bigcap_{x\in\mathcal{X}} s(x)\mathcal{D}_x s(x)^{-1} = \mathcal{D}_\mathcal{L}(\mathcal{X}),$$

as required.

Given the isomorphism of the previous proposition one is led to ask: when are $\mathcal{D}(\mathcal{X})$ and $\mathcal{D}_\mathcal{L}(\mathcal{X})$ morita equivalent? In particular: when is $\Gamma(\mathcal{X}, \mathcal{D}_\mathcal{X} \otimes \mathcal{L}^{-1})$ a progenerator? In the next section we give a complete answer to this question in the special case when $\mathcal{X}$ is the projective space.

4. Differential operators on Projective Space.

We are now ready to study the rings of differential operators on projective space, including the twisted case. In view of Proposition 3.8 it will be necessary to consider in detail some of the properties of locally free sheaves of rank one over $\mathcal{D}_{\mathbf{P}^n}$. In particular we calculate the modules of global sections of these sheaves and determine when they are projective.

Let us fix a system of homogeneous coordinates $(x_0 : \ \dots \ : x_n)$ for $\mathbf{P}^n$. Now if we put $V_i = \{(x_0 : \ \dots \ : x_n) \in \mathbf{P}^n : x_i \neq 0\})$ then V_i is an open affine subset of $\mathbf{P}^n$. The isomorphism between V_i and $\mathbb{A}^n$ is given by the map

$$(x_0 : \ \dots \ : x_n) \longmapsto (x_0/x_i, \ \dots \ , \widehat{x_i/x_i}, \ \dots \ , x_n/x_i),$$

where ^ denotes omission. We call $\{V_0, \ \dots \ V_n\}$ the *standard* (open affine) cover of $\mathbf{P}^n$. With this notation:

$$\mathcal{D}(V_i) = \mathbb{C}[x_0/x_i, \ \dots \ , \widehat{x_i/x_i}, \ \dots \ , x_n/x_i, x_i\partial/\partial x_0, \ \dots \ , \widehat{x_i\partial/\partial x_i}, \ \dots \ , x_i\partial/\partial x_n]$$

is isomorphic to the n-th Weyl algebra, A_n. In order to simplify the notation, let $t_j = x_j/x_0$ and $\partial_j = \partial/\partial t_j$. An easy calculation shows that $\mathcal{D}(V_0) = \mathbb{C}[t_1, \ \dots \ , t_n, \partial_1, \ \dots \ , \partial_n]$ and, for $1 \leq i \leq n$,

$$\mathcal{D}(V_i) = \mathbb{C}[t_1 t_i^{-1}, \ \dots \ , t_n t_i^{-1}, t_i^{-1}, t_i\partial_1, \ \dots \ , t_i\partial_n, t_i\pi],$$

where $\pi = \sum_1^n t_j\partial_j$. We want to write $\mathcal{D}(\mathbf{P}^n) \subseteq \bigcap_0^n \mathcal{D}(V_i)$ as a subring of $\mathcal{D}(V_0)$. A brute force calculation of the intersection is out of the question; we resort to a more roundabout method.

Recall, from 2.10, that there is a surjective homomorphism $\varphi : U(\mathfrak{sl}_{n+1}) \twoheadrightarrow \mathcal{D}(\mathbf{P}^n)$. This map is most easily described when we consider $\mathcal{D}(\mathbf{P}^n)$ as a subalgebra of $\mathcal{D}(\mathbf{C}(\mathbf{P}^n))$. Recall, using the notation of Example 2.10, that $\mathfrak{sl}_{n+1}$ has a basis consisting of the matrices:

$$E_{ij} \quad \text{for } 1 \le i \ne j \le n+1 \quad \text{and} \quad H_i = E_{i+1,i+1} - E_{ii} \quad \text{for } 1 \le i \le n.$$

One has that $\varphi(E_{ij}) = x_{i-1}\partial/\partial x_{j-1}$ (see [Borho and Brylinski 1982]). In $\mathcal{D}(V_0)$ we have:

$$\begin{aligned} x_0\partial/\partial x_j &= \partial_j & j \ne 0 \\ x_i\partial/\partial x_j &= (x_i/x_0)x_0\partial/\partial x_j = t_i\partial_j & 0 \ne i, j \ne 0. \end{aligned}$$

But $x_i\partial/\partial x_0, i \ne 0$ is slightly harder. Note first that the *Euler* relation $\sum_0^n x_i\partial/\partial x_i = 0$ holds in $\mathcal{D}(\mathbf{C}(\mathbf{P}^n))$. It implies that $x_0\partial/\partial x_0 = -\sum_1^n x_j\partial/\partial x_j = -\pi$. Hence

$$x_i\partial/\partial x_0 = -(x_i/x_0)\sum_1^n (x_j/x_0)x_0\partial/\partial x_j = -t_i\pi.$$

Putting all this together, $\mathcal{D}(\mathbf{P}^n)$ is the $\mathbf{C}$-subalgebra of $\mathcal{D}(V_0)$ generated by $\partial_j, t_i\partial_j, t_i\pi$ for $1 \le i \ne j \le n$, $t_1\partial_1 + \pi$ and $t_i\partial_i - t_{i-1}\partial_{i-1}$ for $2 \le i \le n$. Notice that $[\partial_i, t_i\pi] = \pi + t_i\partial_i$ and so $\sum_1^n[\partial_i, t_i\pi] = (n+1)\pi$. We deduce that $t_i\partial_i \in \mathcal{D}(\mathbf{P}^n)$ for $1 \le i \le n$. Therefore

$$\mathcal{D}(\mathbf{P}^n) = \mathbf{C}[\partial_i, t_i\partial_j, t_i\pi : 1 \le i, j \le n].$$

A number of important ring-theoretic properties of $\mathcal{D}(\mathbf{P}^n)$ can be deduced from the homomorphism φ mentioned above. For example, $\mathcal{D}(\mathbf{P}^n)$ is left and right noetherian, since so is $U(\mathfrak{sl}_{n+1})$. Using deeper results from enveloping algebra theory, the two-sided ideal structure of $\mathcal{D}(\mathbf{P}^n)$ can be explicitly determined. Let us do this.

Let $\mathfrak{g}$ be a semisimple Lie algebra over $\mathbf{C}$. Recall that if J is a primitive ideal of $U(\mathfrak{g})$ then the Gelfand-Kirillov dimension of $U(\mathfrak{g})/J$ is the dimension of the nilpotent orbit which corresponds to J. See, for example [Borho 1986]. Since the dimension of the minimal nilpotent orbit of $\mathfrak{sl}_{n+1}$ is $2n$ then $\mathrm{GK}(U(\mathfrak{sl}_{n+1})/J)$ is either zero or greater than or equal to $2n$. The next result is well-known to experts but this is apparently its first appearance in print.

PROPOSITION 4.1. *$\mathcal{D}(\mathbf{P}^n)$ has exactly one non-zero, proper two-sided ideal, $\mathbf{m}$, namely the image of the augmentation ideal of $U(\mathfrak{sl}_{n+1})$ under the map φ above.*

PROOF: We begin by determining all the primitive ideals of $\mathcal{D}(\mathbf{P}^n)$. Recall that $\mathcal{D}(\mathbf{P}^n)$ is primitive (Example 2.10), hence $\{0\}$ is a primitive ideal. For the remainder of this proof let $\varphi : U(\mathfrak{sl}_{n+1}) \twoheadrightarrow \mathcal{D}(\mathbf{P}^n)$ be the surjection described above.

Let $\mathbf{a}$ be the augmentation ideal of $U(\mathfrak{sl}_{n+1})$. Then $\varphi(\mathbf{a})$ is generated as a left (or right) ideal of $\mathcal{D}(\mathbf{P}^n)$ by $t_i\partial_j, \partial_j$ and $t_i\pi$ for $1 \le i, j \le n$. If $\varphi(\mathbf{a}) = \mathcal{D}(\mathbf{P}^n)$ then $\mathcal{D}(V_0)\varphi(\mathbf{a}) = \mathcal{D}(V_0)$ but, clearly, $\mathcal{D}(V_0)\varphi(\mathbf{a}) = \sum_1^n \mathcal{D}(V_0)\partial_i \ne \mathcal{D}(V_0)$. Hence $\mathbf{m}$ is a proper ideal of $\mathcal{D}(\mathbf{P}^n)$; it is also primitive since $\mathcal{D}(\mathbf{P}^n)/\mathbf{m} \simeq \mathbf{C}$.

If Z is the centre of $U(\mathfrak{sl}_{n+1})$ then $\mathfrak{a} \cap Z = \ker(\varphi) \cap Z$. Since $\mathfrak{a}$ has *trivial central character*, then so does $\ker(\varphi)$. Now let J be a primitive ideal of $U(\mathfrak{sl}_{n+1})$ and assume that $\varphi(J)$ is non-zero and proper. Then

$$\mathrm{GK}(U(\mathfrak{sl}_{n+1})/J) = \mathrm{GK}(\mathcal{D}(\mathbf{P}^n)/\varphi(J)) < \mathrm{GK}(\mathcal{D}(\mathbf{P}^n)) = 2n.$$

By the observations preceding the proposition we conclude that $\mathrm{GK}(\mathcal{D}(\mathbf{P}^n)/\varphi(J)) = 0$, that is, $\mathcal{D}(\mathbf{P}^n)/\varphi(J)$ is finite-dimensional. Thus $\varphi(J)$ is maximal, and has trivial central character. By [Borho 1986, Proposition 3.2(b)] this implies that $\mathbf{m} = \varphi(J)$. Consequently the only primitive ideals of $\mathcal{D}(\mathbf{P}^n)$ are $\{0\}$ and $\mathbf{m}$.

By [Dixmier 1974,Proposition 3.1.15], and the above, the only semiprime ideals of $\mathcal{D}(\mathbf{P}^n)$ are $\{0\}$ and $\mathbf{m}$. Now let $\{0\} \neq I$ be a proper two-sided ideal of $\mathcal{D}(\mathbf{P}^n)$. The prime radical of I must be $\mathbf{m}$. Hence there exists $t > 0$ such that $\mathbf{m}^t \subseteq I$. But, as one easily checks, $\mathbf{m}^t = \mathbf{m}$ for every $t \in \mathbf{N}$. Thus $I = \mathbf{m}$, and the proposition is proved.

The *Chevalley* anti-automorphism of $U(\mathfrak{sl}_{n+1})$ appears as an ingredient in the proof of the next proposition. We recall it briefly. The Chevalley anti-automorphism $u \longmapsto {}^t u$ is defined by setting ${}^t E_{ij} = E_{ji}$ (see [Joseph 1983]). This induces an anti-automorphism of $\mathcal{D}(\mathbf{P}^n)$. Keeping to the enveloping algebra notation we have ${}^t(x_i \partial/\partial x_j) = x_j \partial/\partial x_i$. Note that ${}^t\mathcal{D}(\mathbf{P}^n) = \mathcal{D}(\mathbf{P}^n)$ and ${}^t\mathbf{m} = \mathbf{m}$. This will be called the Chevalley anti-automorphism of $\mathcal{D}(\mathbf{P}^n)$.

PROPOSITION 4.2. *$\mathcal{D}(\mathbf{P}^n)$ is a maximal order.*

PROOF: Write Λ for the dual of $\mathbf{m}$ as a left ideal and R for its dual as a right ideal. It is easily verified that $\mathbf{m}\mathcal{D}(V_i) = \mathcal{D}(V_i)$ for $0 \leq i \leq n$. Thus if $x \in R$ then

$$x\mathbf{m}\mathcal{D}(V_i) = x\mathcal{D}(V_i) \subseteq \mathcal{D}(V_i).$$

We conclude that $x \in \bigcap_0^n \mathcal{D}(V_i) = \mathcal{D}(\mathbf{P}^n)$. Therefore $R = \mathcal{D}(\mathbf{P}^n)$. Since the Chevalley anti-automorphism leaves $\mathbf{m}$ invariant we also have $\Lambda = \mathcal{D}(\mathbf{P}^n)$.

Now suppose that S is an order in $Q(\mathbf{P}^n)$ equivalent to $\mathcal{D}(\mathbf{P}^n)$ and $\mathcal{D}(\mathbf{P}^n) \subseteq S$. Suppose also that $aSb \subseteq \mathcal{D}(\mathbf{P}^n)$ for some non-zero $a, b \in \mathcal{D}(\mathbf{P}^n)$. Then, without loss of generality, $\mathbf{m}S\mathbf{m} \subseteq \mathcal{D}(\mathbf{P}^n)$. It follows that $S\mathbf{m} \subseteq \Lambda = \mathcal{D}(\mathbf{P}^n)$. But then $S \subseteq R = \mathcal{D}(\mathbf{P}^n)$, and so $S = \mathcal{D}(\mathbf{P}^n)$, as required.

The following properties of $\mathcal{D}(\mathbf{P}^n)$ are immediate consequences of Proposition 2.7:

(a) $\mathcal{D}(\mathbf{P}^n)$ has Krull dimension n.

(b) $n \leq$ global dimension $\mathcal{D}(\mathbf{P}^n) \leq 2n$.

We now turn to the study of $\mathcal{D}(\mathbf{P}^n)$-modules. A $\mathcal{D}(\mathbf{P}^n)$-module, M, is said to be *locally free* if the sheaf $\mathcal{D}_{\mathbf{P}^n} \otimes_{\mathcal{D}(\mathbf{P}^n)} M$ is a locally free $\mathcal{D}_{\mathbf{P}^n}$-module in the sense of section 3. In Proposition 3.4 we classified all locally free sheaves of $\mathcal{D}_{\mathbf{P}^n}$-modules of rank one. By Theorem 2.6 this amounts to a classification of all locally free $\mathcal{D}(\mathsf{P}^n)$-modules of uniform dimension (rank) one. Combining this with the fact that $H^1(\mathsf{P}^n, \mathcal{O}^*_{\mathsf{P}^n}) \simeq \mathrm{Pic}(\mathsf{P}^n) \simeq \mathbf{Z}$ one obtains the following result.

PROPOSITION 4.3. *To each integer there corresponds a unique isomorphism class of a rank one locally free $\mathcal{D}(\mathsf{P}^n)$-module.*

We want to determine a representative for each of these isomorphism classes. Let k be an integer and let $\mathcal{O}(k)$ be a locally free $\mathcal{O}_{\mathsf{P}^n}$-module, of rank one, corresponding to k. A standard way to define such a sheaf is by setting:

$$\Gamma(V_0, \mathcal{O}(k)) = \mathcal{O}(V_0)$$
$$\Gamma(V_i, \mathcal{O}(k)) = \mathcal{O}(V_i)t_i^k \qquad \text{for } 1 \le i \le n.$$

We write

$$D(k) = \Gamma(\mathsf{P}^n, \mathcal{D}_{\mathsf{P}^n} \otimes_{\mathcal{O}} \mathcal{O}(k)).$$

It is a $\mathcal{D}(\mathsf{P}^n)$-module in the isomorphism class corresponding to k. A more explicit desription of $D(k)$ will be one of the goals of this section.

This is a good place to mention that following [Hodges 1988] one can show that there are isomorphisms between the K-groups $K_i(\mathcal{D}(\mathsf{P}^n)) \simeq K_i(\mathsf{P}^n)$ for $i \ge 0$. In particular $K_0(\mathcal{D}(\mathsf{P}^n))$ is a free abelian group, of rank $n+1$, generated by $[D(k)]$ for $0 \le k \le n$.

According to the theory developed in section 3, to the sheaf $\mathcal{O}(m)$ there corresponds a ring of twisted differential operators $\mathcal{D}_{\mathcal{O}(m)}(\mathsf{P}^n)$. From now on assume that n has been fixed. For $m \in \mathbf{Z}$, let $\mathcal{D}_m$ be the sheaf $\mathcal{D}_{\mathcal{O}(m)}$ and $D_m = \Gamma(\mathsf{P}^n, \mathcal{D}_m)$. Let $\mathcal{D} = \mathcal{D}_0$ and $D = D_0$. By Proposition 3.8, $D_m = \mathrm{End}_D D(-m)$. Thus it is convenient to study the rings D_m and the D-modules $D(-m)$ together.

Let us start with D_m. Recall that $D_m = \bigcap_{i=0}^{n} \mathcal{D}_m(V_i)$. From above one obtains:

$$\mathcal{D}_m(V_0) = \mathcal{D}(V_0)$$
$$\mathcal{D}_m(V_i) = t_i^m \mathcal{D}(V_i) t_i^{-m} = \mathbf{C}[t_i^{-1}, t_i^{-1}t_1, \ \ldots \ , t_i^{-1}t_n, t_i\partial_1, \ \ldots \ , t_i\partial_n, t_i(\pi - m)].$$

Hence $E_m = \mathbf{C}[\partial_j, t_i\partial_j, t_i(\pi - m) \ : \ 1 \le i,j \le n] \subseteq D_m$. We want to show that $E_m = D_m$ for every $m \in \mathbf{Z}$. At the beginning of this section we showed that this equality holds if $m = 0$. The following properties of E_m will be useful in the sequel.

PROPOSITION 4.4.

(a) *E_k is a right and left noetherian ring for every $k \in \mathbf{Z}$.*

(b) *$E_k \simeq E_{-(k+n+1)}$ for every $k \in \mathbf{Z}$.*

PROOF: To prove (a) use the fact that there is a surjection from $U(\mathfrak{sl}_{n+1})$ to E_k. This map is analogous to the map φ defined at the beginning of this section. It is given explicitly in [Hodges and Smith 1985c].

Now consider the anti-automorphism of $Q(\mathbf{P}^n)$ induced by $t_i \longmapsto t_i$ and $\partial_i \longmapsto -\partial_i$ for $1 \le i \le n$. It restricts to an anti-isomorphism $E_k \simeq E_{-(k+n+1)}$. The isomorphism of (b) is obtained by composing this anti-isomorphism with the Chevalley anti-automorphism of E_k.

The explicit description of $D(k)$, which is one of our aims, is given as a product of E_m-modules contained in D. We proceed to define these modules. For $k \in \mathbb{Z}$ set:

$$M(k) = \sum_{i=1}^{n} E_k\partial_i + E_k(\pi - k - 1) = \sum_{j=1}^{n} \partial_j E_{k+1} + (\pi - k - 1)E_{k+1}$$

$$N(k) = \sum_{i=1}^{n} t_i E_k + E_k = \sum_{j=1}^{n} E_{k+1}t_j + E_{k+1}.$$

The equalities follow from an easy calculation.

LEMMA 4.5. *Let* $k \in \mathbb{Z}$.
(a) $M(k)N(k) = E_k$ *if* $k \neq -(n+1)$.
(b) $N(k)M(k) = E_{k+1}$ *if* $k \neq -1$.

PROOF: We check (a); (b) is analogous. An explicit calculation with generators shows that $M(k)N(k)$ is a two-sided ideal of E_k. But

$$\sum_{1}^{n} \partial_i t_i - (\pi - k - 1) = k + n + 1$$

belongs to $M(k)N(k)$. Therefore $M(k)N(k) = E_k$ if $k \neq -(n+1)$.

The next technical lemma is a consequence of Theorem 2.6 and its proof is left to the reader.

LEMMA 4.6. *Let J be a finitely generated right D-submodule of $Q(\mathbf{P}^n)$ and let J^* denote its dual (regarded as a submodule of $Q(\mathbf{P}^n)$). Then*
$J^* = \bigcap_{i=0}^{n} (J\mathcal{D}(V_i))^*$.

We now come to the main theorem. In it we give explicit descriptions of the modules $D(k)$ and their duals. If $L_1, \ldots, L_t$ are a collection of additive abelian subgroups of $Q(\mathbf{P}^n)$ we shall denote $\prod_{j=1}^{t} L_j = L_1 \ldots L_t \subseteq Q(\mathbf{P}^n)$.

THEOREM 4.7. *Let $k > 0$ be an integer.*
(a) $D(-k) \simeq \prod_{j=0}^{k-1} M(j)$ *and* $D(-k)^* \simeq \prod_{j=0}^{k-1} N(k-1-j)$.
(b) *Let $F(k)$ denote the set of polynomials in $\mathbb{C}[t_1, \ldots, t_n]$ of total degree less than or equal to k. Then $D(k) \simeq D.F(k) = \prod_{j=1}^{k} N(-j)$ and, if $1 \le k \le n$, then $D(k)^* \simeq \prod_{j=1}^{k} M(j-k-1)$.*

PROOF: (a) Let $A(-k) = \prod_{j=0}^{k-1} M(j)$. By Theorem 2.6,

$$A(-k) \simeq D(-k) \quad \text{if and only if} \quad \mathcal{D} \otimes A(-k) \simeq \mathcal{D} \otimes D(-k).$$

This can be checked locally. It is easy to verify that $\mathcal{D}(V_0)A(-k) = \mathcal{D}(V_0)$. We show that if $1 \le i \le n$, then $\mathcal{D}(V_i)A(-k) = \mathcal{D}(V_i)t_i^{-k}$. From

$$t_i^{-1}.t_i.(\pi - k - 1) - \sum_{j \neq i}(t_i^{-1}t_j)t_i.\partial_j = t_i\partial_i - k - 1,$$

we get that $\mathcal{D}_k(V_i)t_iM(k) = \mathcal{D}_k(V_i)$. Hence

$$\begin{aligned} t_i^{-k}\mathcal{D}_k(V_i)M(k) &= t_i^{-k}\mathcal{D}_k(V_i)t_i^{-1}t_iM(k) \\ &= t_i^{-k-1}\mathcal{D}_{k+1}(V_i) \\ &= \mathcal{D}(V_i)t_i^{-(k+1)}, \end{aligned}$$

since $\mathcal{D}_{k+1}(V_i) = t_i^{k+1}\mathcal{D}(V_i)t_i^{-(k+1)}$. Therefore, by induction,

$$\mathcal{D}(V_i)A(-k) = \mathcal{D}(V_i)\prod_{j=0}^{k-1} M(j) = \mathcal{D}(V_i)t_i^{-k}$$

and thus $A(-k) \simeq D(-k)$.

We want to determine $A(-k)^*$. An easy calculation shows that

$$B(-k) = N(k-1) \ \ldots \ N(0) \subseteq A(-k)^*.$$

By Lemma 4.6

$$B(-k)^* = \bigcap_{i=0}^{n}(B(-k)\mathcal{D}(V_i))^*.$$

But, proceeding as we did for $A(-k)$, it can be shown that

$$B(-k)\mathcal{D}(V_0) = \mathcal{D}(V_0) \qquad \text{and} \qquad B(-k)\mathcal{D}(V_i) = t_i^k\mathcal{D}(V_i) \qquad \text{for } 1 \le i \le n,$$

and so

$$B(-k)^* = \mathcal{D}(V_0) \cap (\bigcap_{i=1}^{n}\mathcal{D}(V_i)t_i^{-k}).$$

From Lemma 3.7(a) one concludes that $B(-k)^* = A(-k)$. Thus $A(-k)^* = B(-k)$, and the proof of (a) is complete.

The proof of (b) is along the same lines and will be left to the reader.

COROLLARY 4.8. *Let $k > 0$ be an integer.*

(a) *$D(-k)$ is a progenerator for every k.*

(b) *$D(k)$ is projective if $1 \le k \le n$.*

(c) *$\mathrm{m}D(k) = D(k)$, and so $D(k)$ is never a generator.*

PROOF: (a) From Theorem 4.7(a), one has $D(-k)^* = \prod_{j=0}^{k-1} N(k-1-j)$. Now from Lemma 4.5 and induction it is easy to conclude that $D(-k)$ is always a progenerator. Similarly one obtains (b).

As for (c), it is easy to check that $\mathbf{m}N(-1) = N(-1)$. Then

$$\mathbf{m}D(k) = \mathbf{m}N(-1) \ \dots \ N(-k) = N(-1) \ \dots \ N(-k) = D(k),$$

as required.

COROLLARY 4.9.

(a) $D_k \simeq E_k = \mathbf{C}[\partial_i, t_i\partial_j, t_i(\pi - k) \ : \ 1 \leq i,j \leq n]$ *for every* $k \in \mathbf{Z}$.

(b) D_k *is Morita equivalent to* D *whenever* $k \in \mathbf{Z}\backslash\{-1, \ \dots \ ,-n\}$.

(c) D_k *is simple whenever* $-n \leq k \leq -1$, *and any two such rings are Morita equivalent.*

PROOF: (a) From Lemma 4.5(b), it follows that $N(j)M(j) = E_{j+1}$ if $j \neq -1$. If $k > 0$, then

$$\begin{aligned} \operatorname{End} D(-k) &= D(-k)^* D(k) \\ &= N(k-1) \dots \ N(0)M(0) \ \dots \ M(k-1). \end{aligned}$$

Using the above equality and induction, one obtains: $\operatorname{End} D(-k) = E_k$. But from Proposition 3.7, $\operatorname{End} D(-k) \simeq D_k$. Thus $D_k \simeq E_k$, as claimed. Similarly $D_{-k} \simeq E_{-k}$ for $1 \leq k \leq n$. Suppose that $k \geq n+1$. It is now easy to see that the isomorphism $E_{-k} \simeq E_{k-n-1}$ extends to one $D_{-k} \simeq D_{k-n-1}$ and so $E_{-k} \simeq D_{-k}$.

(b) is clear, since $D(-k)$, $k > 0$, is a progenerator. To prove (c) let $-n \leq k \leq -1$, and let $I \neq 0$ be a two-sided ideal of D_k. Since $D(-k)ID(-k)^*$ is a non-zero two-sided ideal of D, we have that either $D(-k)ID(-k)^* = \mathbf{m}$ or $D(-k)ID(-k)^* = D$. But $D(-k)^*D(-k) = D_k$. Therefore, either $I = D(-k)^*\mathbf{m}D(-k)$ or $I = D(-k)^*D.D(-k)$. Since $\mathbf{m}D(-k) = D(-k)$, in either case $I = D_k$. The rest is clear.

Part (c) of the previous corollary implies that Corollary 4.7(b) can be strengthened to: If $k > 0$, then $D(k)$ is projective if and only if $1 \leq k \leq n$. One should also note that Corollary 4.9 is really part of the translation principle (see [Bernstein and Gelfand 1980]). The reader should also compare this result with those in [Joseph and Stafford 1984].

References.

A. Beilinson and I.N. Bernstein 1981, *Localisation de* $\mathfrak{g}$*-modules*, C. R. Acad. Sci. Paris Ser. I 292, pp. 15-18.

A. Beilinson and I.N. Bernstein 1983, *A generalisation of Casselman's submodule theorem* in *Representation theorey of reductive Lie groups*, P.C. Trombi, editor, Progress in Mathematics 40, Birkhäuser Boston 1983.

J.N. Bernstein and S.I. Gelfand 1980, *Tensor products of finite and infinite dimensional representations of semisimple Lie algebras*, Compositio. Math. 41, pp. 245-285.

A. Borel 1987, *Algebraic D-modules*, Perspectives in Mathematics 2, Academic Press Boston.

W. Borho 1986, *A survey on enveloping algebras of semi-simple Lie algebras I*, Canadian Math. Soc. Conf. Proc. 5, pp. 19-50.

W. Borho and J.-L. Brylinski 1982, *Differential operators on homogeneous spaces I*, Invent. Math. 69, pp. 437-476.

W. Borho and J.-L. Brylinski 1985, *Differential operators on homogeneous spaces III*, Invent. Math. 80, pp. 1-68.

J.-L. Brylinski 1981, *Differential operators on the flag varieties* in *Tableaux de Young et foncteurs de Schur en algèbre et géométrie*, Astérisque 87-88, pp. 43-60.

S.C. Coutinho and M.P. Holland, *Module structure of rings of differential operators*, Proc. London Math. Soc., to appear.

S.C. Coutinho and M.P. Holland 1988, *Module structure of the ring of differential operators on projective space*, Leeds University preprint.

J. Dixmier 1974, *Algèbres enveloppantes*, Gauthier-Villars Paris.

P. Gabriel 1962, *Des catégories abéliennes*, Bull. Soc. Math. France 90, pp. 323-448.

R. Hartshorne 1977, *Algebraic Geometry*, Graduate texts in Mathematics 52, Springer-Verlag Berlin/New York.

T.J. Hodges 1988, *K-theory of Noetherian rings*, University of Cincinnati preprint.

T.J. Hodges and S.P. Smith 1985a, *Rings of differential operators and the Beilinson-Bernstein equivalence of categories*, Proc. Amer. Math. Soc. 93, pp. 379-386.

T.J. Hodges and S.P. Smith 1985b, *The global dimension of certain primitive factors of the enveloping algebra of a semi-simple Lie algebra*, J. London Math. Soc. (2) 82, pp. 411-418.

T.J. Hodges and S.P. Smith 1985c, *Differential operators on projective space*, University of Cincinnati preprint.

S. Iitaka 1982, *Algebraic Geometry*, Graduate texts in Mathematics 76, Springer-Verlag Berlin/New York.

A. Joseph 1983, *On the classification of primitive ideals in the enveloping algebra of a semi-simple Lie algebra*, Lecture notes in Mathematics 1024, Springer-Verlag Berlin/New York.

A. Joseph and J.T. Stafford 1984, *Modules of $\mathfrak{k}$-finite vectors over semi-simple Lie algebras*, Proc. London Math. Soc. (3) 49, pp. 361-384.

E. Kunz 1985, *Introduction to Commutative Algebra and Algebraic Geometry*, Birkhäuser Boston.

J.C. McConnell and J.C. Robson 1987, *Noncommutative Noetherian rings*, John Wiley Chichester/New York.

P. Schapira 1985, *Microdifferential Systems in the Complex Domain*, Die Grundlehren der Mathematischen Wissenschaften 269, Springer-Verlag Berlin/New York.

J.A. Seebach, L.A. Seebach and L.A. Steen 1970, *What is a sheaf?*, Amer. Math. Monthly 77ii, pp. 681-703.

J.-P. Serre 1955, *Faisceaux Algébriques Cohérents*, Ann. of Math. 61, pp. 197-278.

J.T. Stafford 1982, *Homological properties of the enveloping algebra $U(\mathfrak{sl}_2)$*,Math. Proc. Camb. Phil. Soc. 91, pp. 29-37.

B. Stenström 1975, *Rings of Quotients*, Die Grundelhren der Mathematischen Wissenschaften 217, Springer-Verlag Berlin/New York.

THE AUSLANDER CONDITION ON GRADED AND FILTERED NOETHERIAN RINGS

Eva Kristina Ekström
Department of Mathematics
University of Stockholm
Box 670 1 , 11385 Stockholm Sweden

[*This paper is in final form and no version of it will be submitted for publication elsewhere*].

0. Introduction.

In this work we study the so called Auslander condition on noetherian rings having finite injective, resp. global homological, dimension. Let us briefly describe the main results of this article. In the first sections we study a graded noetherian ring $A = \oplus A_v$ and impose the graded Auslander condition by the following :

For every finitely generated graded left or right A-module M, every integer v and each graded submodule N of $\mathrm{Ext}_A^v(M,A)$, it follows that $\mathrm{Ext}_A^i(N,A) = 0$ for every $i < v$.

0.1. THEOREM. Let A be a graded noetherian ring satisfying the graded Auslander condition. Then, if A has a finite injective dimension, (resp. finite global homological dimension), it follows that the ring A is Auslander-Gorenstein, (resp. Auslander regular).

We refer to the article by Björk [Bj:2] in this volume for the definition of Auslander-Gorenstein, (resp. Auslander regular) rings. We also establish a result about graded noetherian rings which has an interest in its own :

0.2. THEOREM. Let A be a noetherian graded ring. Then A has a finite injective, (resp. global homological), dimension if (1), (resp. (2)) hold below :

(1) There exists an integer μ such that : $\mathrm{Ext}_A^v(M,A) = 0$ for every $v > \mu$ and each graded and finitely generated A-module M.

(2) There exists an integer μ such that : $\mathrm{Ext}_A^v(M,N) = 0$ for

every $v > \mu$ and each pair of graded and finitely generated left or right A-modules.

In section 4 we prove that if a ring A is Auslander-Gorenstein (resp. Auslander regular) so is the polynomial ring A[t]. This result was originally obtained in my report [E.K] . In § 5 we study Rees rings associated with filtered noetherian rings. In the case when A is a ring with a positive filtration and the associated graded ring G(A) is Auslander-Gorenstein (resp. Auslander regular) then it was proved in [E.K] that the Rees ring $\tilde{A}$ is so. This result was generalized in [H.O] to the case when the filtration is Zariskian.

Here we treat the case when we only assume the so called comparison condition in which case we have to assume that both A and G(A) are Auslander-Gorenstein,(resp. Auslander regular).We mention that in the Zariskian case we only have to assume this for G(A) , by Theorem 4.1 in [Bj:1] . See also Björk's article [Bj:2] .

So in Theorem 5.3 we prove that if A is a filtered ring such that A and G(A) are Auslander-Gorenstein, (resp. Auslander regular), then the Rees ring $\tilde{A}$ is so under the hypothesis that the filtration satisfies the comparison condition. Notice that the comparison condition is equivalent to the condition that the Rees ring $\tilde{A}$ is noetherian by Proposition 2.17 in [Bj:2] .

It turns out that a quite involved part of the proof of Theorem 5.3 occurs for the case of Auslander regularity where we need that the Rees ring has finite global homological dimension.
So actually we establish a result in Lemma 5.5 which has an interest in its own and asserts the following :

THEOREM. Let A be a filtered ring such that the filtration satisfies the comparison condition. Then, if both A and G(A) have finite global homological dimension, it follows that :

$$\text{gl.dim}(\tilde{A}) \leq 1 + \sup\{\text{gl.dim}(A), \text{gl.dim}(\text{gr}(A))\}.$$

Acknowledgment. This article is the outcome of my work during the academic years 1986-88 towards a Licentiatexamen at University of Stockholm. I want to thank warmly J.E.Björk, T.Ekedahl and C.Löfwall for their kind interest and advice.

1. External homogenization.

Let A be a $\mathbb{Z}$-graded ring. Thus, $A = \oplus A_v$, where the indices are integers and $A_v A_k \subset A_{v+k}$. A graded left (respectively right) A-module is a left (resp. right) A-module M which has a decomposition $M = \oplus M_v$ as an additive group. Moreover, the decomposition is compatibl with the A-module structure by the inclusions $A_k M_v \subset M_{k+v}$ (resp. $M_v A_k \subset M_{v+k}$) .

Let t be an indeterminate and consider the polynomial ring $A[t]$. The graded structure on A gives a graded ring structure on $A[t]$ by :

$$A[t]_m = \bigoplus_{v+k = m} A_v t^k \tag{1.1}$$

Following methods from the book [N.O] we are going to relate A-modules with graded $A[t]$-modules. We explain this in the categories of left modules and remark only that everything below remains valid with left replaced by right. Denote by Mod(A) the category of left A-modules. By $\underline{\text{Mod}}_g(A)$ we denote the family of graded left A-modules. If we consider the ring $A[t]$ we notice that $1-t$ is a central element and if $(1-t)$ denotes the two-sided ideal in $A[t]$ generated by $1-t$, then the rings $A[t]/(1-t)$ and A are isomorphic. It follows that if M is a left $A[t]$-module, then $M/(1-t)M$ carries a left A-module struc ture. This leads to :

1.2. DEFINITION. If M is a left $A[t]$-module, then $\varepsilon(M)$ denotes the A-module $M/(1-t)M$.

Now we announce a result which will be used several times later on

1.3. PROPOSITION. Let $M \in \text{Mod}(A)$. Then there exists a graded $A[t]$-module L such that $\varepsilon(L) = M$.

Before we prove Proposition 1.3 we shall need some preliminary con tructions. First, if $M \in \underline{\text{Mod}}_g(A)$, then we obtain the left $A[t]$-modul $A[t] \otimes_A M$. It is denoted by $M[t]$ and has a graded $A[t]$-module structure defined by $M[t]_m = \bigoplus_{v+k=m} M_v t^k$.

The equality $M = \varepsilon(M[t])$ is obvious. Let us now consider an A-submodule K of the graded A-module M . Here we do not assume that the A-module K is graded. Using the so called external homogenization we construct a graded $A[t]$-submodule K^* of $M[t]$ as follows :

First, if $x \in M$, we decompose x into homogeneous components, i.e. we write $x = \Sigma x_v$ where $x_v \in M_v$. Assume that x is non-zero and choose the largest integer v_o such that $x_{v_o} \neq 0$. Then we see that the element $\Sigma t^{v_o - v} x_v$ is a homogeneous element of degree v_o in the graded $A[t]$-module $M[t]$. We denote this element with x^* and refer to x^* as the external homogenization of x. Now, if K^* is a left A-submodule of M, then K^* denotes the $A[t]$-submodule of $M[t]$ generated by $\{k^* : k \in K\}$.

1.4. LEMMA. Let $M \in \underline{\mathrm{Mod}}_g(A)$. Then, for each A-submodule K of M, it follows that $\varepsilon(K^*) = K$.

Proof. Notice that the equality $\varepsilon(M[t]) = M$ can be interpreted by the decomposition $M[t] = M \oplus (1-t)M[t]$. Next, the $A[t]$-module $M[t]/K^*$ is graded. Since the $A[t]$-element t is homogeneous of degree one in the graded ring $A[t]$, it follows that the $A[t]$-element $1-t$ induces an injective map on every graded $A[t]$-module. In particular $1-t$ is injective on $A[t]/K^*$ and then we conclude that $(1-t)K^* = K^* \cap (1-t)M[t]$. This yields :

$$\varepsilon(K^*) = K^*/K^* \cap (1-t)M[t] = (K^* + (1-t)M[t])/(1-t)M[t] .$$

Identifying M with $M[t]/(1-t)M[t]$, it follows easily from $\varepsilon(K^*) = (K^* + (1-t)M[t])/(1-t)M[t]$, that $\varepsilon(K^*) = K$.

Proof of Proposition 1.3. The A-module M is the quotient of a free A-module F. Thus, $M = F/K$, where F is a direct sum of copies of the left A-module A. The graded left A-module structure on A gives a graded A-module structure on F. Now K is a left A-submodule of F and we obtain $\varepsilon(K^*) = K$. Then, since $1-t$ is injective on the graded $A[t]$-module $F[t]/K^*$, it follows that the left A-modules M and $\varepsilon(F[t]/K^*)$ are isomorphic. So Proposition 1.3 follows with $L = F[t]/K^*$.

1.5. Remark. Let $M \in \mathrm{Mod}(A)$ and let N be a left A-submodule of M. Then we can find a graded $A[t]$-module L and a graded submodule L' of L such that $\varepsilon(L) = M$ and $\varepsilon(L') = N$. To see this we take $M = F/K$ and then $N = K_1/K$ where $K \subset K_1 \subset F$. Then we can use

$L = \varepsilon(F[t]/K^*)$ and $L_1 = \varepsilon(K_1^*/K^*)$.

<u>Special graded</u> $A[t]$-<u>modules</u> . Let $M \in \underline{Mod}_g(A[t])$. We say that M is <u>special graded</u> if we have a decomposition $M = \oplus M(v)$, where $tM(v) \subset M(v+1)$ holds for every v . Moreover, each $M(v)$ is a graded A-module and if $M(v) = \oplus M(v)_k$, then the graded $A[t]$-module structure on M is given by $M_m = \bigoplus_{v+k=m} M(v)_k$.

<u>Remark</u>. Let M be a special graded $A[t]$-module. So we have a decomposition $M = \oplus M(v)$ where $tM(v) \subset M(v+1)$. Moreover, every A-module $M(v)$ is graded and we set $M(v) = \bigoplus_k M(v)_k$. On the other hand the graded $A[t]$-module M has homogeneous components given by $M_m = \bigoplus_{v+k=m} M(v)_k$. It follows that t maps $M(v)_k$ into $M(v+1)_k$ for every pair v and k . In particular, the A-linear map t from $M(v)$ into $M(v+1)$ is homogeneous of degree zero for every v .

Let us also consider a graded A-module $N = \oplus N_k$. We get the $A[t]$-module $M = A[t] \otimes_A N$ and M becomes a special graded $A[t]$-module with $M_m = \bigoplus_{j=k=m} t^j \otimes N_k$ while $M(v) = t^v \otimes N$.

Denote by <u>SG</u> the family of special graded $A[t]$-modules. If $M \in \underline{Mod}_g(A)$, then it is obvious that $M[t]$ is special graded. Another family of special graded $A[t]$-modules is obtained as follows :

1.6. DEFINITION. If $M \in \underline{Mod}(A)$, then $\psi(M)$ denotes the left $A[t]$-module for which $t\psi(M) = 0$,while the underlying A-module ${}_A\psi(M)$ is equal to the A-module M .

<u>Remark</u>. Let us say that a left $A[t]$-module S is <u>concentrated in degree zero</u> if the $A[t]$-element t annihilates S . Using the identification between the rings A and $A[t]/(t)$ we see that if $M \in \underline{Mod}(A)$, then $\psi(M)$ is concentrated in degree zero and conversely, if S is a left $A[t]$-module which is concentrated in degree zero, then $S = \psi({}_AS)$, where ${}_AS$ denotes the underlying A-module, derived from the inclusion of A as a subring of $A[t]$.

Now, if $M \in \underline{Mod}_g(A)$, then it is obvious that $\psi(M)$ is special graded. So in this way we obtain another subfamily of <u>SG</u> .

<u>From now on we shall assume that the ring A is Noetherian</u>, i.e. every left or right ideal of A is a finitely generated A-module. It follows that the ring $A[t]$ is Noetherian and we are going to study the class of special graded and finitely generated $A[t]$-modules.

Denote by $\underline{SG}_f$ the family of special graded and finitely generated A[t]-modules. If $M \in \underline{SG}_f$, we say that M has a <u>finite amplitude</u> if only finitely many M(v)'s are non-zero in the decomposition $M = \oplus M(v)$. Since the ring A[t] is Noetherian, we easily obtain :

1.7. LEMMA. Let $M \in \underline{SG}_f$. Then M has a finite amplitude if and only if there exists an integer w such that $t^w M = 0$.

In particular we notice that if M is finitely generated and graded A-module, then $\psi(M)$ has a finite amplitude. In fact, by definition, we notice that $\psi(M) = \oplus \psi(M)(v)$, where $\psi(M)(0) = M$ and $\psi(M)(v) = 0$ for every $v \neq 0$. Next, if M belongs to $\underline{SG}_f$ and the A[t]-module M is concentrated in degree zero, then $M = \oplus M(v)$ where only finitely many M(v)'s are non-zero and moreover, $tM(v) = 0$ for every v. It follows that each M(v) is a left A[t]-submodule of M. So the A[t]-module M is a direct sum of A[t]-modules of the form M(v), where M(v) is a graded A-module. Of course, this means just that the A[t]-module M is isomorphic to the direct sum of A[t]-modules of the form $\psi(M(v))$.

Denote by $\underline{SG}_f(0)$ the family of modules in $\underline{SG}_f$ of the form $\psi(M)$, where M is a finitely generated and graded A-module. By an induction over the integer w which yields $t^w M = 0$ for some M in the family $\underline{SG}_f$ having finite amplitude, it follows that $\underline{SG}_f(0)$ generates the family of modules in $\underline{SG}_f$ having finite amplitude within the category of finitely generated left A[t]-modules.

1.8. LEMMA. Let $M \in \underline{SG}_f$. Then there exists an integer w and an exact sequence $0 \to A[t] \otimes_A M(w) \to M \to M' \to 0$ where M' is zero or a special graded A[t]-module with finite amplitude.

<u>Proof of Lemma</u> 1.8. Set $K = \mathrm{Ker}_t(M)$ and $\bar{M} = M/tM$. Since t is a central element in the noetherian ring A[t] it first follows that K and $\bar{M}$ are finitely generated A[t]-modules. Moreover, since $tK = t\bar{M} = 0$ we conclude that the A-modules K and $\bar{M}$ are finitely generated. Now we consider the decomposition $M = \oplus M(v)$. It is obvious that

$$K = \bigoplus_v K \cap M(v) \quad \text{and} \quad \bar{M} = \bigoplus_v M(v)/tM(v-1) .$$

Next, since the A[t]-module M is finitely generated we see that there exists an integer v_o such that $v < v_o$ gives $M(v) = 0$. Next, since the A-modules K and $\bar{M}$ are finitely generated, it follows from the two decompositions above that there exists an integer w such that $v \geq w$ gives $K \cap M(v) = 0$ and $M(v)/tM(v-1) = 0$. Thus, if $v \geq w$ then the map $t : M(v) \to M(v+1)$ is bijective.

Denote by N the $A[t]$-submodule of M defined by $\bigoplus_{v \geq w} M(v)$. Then N is isomorphic to the special graded $A[t]$-module $A[t] \otimes_A M(w)$ and the quotient module M/N is a special graded $A[t]$-module with a finite amplitude. This proves Lemma 1.8.

1.9. Conclusion. Lemma 1.8 and the previous discussion shows that the family $\underline{SG}_f(0)$ and the family of special graded $A[t]$-modules of the form $\{M[t] : M$ is a graded and finitely generated A-module$\}$ together generate $\underline{SG}_f$ in the category of finitely generated left $A[t]$-modules

The Σ-filtration on $A[t]$. We have a positive filtration on the ring $A[t]$, defined by $\Sigma_v = A + At + \ldots + At^v$. Of course, the associated graded ring $gr_\Sigma(A[t]) = \oplus \Sigma_v/\Sigma_{v-1}$ is equal to the ring $A[t]$ Notice also that the Σ-filtration commutes with the graded structure on $A[t]$ inherited from A . In other words, if we put $\Gamma_m(A[t]) = A_m + A_m t + \ldots$, then $\Sigma_v = \bigoplus_m \Sigma_v \cap \Gamma_m(A[t])$ for every m .

Let us now consider a finitely generated left $A[t]$-module M . Recall that a good Σ-filtration on M consists of an increasing sequence $\{\Pi_v\}$ such that there exists a finite set $m_1,\ldots,m_s$ of generators of the $A[t]$-module M and integers $v_1,\ldots,v_s$ so that:

$$\Gamma_v = \Sigma_{v-v_1} m_1 + \ldots + \Sigma_{v-v_s} m_s .$$

In the case when M is a graded and finitely generated $A[t]$-module we say that a good Σ-filtration $\{\Gamma_v\}$ is homogeneous if we describe $\{\Gamma_v\}$ as in (1.10) using homogeneous generators $m_1,\ldots,m_s$ of the graded module M .

Obviously every graded and finitely generated $A[t]$-module has some homogeneous good Σ-filtration. Now we have

1.11. LEMMA. Let $\{\Gamma_v\}$ be a homogeneous good Σ-filtration on a graded and finitely generated $A[t]$-module. Then the associated graded module $gr_\Gamma(M) = \oplus \Gamma_v/\Gamma_{v-1}$ is special graded .

Proof of Lemma 1.11. We have the graded A[t]-module $M = \oplus M_m$. Since $m_1,\dots,m_s$ are homogeneous we see that

$$\Gamma_v = \bigoplus_m \Gamma_v \cap M_m .$$

Now $gr_\Gamma(M) = \oplus \Gamma_v/\Gamma_{v-1} = \oplus gr_\Gamma(M)(v)$.

Then $tgr_\Gamma(M)(v) \subset gr_\Gamma(M)(v+1)$ for every v . Next, for every v we construct a graded A-module structure on $gr_\Gamma(M)(v)$ as follows :

Set $gr_\Gamma(M)(v)_m = (\Gamma_v \cap M_{v+m})/(\Gamma_{v-1} \cap M_{m+v})$.

we see that $\bigoplus_m gr_\Gamma(M)(v)_m$ gives a graded A-module structure on $gr_\Gamma(M)(v)$. Finally , since $tM_{v+m} \subset M_{v+1+m}$ for every pair v,m, it follows from the Remark above Definition 1.6 that we have constructed a special graded A[t]-module structure on $gr_\Gamma(M)$.

Remark about homogeneous good Σ-filtrations. Let $M = \oplus M_v$ be a finitely generated and graded A[t]-module. It is easily seen that a good Σ-filtration $\{\Gamma_v\}$ on M is homogeneous if and only if $\Gamma_i = \oplus_v \Gamma_i \cap M_v$ for every i . From this it easily follows that if Γ is a homogeneous good Σ-filtration on M and N is a graded submodule of M , then the induced filtration $\{\Gamma_v \cap N\}$ is a homogeneous good Σ-filtration on N .

. On graded global and injective dimensions.

In this section we prove Theorem 0.2 from the introduction. So let A be a graded ring. We assume also that the ring A is Noetherian and obtain the graded ring A[t] as in Section 1 .

2.1. LEMMA. Assume that A has a finite graded injective dimension μ . Then A[T] has a finite injective graded dimension equal to $\mu + 1$.

Proof. First, if M is a finitely generated and graded A-module such that $Ext^\mu_A(M,A)$ is non-zero, then a standard result in homological algebra yields $Ext^{\mu+1}_{A[t]}(\psi(M),A[t]) \neq 0$. Indeed, the underlying right A-module of $Ext^{\mu+1}_{A[t]}(\psi(M),A[t])$ is equal to the right A-module $Ext^\mu_A(M,A)$.

It follows that the graded injective dimension of the ring A[t] is at least $\mu + 1$. Hence Lemma 2.1 follows if we can show that if

$v > \mu + 1$ and M is a finitely generated and graded $A[t]$-module, then $Ext^v_{A[t]}(M,A[t]) = 0$.

To prove this we choose a homogeneous good Σ-filtration Γ on M. Since Σ is a positive filtration on the Noetherian ring $A[t]$ it follows that there exists an induced good Σ-filtration on $Ext^v_{A[t]}(M,A[t])$ such that its associated graded module is a subquotient of $Ext^v_{A[t]}(gr_\Gamma(M),A[t])$. For this fact we refer to section 3 in [Bj : 2] .

We conclude that it suffices to show that $v > \mu + 1$ implies that $Ext^v_{A[t]}(gr_\Gamma(M), A[t]) = 0$. To show this we use that $gr_\Gamma(M)$ is <u>special</u> graded by Lemma 1.11 . Hence $gr_\Gamma(M)$ belongs to the family SG_f which is generated as in section 1.9 . Using long exact Ext-sequences we therefore reduce the proof of the vanishing of Ext-groups in degree $> \mu + 1$ to the cases when M is $\psi(N)$ for some graded A-module N, or $M = A[t] \otimes_A N$. In both these cases the hypothesis that $v > \mu$ gives $Ext^v_A(N,A) = 0$ imply that $Ext^v_{A[t]}(M,A[t]) = 0$ if $v > \mu + 1$ by standard homological algebra.

2.2. LEMMA. Let A be a graded Noetherian ring and assume there exists an integer μ such that $Ext^v_A(M,N) = 0$ for every $v > \mu$ and each pair of graded and finitely generated left or right A-modules. Then $Ext^v_{A[t]}(M,N) = 0$ for every $v > \mu + 2$ and any pair of graded and finitely generated $A[t]$-modules .

<u>Proof</u>. Let M and N be finitely generated and graded $A[t]$-modules. On each of them we choose a homogeneous good Σ-filtration. Since the Σ-filtration is positive on the Noetherian ring $A[t]$, it follows from [Bj : 2 , Proposition 3.2] that $Ext^v_{A[t]}(gr(M),gr(N)) = 0$ yields $Ext^v_{A[t]}(M,N) = 0$ for any given integer v . Hence it suffices to show that if $v > \mu + 2$, then $Ext^v_{A[t]}(gr(M),gr(N)) = 0$. Here $gr(M)$ and $gr(N)$ belong to the class <u>SG_f</u> . Using the generating family of <u>SG_f</u> in section 1.9 and standard homological algebra we obtain Lemma 2.2 .

Recall the definition of the ε-functor in Definition 1.2 and let us remark that we also have the ε-functor from right $A[t]$-modules into right A-modules, defined by $\varepsilon(N) = N/N(1-t)$ for every right $A[t]$-module N.

2.3 LEMMA. Let M be a finitely generated left or right $A[t]$-module. Then :

$$\varepsilon(\mathrm{Ext}^{v}_{A[t]}(M,A[t])) = \mathrm{Ext}^{v}_{A}(\varepsilon(M),A) \quad \text{holds for every } v .$$

Proof. The element $1-t$ is a central non - zero divisor in $A[t]$. Since $A = A[t]/A[t](1-t)$, we obtain Lemma 2.3 by standard homological algebra.

Proof of Theorem 0.2. First we prove (i) which amounts to show that the ring A has a finite injective dimension under the hypothesis that $\mathrm{Ext}^{v}_{A}(M,A) = 0$ for every $v > \mu$ and every finitely generated and graded A-module M. To show that $\mathrm{inj.dim}(A)$ is finite, we then consider a finitely generated A-module M. Proposition 1.3 yields a graded and finitely generated $A[t]$-module L such that $M = \varepsilon(L)$. Now Lemma 2.1 and Lemma 2.3 show that $\mathrm{Ext}^{v}_{A}(M,A) = 0$ for every $v > \mu + 1$ and hence $\mathrm{inj.dim}(A) \leq \mu + 1$.

To prove (ii) in Theorem 0.2, we use Lemma 2.2 and Lemma 2.3 as follows : Let M and N be a pair of finitely generated left A-modules. Choose graded and finitely generated $A[t]$-modules L and L' with $M = \varepsilon(L)$ and $N = \varepsilon(L')$. Since L' is graded, it follows that $1-t$ is injective on L' and we have the exact sequence :

$$0 \longrightarrow L' \xrightarrow{\;1-t\;} L' \longrightarrow L'/(1-t)L' \longrightarrow 0 .$$

By the associated long exact Ext-sequence and Lemma 2.2., we see that $\mathrm{Ext}^{v}_{A[t]}(L,L'/(1-t)L') = 0$ for every $v > \mu + 3$. Next, by standard homological algebra, it follows that the underlying right A-module of

$Ext^{v}_{A[t]}(L,L'(1-t)L')$ is equal to $Ext^{v}_{A}(M,N)$ and this proves that gl.dim(A) is finite, i.e. we have obtained $gl.dim(A) \leq \mu + 3$ with μ as in Lemma 2.2 .

Finally, we recall that since A is a Noetherian ring, it follows that if gl.dim(A) is finite, then gl.dim(A) is equal to inj.dim(A). Then we see that the proof of (1) in Theorem 0.2 yields $gl.dim(A) \leq \mu + 1$, where μ is the integer from Lemma 2.2 .

3. Graded Auslander conditions.

In this section we prove Theorem 0.1 from the introduction. But first we need some preliminaries. Let $A = \oplus A_v$ be a graded ring. Assume that A is Noetherian and denote by $\underline{Mod}^f_g(A)$ the family of finitely generated and graded left A-modules . Let $M = \oplus M_v$ and $N = \oplus N_v$ belong to $\underline{Mod}^f_g(A)$. A left A-linear map $\varphi : M \to N$ is homogeneous of degree zero if $\varphi(M_v) \subset N_v$ for every v . By a finitely generated and graded free left A-module we mean a finite direct sum of shifted A-modules of the form A[w] , where w is an integer and A[w] is graded by $A[w]_v = A_{v+w}$ for every v .

It is wellknown that if $M \in \underline{Mod}^f_g(A)$, then there exists a graded free resolution $\longrightarrow F_1 \longrightarrow F_0 \longrightarrow M \longrightarrow 0$. That is, each F_i is graded free and the differentials are homogeneous of degree zero. Moreover, if $F_.$ and $F'_.$ is a pair of two graded free resolutions of M, then the two complexes $F_.$ and $F'_.$ are homotopic and the A-linear maps $k_v : F_v \to F_{v+1}$ which yield the homotopy can be taken to be homogeneous of degree zero.

If $F_.$ is a graded free resolution of some M in $\underline{Mod}^f_g(A)$, then we see that $Hom_A(F_.,A)$ is a complex of right A-modules whose differentials are right A-linear and homogeneous of degree zero. We conclude that the cohomology modules $\mathrm{Ext}_A^v(M,A)$ are graded right A-modules. Moreover, the graded structure on $\mathrm{Ext}_A^v(M,A)$ is <u>canonical</u> since the previous homotopy relation between pairs of graded free resolutions of M show that the graded structure on $\mathrm{Ext}_A^v(M,A)$ does not depend on the special graded free resolution.

Using graded free resolutions over an exact sequence $0 \longrightarrow M' \longrightarrow M \longrightarrow M'' \longrightarrow 0$ in $\underline{\mathrm{Mod}}_g^f(A)$, where we assume that the A-linear maps are homogeneous of degree zero, it follows that the long exact Ext-sequence $\ldots \to \mathrm{Ext}^v(M'',A) \to \mathrm{Ext}^v(M,A) \to \mathrm{Ext}^v(M',A) \to \mathrm{Ext}_A^{v+1}(M'',A) \to \cdots$ has arrows given by right A-linear maps which are homogeneous of degree zero, provided that the <u>canonical</u> graded A-module structure is used on every Ext-group.

The previous discussion has been inserted in order to clarify the subsequent discussion about the <u>graded</u> Auslander condition.

Let A be a graded noetherian ring. Then we give :

3.1. DEFINITION. Let $M \in \underline{\mathrm{Mod}}_g^f(A)$. We say that M <u>satisfies the graded Auslander condition</u> if we have : for every integer v and every graded submodule N of $\mathrm{Ext}_A^v(M,A)$, it follows that $\mathrm{Ext}_A^i(N,A) = 0$ for every $i < v$.

3.2. **LEMMA.** Let $0 \to M' \to M \to M'' \to 0$ be an exact sequence of graded and finitely generated A-modules where the A-linear maps are homogeneous of degree zero. Then, if both M' and M" satisfy the graded Auslander condition, it follows that M does.

<u>Proof</u>. Let N be a graded submodule of $\mathrm{Ext}_A^v(M,A)$. By the long exact

Ext-sequence associated with $0 \to M' \to M \to M'' \to 0$, we obtain an exact sequence

$$0 \to N' \to N \to N'' \to 0$$

where $N'' \subset Ext^v_A(M',A)$ and N'' is a subquotient of $Ext^v_A(M'',A)$. The hypothesis gives $Ext^j_A(N',A) = Ext^j_A(N'',A) = 0$ for every $j < v$. By th long exact Ext-sequence associated with $0 \to N' \to N \to N'' \to 0$ we get $Ext^j_A(N,A) = 0$ for every $j < v$.

Let us now assume that A is a graded ring of finite injective dimension and every graded module satisfies the graded Auslander condition. In particular A is a Noetherian ring with a finite injective dimension. If M is a graded and finitely generated A-module we construct a <u>graded free</u> resolution $P_{\cdot}$ of M and then a <u>graded projective</u> resolution $Q_{\cdot\cdot}$ exists of the complex $Hom_A(P_{\cdot},A)$. Thus, $Q_{\cdot\cdot}$ is a double complex where every Q_{iv} is a graded A-module and a direct summan of a graded free A-module of finite rank. Moreover the differentials in $Q_{\cdot\cdot}$ are homogeneous of degree zero. Moreover, the conditions explained in § 1 from [Bj:2] hold. Then $Hom_A(Q_{\cdot\cdot},A)$ represents the <u>bidualizing complex</u> of M.

By the result of Th.Levasseur described in § 1 of [Bj:2], the spectral sequence associated with the first filtration is regular and we obtain the B-filtration from Theorem 1 in [Bj:2], because we have assumed that A is <u>graded Auslander-Gorenstein</u> which implies that the <u>graded</u> A-modules $Ext^v_A(Ext^j_A(M,A),A) = 0$ when $v < j$.

Moreover, the B-filtration on M consists of <u>graded</u> submodules because they are obtained from the associated spectral sequence which by the construction consists of complexes of graded A-modules where the differentials are homogeneous of degree zero.

Summing up, we have

3.3. PROPOSITION. Assume that A is graded Auslander-Gorenstein. Then, if M is a finitely generated and graded A-module, there exists a sequence $M = B_0(M) \supset B_{-1}(M) \supset \cdots \supset B_{-\mu}(M) \supset 0$ of graded submodules and for every v an exact sequence :

$$0 \to B_{-v}(M)/B_{-v-1}(M) \to \mathrm{Ext}^v(\mathrm{Ext}^v(M,A),A) \to Q(v) \to 0$$

of graded A-modules where the A-linear maps are homogeneous of degree zero.

Let us now draw some conclusions from Proposition 3.3. The graded Auslander condition gives $j_A(B_{-v}(M)/B_{-v-1}(M)) \geq v$ for every v, where j_A refers to the <u>grade number</u>, i.e. if N is a finitely generated and non-zero A-module then $j_A(N)$ is the smallest integer such that $\mathrm{Ext}_A^{j_A(N)}(N,A)$ is non-zero .

Then, exactly as in the proof of Proposition 1.7 in [Bj:2], an induction over v gives $j_A(B_{-v}(M)) \geq v$ for every v. Then we conclude that $B_{-j(M)}/B_{-j(M)-1}$ cannot be zero and hence the exact sequence in Proposition 3.3 implies that $\mathrm{Ext}^{j_A(M)}(\mathrm{Ext}^{j_A(M)}(M,A),A)$ is non-zero. This conclusion is what is needed later on for the proof of Theorem 0.1 . So we announce this by the following:

Conclusion. <u>Let</u> A <u>be a graded Auslander-Gorenstein ring</u>. <u>Then, if</u> M <u>is a graded and finitely generated</u> A-<u>module</u>, <u>it follows that</u> :

$$j_A(\mathrm{Ext}_A^{j_A(M)}(M,A)) = j_A(M).$$

At this stage we are prepared to begin the proof of Theorem 0.1 where the first step is to establish the following :

3.4. LEMMA. Assume that the graded ring A is graded Auslander-Gorenstein. Then $A[t]$ is graded Auslander-Gorenstein.

Proof. Of course, we have $\text{inj.dim}(A[t]) = \text{inj.dim}(A)+1$ so there remain only to verify that every graded and finitely generated $A[t]$-module satisfies the graded Auslander condition. So let $M \in \underline{\text{Mod}}_g^f(A[t])$ and choose a homogeneous good Σ-filtration Γ on M.

Let N be a graded submodule of $\text{Ext}^v_{A[t]}(M,A[t])$. Since the Σ-filtration is discrete and $gr_\Sigma(A[t]) = A[t]$, it follows from the general result in [Bj : 2, Proposition 3.1] combined with the remarks after Lemma 1.11, that there exists a homogeneous good Σ-filtration on N such that $gr(N)$ is a subquotient of $\text{Ext}^v_{A[t]}(gr_\Gamma(M),A[t])$. So if we have proved that $gr_\Gamma(M)$ satisfies the graded Auslander condition, then the grade number $j_{A[t]}(gr(N)) \geq v$. We also note that the inequality $j_{A[t]}(N) \geq j_{A[t]}(gr(N))$ holds because the Σ-filtration on the ring $A[t]$ is Zariskian, i.e. see Proposition 3.1 in [Bj:2].

We conclude that if $gr_\Gamma(M)$ satisfies the graded Auslander condition so does M. Now $gr_\Gamma(M)$ is a special graded $A[t]$-module and using Lemma 3.2 and Conclusion 1.9, it follows that there only remain to show that if $M \in \underline{\text{Mod}}_g^f(A)$, then the $A[t]$-modules $\psi(M)$ and $A[t] \otimes_A$ both satisfy the graded Auslander condition.

The proof that $\psi(M)$ satisfies the graded Auslander condition goe as follows : Let N be a graded $A[t]$-submodule of $\text{Ext}^v_{A[t]}(\psi(M),A[t])$ By standard homological algebra, it follows that $N = \psi(P)$ for a graded right A-submodule P of the right A-module $\text{Ext}^{v-1}_A(M,A)$. Then $j_A(P) \geq v-1$ and again by standard homological algebra we get $j_{A[t]}(N) = j_A(P) + 1 \geq v$. This proves that $\psi(M)$ satisfies the graded Auslander condition.

Next, let N be a graded $A[t]$-submodule of $\text{Ext}^v_{A[t]}(M[t],A[t])$. Let us put $\mathcal{M} = \text{Ext}^v_A(M,A)$. We notice that the graded right $A[t]$- module $\text{Ext}^v_{A[t]}(M[t], A[t]) = \mathcal{M} \otimes_A A[t] = \mathcal{M}[t]$. We have the homogeneous good Σ-filtration on $\mathcal{M}[t]$, defined by $\mathcal{M}[t]_v = \mathcal{M} + \mathcal{M}t + \dots + \mathcal{M}t^v$. Then $gr(\mathcal{M}[t]) = \mathcal{M}[t]$ and we obtain an induced homogeneous good Σ-filtration on N . The inequality $j_{A[t]}(N) \geq j_{A[t]}(gr(N))$ reduces the proof that $\mathcal{M}[t]$ satisfies the Auslander condition to show that $j_{A[t]}(gr(N)) \geq v$. To show this we decompose the special graded $A[t]$-module $gr(N)$, i.e. $gr(N) = \oplus N(i)$. We observe that $N(i)$ is a graded right A-submodule of $Ext^v_A(M,A)$ for every i . So the hypothesis that A is graded Auslander-Gorenstein implies that $j_A(N(i)) \geq v$ for every i . The required inequality $j_{A[t]}(N) \geq v$ now follows from the following :

SUBLEMMA. Let $N = \oplus N(i)$ be a special graded $A[t]$-module . Let k be an integer such that $j_A(N(i)) \geq k$ for every i . Then $j_{A[t]}(N)) \geq k$.

<u>Proof</u>. There exists an integer v such that $t : N(i) \to N(i+1)$ is bijective for every $i \geq w$. Then we obtain the exact sequence :

$$0 \longrightarrow A[t] \otimes_A N(w) \longrightarrow N \longrightarrow N' \longrightarrow 0$$

where $N' = \oplus N'(i)$ has a finite amplitude, i.e. $N'(i) = N(i)$ if $i < w$ while $i \geq w$ gives $N'(i) = 0$. Using long exact Ext-sequences we easily get $j_{A[t]}(N) \geq k$.

The next result is essential for the proof of Theorem 0.1 .

3.6. LEMMA. Let M be a graded $A[t]$-module and assume that $(0:t)_M = 0$. Then $j_A(\varepsilon(M)) = j_{A[t]}(M)$.

<u>Proof</u>. Since the $A[t]$-module M is graded, it follows that $1-t$ is injective on M and then $0 \to M \xrightarrow{1-t} M \to M/(1-t)M \to 0$ is an exact sequence of $A[t]$-modules. Moreover, $1-t$ is injective on the graded modules $Ext^v_{A[t]}(M,A[t])$ and hence the long exact Ext-sequence gives with $\mathcal{M}_v = Ext^v_{A[t]}(M,A[t])$;

(1) $\quad Ext^{v+1}_{A[t]}(M/(1-t)M,A[t]) = \mathcal{M}_v/\mathcal{M}_v(1-t)$ for every v .

This shows that $j_{A[t]}(M/(1-t)M) \geq j_{A[t]}(M) + 1$. Standard homological algebra gives $j_A(\varepsilon(M)) = j_{A[t]}(M/(1-t)M)$ and hence we have

$j_A(\varepsilon(M)) \geq j_{A[t]}(M)$. By (1) we get the equality if we have proved the following :

SUBLEMMA. Let $k = j_A(M)$. Then $\mathcal{M}_k/\mathcal{M}_k(1-t)$ is non-zero.

Proof of Sublemma. If K is a graded A[t]-module it is easily seen that $K/(1-t)K = 0$ holds if and only if K is equal to its t-torsion module. So if $\mathcal{M}_k/\mathcal{M}_k(1-t) = 0$, then the noetherian right A[t]-module $\mathcal{M}_k$ is annihilated by some power of t . Suppose then that $\mathcal{M}_k t^w = 0$. We are going to derive a contradiction from this and then conclude that $\mathcal{M}_k/\mathcal{M}_k(1-t)$ cannot be zero .

To get the contradiction we use the injectivity of t on M and consider $0 \to M \xrightarrow{t^w} M \to M/t^wM \to 0$. The long exact Ext-sequence contains

$$\mathcal{M}_k \xrightarrow{t^w} \mathcal{M}_k \longrightarrow \mathrm{Ext}^{k+1}_{A[t]}(M/t^wM, A[t]) .$$

So if $\mathcal{M}_k t^w = 0$, then $\mathcal{M}_k$ is a submodule of $\mathrm{Ext}^{k+1}_{A[t]}(M/t^wM,A[t])$ The graded Auslander condition gives $j_{A[t]}(\mathcal{M}_k) \geq k+1$. But now we recall that $k = j_{A[t]}(M)$. So the conclusion after Proposition 3.3 gives $j_{A[t]}(\mathcal{M}_k) = k$. This gives the required contradiction.

Proof of Theorem 0.1. Since Theorem 0.2 already is proved,we see that there only remains to verify the Auslander conditions under the hypothesis that A is graded Auslander-Gorenstein. So let M be a finitely generated A-module and $N \subset \mathrm{Ext}^v_A(M,A)$. First we choose a graded A[t]-module L such that $M = \varepsilon(L)$. Lemma 2.3 gives :

$$\mathrm{Ext}^v_A(M,A) = \varepsilon(\mathrm{Ext}^v_{A[t]}(L,A[t])) .$$

Now we obtain $j_A(\mathrm{Ext}^v_A(M,A)) \geq j_{A[t]}(\mathrm{Ext}^v_{A[t]}(L,A[t]) \geq v$, where the first inequality follows from the proof of Lemma 3.6 and the second since A[t] is graded Auslander-Gorenstein.

Next, we can apply the Remark in 1.5 and find a graded A[t]-module S and a graded submodule S' so that $N = \varepsilon(S')$ and $\mathrm{Ext}^v_A(M,A) = \varepsilon(S)$. We notice that if S_o is the t-torsion part of S , then $\varepsilon(S) = \varepsilon(S/S_o)$ and similarly $N = \varepsilon((S' + S_o)/S_o)$. So we may assume that $(O:t)_S = (O:t)_{S'} = O$. Then Lemma 3.6 gives

$$j_A(N) = j_{A[t]}(S') \geq j_{A[t]}(S) = j_A(\mathrm{Ext}^v_A(M,A)) \geq v$$

and hence Theorem 0.1 is proved.

4. On the preservation of the Auslander condition.

First we show that polynomial rings over Auslander-Gorenstein, (respectively Auslander-regular) rings remain Auslander-Gorenstein, (resp. Auslander regular). Recall first that if A is a Noetherian ring and if inj.dim(A) is finite so is inj.dim(A[t]) . In fact, we have inj.dim(A[t]) = inj.dim(A) + 1 . Similarly, gl.dim(A[t]) = gl.dim(A) + 1.

So there remains only to verify the Auslander condition for finitely generated A[t]-modules, assuming that the ring A is Auslander-Gorenstein. Notice that if we instead assume that A is Auslander regular, then the finiteness of gl.dim(A[t]) and the Auslander condition entail that A[t] is Auslander regular.

So from now on we assume that A is Auslander-Gorenstein. We must show that every finitely generated A[t]-module satisfies the Auslander condition. Denote by $\mathcal{A}$ the family of finitely generated A[t]-modules satisfying the Auslander condition. The ring A[t] has the positive Σ-filtration defined by $\Sigma_v = A + \ldots + At^v$ and A[t] is equal to $gr_\Sigma(A[t])$.

4.1. LEMMA. Let M be a finitely generated A[t]-module such that $gr_\Gamma(M) \in \mathcal{A}$ for some good Σ-filtration Γ on M . Then $M \in \mathcal{A}$.

Proof of Lemma 4.1. Let $N \subset \mathrm{Ext}^v_{A[t]}(M,A[t])$ for some $v \geq 0$. We must show that $j_{A[t]}(N) \geq v$. To prove this we construct a filtered free resolution of the Γ-filtered A[t]-module M and obtain a good filtration on $\mathrm{Ext}^v_{A[t]}(M,A[t])$ such that its associated graded module is a subquotient of $\mathrm{Ext}^v_{gr_\Sigma(A[t])}(gr_\Gamma(M),gr_\Sigma(A[t]))$. See also Proposition 3.1 in [Bj : 2] . It follows that N has a good Σ-filtration such that gr(N) is a subquotient of $\mathrm{Ext}^v_{gr_\Sigma(A[t])}(gr_\Gamma(M),gr_\Sigma(A[t]))$. Recall that the rings A[t] and $gr_\Sigma(A[t])$ are the same. The hypothesis that $gr_\Gamma(M)$ belongs to $\mathcal{A}$ gives $j_{gr_\Sigma(A[t])}(gr(N)) \geq v$. Next , since the Σ-filtration is positive on the Noetherian ring $A[t]$, it follows from Proposition 3.2 in [Bj : 2] that $j_A(N) \geq v$ and hence Lemma 4.1 is proved.

Next, a <u>t-graded</u> <u>A[t]-module</u> is of the form $M = \oplus M(v)$, where $tM(v) \subset M(v+1)$ and every $M(v)$ is a left A-module. If Γ is a good Σ-filtration on some finitely generated $A[t]$-module M, then $gr_\Gamma(M)$ is t-graded. Also, if $M = \oplus M(v)$ is t-graded, then a similar proof as in Lemma 1.8 shows the existence of an integer w such that we have an exact sequence :

$$0 \longrightarrow A[t] \otimes_A M(w) \longrightarrow M \longrightarrow M' \longrightarrow 0$$

where $M' = \oplus M'(v)$, with $M'(v) = 0$ if $v \geq w$ and $M'(v) = M(v)$ if $v \leq w$. Using this we see that similar methods as in the proof of Lemma 3.4 show that every finitely generated and t-graded $A[t]$-module belongs to $\mathcal{A}$. Combined with Lemma 4.1 we have then proved that every finitely generated $A[t]$-module satisfies the Auslander condition.

<u>Skew polynomial rings</u>. Let A be a ring and $\rho: A \to A$ a ring automorphism. Let δ be a ρ-derivation, i.e. we have $\delta(ab) = \delta(a)b + \rho(a)\delta(b)$. Then the skew polynomial ring $A[x,\rho,\delta]$ is the ring whose additive group is $\oplus Ax^v$ and the ring structure is obtained by $xa = \rho(a)x + \delta(a)$ for all $a \in A$.

4.2. THEOREM. Let A be Auslander-Gorenstein, (resp. Auslander-Gorenstein). Then every skew polynomial ring over A is Auslander-Gorenstein, (resp. Auslander regular).

<u>Proof</u>. The ring $A[x,\rho,\delta]$ has the positive Σ-filtration defined by $\Sigma_v = \bigoplus_{j \leq v} Ax^j$. We see that $gr_\Sigma(A[x,\rho,\delta]) = A[x,\rho]$.

The ring $A[x,\rho]$ is Noetherian and hence the Σ-filtration on $A[x,\rho,\delta]$ is Zariskian. So by Theorem 3.9 in [Bj:2] we see that it suffices to show that $A[x,\rho]$ is Auslander-Gorenstein (respectively Auslander-regular). Thus, we may assume that the ρ-derivation δ is zero. The proof that $A[x,\rho]$ is Auslander-Gorenstein if A is, follows by similar methods as in the case then ρ is the identity map.

<u>T-rings</u>. Let A be a ring which contains an element T satisfying : $AT = TA$ and T is neither a left nor a right zero-divisor on A.

Then we denote by (T) the two-sided ideal generated by T and get the ring $A/(T)$. Now we have the following :

4.3. THEOREM. If $A/(T)$ is Auslander-Gorenstein, (resp. Auslander regular) and if T belongs to the Jacobson radical of A, it follows that A is Auslander-Gorenstein, (resp. Auslander regular) .

Proof. Put $A_{-v} = (T)^v$ for every $v \geq 1$. With $A_0 = A$ we see that $A_0 \supset A_{-1} \supset \dots$ is a filtration where we notice that $\cap A_{-v} = 0$ follows because T belongs to the Jacobson radical.

The associated graded ring $\oplus (T)^v/(T)^{v+1}$ is denoted by $gr_T(A)$. Since T has no zero-divisors we see that $gr_T(A)$ is the skewpolynomial ring $A/(T)[X, \bar{\rho}]$ where $\bar{\rho}$ is the automorphism on $A/(T)$ induced by the automorphism ρ on A, defined by $\rho(a)T = Ta$ for every a in A.

Hence Theorem 3.9 in [Bj:2] gives Theorem 4.3 if the T-filtration is Zariskian. To show this we notice that the Rees ring $\oplus(T)^v$ is isomorphic to $A[X, \rho]$ and hence is Neotherian. Finally, since T belongs to the Jacobson radical it follows from Nakayama's Lemma that good filtrations on finitely generated A-modules are separated.

5. A study of Rees rings.

Let A be a filtered ring. The associated Rees ring is the graded ring $\oplus A_v$. It is denoted by $\tilde{A}$ and we refer to [Bj:2] for further comments about the ring $\tilde{A}$. In particular $\tilde{A}$ contains the central element $j_1(1)$ which we denote by T. We obtain the T-filtration on $\tilde{A}$, defined by $\tilde{A}_v = \tilde{A}$ if $v \geq 0$ and $\tilde{A}_{-v} = (T)^v$ if $v \geq 1$. Recall that $\tilde{A}/(T)$ is isomorphic with the associated graded ring $gr(A)$. It follows easily that $gr_T(\tilde{A}) = \oplus \tilde{A}_v/\tilde{A}_{v-1}$ is isomorphic with the polynomial ring in one variable over $gr(A)$.

Next, the central $\tilde{A}$-element T is a non-zero divisor and hence we can construct the localisation $\tilde{A}_T$, i.e. the so-called universal T-inverting ring.

5.1. LEMMA. The ring $\tilde{A}_T$ is isomorphic to the ring $A[x,x^{-1}]$ of finite Laurent series in one variable over A.

For the proof we refer to [Bj:2, Proposition 2.14] .

Concerning the T-filtration on A we shall need the following

5.2. LEMMA. Let $\mathscr{R}$ be the associated Rees ring of $\tilde{A}$ with respect to the T-filtration. Then $\mathscr{R}$ is a Noetherian ring.

Proof. The graded ring $\mathscr{R} = \oplus\, gr_v(\mathscr{R})$, where $gr_{-v}(\mathscr{R}) = (T)^v$ if $v \geq 1$ while $gr_v(\mathscr{R}) = \tilde{A}$ if $v \geq 0$. Since $\tilde{A}$ is assumed to be Noetherian it is easy to show that every graded left or right ideal in $\mathscr{R}$ is finitely generated. Then Theorem 2.18 in [Bj:2] shows that the ring $\mathscr{R}$ is Noetherian.

After these preliminaries we are prepared to enter the proof of the following:

5.3. THEOREM. Let A be a filtered ring such that $\tilde{A}$ is Noetherian. Then, if A and gr(A) both are Auslander-Gorenstein, (resp. Auslander regular) it follows that $\tilde{A}$ is Auslander-Gorenstein, (resp. Auslander regular).

The proof of Theorem 5.3 requires several steps. First we establish results concerned with injective and global dimensions where the Auslander condition does not appear.

5.4. LEMMA. Let A be a filtered ring such that $\tilde{A}$ is Noetherian. Then, if both inj.dim(A) and inj.dim(gr(A)) are finite, it follows that

$$\text{inj.dim}(\tilde{A}) \leq 1 + \sup\{\text{inj.dim}(A),\ \text{inj.dim}(gr(A))\},$$

Proof. Let M be a finitely generated left $\tilde{A}$-module. The localization with respect to T commutes with Ext and hence we have

(1) $$\text{Ext}^v_{\tilde{A}}(M,\tilde{A})_T = \text{Ext}^v_{\tilde{A}_T}(M_T,\tilde{A}_T) \quad \text{for every } v.$$

Now $\tilde{A}_T = A[T,T^{-1}]$ so $\text{inj.dim}(\tilde{A}_T) \leq 1 + \text{inj.dim}(A)$. We conclude that if $\mu = \sup\{\text{inj.dim}(A), \text{inj.dim}(gr(A))\}$, then $v > \mu + 1$ gives $\text{Ext}^v_{\tilde{A}}(M,\tilde{A})_T = 0$ which means that the right $\tilde{A}$-module $\text{Ext}^v_{\tilde{A}}(M,\tilde{A})$ has T-torsion, i.e. every element is annihilated by some power of T. Since $\text{Ext}^v_{\tilde{A}}(M,\tilde{A})$ is a finitely generated A-module we also obtain an integer w such that T^w annihilates $\text{Ext}^v_{\tilde{A}}(M,\tilde{A})$, where v is a fixed integer $> \mu + 1$.

Next, by Lemma 5.2 the T-filtration satisfies the comparison condition as explained in section 2 of [Bj:2].

Let Γ be a good T-filtration on M . Then, by the general result in [Bj:2, Proposition 3.1]

(2) $gr_T(Ext^v_{\tilde{A}}(M,\tilde{A}))$ is a subquotient of $Ext^v_{gr_T(\tilde{A})}(gr_{\tilde{T}}(M), gr_T(\tilde{A}))$.

Now $gr_T(\tilde{A})$ is the polynomial ring in one variable over $gr(A)$. Hence (2) shows that $v > \mu + 1$ gives $gr_T(Ext^v_{\tilde{A}}(M,\tilde{A}) = 0$.

At this stage we conclude that $v > \mu + 1$ gives $Ext^v_{\tilde{A}}(M,\tilde{A}) = 0$. Indeed, we notice that if K is some A-module such that T^w annihilates K for some integer w , then $gr_T(K)$ cannot be zero . This applies to the A-module $Ext^v_{\tilde{A}}(M,\tilde{A})$ and Lemma 5.4 is proved.

5.5 LEMMA. Let A be a filtered ring such that $\tilde{A}$ is Noetherian . Then, if both gl.dim(A) and gl.dim(gr(A)) are finite, it follows that

$$gl.dim(\tilde{A}) \leq 1 + \sup\{gl.dim(A), gl.dim(gr(A))\} .$$

<u>Proof</u>. Recall first that if B is a Noetherian ring with a finite global dimension, then gl.dim(B) = inj.dim(B) . We conclude that Lemma 5.4 gives the inequality in Lemma 5.5 if we have proved that $gl.dim(\tilde{A})$ is <u>finite</u>. So now we are going to establish the existence of some integer ω such that every finitely generated $\tilde{A}$-module has a projective resolution of length ω at most. To obtain ω we proceed as follows. First $\tilde{A}$ is a graded ring and Lemma 2.2 shows that if we have found an integer ω_o such that the projective dimension of every finitely generated and graded $\tilde{A}$-module is at most ω_o , then $pd_{\tilde{A}}(M) \leq \omega + 2$ for every finitely generated $\tilde{A}$-module. So it suffies to consider graded $\tilde{A}$-modules .

If M is a graded $\tilde{A}$-module we denote by M^t the T-torsion part. Thus, $M^t = \{m \in M : T^v m = 0$ for some $v \geq 1\}$. We see that M^t is a graded submodule of M . Also, there exists an integer w such that $T^w M^t = 0$. In the case when M is a graded $\tilde{A}$-module such that $TM = 0$ we say that M is concentrated in degree zero. If M is concentrated in degree zero we consider the A-module $\varepsilon(M)$ given by $M/(1-t)M$, where the A-module structure on $M/(1-t)M$ comes from the equality between the rings $\tilde{A}$ and $A/(1-T)$.

By standard homological algebra we get $pd_{\tilde{A}}(M) = pd_A(\varepsilon(M)) + 1$ for every graded $\tilde{A}$-module which is concentrated in degree zero. By an induction over the integer w we obtain $pd_{\tilde{A}}(M) \leq 1 + gl.dim(A)$ for every graded finitely generated $\tilde{A}$-module with $M = M^t$.

Hence there only remains to find an integer ω such that $pd_{\tilde{A}}(M) \leq \omega$ for every T-torsion free and graded A-module M . To prove the existence of ω, we recall that if $M^t = 0$ and M is graded, then $M = \varepsilon(M)^{\sim}$, where $\varepsilon(M)$ has a good filtration and $\varepsilon(M)^{\sim}$ denotes the associated Rees module. See also section 2 in [Bj:2] for details .

So there remains to show the existence of an integer such that $pd_{\tilde{A}}(\tilde{M}) \leq \omega$, where $\tilde{M}$ is the Rees module of some finitely generated A-module M with a good filtration. To obtain ω we first treat a special case .

SUBLEMMA 1. Let M be a finitely generated A-module. Suppose that $gr_{\Gamma}(M) = 0$ for every good filtration on M . Then $Ext^{v}_{\tilde{A}}(\tilde{N},\tilde{M}) = 0$ for every $v > gl.dim(A)$ and every Rees module $\tilde{N}$ of a finitely generated A-module N with a good filtration.

Proof of Sublemma 1. The hypothesis that $gr(M) = 0$ means that the A-element T is bijective on the Rees module $\tilde{M}$. Next, using a graded free resolution $F_{\bullet}$ of N we see that $Hom_{\tilde{A}}(F_{\bullet},\tilde{M})$ is a complex of graded abelian groups. The bijective $\tilde{A}$-linear map T on $\tilde{M}$ induces a map denoted by $\rho_v(T)$ on the abelian group $Ext^{v}_{\tilde{A}}(\tilde{N},\tilde{M})$ for each v . We see that $\rho_v(T)$ must be injective for every v . Next, consider the exact sequence $0 \longrightarrow \tilde{M} \xrightarrow{1-T} \tilde{M} \longrightarrow \tilde{M}/(1-T)\tilde{M} \longrightarrow 0$. This gives a long exact sequence which contains

$$\longrightarrow Ext^{v}_{\tilde{A}}(\tilde{N},\tilde{M}) \xrightarrow{1-\rho_v(T)} Ext^{v}_{\tilde{A}}(\tilde{N},\tilde{M}) \longrightarrow Ext^{v+1}_{\tilde{A}}(\tilde{N},\tilde{M}/(1-T)\tilde{M}) \longrightarrow$$

Notice that $\rho_v(T)$ is a homogeneous map of degree one on the graded group $Ext^{v}_{\tilde{A}}(\tilde{N},\tilde{M})$ and therefore injective. So for each integer v we have the following equality between abelian groups :

$$Ext^{v+1}_{\tilde{A}}(\tilde{N},\tilde{M}/(1-T)\tilde{M}) = Ext^{v}_{\tilde{A}}(\tilde{N},\tilde{M})/(1-\rho_v(T)(Ext^{v}_{\tilde{A}}(\tilde{N},\tilde{M})) .$$

Next, since $\rho_v(T)$ is injective on $Ext^{v}_{\tilde{A}}(\tilde{N},\tilde{M})$ we see that $(1-\rho_v(T))Ext^{v}_{\tilde{A}}(\tilde{N},\tilde{M}))$ cannot contain homogeneous elements of the graded group $\mathrm{Ext}^{v}_{\tilde{A}}(\tilde{N},\tilde{M})$. So $1 - \rho_v(T)$ cannot be surjective, unless $\mathrm{Ext}^{v}_{\tilde{A}}(\tilde{N},\tilde{M}) = 0$. We conclude that Sublemma 1 follows if we have shown that $v > \mathrm{gl.dim}(A)$ gives $\mathrm{Ext}^{v}_{\tilde{A}}(\tilde{N},\tilde{M}/(1-t)\tilde{M}) = 0$. To prove this we notice that standard homological algebra identifies the abelian group $\mathrm{Ext}^{v}_{\tilde{A}}(\tilde{N},\tilde{M}/(1-T)\tilde{M})$ with $\mathrm{Ext}^{v}_{A}(N,M)$. The last Ext-group vanishes if $v > \mathrm{gl.dim}(A)$.

Proof continued. Let us now consider an arbitrary finitely generated A-module M . So here gr(M) can be non-zero for a good filtration. We choose a good filtration on M and construct a so called filt-free resolution $\ldots \to F_1 \to F_o \to M \to 0$. See [Bj:2].

Let $\mu = \mathrm{gl.dim}(\mathrm{gr}(A))$. Passing to the associated graded complex we see that if K is the kernel of the map $F_\mu \to F_{\mu-1}$, then $0 \to \mathrm{gr}(K) \to \mathrm{gr}(F_\mu) \to \cdots \to \mathrm{gr}(F_o) \to \mathrm{gr}(M) \to 0$ is an exact sequence of gr(A)-modules and gr(K) is a projective gr(A)-module. Then we find a gr(A)-linear map $\varphi: \mathrm{gr}(F_\mu) \to \mathrm{gr}(K)$ which is homogeneous of degree zero and restricts to the identity map on gr(K) . Now we obtain a filter preserving A-linear map $\Phi: F_\mu \to K$ whose associated graded map is φ. Since $\varphi(\mathrm{gr}(K)) = \mathrm{gr}(K)$, it follows that $K/\Phi(K)$ is a left A-module such that $\mathrm{gr}_\Gamma(K/\Phi(K)) = 0$ for every good filtration Γ on $K/\Phi(K)$.

Next, let S denote the kernel of the map Φ . We notice that since K is a submodule of the free A-module F , it follows that the induced good filtration on K is separated. Then, since φ is the identity map on gr(K), it follows that the Φ-kernel cannot intersect K . Thus , we have the submodule of F_μ given by the direct sum $\mathrm{Ker}(\Phi) \oplus K$. We notice that $F_\mu/(\mathrm{Ker}(\Phi) \oplus K)$ is isomorphic with a submodule of $K/\Phi(K)$.

Hence $\mathrm{gr}_\Gamma(F_\mu/(\mathrm{Ker}(\Phi) \oplus K)) = 0$ for every good filtration and then Sublemma 1 applies to the A-module $F_\mu/(\mathrm{Ker}(\Phi) \oplus K)$. Using Sublemma 1 together with Lemma 2.2 and the previous discussion about graded $\tilde{A}$-modules with T-torsion, it follows that there exists a fixed integer ω such that $\mathrm{pd}_{\tilde{A}}(F/(\mathrm{Ker}(\Phi) \oplus K)) \leq \omega$. But then $\mathrm{pd}_{\tilde{A}}(\mathrm{Ker}(\Phi) \oplus K) \leq \omega$ and hence also $\mathrm{pd}_{\tilde{A}}(K) \leq \omega$. Now we obtain

$pd_{\widetilde{A}}(\widetilde{M}) \leq pd_{\widetilde{A}}(K) + \mu \leq \omega + \mu$. This proves the required uniform bound and Lemma 5.5 follows.

5.6. Remark. In the special case when the filtration on the ring A is Zariskian the previous proof of Lemma 5.5 is much simpler. In fact, then it suffices to observe that if $\widetilde{M}$ is a Rees module of some finitely generated A-module with a good filtration and $F_{\bullet}$ is a filt-free resolution, it follows that the kernel K of the map $F_{\mu} \to F_{\mu-1}$, with $\mu = \text{gl.dim}(gr(A))$, enables us to construct the map $\Phi : F \to K$ which in the Zariskian case is surjective. This gives a decomposition $F = \text{Ker}(\Phi) \oplus K$ and then it follows that the Rees module $\widetilde{K}$ is projective and so $pd_{\widetilde{A}}(\widetilde{M}) \leq 1 + \text{gl.dim}(gr(A))$. We remark that this device was used in [H-O] .

Proof of Theorem 5.3. By Lemma 5.4 and Lemma 5.5 we see that there only remains to verify Auslander conditions. So let M be a finitely generated $\widetilde{A}$-module. Let N be an $\widetilde{A}$-submodule of $\text{Ext}^{v}_{\widetilde{A}}(M,\widetilde{A})$. Then the localization $N_T \subset \text{Ext}^{v}_{\widetilde{A}}(M_T,\widetilde{A}_T)$. The ring $\widetilde{A}_T = A[T,T^{-1}]$ and from section 4 we see that if A is Auslander-Gorenstein so is $A[T,T^{-1}]$. We conclude that $i < v$ gives $\text{Ext}^{v}_{\widetilde{A}_T}(N_T,\widetilde{A}_T) = 0$. Since Ext and localization commute, it follows that

(1) $\text{Ext}^{i}_{\widetilde{A}}(N,\widetilde{A})_T = 0$ for every $i < v$.

Next, choose a good T-filtration on M . We obtain a good T-filtration on N such that $gr_T(N)$ is a subquotient of $\text{Ext}^{v}_{gr_T(\widetilde{A})}(gr_T(M),gr_T(\widetilde{A}))$. Now $gr_T(\widetilde{A})$ is Auslander-Gorenstein since this ring is the polynomial ring in one variable over $gr(A)$.

So $i < v$ gives $\text{Ext}^{i}_{gr_T(\widetilde{A})}(gr_T(N),gr_T(\widetilde{A})) = 0$. Next, we find a good T-filtration on $\text{Ext}^{i}_{\widetilde{A}}(N,\widetilde{A})$ such that its associated graded module is a subquotient of $\text{Ext}^{i}_{gr_T(\widetilde{A})}(gr_T(N),gr_T(\widetilde{A}))$. We conclude that if $i < v$ and $K = \text{Ext}^{i}_{\widetilde{A}}(N,\widetilde{A})$, then $K_T = 0$ and $gr_T(K) = 0$ for some good T-filtration on K . But we have seen that $K_T = gr_T(K) = 0$ yield $K = 0$. This proves that the $\widetilde{A}$-module M satisfies the Auslander condition and Theorem 5.3 follows.

R E F E R E N C E S

[Bj:1] BJÖRK,J.-E., Noetherian rings and their applications. Math. surveys and monographs, 24, A.M.S., 1987.

[Bj:2] BJORK,J.-E., The Auslander condition on noetherian rings (in this proceeding).

[E] EKSTRÖM, E.K. , A study of non-commutative noetherian rings. Stockholm University Report, serie N° 18, 1988.

[F-F] FOSSUM,R., FOXBY, H. , The category of Graded Modules. Math. Scand. 35, 1974.

[N-O] NASTASESCU, C., Van OYSTAEYEN, F., Graded ring theory. North Holland, Math. Libr. series, 1982.

[H-O] HUISHI LI - Van OYSTAEYEN, F. , Global dimension and Auslander regularity of Rees rings (to appear).

K-THEORY OF NOETHERIAN RINGS

Timothy J. Hodges[1]

University of Cincinnati

Cincinnati, OH 45221 U.S.A.

"This paper is in final form and no version of it will be submitted for publication elsewhere".

Introduction.

This article is a survey of some aspects of the K-theory of Noetherian rings. Mainly we shall be concerned with calculating the (Quillen) K-groups for specific Noetherian rings but we shall also mention a number of applications (old and new) and describe some open problems and possible directions for futher research in this area. The kind of examples we shall be concerned with include differential operators on algebraic varieties, factors of enveloping algebras of semisimple Lie algebras, HNP rings and fixed rings and skew group rings with respect to finite groups of automorphisms. The main theme is the use of categorical techniques (in particular categorical or torsion-theoretic localization) together with Quillen's generalizations of the classical theorems (due to Serre, Grothendieck, Bass et al.) on devissage, resolution, localization and filtered and graded rings.

[1] The author was partially supported by United States National Science Foundation grant DMS 8602291.

In the first section we discuss Quillen's filtered rings theorem, its well-known applications and some less well-known applications to fixed rings and rings of differential operators on quotient varieties. The next section consists of a preliminary discussion of the localization theorem together with a review of some of its applications, in particular to HNP rings and certain skew group rings. In the third section we show how to calculate the K-groups for $U(\mathfrak{g})/P_\lambda$ where $U(\mathfrak{g})$ is the enveloping algebra of a finite dimensional semi-simple Lie algebra and P_λ is a minimal primitive corresponding to a regular weight $\lambda \in \mathfrak{h}^*$. Our technique is to use the localization theorem again to calculate the groups K_i for sheaves $\mathcal{D}_\lambda$ of twisted differential operators on the associated flag variety G/B and then to use a theorem of Bernstein and Beilinson to deduce that $K_i(\mathcal{D}_\lambda) \cong K_i(U(\mathfrak{g})/P_\lambda)$.

The fourth section is concerned with local-global questions for K-theory of noncommutative Noetherian rings. We reproduce an unpublished result of S.P.Smith on a Mayer-Vietoris sequence connected to a faithfully flat pair of localizations and discuss a possible generalization of this result to a spectral sequence along the lines of results of Brown, Gersten and others in the commutative case. Finally in the last section we discuss very briefly the map between K-theory and cyclic homology which is familiar to ring theorists on the 0-th level as the map from $K_0(R)$ to the trace group $R/[R,R]$.

Of course this survey reflects heavily the author's own interests and is not meant to be a comprehensive review of the subject. One particularly notable ommission is the very interesting recent work of Moody on group algebras of poly-cyclic-by-finite groups.

Our notation will usually follow that of Quillen [32]. If C is a full subcategory of an abelian category, then we denote by $K_i(C)$ for $i \geq 0$, the K-groups as defined by Quillen. If R is a Noetherian ring, we denote by Mod_f-R the category of finitely generated left R-modules and by P(R) the full subcategory of finitely generated projective left R-modules. We set $K'_i(R) = K_i(Mod_f\text{-}R)$ and $K_i(R) = K_i(P(R))$.

1 Filtered Rings.

1.1. We begin with one of the most familiar results in this subject, Quillen's theorem on the K-theory of filtered rings, generalising earlier results of Grothendieck and Bass. By a filtered ring we shall mean a ring S equipped with a chain of additive subgroups, $S_0 \subseteq S_1 \subseteq \ldots \subseteq \bigcup S_i = S$ such that $S_i S_j \subseteq S_{i+j}$. The associated graded ring is then defined to be $gr(S) = \oplus (S_i/S_{i+1})$.

Theorem. Let S be a filtered ring such that S_0 and $gr(S)$ are Noetherian rings of finite global dimension. Then the functor $S\otimes_{S_0}$ - induces isomorphisms:

$$K_i(S_0) \cong K_i(S)$$

for all $i \geq 0$.

As Quillen himself observes, this implies that if $A_n(k)$ is the n-th Weyl algebra over a field (or in fact any Noetherian ring of finite global dimension) k, then $K_i(A_n(k)) \cong K_i(k)$. Similarly, if $\mathfrak{g}$ is a finite dimensional Lie algebra over k and $U(\mathfrak{g})$ is the universal enveloping algebra of $\mathfrak{g}$, then $K_i(U(\mathfrak{g})) \cong K_i(k)$. On the other hand, if V is a smooth affine variety over a field k of characteristic zero, let us denote by k[V] the coordinate ring of V and by D[V], the ring of differential operators on D. Then D[V] may be filtered by the degree of the differential operators. In this case $D[V]_0 = k[V]$ and the associated graded ring is isomorphic to $k[T^*V]$, the coordinate ring of the cotangent bundle. Since the cotangent bundle is again smooth, the theorem applies and yields that $K_i(D[V]) \cong K_i(k[V])$.

1.2. Another interesting application of the above theorem is to skew group rings. Suppose that S is a filtered ring as above and that G is a finite group of automorphisms of S. We say that the action of G is *linear* if $S_i^g = S_i$ for all $g \in G$. In this case S*G may be given the filtration $(S{*}G)_i = \{\Sigma s_g g \mid s_g \in S_i\}$ and the associated graded ring is naturally isomorphic to gr(S)*G where the action of G on gr(S) is the natural induced action. Suppose further that S and gr(S) are Noetherian rings of finite global dimension and that $|G|^{-1} \in S_0$. Then by some standard results, $S_0{*}G$ and gr(S)*G are again Noetherian rings of finite global dimension and we may apply the theorem. Thus we have proved:

Theorem. Let S be a filtered ring and let G be a finite group of automorphisms acting linearly with respect to this filtration. Assume also that S_0 and gr(S) are Noetherian rings of finite global dimension and that $|G|^{-1} \in S_0$. The the induced maps,

$$\phi_i \; : \; K_i(S_0{*}G) \longrightarrow K_i(S{*}G)$$

are isomorphisms for all $i \geq 0$.

1.3. This result was first discussed by Goodearl in the context of an example of Zalesskii and Neroslavski of a simple Noetherian ring not isomorphic to a matrix ring over a domain. Let us mention in passing the following interesting generalisation of this result.

Corollary. Let G be *any* finite group of automorphisms of the Weyl algebra $A_1(\mathbb{C})$. Then the skew group ring $A_1(\mathbb{C}){*}G$ is an hereditary simple Noetherian ring not isomorphic to a matrix ring over a domain.

Proof. It was shown by Alev [1] that any finite group of automorphisms of $A_1(\mathbb{C})$ is conjugate to a group which is linear with respect to the standard filtration. Thus we may assume that G is linear and Goodearl's argument may be used. Applying the theorem yields that $K_0(A_1(\mathbb{C}){*}G) \cong K_0(\mathbb{C}G)$ via an isomorphism sending $[A_1(\mathbb{C})*G]$ to $[\mathbb{C}G]$. If $A_1(\mathbb{C})*G$ were a matrix ring, then $[A_1(\mathbb{C})*G]$ would be divisible in $K_0(A_1(\mathbb{C})*G)$ by the size of the matrices. But it is easily seen that $[\mathbb{C}G]$ cannot be divisible in $K_0(\mathbb{C}G)$ by any integer greater than one.

One intuitive reason why this result works is that $A_1(\mathbb{C})*G$ is isomorphic to the endomorphism ring of $A_1(\mathbb{C})$ as a module over the fixed ring $A_1(\mathbb{C})^G$ and that $A_1(\mathbb{C})$ is never free over the fixed ring. This raises the following interesting question: Suppose that A is projective over A^G. When is A free over A^G and why? Presumably the answer will involve further K-theoretical properties of A and A^G. An example of the situation where A is free over A^G is given at the end of the next paragraph. The crucial difference seems to be the size of $K_1(A)$.

1.4. Another interesting application of this theorem is to the general question of the structure of fixed rings of the Weyl algebra. Let A_n denote the n-th Weyl algebra over a field k of characteristic zero and let G be a finite group of automorphisms linear with respect to the standard filtration. Since A_n is simple, the fixed ring A_n^G is morita equivalent to the skew group ring A_n*G [30; Theorem 2.4]. Thus $K_0(A_n^G) \cong K_0(kG)$. In particular, A_n^G is never isomorphic to A_n (see [2] for more on this question). It is illuminating to compare this result with the analogous geometric problem which is answered in the case of curves by the classical theorem of Hurwitz. Let X be a projective curve, G be a finite group of automorphisms of X and Y be the quotient variety X/G (the geometric version of the fixed ring). If X is unramified over Y then the genera of X and Y are related by the formula: $g(Y) = g(X) + (|G|-1)$. Thus again we see that as the order of G increases, the structure of Y gets more complicated. Notice that the assumption of projectivity is vitally important here. For instance, let $A = k[q,q^{-1}]$ and let G be the group generated by the automorphism sending q to -q. Then $A^G = k[q^2,q^{-2}] \cong A$.

Taking differential operators yields another interesting example. Let $B = k[q,q^{-1},p]$ where $[p,q] = 1$ and let G be the group generated by the automorphism sending q to $-q$ and p to $-p$. Then $B^G = k[q^2,q^{-2},q^{-1}p] \cong B$. Since B is a localisation of $A_1(k)$, it is again a simple Noetherian hereditary domain with trivial K_0. Perhaps the significant point here is that B contains nontrivial units (i.e. $K_1(B) \neq k^*$). It would be interesting to see if and how the structure of a fixed ring is related to K-theoretical properties of the original ring.

1.5. One final observation concerning Theorem 1.3 is that it may be used to calculate the K-groups for rings of differential operators on certain singular varieties. Let X be complex affine n-space and let G be a finite group of linear automorphisms of X containing no pseudo-reflections. Then it was shown in [26] that G induces an action on the ring of differential operators on X (again $A_n(\mathbb{C})$) and that the ring of differential operators $D(X/G)$ on the corresponding quotient singularity X/G is just the fixed ring $A_1(\mathbb{C})^G$. Using the ideas described above we get that $K_i(D(X/G)) \cong K_i(\mathbb{C}G)$. On the other hand, if we use the filtration of $A_n(\mathbb{C})$ with respect to degree of the differential operator, then we obtain a natural isomorphism from $K_i(D(X/G))$ to $K_i^G(X)$, the corresponding equivariant K-theory.

2. The Long Exact Localization Sequence.

2.1. In this section we discuss various applications of Quillen's localization theorem to Noetherian rings. We first state the theorem in its full generality.

Localization Theorem. Let A be an abelian category having a set of isomorphism classes of objects. Let B be a Serre subcategory of A and let A/B be the quotient category. Then there exists a long exact sequence,

$$\cdots \to K_{i+1}(A/B) \to K_i(B) \to K_i(A) \to K_i(A/B) \to \cdots$$

The most well-known situation in which the theorem may be applied is classical localization with respect to an Ore set. Let $\mathcal{S}$ be a left Ore set in the ring R and let $R_{\mathcal{S}}$ be the localized ring. If B is the set of finitely generated $\mathcal{S}$-torsion left R-modules, then B is a Serre subcategory of Mod_f-R and Mod_f-$R_{\mathcal{S}}$ is equivalent to Mod_f-R/B. However this is not the sort of situation we will generally be considering. If $\phi : R \longrightarrow S$ is an epimorphism of rings which makes S into a flat right R-module, then we call S a *perfect left localization* of R (see [37] for further details on this subject). Again, if B is the subcategory of finitely generated S-torsion left R-modules, then B is a Serre subcategory of Mod_f-R, and Mod_f-S is equivalent to Mod_f-R/B. For instance, if R is a semiprime Noetherian ring and S is a ring contained between R and its classical full quotient ring such that S is a flat right R-module, then S is a perfect localization of R. In particular, if R is an hereditary Noetherian prime (or HNP) ring then all rings between R and its quotient ring are flat as right R-modules and hence are perfect right (or left) localizations of R.

2.2. One easy corollary of this theorem is an analogy for HNP rings of the long exact sequence of K-groups for Dedekind domains. The proof is straightforward. Details may be found in [15]. A preliminary version of this theorem was proved directly by LeBruyn in [25].

Corollary. Let R be an HNP ring with quotient ring Q. Then there exists a long exact sequence:

$$\cdots \rightarrow K_{i+1}(Q) \rightarrow \coprod_{V} K_i(\mathrm{End}_R V) \rightarrow K_i(R) \rightarrow K_i(Q) \rightarrow \cdots$$

where the summation is over the isomorphism classes of simple R-modules.

This corollary suggests the following problem. Suppose that R is an HNP ring with a single simple module. Then it follows from the above corollary that $K_0(R) \cong \mathbb{Z} \oplus \mathbb{Z}_n$ for some positive integer n. Do there exist examples where $K_0(R)$ is not isomorphic to $\mathbb{Z}$?

2.3. A very interesting application of the Localization Theorem due to Schofield is his proof of the fact that the quotient division ring of the ring of differential operators on a non-rational affine curve is not isomorphic to the quotient division ring of the Weyl algebra. The essence of the proof is the following interesting observation.

Proposition. Let Q be a simple artinian algebra over a field k and let R and S be two hereditary Noetherian orders in Q which are finitely generated as k-algebras. Then $K_0(R)$ is finitely generated if and only if $K_0(S)$ is finitely generated.

Proof. Let T be the subring of Q generated by R and S. Then R,S and T are all HNP rings. As noted above, T is a perfect left localization of R. Since S is a finitely generated k-algebra, T is generated over R by finitely many elements and hence the subcategory of finitely generated T-torsion R-modules is generated by finitely many simple modules. Thus the Localization Theorem implies that $K_0(T) \cong K_0(R)/G$ for some finitely generated group G. Since a similar result obviously holds for S, the proposition follows.

Now let X be a smooth non-rational complex affine curve and let D[X] be the ring of differential operators on X. By Quillen's theorem on filtered rings, $K_0(D[X]) \cong K_0(X)$. But by [12;p447] and Quillen's localization theorem for schemes [32;Proposition 3.1], $K_0(X)$ is certainly not finitely generated as an abelian group. On the other hand $K_0(A_1(\mathbb{C})) \cong \mathbb{Z}$. Hence the quotient division rings of D[X] and $A_1(\mathbb{C})$ cannot be isomorphic.

2.4. A similar argument can be used to prove a result on endomorphisms of the Weyl algebra. Interestingly, we now need to know about K_1 also. We show that if $\phi : A_1(\mathbb{C}) \longrightarrow A_1(\mathbb{C})$ is a strict embedding, then the induced endomorphism of the quotient division ring $D_1(\mathbb{C})$ is also a strict embedding.

For suppose not; then we would have a copy of $A_1(\mathbb{C})$ lying strictly between $A_1(\mathbb{C})$ and $D_1(\mathbb{C})$. As in the previous paragraph this implies that we have an exact sequence,

$$K_1(A_1(\mathbb{C})) \rightarrow K_1(A_1(\mathbb{C})) \rightarrow K_0(B) \rightarrow K_0(A_1(\mathbb{C})) \rightarrow K_0(A_1(\mathbb{C})) \rightarrow 0$$

where B, and hence $K_0(B)$, is nontrivial. But by Quillen's filtered ring theorem, $K_i(A_1(\mathbb{C}))$ is naturally isomorphic to $K_i(\mathbb{C})$. Thus, since ϕ is a $\mathbb{C}$-algebra map, the first and fourth arrows above must be isomorphisms, a contradiction.

2.5. As a final application of this technique, we sketch a calculation of K_0 for primitive factors of the enveloping algebra of the Lie algebra $\mathfrak{sl}(2,\mathbb{C})$. Let $R = U(\mathfrak{sl}(2,\mathbb{C}))/P_\lambda$ where P_λ is a minimal primitive ideal and λ is a regular weight (see section four for any unexplained terminology). Then R can be embedded in $A_1(\mathbb{C})$ so that $A_1(\mathbb{C})$ is a perfect left localization of R with respect to the subcategory B generated by a single simple Verma module (this fact is a consequence of the Bernstein-Beilinson theorem but may also be established easily and directly [19]). As in the previous section, one sees that the map $K_1(R) \longrightarrow K_1(A_1(\mathbb{C}))$ is surjective. Hence we have an exact sequence:

$$0 \longrightarrow K_0(B) \longrightarrow K_0(R) \longrightarrow K_0(A_1(\mathbb{C})) \longrightarrow 0$$

from which one may immediately deduce that $K_0(R) \cong \mathbb{Z}\oplus\mathbb{Z}$. This result is in fact a very special case of Corollary 4.3.

3. Short Exact Localization Sequences.

3.1. Assume as before that S is a perfect left localization of R. There are a number of interesting situations in which the long exact localization sequence splits into split short exact sequences of the form:

$$0 \rightarrow K_i(B) \rightarrow K_i'(R) \rightarrow K_i'(S) \rightarrow 0.$$

This seems to have been first noticed by Keating in [24]. However the general statement proved in section one of that paper contains an error. The following proposition covers two situations where the result still holds.

Proposition. Let R be a Noetherian ring and let S be a perfect left localization of R. Let $\mathcal{B}$ be the category of finitely generated S-torsion left R-modules. Suppose that either:

(i) S is finitely generated as a left R-module

or (ii) S is finitely generated as a right R-module and R has finite global dimension.

Then the long exact localization sequence splits up into spit short exact sequences:

$$0 \to K_i(\mathcal{B}) \to K'_i(R) \to K'_i(S) \to 0.$$

Remark. When S is finitely generated as a *right* R-module, it will in fact be a finitely generated *projective* right R-module. Such localizations are known as *finite left localizations* of R. They were first studied by Silver in [35].

Proof. (i) In this case we have a well-defined restriction functor G: Mod_f-S $\to \mathrm{Mod}_f$-R. Let F be the localization functor $S\otimes_R$- : Mod_f-R $\to \mathrm{Mod}_f$-S. Since $S\otimes_R S$ is naturally isomorphic to S [37; XI.1], the composition FG is naturally equivalent to the identity. Hence the induced maps $K_i(G)$ split the long exact sequence.

(ii). Let $S^* = \mathrm{Hom}_R(S,R)$ be the dual of S as a right R-module. Then S^* is a finitely generated projective left R-module and we may then define the functor $H = S^*\otimes_S$- : $\mathcal{P}(S) \to \mathcal{P}(R)$ (where $\mathcal{P}(-)$ denotes the category of finitely generated projective modules). Now $S\otimes_R S^* \cong \mathrm{End}_R(S) \cong S$ by [35; p58], so again FH is equivalent to the identity. Since R has finite global dimension, S has finite left global dimension and in both cases we may identify K_i with K'_i. But then $K_i(H) : K_i(S) \to K_i(R)$ again splits the long exact sequence.

Neither of the techniques in the above proof can be extended to prove (ii) without the additional hypothesis on the finiteness of the global dimension. Examples exist to show that S may be finitely generated on the right but not on the left (for instance, idealisers in $A_1(\mathbb{C})$). Further, S^* need not be flat as a right S-module, as may be easily illustrated using tiled matrix orders. Hence the functor H above will not generally be exact on Mod_f-S.

3.2. When R is a Noetherian prime ring, the perfect left localizations of R are precisely the rings S between R and its quotient ring such that S is flat as a right R-module. If S is further a finite left localization, then S^* is naturally isomorphic to the trace ideal $\tau(S) = \{r \in R : rS \subseteq R\}$. Further the S-torsion modules are precisely those annihilated by $\tau(S)$ [37;XI.7].

Corollary 1. Let R be a Noetherian prime ring with finite global dimension and let S be a finite left localization of R. Then:

$$K_i(R) \cong K_i(S) \oplus K'_i(R/\tau(S)).$$

When R is an HNP ring, the finite left localizations are precisely the rings right equivalent to R (in the sense that there exists a nonzero element element $x \in R$ such that $xS \subseteq R$). Furthermore, in this case $R/\tau(S)$ is artinian so by devissage we may replace $K'_i(R/\tau(S))$ by $\coprod_{P \supseteq \tau(S)} K_i(R/P)$, where the direct sum is over the prime ideals containing $\tau(S)$.

Corollary 2. Let R be an HNP ring and let S be a ring right equivalent to R. Then:

$$K_i(R) \cong K_i(S) \oplus \coprod_{P \supseteq \tau(S)} K_i(R/P).$$

This result was proved for hereditary orders by Keating [24] and independently for HNP rings by the author [13].

3.3. This technique may also be applied to calculate the K-theory of certain skew group rings. The following theorem appears in [14]. The hypotheses have been simplified a little for the sake of clarity.

Theorem. Let S be a semiprime Noetherian ring of finite global dimension, let G be a finite group of outer automorphisms of S such that $|G|^{-1} \in S$ and let R be the fixed ring. If S is a projective left R module, then

$$K_i(S*G) \cong K_i(R) \oplus K'_i(S*G/S\pi S)$$

where $\pi = \sum_G g$.

The idea of the proof is that S*G embeds naturally in $End_R S$ and that because of the projectivity assumption $End_R S$ is a finite right localization of S*G. Further $End_R S$ is Morita equivalent to R so $K_i(End_R S)$ may be replaced by $K_i(R)$. The final step is to identify the trace ideal as the ideal $S\pi S$. Of course this leaves the problem of calculating $K'_i(S*G/S\pi S)$. The first step is to note that $Rad((S\pi S)\cap S)$ is the intersection of the ramified primes (where the concept of ramification is suitably defined so as to correspond with the usual definition in the commutative case [3]). Thus $T = S/Rad((S\pi S)\cap S)$ is an analogue of the coordinate ring of the branch locus and the torsion modules are the "trace-zero modules" over T*G. In general, however it is still difficult to calculate explicitly this term. Some specific examples are calculated in [14,16] (when S is an HNP ring or when G is generated by a "product of orthogonal pseudo-reflections"). These results, suggest that one should expect contributions to $K'_i(S*G/S\pi S)$ to come from all components of the natural "stratification of the branch locus". Rather than getting into further details here we refer the reader to [14] for more information. These ideas may be globalized to cover the case of schemes [16]. In particular this provides a generalization of a geometric result of Ellingsrud and Lonsted on the equivariant K-theory of smooth curves.

4. Primitive Factors of enveloping algebras / $\mathcal{D}$-Modules.

4.1. Let G be a connected semisimple algebraic group over an algebraically closed field k of characteristic zero and let B be a Borel subgroup of G. Let $\mathfrak{g}$ be the Lie algebra of G and let $\mathfrak{h}$ be a fixed Cartan subalgebra of $\mathfrak{g}$ contained in the Lie algebra of B. For each $\lambda \in \mathfrak{h}^*$, define D_λ to be the ring $U(\mathfrak{g})/\mathrm{ann}_{U(\mathfrak{g})}M(\lambda)$ where $U(\mathfrak{g})$ denotes the enveloping algebra and $M(\lambda)$ is the Verma module associated to λ as defined in [9]. In [22], Joseph and Stafford show that the Grothendieck group $K_0(D_\lambda)$ contains a free abelian group with rank the cardinality of Spec D_λ. On the other hand, in [21] Joseph describes a certain subgroup K_0' of K_0 and states that he was informed by Bernstein that this subgroup is free with rank equal to the cardinality of the associated Weyl group. Joseph then conjectures that $K_0' = K_0$. In this section we outline a proof of the fact that $K_i(D_\lambda) \cong K_i(G/B)$ for regular λ. After suitable identifications have been made, this proves both the above statements.

4.2. Our approach will be geometric, relying heavily on the now famous theorem [4] of Bernstein and Beilinson concerning $\mathcal{D}$-modules over G/B. Another result which plays a key role is the following theorem due to Kashiwara. Most of the rest of this section is taken from [5]. The reader may also refer to [29] for a more detailed explanation. Let X be a smooth algebraic variety, let $\mathcal{O}_X$ be the sheaf of regular functions on X and let $\mathcal{D}$ be a sheaf of twisted differential operators on X as described, for instance, in [4]. Let Y be a smooth closed subvariety of X. Then a sheaf of differential operators may be constructed on Y in the following fashion. Let $\mathcal{I}_Y$ be the ideal sheaf in $\mathcal{O}_X$ corresponding to Y. Consider $\mathcal{E}nd(\mathcal{D}/\mathcal{I}_Y\mathcal{D})$, the sheaf of local endomorphisms of the right $\mathcal{D}$-module $\mathcal{D}/\mathcal{I}_Y\mathcal{D}$. Then $\mathcal{E}nd(\mathcal{D}/\mathcal{I}_Y\mathcal{D})$ is certainly a sheaf of rings supported on Y, so we may define $\mathcal{D}^Y$ to be the sheaf of rings on Y such that $i_*(\mathcal{D}^Y) = \mathcal{E}nd(\mathcal{D}/\mathcal{I}_Y\mathcal{D})$ where $i: Y \longrightarrow X$ is the natural inclusion. It can be shown (see for instance [5]) that $\mathcal{D}^Y$ is a sheaf of twisted differential operators on Y. For an open set U of X we define $\mathcal{D}_U$ to be $\mathcal{D}|_U$. Then again $\mathcal{D}_U$ is a sheaf of twisted differential operators on U. Finally let $\mathcal{A}$ be a sheaf of k-algebras and let j: $\mathcal{O}_X \longrightarrow \mathcal{A}$ be a morphism of k-algebras. An $\mathcal{A}$-module is said to be quasi-coherent if it is quasi-coherent as an $\mathcal{O}_X$-module. Such a module $\mathcal{M}$ is said to be a coherent as an $\mathcal{A}$-module if there is a local cover of X on which $\mathcal{M}|_U \cong \mathcal{O}_U \otimes \Gamma(U,\mathcal{M})$ and $\Gamma(U,\mathcal{M})$ is a finitely generated $\Gamma(U,\mathcal{A})$-module. The subcategory of such modules is denoted by Coh-$\mathcal{A}$. When $\mathcal{A}$ is itself quasicoherent and locally Noetherian (for instance when $\mathcal{A}$ is a sheaf of twisted differential operators), Coh-$\mathcal{A}$ is an abelian category and we may consider $K_i(\text{Coh-}\mathcal{A})$.

Theorem (Kashiwara). Let Y be a smooth closed subvariety of the smooth algebraic variety X and let $\mathcal{D}$ be a sheaf of twisted differential operators on X. Then there exists a direct image functor i_+: Coh-$\mathcal{D}^Y \longrightarrow$ Coh-$\mathcal{D}$, defining an equivalence of categories between the category of coherent left $\mathcal{D}^Y$-modules and the full subcategory of coherent left $\mathcal{D}$-modules supported on Y.

4.3. **Theorem**. Let $\mathcal{D}$ be a sheaf of twisted differential operators on G/B. Then for all $i \geq 0$, the functor $F = \mathcal{D}\otimes_{\mathcal{O}}-$ induces isomorphisms $K_i(F)$: $K_i(G/B) \longrightarrow K_i(\text{Coh-}\mathcal{D})$.

Proof. We prove the result in a more general setting. Let X be a smooth quasi-projective variety and let $\mathcal{D}$ be a sheaf of twisted differential operators on X. We claim that if X has a chain of closed subspaces:

$$\emptyset = X_0 \subseteq X_1 \subseteq \ldots \subseteq X_t = X$$

such that $X_i \backslash X_{i-1}$ is a smooth affine subset of $X \backslash X_{i-1}$, then tensoring with $\mathcal{D}$ induces an isomorphism from $K_i(X)$ to $K_i(\text{Coh-}\mathcal{D})$. When X = G/B the existence of such a filtration follows from the usual Bruhat decomposition [6].

We go by induction on t. If t=1, then X is affine and the result follows from Quillen's result (Theorem 1.1). Suppose that t>1 and let $U = X \backslash X_1$ and $Z = X_1$. Then clearly U satisfies the hypotheses for smaller t and hence the theorem is true for U. Let B_0 be the category of coherent $\mathcal{O}_X$-modules supported on Z. It is well-known [10; III.5] that there is a natural equivalence of categories between Coh-$\mathcal{O}_U$ and the quotient category Coh-$\mathcal{O}_X/B_0$ and hence that there is a localization sequence for the corresponding K-groups:

$$\ldots \to K_{i+1}(U) \to K_i(B_0) \to K_i(X) \to K_i(U) \to \ldots$$

Let B be the category of coherent left $\mathcal{D}$-modules supported on Z. In an analogous fashion it may be shown that Coh-$\mathcal{D}_U$ is equivalent to Coh-$\mathcal{D}/B$. Thus we have an analogous long exact sequence in this situation also.

$$\ldots \to K_{i+1}(\text{Coh-}\mathcal{D}_U) \to K_i(B) \to K_i(\text{Coh-}\mathcal{D}) \to K_i(\text{Coh-}\mathcal{D}_U) \to \ldots$$

Tensoring with $\mathcal{D}$ is an exact functor which takes B_0 to B and Coh-$\mathcal{O}_U$ to Coh-$\mathcal{D}_U$. Thus it induces a map between the two above complexes. By devissage we may identify $K_i(B_0)$ with $K_i(Z)$ and by Kashiwara's theorem, we may identify $K_i(B)$ with $K_i(\mathcal{D}^Z)$. Furthermore the induced map from $K_i(Z)$ to $K_i(\mathcal{D}^Z)$ can be seen to be the natural one, and therefore must be an isomorphism since Z is smooth and affine. Thus the map between the complexes is an isomorphism at the Z- and (by induction) U-levels. Hence it must be an isomorphism of complexes and in particular, $K_i(X) \cong K_i(\text{Coh-}\mathcal{D})$ for all i.

Corollary. If λ is a regular weight, then $K_i(D_\lambda) \cong K_i(G/B)$ for all $i \geq 0$. In particular, $K_0(D_\lambda)$ is a free abelian group whose rank is the cardinality of the Weyl group W.

Proof. Since $D_\lambda \cong D_{w\lambda}$ for all $w \in W$, we may assume that λ is dominant regular in the sense of [4]. In [4] Bernstein and Beilinson construct, for each $\lambda \in \mathfrak{h}^*$, a sheaf of twisted differential operators $\mathcal{D}_\lambda$ on G/B. They then show that when λ is dominant and regular, the category of coherent $\mathcal{D}_\lambda$-modules is equivalent to the category of finitely generated left D_λ-modules. Since D_λ has finite global dimension [18; Theorem 3.9], it then follows that $K_i(G/B) \cong K_i(\text{Coh-}\mathcal{D}_\lambda) \cong K'_i(D_\lambda) \cong K_i(D_\lambda)$. The second claim then follows from the analogous statement for $K_0(G/B)$ which is well-known.

5 A Mayer-Vietoris Sequence.

5.1. We now come to consider the question: to what extent is $K_i(R)$ determined by 'local data'? Of course, when we are considering noncommutative rings it is not clear exactly what should be meant by 'local data'. One natural scenario is the following. Let R be a ring and let $R_1,\ldots,R_n$ be localizations of R such that $R_1\oplus\ldots\oplus R_n$ is faithfully flat as a right R-module. Then one might expect certain invariants of R to be determined by those of $R_1,\ldots,R_n$. For instance, the Krull dimension of R (in the sense of Gabriel and Rentschler) is equal to the maximum of the Krull dimensions of the R_i. On the other hand, the analogous statement for global dimension is not true and an additional "compatibility" assumption (see below) is required for the result to hold [18]. Of course this process is analogous to passing to a (finite) open affine cover in the geometric situation. Thus we look to geometry for a model of the kind of result we might expect.

5.2. When X is a Noetherian scheme, it was shown by Brown and Gersten that there is a 'local to global' spectral sequence:

$$E^{pq} = H^p(X,\mathcal{K}'_{-q}) \Rightarrow K'_{-p-q}(X)$$

where $\mathcal{K}'_{-q}$ is the sheaf defined by the K-groups of the rings of local sections. This result can be reexpressed in terms of an open affine cover rather than the whole sheaf using the kind of approach described in [8]. It would be very interesting to know if an analogous result holds in the noncommutative case described above. In particular, such a result would provide an alternative proof of Theorem 4.3. However its theoretical significance is much more important.

5.3. When n = 2 (in the notation of 4.1), the problem becomes that of constructing a Mayer-Vietoris sequence. This was done for semi-prime Noetherian rings by S. P. Smith and we are grateful to him for letting us reproduce his (unpublished) result here. Notice that the "compatibility conditions" that are used here are precisely those that were needed to prove a local-global result for global dimension [18]. It is not yet whether such conditions are necessary, though it seems likely that they are.

Theorem. Let R be a semiprime Noetherian ring with quotient ring Q. Let R_1 and R_2 be perfect left localizations of R (lying between R and Q) and let R′ be the subring of Q generated by R_1 and R_2. Suppose that:

(a) $R_1 \oplus R_2$ is a faithfully flat right R-module

and (b) the natural maps $R_i \otimes_R R_j \to R'$ for $i \neq j$ are R-bimodule isomorphisms.

Then there is a Mayer-Vietoris sequence,

$$\cdot \to K_1'(R) \to K_1'(R_1) \oplus K_1'(R_2) \to K_1'(R') \to K_0'(R) \to K_0'(R_1) \oplus K_0'(R_2) \to K_0'(R') \to 0.$$

Proof. By (b) and some standard localization theory [37], R′ is a perfect left localization of R_1. Denote by $\mathcal{T}_2'$ the torsion theory associated to this localization and by $\mathcal{T}_i$ the torsion theory associated to the localization R_i of R for i = 1,2. Condition (b) implies that the functor $R_1 \otimes$- carries $\mathcal{T}_2$ into $\mathcal{T}_2'$. Because of the faithfulness condition (a), M embeds in $R_1 \otimes M$ whenever M is in $\mathcal{T}_2$. On the other hand $R_1 \otimes M/M$ is torsion with respect to both R_1 and R_2 so that, again by (a), we must have that M is naturally isomorphic to $R_2 \otimes M$ as an R-module. Thus $R_1 \otimes$- induces an equivalence of categories between $\mathcal{T}_2$ and $\mathcal{T}_2'$. Hence tensoring with R_1 induces a map from the long exact sequence for the localization $R \to R_2$ to the corresponding one for $R_1 \to R'$:

$$\begin{array}{ccccccccccccc}
\cdots \to K_1'(\mathcal{T}_2) & \to & K_1'(R) & \to & K_1'(R_2) & \to & K_0'(\mathcal{T}_2) & \to & K_0'(R) & \to & K_0'(R_2) & \to & 0 \\
\downarrow \gamma & & \downarrow & & \downarrow & & \downarrow \gamma & & \downarrow & & \downarrow & & \\
\cdots \to K_1'(\mathcal{T}_2') & \to & K_1'(R_1) & \to & K_1'(R') & \to & K_0'(\mathcal{T}_2') & \to & K_0'(R_1) & \to & K_0'(R') & \to & 0
\end{array}$$

Since the maps γ are isomorphisms, the required sequence then follows from a standard diagram chasing argument [34;Lemma 6.6].

6. K-Theory, the Trace Group and Cyclic Homology

6.1 There seems to be much interest at the present time in the relationship between K-theory and cyclic homology. It is well-known that there exist maps from the higher K-theory $K_i(R)$ to the cyclic homology groups $HC_j(R)$ (see for instance, [23]). In the setting of Noetherian rings it is not yet clear whether these maps will prove interesting. However the map at the zero-th level is known to a certain group of ring theorists - it is the map from $K_0(R)$ to the "Trace Group" $HC_0(R) = R/[R,R]$ (where $[R,R]$ is the subgroup generated by additive commutators). The purpose of this section is to bring this map to more general attention and to suggest some areas where it may be interesting.

6.2. First we need to define this map and show that it is nontrivial. For this we basically follow Passman's exposition in [31]. Let P be a finitely generated projective module. Represent P as a summand of a free module, say $F \cong P \oplus Q$. Then P defines a projection e in $\mathrm{End}_R F$. With respect to some basis $\mathrm{End}_R F$ is a matrix ring and one may define a trace map $\mathrm{Tr}\colon \mathrm{End}_R F \longrightarrow R/[R,R]$ by letting $\mathrm{Tr}(\phi)$ equal the image of the usual matrix trace. It is easily verified that $\mathrm{Tr}(\phi\theta) = \mathrm{Tr}(\theta\phi)$ and hence that Tr does not depend on the choice of basis. We now define $\chi(P)$ to be Tr(e) and claim that Tr(e) depends only on P and not on the choice of F or e. Let (F_1, e_1) and (F_2, e_2) be two such choices and let $F = F_1 \oplus F_2$. Then there exists an automorphism γ of F given by switching the copies of P. We may extend the e_i trivially to F (that is, by letting them annihilate the other summand). These e_i are now conjugate under γ and hence $\mathrm{Tr}(e_i) = \mathrm{Tr}(e_j)$. Thus $\chi(P)$ is well-defined. Furthermore it is clear that $\chi(P \oplus Q) \cong \chi(P) \oplus \chi(Q)$ so by the universal property of the Grothendieck group, we may extend χ to a map $\chi : K_0(R) \longrightarrow R/[R,R]$.

6.3. **Example.** As an illustration of a situation where this map is significantly nontrivial, let $R = A_1(\mathbb{C})^G$ the fixed ring of the Weyl algebra with respect to the action of a cyclic group G of order n. It is shown in [2] that $K_0(R) \cong \mathbb{Z}^n$ and that $\dim_{\mathbb{C}} R/[R,R] = n-1$. A careful examination reveals that in this case $\mathrm{Ker}\chi \cong \mathbb{Z}$ and $\mathrm{Im}\chi$ is a dense lattice inside $R/[R,R]$. This example raises a number of interesting questions. For instance, although the K_0 calculation follows from a very general result (see 1.4) and the behaviour of the trace group is similar, no general result is known in this case. The results in [2] all depend on explicit calculations. In particular, it was not possible to calculate $HC_0(A_1(\mathbb{C})^G)$ when G was an exceptional group. It seems very likely that there is a better approach to this problem that would provide a description of $HC_0(A_1(\mathbb{C})^G)$ in all cases without the need for explicit calculations.

6.4. As an illustration of an application of this theory to a structural question in the theory of Noetherian rings, let us look at a key part of Lorenz's elegant proof that the "second Zalesskii-Neroslavski example" has no idempotents. Recall that this example of Zalesskii and Neroslavski [38] was the first example of a simple Noetherian ring with zero-divisors but without idempotents. Their original proof was an extremely complicated direct verification. The crucial step in Lorenz's proof is the following result which reduces the problem of showing that there are no idempotents to two other conditions which are more easily verified. This result first appeared in [27]. A more elegant proof using Frobenius reciprocity is given in [29].

By the *rank* of a module M over a semiprime Noetherian ring S, we mean the length of $Q \otimes_S M$ where Q is the classical qotient ring of S.

Theorem. Let S be a semiprime Noetherian algebra over a field of characteristic p. Suppose that:

(a) $\mathrm{rank}(S) = p$

(b) $1 \notin [S,S]$

(b) for all projective modules P, $P^{\oplus p}$ is stably free.

Then S has no idempotents.

Proof. We claim that $p | \mathrm{rank}(P)$ for all projective modules P. It then follows easily that S has no idempotents. Suppose to the contrary that there exists a projective P such that $p \nmid \mathrm{rank}(P)$. By assumption (b) there exists a positive integer n such that $p[P] = n[S]$ in $K_0(S)$. Taking ranks of both sides yields that $n = \mathrm{rank}(P)$. Now look at the images of both sides under $\chi : K_0(S) \longrightarrow S/[S,S]$. On the one hand, $\chi(n[S]) = \bar{n} \neq 0$ while on the other hand $\chi(p[P]) = p\chi([P]) = 0$, a contradiction.

An interesting consequence of this result is the existence of a Noetherian domain R of Krull dimension one with an indecomposable projective module of rank two [20]. One important gap in the theory of Noetherian rings is the lack of examples of indecomposable projective modules to complement the results of Stafford [36] and it is possible that such K-theoretical techniques may help surmount the considerable technical difficulties involved in proving that certain modules are indecomposable.

References

1. J.Alev, "Actions de groupes sur $A_1(\mathbb{C})$", in *Ring Theory* LNM 1197, Springer New York 1986.

2. J.Alev, T.J.Hodges and J.Velez "Fixed rings of Weyl algebras", preprint, University of Cincinnati, 1987.

3. M. Auslander and D.A. Buchsbaum, "On ramification theory in Noetherian rings", *Amer. J. Math.*, 81 (1959), 749-764.

4. A.Beilinson and I.N.Bernstein, "Localization de $\mathfrak{g}$-modules" *C.R. Acad. Sci. Paris Ser. I* 292 (1981), 15-18.

5. A. Beilinson and I.N.Bernstein, "A generalisation of Casselman's submodule theorem", *Representation theory of reductive groups*, Birkhauser, Boston 1983.

6. A. Borel, *Linear Algebraic Groups*, Benjamin, New York 1969.

7. K.S.Brown and S.M.Gersten, "Algebraic K-theory as generalized sheaf cohomology" in *Algebraic K-theory 1*, LNM 341 Springer, New York 1973.

8. B.H.Dayton and C.A.Weibel, "A Spectral Sequence for the K-theory of Affine Glued Schemes" in *Algebraic K-theory, Evanston 1980*, LNM 854 Springer, New York 1981.

9. J.Dixmier, "Sur les algebres de Weyl", *Bull. Soc. Math. France*, 96 1968, 209-242.

10. P.Gabriel, "Des catégories abéliennes", *Bull. Soc. Math. France*, 90 (1962), 323-448.

11. K. R. Goodearl, "Simple Noetherian rings not isomorphic to matrix rings over domains", *Comm. in Alg.*, 12 1984, 1421-1435.

12. R.Hartshorne, *Algebraic Geometry*, Springer-Verlag, New York 1977.

13. T.J.Hodges, "K-theory and right ideal class groups for HNP rings", *Trans. Amer. Math. Soc.*, 302 (1987), 751-767.

14. T.J.Hodges, "Equivariant K-theory of Noetherian rings," *J. London Math. Soc.*, to appear.

15. T.J.Hodges, "A noncommutative Bass-Tate sequence," *Bull. Soc. Math. Belg.*, to appear.

16. T.J.Hodges, "Equivariant K-theory of Noetherian schemes", preprint, University of Cincinnati, 1987.

17. T.J.Hodges and S.P.Smith, "Rings of differential operators and the Bernstein-Beilinson equivalence" of categories, *Proc. Amer. Math. Soc.*, 93 (1985), 379-386.

18. T.J.Hodges and S.P.Smith, "The global dimension of certain primitive factors of the enveloping algebra of a semi-simple Lie algebra", *J. London Math. Soc.* (2), 82 (1985),411-418.

19. T.J.Hodges and S.P.Smith, ""Differential operators on projective space", preprint, University of Cincinnati 1985.

20. T.J.Hodges and J.Osterburg, "An indecomposable projective module over a Noetherian domain of Krull dimension one", *Bull. London Math. Soc.*, 19 (1987), 139-144.

21. A.Joseph, "On the classification of primitive ideals in the enveloping algebra of a semisimple Lie algebra", in *Lie Group Representations I*, LNM 1024 Springer, New York.

22. A.Joseph and J.T.Stafford, "Modules of $\mathfrak{k}$-finite vectors over semi-simple Lie algebras", *Proc. London Math. Soc.* (3), 49 (1984), 361-384.

23. M. Karoubi, "Homologie cyclique et K-théorie algébrique 1", *C. R. Acad. Sci. Paris, Ser.I*, 297 (1983), 447-450.

24. M.E.Keating, "On the theory of tiled orders", *J. Algebra*, 43 (1976), 193-197.

25. L.LeBruyn, "Towards a noncommutative version of the Bass-Tate sequence", *Bull. Soc. Math. Belg. Ser. B*, 35 (1983), 45-67.

26. T.Levasseur, "Anneaux d'opérateurs differentiels", *Seminaire d'algebre Malliavin*, LNM 867 Springer, New York 1981.

27. M.Lorenz, "K_0 of skew group rings and simple Noetherian rings without idempotents", *J. London Math. Soc.* (2), 32 (1985), 41-50.

28. M. Lorenz, "Frobenius reciprocity and G_0 of skew group rings", preprint, Max-Planck-Institut 1986.

29. D. Milicic, "Localization and representation theory of reductive Lie groups", lecture notes, University of Utah 1987, to appear.

30. S. Montgomery, *Fixed Rings of Finite Automorphism Groups of Associative Rings*, LNM 818, Springer, Berlin 1980.

31. D.S.Passman, *The algebraic structure of group rings*, Wiley, New York 1977.

32. D. Quillen, "Higher Algebraic K-theory" in *Algebraic K-Theory*, LNM 341, Springer, Berlin 1973.

33. J.C.Robson, "Idealisers and Hereditary Noetherian Prime Rings", *J. Algebra*, 22 (1972), 45-81.

34. J.J. Rotman, *An introduction to homological algebra*, Academic Press, New York 1979.

35. L. Silver, "Non-commutative localization and applications", *J. Algebra*, 7 (1967), 44-76.

36. J.T. Stafford, "Generating modules efficiently: algebraic K-theory for noncommutative Noetherian rings", *J. Algebra*, 69 (1981), 312-335.

37. B. Stenstrom, *Rings of Quotients*, Springer Verlag, Berlin 1975.

38. A.E.Zalesskii and O.M.Neroslavskii, "There exists a simple Noetherian ring with divisors of zero but without idempotents", *Comm. Alg.* 5 (1977), 231-234.

OPERATEURS DIFFERENTIELS SUR LES SURFACES MUNIES D'UNE BONNE $\mathbb{C}^*$ -ACTION

Par Thierry LEVASSEUR

Département de Mathématiques et Informatique

Université de Bretagne Occidentale

6 Av. V. Le Gorgeu

29287 BREST CEDEX - FRANCE -

"This paper is in final form and no version of it will be submitted for publication elsewhere".

__Abstract__ : Let $R = \bigoplus_{n \in \mathbb{N}} R_n$ be a finitely generated normal graded $\mathbb{C}$-algebra of dimension 2. One studies the ring $\mathcal{D}(R)$ of differential operators over R ; in particular one shows that, most of the time, $\mathcal{D}(R)$ is neither simple nor noetherian. This generalizes results by J. Bernstein - I.N. Gelfand - S.I. Gelfand and J.P. Vigué.

INTRODUCTION

Soient $\mathfrak{X}$ une variété algébrique affine (irréductible) définie sur $\mathbb{C}$, R l'anneau des fonctions régulières sur $\mathfrak{X}$. On désigne par $\mathcal{D}(\mathfrak{X}) = \mathcal{D}(R)$ l'anneau des opérateurs différentiels sur $\mathfrak{X}$, cf. § O pour une description de $\mathcal{D}(\mathfrak{X})$.
Rappelons que si $\mathfrak{X}$ est lisse, $\mathcal{D}(\mathfrak{X})$ est un anneau qui possède d'agréables propriétés, à savoir :

(i) $\mathcal{D}(\mathfrak{X})$ est un anneau noethérien (à droite et à gauche).

(ii) R, qui est naturellement un $\mathcal{D}(R)$-module, est simple.

En fait, $\mathcal{D}(R)$ est un anneau simple (i.e. sans idéal bilatère non trivial).

(iii) $\mathcal{D}(\mathfrak{X})$ est une $\mathbb{C}$-algèbre de type fini.

Le problème se pose donc de déterminer lesquelles des propriétés (i) à (iii) ci-dessus subsistent lorsque $\mathfrak{X}$ n'est pas lisse.

Si $\mathfrak{X}$ est une courbe, on dispose d'une bonne connaissance de $\mathcal{D}(\mathfrak{X})$ grâce à [S.S.].

Si $\mathfrak{X}$ est une surface normale, J. Bernstein, I. et S. Gelfand, **[B.G.G.]** , ont montré que pour le cône cubique $\{X^3+Y^3+Z^3=O\}$ aucune de ces propriétés n'est vérifiée.

Cet exemple a été généralisé par J.P. Vigué, **[V]**, au cas d'une surface $\mathfrak{X} = \{F(X,Y,Z) = O\}$, F polynôme homogène de degré ≥ 3 ayant une singularité isolée à l'origine.

D'autre part on a maintenant nombre d'exemples où (i) à (iii) ont lieu : cf **[L]**, **[L-Sm]**, **[L.St]**, **[L.S.S]**, **[Mu]**, **[S.S]**. On trouvera dans **[L.Sm]**, **[L.St]** et **[L.S.S.]** les raisons qui nous ont amené à étudier les opérateurs différentiels sur dés variétés singulières ; elles sont issues de la théorie des représentations des algèbres de Lie (semi-simples), leur développement augmenterait par trop la longueur de cette introduction.

Le but principal des présentes notes est de revenir sur l'étude de $\mathcal{D}(\mathfrak{X})$ pour des surfaces normales munies d'une bonne $\mathbb{C}^*$-action, voir **[P.2]** ou § 5 pour cette notion, et d'inscrire les résultats de **[B.G.G]** et **[V]** dans un cadre général.

Plutôt que de détailler les résultats qui suivent nous indiquerons simplement l'idée directrice de ce travail : pour une surface du type précédent $\mathcal{D}(\mathfrak{X})$ n'a de bonnes propriétés que si $\mathfrak{X}$ est une singularité quotient, $\mathbb{C}^2/G$, du plan par un groupe fini.

On trouvera les énoncés précis des principaux résultats en (3.2.2), (4.3.3), (4.3.4), (4.4.1O), (4.4.11), (5.2.4 à 7).

Les preuves ne sont pas originales, elles se déduisent facilement de celles de **[B.G.G]** et **[V]**, déjà réécrites dans **[I.1]**.

Une conséquence de cette étude est de répondre à une question qui se pose naturellement en examinant les exemples positifs cités ci-dessus. Dans tous les cas la variété $\mathfrak{X}$ n'a que des singularités rationnelles, e.g. $\mathfrak{X} = \{X^2+Y^2+Z^2=O\}$ ou certains anneaux classiques d'invariants ; on montre ici que (malheureusement) une telle condition géométrique est insuffisante pour assurer un bon comportement de $\mathcal{D}(\mathfrak{X})$, cf. (5.3.1) et (5.3.2). On pourra également à ce propos consulter l'introduction de **[L.St]** où certaines questions connexes sont discutées.

O. NOTATIONS

O.1. Les définitions et notations concernant les variétés et les notions géométriques sont celles de **[H]** . Signalons en particulier :

(O.1.1.) $\mathbb{P}^n$ est l'espace projectif de dimension n sur $\mathbb{C}$.

(O.1.2) Soit R une $\mathbb{C}$-algèbre de type fini graduée $R = \bigoplus_{k\geq 0} R_k$, $R_o = \mathbb{C}$.

On note Proj R la variété projective associée. Rappelons que si $f \in R_q$ on dit que $\deg f = q$ et le localisé R_f est gradué par $[R_f]_k = \left\{\frac{a}{f^t} / \deg a = t \deg f + k\right\}$.

Une base d'ouverts affines de Proj R est alors donnée par les $D_+(f) = \mathrm{Spec}\,[R_f]_o$.

Si $\ell \in \mathbb{Z}$ on désigne par $R(\ell)$ le R-module gradué $\bigoplus_{i\in\mathbb{Z}} R_{\ell+i}$. Le $\mathcal{O}_{\mathrm{Proj}\,R}$-module associé est noté $\mathcal{O}_{\mathrm{Proj}\,R}(\ell)$.

(O.1.3.) Soit $\tilde{\mathfrak{X}}$ une courbe projective lisse ; tout faisceau inversible $\mathfrak{F}$, également appelé fibré de rang 1, s'écrit $\mathfrak{F} = \mathcal{O}_{\tilde{\mathfrak{X}}}(D)$ pour un diviseur $D = \sum_p n_p P$ sur $\tilde{\mathfrak{X}}$. Ici $\mathcal{O}_{\tilde{\mathfrak{X}}}(D)$ est le $\mathcal{L}(D)$ de **[H]** . Le dual de $\mathfrak{F}$ sera noté $\mathfrak{F}^\vee$.

0.2. En ce qui concerne les anneaux d'opérateurs différentiels nous utiliserons les notations introduites dans [L], d'après [E.G.A.]. Indiquons quelques notations et propriétés utiles.

(0.2.1.) Soit R une $\mathbb{C}$-algèbre et $\mathfrak{X}$ = Spec R, on désignera par $\mathcal{D}(\mathfrak{X})$ ou $\mathcal{D}(R)$ l'anneau des opérateurs différentiels associé. Rappelons que si R est une $\mathbb{C}$-algèbre de type fini écrite sous la forme $\frac{\mathbb{C}[X_1...X_n]}{I}$, X_i indéterminées et I un idéal, alors on a :

$$\mathcal{D}(R) = \frac{\{P=\sum_{\alpha} a_{\alpha}\left(\frac{\partial}{\partial X}\right)^{\alpha} / a_{\alpha} \in \mathbb{C}[X_1,\dots,X_n] \text{ et } P(I) \subset I\}}{\{P=\sum_{\alpha} a_{\alpha}\left(\frac{\partial}{\partial X}\right)^{\alpha} / a_{\alpha} \in I \text{ pour tout } \alpha\}}$$

où comme d'habitude $\left(\frac{\partial}{\partial X}\right)^{\alpha} = \left(\frac{\partial}{\partial X_1}\right)^{\alpha_1} \dots \left(\frac{\partial}{\partial X_n}\right)^{\alpha_n}$.

(0.2.2) Il existe sur $\mathcal{D}(R)$ une filtration par l'ordre des opérateurs ; ainsi si R est une $\mathbb{C}$-algèbre de type fini comme ci-dessus on pose :

$$\mathcal{D}^m(R) = \{\text{classes des } \sum_{|\alpha|\leq m} a_{\alpha}\left(\frac{\partial}{\partial X}\right)^{\alpha},\ |\alpha| = \alpha_1 + \dots\dots + \alpha_n\}$$

L'anneau gradué associé gr $\mathcal{D}(R) = \bigoplus_m \frac{\mathcal{D}^m(R)}{\mathcal{D}^{m-1}(R)}$ est commutatif ; si $P \in \mathcal{D}^m(R) \backslash \mathcal{D}^{m-1}(R)$ on dira que ord P = m et on notera $\sigma(P)$ la classe de P dans $\frac{\mathcal{D}^m(R)}{\mathcal{D}^{m-1}(R)}$, c'est le symbole principal de P.

(0.2.3) Si R est une $\mathbb{C}$-algèbre de type fini intègre alors $\mathcal{D}(R)$ et gr $\mathcal{D}(R)$ sont des $\mathbb{C}$-algèbres intègres.

Pour la première assertion cf [L], et pour la seconde on raisonne comme suit. Soit K le corps des fractions de R, alors $\mathcal{D}^m(R) = \{P \in \mathcal{D}^m(K) / P(R) \subset R\} = \mathcal{D}^m(K) \cap \mathcal{D}(R)$ et ainsi gr $\mathcal{D}(R)$ est une sous-algèbre de gr $\mathcal{D}(K)$, on montre aisément que cette dernière est intègre.

(0.2.4) Comme $\mathcal{D}(R) \subset \mathrm{End}_{\mathbb{C}}(R)$, l'anneau R est naturellement un $\mathcal{D}(R)$-module à gauche ; pour $D \in \mathcal{D}(R)$ et $f \in R$ on notera $D * f$ à la place de D(f).

(0.2.5) Soit $\mathfrak{X}$ un schéma sur $\mathbb{C}$, de faisceau structural $\mathcal{O}_{\mathfrak{X}}$, on construit un faisceau d'anneaux d'opérateurs différentiels, noté $\mathcal{D}_{\mathfrak{X}}$, tel que la fibre en $p \in \mathfrak{X}$ soit :

$$\mathcal{D}_{\mathfrak{X},p} = \mathcal{D}(\mathcal{O}_{\mathfrak{X},p})$$

où $\mathcal{O}_{\mathfrak{X},p}$ est l'anneau local de $\mathfrak{X}$ en p.

1. CÔNES ASSOCIES A UN FIBRE AMPLE

1.1 Fixons une variété projective (intègre) lisse $\bar{\mathfrak{X}}$ et un faisceau inversible ample $\mathfrak{L}$ sur $\bar{\mathfrak{X}}$.

On notera $\mathfrak{L}^i$ la puissance tensorielle $\mathfrak{L}^{\otimes i}$, $i \in \mathbb{Z}$.

(1.1.0.) Nous utiliserons les notations suivantes :

$$R = R(\bar{\mathfrak{X}}, \mathfrak{L}) = \bigoplus_{i \geq 0} H^\circ(\bar{\mathfrak{X}}, \mathfrak{L}^i)$$

$$R_i = H^\circ(\bar{\mathfrak{X}}, \mathfrak{L}^i)$$

$$\mathfrak{X} = \mathfrak{X}(\bar{\mathfrak{X}}, \mathfrak{L}) = \mathrm{Spec}\, R$$

$$\mathfrak{M} = \bigoplus_{i>0} R_i.$$

Si $\mathfrak{F}$ est un faisceau sur $\bar{\mathfrak{X}}$ on abrègera $H^i(\bar{\mathfrak{X}}, \mathfrak{F})$ en $H^i(\mathfrak{F})$ et on posera $h^i(\mathfrak{F}) = \dim_{\mathbb{C}} H^i(\mathfrak{F})$.

Il est clair que l'anneau R est gradué ; on dit que $\mathfrak{X}$ est le <u>cône associé</u> à $\mathfrak{L}$, le point de $\mathfrak{X}$ correspondant à l'idéal maximal $\mathfrak{M}$ est appelé <u>l'origine</u> du cône et sera noté $\{O\}$.

Rappelons quelques faits bien connus, cf ([E.G.A.] ou [W.3])

(1.1.1) Comme $\bar{\mathfrak{X}}$ est normale l'anneau R est intègre et intégralement clos.

(1.1.2) La $\mathbb{C}$-algèbre graduée R est de type fini et la variété projective associée, Proj R, vérifie : $\bar{\mathfrak{X}} \cong \mathrm{Proj}\, R$.

(1.1.3) On a prof $R_{\mathfrak{M}} \geq 2$, où prof est la profondeur d'un anneau local et $\dim \mathfrak{X} = \dim \bar{\mathfrak{X}} + 1$.

(1.1.4) Si $\dim \bar{\mathfrak{X}} = 1$, i.e. si $\bar{\mathfrak{X}}$ est une courbe projective lisse, alors le lieu singulier de $\mathfrak{X}$ vérifie

$\mathrm{Sing}\, \mathfrak{X} \subset \{O\}$.

On a $\mathrm{Sing}\, \mathfrak{X} = \emptyset$ si et seulement si R est un anneau de polynômes, auquel cas $\bar{\mathfrak{X}} \cong \mathbb{P}^1$ (on verra qu'alors $\mathfrak{L} = \mathcal{O}_{\mathbb{P}^1}(1)$).

(1.1.5) Notons $\mathbb{V}(\bar{\mathfrak{X}}, \mathfrak{L}) = \mathbb{V}$ le schéma *Spec* $S(\mathfrak{L})$, (cf. [H] ex. II. 5.18), c'est un fibré au dessus de $\bar{\mathfrak{X}}$ ayant $\mathfrak{L}^{-1}$ pour faisceau de sections ; on note p le morphisme (affine) de $\mathbb{V}$ vers $\bar{\mathfrak{X}}$.

Comme $\mathfrak{L}$ est ample il existe un morphisme de contraction (dominant birationnel et propre), cf ([E.G.A] chap. II 8.9.1) :

$$\pi : \mathbb{V} \to \mathfrak{X}$$

tel que $\pi^{-1}(O) \cong \bar{\mathfrak{X}}$ lorsque $\bar{\mathfrak{X}}$ est identifié à la section nulle dans $\mathbb{V}$.

En particulier si $\bar{\mathfrak{X}}$ est une courbe et $\mathrm{Sing}\, \mathfrak{X} = \{O\}$ alors $(\mathbb{V}, \pi)$ est une désingularisation de $\mathfrak{X}$.

1.2. Rappelons, qu'avec les notations de 1.1, si $\mathcal{L}$ est ample alors $\mathcal{L}^q$ est très ample pour $q >> O$. Fixons un tel q et posons :

$$R^{(q)} = \bigoplus_{i\geq 0} R_i^{(q)} \quad \text{où} \quad R_i^{(q)} = R_{qi}.$$

$$s+1 = h°(\mathcal{L}^q).$$

Donc par définition il existe une immersion fermée $\bar{\mathfrak{X}} \xrightarrow{\varphi} \mathbb{P}^s$. Notons $\mathcal{Y}$ l'image de cette immersion et $j : \mathcal{Y} \hookrightarrow \mathbb{P}^s$ l'inclusion canonique. On a donc $\mathcal{Y} = \varphi(\bar{\mathfrak{X}})$ et $\mathcal{L}^q \cong j^*\mathcal{O}_{\mathbb{P}^s}(1) = \mathcal{O}_{\mathcal{Y}}(1)$ par l'isomorphisme φ de $\bar{\mathfrak{X}}$ sur $\mathcal{Y}$.

Soit J l'idéal homogène définissant $\mathcal{Y}$ dans $\mathbb{P}^s$ et posons :

$$S = \frac{\mathbb{C}[X_o,..., X_s]}{J}.$$

C'est un anneau gradué ; on note S_i le sous-espace vectoriel formé des éléments homogènes de degré i. Avec des notations évidentes, compte tenu de $\deg X_j = 1$, on a donc :

$$S_i = \frac{\mathbb{C}[X_o,..., X_s]_i}{J \cap \mathbb{C}[X_o,..., X_s]_i}$$

Grâce à $\mathcal{L}^q \cong \mathcal{O}_{\mathcal{Y}}(1)$ il existe $f_j \in H°(\mathcal{L}^q)$ correspondant à $\bar{X}_j \in H°(\mathcal{Y}, \mathcal{O}_{\mathcal{Y}}(1)) = S_1$;on pose $x_j = \bar{X}_j$.

La variété $\mathcal{Y}$ est recouverte par les ouverts affines $\mathcal{Y}_j = \operatorname{Spec}\,[S_{x_j}]_o$, $O \leq j \leq s$,

l'isomorphisme φ donne : $\bar{\mathfrak{X}}_j = \operatorname{Spec}\,[R_{f_j}]_o \cong \mathcal{Y}_j$.

On a :

$$\forall \ell \in \mathbb{Z} \qquad \begin{cases} H°(\mathcal{Y}_j, \mathcal{O}_{\mathcal{Y}}(\ell)) = [S_{x_j}]_o . x_j^{\ell} \\ H°(\bar{\mathfrak{X}}_j, \mathcal{L}^{q\ell}) = [R_{f_j}]_o . f_j^{\ell} \end{cases}$$

1.3. Supposons maintenant que $\bar{\mathfrak{X}}$ soit une courbe de genre g.

Rappelons, ([H] ex. IV, 4.2), que tout faisceau inversible $\mathcal{F}$ de degré $\geq 2g+1$ est normalement engendré : $\mathcal{F}$ est très ample et l'application naturelle

$$S^i(H°(\mathcal{F})) \to H°(\mathcal{F}^i)$$

est surjective pour tout $i \in \mathbb{N}$.

Ainsi l'anneau $\bigoplus_{i\geq 0} H°(\mathcal{F}^i)$ est engendré par ses éléments homogènes de degré 1 et est l'anneau coordonné associé au plongement $\bar{\mathfrak{X}} \hookrightarrow \mathbb{P}^s$, $s+1 = h°(\mathcal{F})$.

Soit maintenant $\mathcal{L}$ un faisceau inversible ample sur $\bar{\mathfrak{X}}$ de degré δ (>O).On a $\deg \mathcal{L}^q = q\delta \geq 2g+1$ dès que $q \geq \frac{2g+1}{\delta}$ et ce qui a été dit ci-dessus s'applique à $\mathcal{F} = \mathcal{L}^q$, utilisant les notations de 1.2, il vient :

$$\forall\ q \geq \frac{2g+1}{\delta} \quad \begin{cases} R_i^{(q)} = R_{qi} = (R_q)^i \\ R_i^{(q)} \equiv S_i \text{ grâce à } \varphi \end{cases}.$$

(1.3.1.) On a ainsi montré que l'anneau gradué $R = R(\bar{\mathfrak{X}}, \mathfrak{L})$ vérifie la condition (#) de [G.W] ou [I.1] c'est à dire :

(#) $\exists\ q_o \in \mathbb{N}^*$ tel que $\forall\ q \geq q_o$ $R^{(q)}$ est engendré par R_q sur $R_o = \mathbb{C}$.

On a vu que $1 + [\frac{2g+1}{\delta}]$ convient, ce n'est évidemment pas le plus petit possible en général.

(1.3.2) Rappelons que $\mathcal{O}_{\bar{\mathfrak{X}}}(\ell)$ est le faisceau sur $\bar{\mathfrak{X}}$ associé au R-module gradué

$$R(\ell) = \bigoplus_{i \in \mathbb{Z}} R(\ell)_i \text{ où } R(\ell)_i = R_{\ell+i}.$$

Par ([G.W.] 5.1.5) on a : $\mathfrak{L}^\ell = \mathcal{O}_{\bar{\mathfrak{X}}}(\ell)$ pour tout $\ell \in \mathbb{Z}$, (on utilise la condition (#))

Notons Ω le faisceau canonique sur $\bar{\mathfrak{X}}$, i.e. le faisceau des 1-formes différentielles, $\vartheta = \Omega^\vee$ le dual de Ω, i.e. le faisceau tangent. On a donc :

$$g = h^\circ(\Omega) = h^1(\mathcal{O}_{\bar{\mathfrak{X}}}),\ \deg \Omega = -\deg \vartheta = 2g-2.$$

Signalons que l'anneau R, qui est Cohen-Macaulay par (1.1.3), (1.1.4), est Gorenstein si et seulement si

$$\Omega = \mathfrak{L}^N = \mathcal{O}_{\bar{\mathfrak{X}}}(N)$$

pour un $N \in \mathbb{Z}$, cf ([G.W.] 5.1.9.). On a alors $N\delta = 2g-2$.

<u>1.4</u> Faisons ici $\bar{\mathfrak{X}} = \mathbb{P}^1$, i.e. g = O. Tout fibré ample de rang 1 s'écrit $\mathfrak{L} = \mathcal{O}_{\mathbb{P}^1}(n)$ avec $n = \deg \mathfrak{L} > 0$. On a $\Omega = \mathcal{O}_{\mathbb{P}^1}(-2)$, donc les anneaux $R(\mathbb{P}^1, \mathcal{O}_{\mathbb{P}^1}(n))$ sont de Gorenstein si et seulement si n = 1 ou 2.

(1.4.1.) Les anneaux $R = \bigoplus_{i \geq 0} H^\circ(\mathcal{O}_{\mathbb{P}^1}(in))$ peuvent se décrire de la façon suivante cf. [P.1.]. $R \cong \frac{\mathbb{C}[X_o,\ldots,X_n]}{I}$

où $X_o,\ldots,X_n$ sont des indéterminées et I est l'idéal engendré par les mineurs 2 x 2 de la matrice

$$\begin{pmatrix} X_o & X_1 & \ldots\ldots & X_{n-1} \\ X_1 & X_2 & \ldots\ldots & X_n \end{pmatrix}$$

Remarquons que R est régulier si et seulement si n = 1.

(1.4.2) Il est utile de donner une autre présentation de R ; soient X, Y deux indéterminées et $\mathbb{Z}_n$ le groupe $\mathbb{Z}/_{n\mathbb{Z}}$ opérant sur $\mathbb{C}[X,Y]$ par $(X,Y) \mapsto (\omega X, \omega Y)$ où ω est une racine primitive n-ième de l'unité. L'anneau des invariants $\mathbb{C}[X,Y]^{\mathbb{Z}_n}$ est égal à

$$\mathbb{C}[X^n,\ldots\ldots, X^{n-i}Y^i,\ldots, Y^n]$$

On remarque alors que l'on a un isomorphisme de $\frac{\mathbb{C}[X_o,\ldots,X_n]}{I}$, introduit en (1.4.1), sur $\mathbb{C}[X,Y]^{\mathbb{Z}_n}$ en envoyant la classe de X_i sur $X^{n-i}Y^i$.

2. LES RESULTATS DE [B.G.G.]

2.1. On fixe une courbe projective lisse $\overline{\mathfrak{X}}$ et un faisceau inversible ample $\mathcal{L}$; on utilise les notations introduites au § 1. En particulier $R = R(\overline{\mathfrak{X}}, \mathcal{L})$, $\mathfrak{X} = \operatorname{Spec} R$.

Nous adopterons les notations de [I.1.].

(2.1.1) Notons $\mathcal{D} = \mathcal{D}(R)$ l'anneau des opérateurs différentiels sur R. Rappelons qu'il est filtré par les

$$\mathcal{D}^m = \mathcal{D}^m(R) = \{D \in \mathcal{D} \ / \ \operatorname{ord} D \leq m\}$$

On a $\mathcal{D}^m = \operatorname{Der}^m(R) \oplus R$ où $\operatorname{Der}^m(R) = \{D \in \mathcal{D}^m / D*1 = O\}$ est le R-module des opérateurs différentiels sans terme constant d'ordre $\leq m$.

On pose $\operatorname{Der}(R) = \bigcup_{m\geq 0} \operatorname{Der}^m(R)$, de sorte que $\mathcal{D} = R \oplus \operatorname{Der}(R)$.

L'anneau $\mathcal{D}$ est gradué : $\mathcal{D} = \bigoplus_{\ell\in\mathbb{Z}} \mathcal{D}_\ell$ où l'on note $\mathcal{D}_\ell = \{D \in \mathcal{D} / D * R_i \subset R_{i+\ell}$ pour tout $i\}$. On pose :

$$\deg D = \ell \quad \text{si } D \in \mathcal{D}_\ell$$

$$\mathcal{D}^m_\ell = \mathcal{D}_\ell \cap \mathcal{D}^m, \ \operatorname{Der}_\ell(R) = \mathcal{D}_\ell \cap \operatorname{Der}(R), \ \operatorname{Der}^m_\ell(R) = \operatorname{Der}_\ell(R) \cap \operatorname{Der}^m(R),$$

ainsi : $\operatorname{Der}^m(R) = \bigoplus_{\ell\in\mathbb{Z}} \operatorname{Der}^m_\ell(R)$, $\operatorname{Der}(R) = \bigoplus_{\ell\in\mathbb{Z}} \operatorname{Der}_\ell(R)$.

(2.1.2.) On a vu que $\operatorname{Sing} \mathfrak{X} \subset \{O\}$, on pose $\mathfrak{X}_o = \mathfrak{X} - \{O\}$ qui est une variété quasi-affine lisse. On note $\mathcal{O}_{\mathfrak{X}_o}$ le faisceau structural et $\mathcal{D}_{\mathfrak{X}_o}$ le faisceau d'anneaux d'opérateurs différentiels sur $\mathfrak{X}_o$; $\mathcal{D}_{\mathfrak{X}_o}$ est filtré par les $\mathcal{D}^m_{\mathfrak{X}_o}$ et on définit $\mathcal{D}er^m_{\mathfrak{X}_o}$, le faisceau des opérateurs différentiels sans terme constant, de manière évidente de sorte que :

$$O \to \mathcal{O}_{\mathfrak{X}_o} \to \mathcal{D}^m_{\mathfrak{X}_o} \to \mathcal{D}er^m_{\mathfrak{X}_o} \to O \ , \ \mathcal{D}er_{\mathfrak{X}_o} = \bigcup_{m\geq 0} \mathcal{D}er^m_{\mathfrak{X}_o}.$$

Par ((1.1.3), (1.1.4) et [L] prop. 2 p. 167), il vient :

$$\forall\, m \geq O,\ H^\circ(\mathfrak{X}_o, \mathcal{D}^m_{\mathfrak{X}_o}) = \mathcal{D}^m(R)\,,\ H^\circ(\mathfrak{X}_o, \mathcal{D}er^m_{\mathfrak{X}_o}) = \operatorname{Der}^m(R).$$

(2.1.3) L'anneau R étant gradué il admet une présentation de la forme :

$$R = \frac{\mathbb{C}[X_o,\dots, X_s]}{(F)} = \mathbb{C}[x_o,\dots, x_s]$$

avec :

X_i indéterminée affectée du degré $d_i > O$.

$(F) = (F_1,\dots, F_t)$ idéal homogène.

x_i = classe de X_i modulo (F).

On posera $d = \min\{d_i, O \leq i \leq s\} \geq 1$.

L'espace vectoriel $\mathbb{C}[X_o,\ldots, X_s]_j$ a pour base les $X_o^{\alpha_o} \ldots\ldots X_s^{\alpha_s}$ tels que $j = \deg X_o^{\alpha_o} \ldots\ldots X_s^{\alpha_s} = \alpha_o d_o + \ldots + \alpha_s d_s$.

Donc en posant $(E)_j = (E) \cap \mathbb{C}[X_o,\ldots, X_s]_j$, on a : $R_j = H^\circ(\mathcal{O}j) = \dfrac{\mathbb{C}[X_o,\ldots, X_s]_j}{(E)_j}$. Insistons sur $R_j = (O)$ si $O < j < d$.

On peut introduire la dérivation d'Euler :

$$I = \sum_{i=O}^{s} d_i X_i \frac{\partial}{\partial X_i}$$

et on note encore $I = \sum_{i=O}^{s} d_i x_i \frac{\partial}{\partial x_i}$ la dérivation qu'elle induit sur R.. On a alors :

$$\forall f \in R_p, I * f = pf.$$

$$\mathcal{D}^m_{\ell} = \mathcal{D}^m_{\ell}(R) = \{D \in \mathcal{D}^m / [I,D] = ID - DI = \ell D\}$$

$$\mathrm{Der}^m_{\ell}(R) = \{D \in \mathrm{Der}^m(R) / [I,D] = \ell D\}$$

Remarquons que $I \in \mathrm{Der}^1_O(R)$.

2.2 Rappelons que $\bar{\mathfrak{X}} \cong \mathrm{Proj}\, R$; le groupe $\mathbb{C}^*$ opère sur R par $\lambda . f = \lambda^d f$ si $f \in R_d$, $\lambda \in \mathbb{C}^*$. Donc $\mathbb{C}^*$ opère sur $\mathfrak{X} = \mathrm{Spec}\, R$ avec un unique point fixe $\{O\}$, on trouve ainsi $\bar{\mathfrak{X}} \cong \mathfrak{X}_o/\mathbb{C}^*$ et l'on note π la projection $\mathfrak{X}_o \rightarrow \bar{\mathfrak{X}} = \mathfrak{X}_o/\mathbb{C}^*$, c'est un morphisme affine.

(2.2.1) Introduisons le faisceau sur $\mathfrak{X}_o$ des germes d'opérateurs différentiels d'ordre $\leq m$ et de degré ℓ sans terme constant, que l'on notera $\mathcal{D}er^m_{\mathfrak{X}_o,\ell}$ ou plus simplement $\mathcal{D}er^m_{\ell}$.

Soit $f \in R_q$, l'anneau R_f étant gradué on définit :

$$\mathrm{Der}^m_{\ell}(R_f) = \{D \in \mathrm{Der}^m(R_f)/D_*[R_f]_i \subset [R_f]_{\ell+i}\}.$$

Notons que la dérivation I de (2.1.3) s'étend à R_f en un élément de $\mathrm{Der}^1_O(R_f)$ et on a :

$$\mathrm{Der}^m_{\ell}(R_f) = \{D \in \mathrm{Der}^m(R_f) / [I,D] = \ell D\}$$

Le faisceau $\mathcal{D}er^m_{\ell}$ sur $\mathfrak{X}_o$ est donné sur l'ouvert $D(f) = \mathrm{Spec}\, R_f$, $f \in R_q$, par :

$\Gamma(D(f), \mathcal{D}er^m_{\ell}) = \mathrm{Der}^m_{\ell}(R_f)$. Signalons que $\Gamma(\mathfrak{X}_o, \mathcal{D}er^m_{\ell}) = \mathrm{Der}^m_{\ell}(R)$, cf. (2.1.2).

(2.2.2) Posons $\Delta^m_{\ell} = \pi_* \mathcal{D}er^m_{\ell}$, c'est un $\mathcal{O}_{\bar{\mathfrak{X}}}$-module ; sur un ouvert affine $D_+(f) = \mathrm{Spec}[R_f]_o$ de $\bar{\mathfrak{X}}$ on a :

$$\Gamma(D_+(f), \mathcal{O}_{\dot{\mathfrak{X}}}) = [R_f]_o$$

$$\Gamma(D_+(f), \Delta^m_{\ell}) = \Gamma(D_+(f), \mathcal{D}er^m_{\ell}) = \mathrm{Der}^m_{\ell}(R_f).$$

Il est donc clair que $[R_f]_o \, \mathrm{Der}^m_{\ell}(R_f) \subset \mathrm{Der}^m_{\ell}(R_f)$. On a en outre :

$$H^o(\bar{\mathfrak{X}}, \Delta^m_{\ell}) = \Gamma(\mathfrak{X}_o, \mathcal{D}er^m_{\ell}) = \mathrm{Der}^m_{\ell}(R).$$

2.3. Nous allons rappeler les résultats de **[B.G.G]** repris dans **[I.1.]** en toute généralité.

(2.3.1.) PROPOSITION. Soit q_o comme dans (#) de (1.3.1). Alors on a :

$$\forall \ell \in \mathbb{Z} \text{ tel que } |\ell| \geq q_o, \ \Delta^m_{\ell} \cong \Delta^m_o \otimes \mathcal{L}^{\ell} \quad .$$

Cet isomorphisme étant compatible à l'inclusion $\Delta^m_{\ell} \hookrightarrow \Delta^p_{\ell}$ pour $p > m$.

<u>**Preuve**</u> On réécrit celle de (**[I.1]** Lemma 6) avec les notations précédentes. Soit $f \in R_q$, $D_+(f) = \mathrm{Spec}\,[R_f]_o$, on a :

$$\Gamma(D_+(f), \mathcal{O}_{\dot{\mathfrak{X}}}) = [R_f]_o, \Gamma(D_+(f), \Delta^m_{\ell}) = \mathrm{Der}^m_{\ell}(R_f).$$

$$\Gamma(D_+(f), \Delta^m_o \otimes \mathcal{L}^{\ell}) = \mathrm{Der}^m_o(R_f) \underset{[R_f]_o}{\otimes} \Gamma(D_+(f), \mathcal{L}^{\ell}).$$

Ecrivons $\ell = \varepsilon|\ell|$, $\varepsilon^2 = 1$, pour $|\ell| \geq q_o$; prenons $f = f_j \in R_{|\ell|}$ comme en <u>1.2</u>, de sorte que les $D_+(f_j) = \bar{\mathfrak{X}}_j$ recouvrent $\bar{\mathfrak{X}}$. On a $\Gamma(\bar{\mathfrak{X}}_j, \mathcal{L}^{\ell}) = [R_{f_j}]_o f_j^{\varepsilon}$ et on peut construire une application $\Gamma(\bar{\mathfrak{X}}_j, \Delta^m_o \otimes \mathcal{L}^{\ell}) \to \Gamma(\bar{\mathfrak{X}}_j, \Delta^m_{\ell})$ en envoyant

$D \otimes f_j^{\varepsilon}$ sur $f_j^{\varepsilon} D$ pour tout $D \in \mathrm{Der}^m_o(R_{f_j})$. L'inverse de cette application étant

$D \in \mathrm{Der}^m_{\ell}(R_{f_j}) \mapsto f_j^{-\varepsilon} D \otimes f_j^{\varepsilon}$.

(2.3.2) Nous utiliserons les notations de **[I.1]**, on pose :

$$\sigma_1 = \Delta^1_o, \quad \sigma_m = \frac{\Delta^m_o}{\Delta^{m-1}_o} \quad \text{si } m \geq 1, \ \mathcal{N} = \mathcal{O}_{\dot{\mathfrak{X}}}.I \cong \mathcal{O}_{\dot{\mathfrak{X}}}, \quad \sigma_0 = \mathcal{O}_{\dot{\mathfrak{X}}}, \sigma_{-1} = (O).$$

Par **[B.G.G]** ou **[I.1.]** on a des suites exactes :

$$(*)_{m,\ell} : \forall m \geq o, \forall \ell \in \mathbb{Z}, O \to \sigma_{m-1} \otimes \mathcal{L}^{\ell} \to \sigma_m \otimes \mathcal{L}^{\ell} \to \vartheta^m \otimes \mathcal{L}^{\ell} \to O.$$

Elles sont déduites de $O \to \sigma_{m-1} \to \sigma_m \to \vartheta^m \to O$ où la première flèche est la mutiplication par $I \in H^\circ(\mathfrak{X}, \Delta^1_o) = \mathrm{Der}^1_o(R)$. Notons que $(*)_{1,O}$ est la suite duale de celle de ([W.3.] (1.1.5)) qui est donnée par la première classe de Chern $c_1(\mathfrak{L}) \in H^1(\Omega) - \{O\}$; en particulier elle n'est pas scindée, cf. ([I.1] preuve du lemme 9).

(2.3.3) Il est clair que l'on a des suites exactes :

$$(**)_{m,\ell} \quad O \to \Delta^{m-1}_\ell \to \Delta^m_\ell \to \frac{\Delta^m_\ell}{\Delta^{m-1}_\ell} \to O \text{ pour } \ell \in \mathbb{Z},\ m \geq 2.$$

Lorsque $|\ell| \geq q_o$ ou $\ell = O$ elles s'écrivent par (2.3.1) :

$$O \to \Delta^{m-1}_\ell \to \Delta^m_\ell \to \sigma_m \otimes \mathfrak{L}^\ell \to O.$$

(2.3.4) Par ([B.G.G.] ou [I.1.]) on a :

$$(***)_m \quad O \to \mathfrak{N}^m \to \sigma_m \to \vartheta \otimes \sigma_{m-1} \to O.$$

(2.3.5) Des suites $(*)_{m,\ell}$ et $(**)_{m,\ell}$ on déduit les longues suites exactes :

$$(*)'_{m,\ell} : O \to H^\circ(\sigma_{m-1} \otimes \mathfrak{L}^\ell) \xrightarrow{\Phi_{m,\ell}} H^\circ(\sigma_m \otimes \mathfrak{L}^\ell) \xrightarrow{\Psi_{m,\ell}} H^\circ(\vartheta^m \otimes \mathfrak{L}^\ell) \longrightarrow$$

$$\longrightarrow H^1(\sigma_{m-1} \otimes \mathfrak{L}^\ell) \longrightarrow H^1(\sigma_m \otimes \mathfrak{L}^\ell) \longrightarrow H^1(\vartheta^m \otimes \mathfrak{L}^\ell) \longrightarrow O$$

pour $m \geq O$, $\ell \in \mathbb{Z}$.

$$(**)'_{m,\ell} : O \to H^\circ(\Delta^{m-1}_\ell) \longrightarrow H^\circ(\Delta^m_\ell) \longrightarrow H^\circ(\sigma_m \otimes \mathfrak{L}^\ell) \longrightarrow$$

$$\longrightarrow H^1(\Delta^{m-1}_\ell) \longrightarrow H^1(\Delta^m_\ell) \longrightarrow H^1(\sigma_m \otimes \mathfrak{L}^\ell) \longrightarrow O$$

pour tout $m \geq 2$, $|\ell| \geq q_o$ ou $\ell = O$.

(2.3.6) En tensorisant la suite $(***)_m$ par ϑ^p, $p \in \mathbb{Z}$, on déduit :

$$(***)'_{m,p} : O \to H^\circ(\mathfrak{N}^m \otimes \vartheta^p) \longrightarrow H^\circ(\sigma_m \otimes \vartheta^p) \longrightarrow H^\circ(\vartheta^{p+1} \otimes \sigma_{m-1}) \longrightarrow$$

$$\xrightarrow{\partial_m} H^1(\mathfrak{N}^m \otimes \vartheta^p) \longrightarrow H^1(\sigma_m \otimes \vartheta^p) \longrightarrow H^1(\vartheta^{p+1} \otimes \sigma_{m-1}) \longrightarrow O$$

3. OPERATEURS DIFFERENTIELS SUR LES CÔNES AU DESSUS D'UNE COURBE DE GENRE ≥1. Cas général.

On considère dans ce paragraphe une courbe $\mathfrak{X}$ de genre $g \geq 1$ et $\mathfrak{L}$ un fibré ample de rang 1 de degré $\delta > O$.

Toutes les notations des § 1 et § 2 seront librement utilisées. Insistons sur $\deg \vartheta = -(2g-2) \leq O$, où $\vartheta = \Omega^\vee$ est le faisceau tangent.

3.1 On va déduire des suites exactes de 2.3. quelques corollaires, cf [B.G.G.] ou [I.1].

(3.1.1) **LEMME** : $\forall \ell < 0$, $\forall m \geq 0$, $H^\circ(\sigma_m \otimes \mathcal{L}^\ell) = 0$.

Preuve : Pour $m = 0 : \sigma_o \otimes \mathcal{L}^\ell = \mathcal{L}^\ell$ avec $\deg \mathcal{L}^\ell = \ell\delta < 0$ si $\ell < 0$, donc le résultat est clair. On fait alors une récurrence sur m. Comme $2g-2 \geq 0$ on a pour $m \geq 0$:

$$\deg \vartheta^m \otimes \mathcal{L}^\ell = \ell\delta + m(2-2g) < 0.$$

De (2.3.5) on tire $H^\circ(\sigma_{m-1} \otimes \mathcal{L}^\ell) \cong H^\circ(\sigma_m \otimes \mathcal{L}^\ell)$ et on termine par récurrence.

(3.1.2) **LEMME** : $\forall m \geq 1$, $\forall \ell \leq -q_o$, $H^\circ(\Delta^m_\ell) \cong H^\circ(\Delta^1_\ell)$.

Preuve : Comme $\ell \leq -q_o < 0$ on a $H^\circ(\sigma_m \otimes \mathcal{L}^\ell) = 0$ par (3.1.1). De (2.3.5) $(**)'_{m,\ell}$ on déduit

$$H^\circ(\Delta^{m-1}_\ell) \cong H^\circ(\Delta^m_\ell) \quad \text{pour } m \geq 2.$$

3.2. Nous allons montrer dans cette section qu'il n'existe pas d'opérateur différentiel sur R de degré < O.

(3.2.1) **PROPOSITION** : $\forall m \geq 1$, $\forall \ell < 0$, $H^\circ(\Delta^m_\ell) = \mathrm{Der}^m_\ell(R) = 0$.

Preuve : Par ([W.2] ou [W.3.]) on a $\mathrm{Der}^1_\ell(R) = 0$ si $\ell < 0$. En particulier : $\forall \ell \leq -q_o$, $H^\circ(\Delta^1_\ell) = \mathrm{Der}^1_\ell(R) = 0$.
Par (3.1.2.) $\forall \ell \leq -q_o$, $\forall m \geq 1$, $H^\circ(\Delta^m_\ell) = \mathrm{Der}^m_\ell(R) = 0$.

Soit $m \geq 1$ et $\ell < 0$, prenons $k \geq 1$ tel que $k\ell < -q_o$; si $D \in \mathrm{Der}^m_\ell(R)$ on a $D^k \in \mathrm{Der}^m_{k\ell}(R)$ dans $\mathcal{D}(R) = \mathcal{D}$ et donc $D^k = 0$ par ce qui précède. Comme $\mathcal{D}$ est intègre il vient $D = 0$.

(3.2.2) **COROLLAIRE** : L'idéal $\mathfrak{M} = \bigoplus_{i>0} R_i$ est un sous-$\mathcal{D}$-module de R.

En particulier $\mathcal{D}$ n'est pas un anneau simple et $\mathbb{C} = \dfrac{R}{\mathfrak{M}}$ est un $\mathcal{D}$-module simple.

Preuve : Par (3.2.1) $\mathcal{D} = \bigoplus_{\ell \geq 0} \mathcal{D}_\ell$; comme $\mathcal{D}_\ell * R_i \subset R_{i+\ell}$ il vient $\mathcal{D} * \mathfrak{M} \subset \mathfrak{M}$. Les autres assertions sont alors évidentes.

3.3 Nous allons décrire $\mathcal{D}_o(R) = \bigcup_{m \geq 0} \mathcal{D}^m_o(R)$ comme dans [B.G.G].

(3.3.1) **LEMME** : On a $H^\circ(\sigma_1) = H^\circ(\Delta^1_O) = \mathrm{Der}^1_O(R) = \mathbb{C}.I$.

Preuve On utilise [W.2.] (se rappeler que $g \geq 1$).

(3.3.2) **LEMME** : Supposons $g \geq 2$, alors :

$$\forall p > 0, \forall m \geq 0 \quad H^\circ(\vartheta^p \otimes \sigma_m) = 0 \text{ et } h^\circ(\sigma_m) = 1.$$

Preuve : L'hypothèse assure que $\deg \vartheta < 0$; donc $H^\circ(\vartheta^p) = 0$ pour $p > 0$. On fait une récurrence sur m. Si $m = 1$, on tire de (2.3.6) :

$$0 \to H^\circ(\vartheta^p) = 0 \to H^\circ(\sigma_1 \otimes \vartheta^p) \to H^\circ(\vartheta^{p+1} \otimes \mathfrak{N}) = 0 \to \cdots$$

(on rappelle que $\mathcal{N} \cong \mathcal{O}_{\bar{\mathfrak{X}}}$).

Pour m quelconque une nouvelle utilisation de (2.3.6) et la récurence donnent $H°(\vartheta^p \otimes \sigma_m) = O$.

Pour calculer $h°(\sigma_m)$ on utilise (2.3.6) avec $p = O$, il vient $H°(\sigma_m) \cong H°(\mathcal{N}^m)$ puisque $H°(\vartheta \otimes \sigma_{m-1}) = O$, d'où $h°(\sigma_m) = 1$.

(3.3.3) LEMME : Supposons $g = 1$, donc $\vartheta \cong \mathcal{O}_{\bar{\mathfrak{X}}}$, alors $\forall\ m \geq O$, $h°(\sigma_m) = 1$.

__Preuve :__ Compte tenu de (3.3.1) c'est la preuve de ([**B.G.G.**] Lemma 5) en utilisant les suites exactes de (2.3.6).

(3.3.4) THEOREME : L'espace vectoriel Der_O^m (R) est de dimension m de base $\{I, I^2, \ldots, I^m\}$. Ainsi $\forall\ m \geq O$

$$\mathcal{D}_O^m(R) = \bigoplus_{i=0}^{m} \mathbb{C}\, I^i.$$

__Preuve__: C'est celle de ([**B.G.G**] Proposition 1. 2°) en notant que $h°(\sigma_m) = 1$ par (3.3.2) et (3.3.3).

__3.4__ Comme il est signalé dans [**B.G.G**] on a des résultats analogues en remplaçant $\mathcal{D}$(R) par l'anneau des germes en O d'opérateurs analytiques sur $\mathfrak{X}$. Nous laissons le soin au lecteur d'énoncer ces résultats.

4. OPERATEURS DIFFERENTIELS SUR LES CÔNES AU DESSUS D'UNE COURBE DE GENRE ≥1. Cas Gorenstein

__4.1.__ On reprend $\bar{\mathfrak{X}}$, $\mathcal{L}$ comme au § 3 et l'on fait l'hypothèse supplémentaire suivante :

$$R = R(\bar{\mathfrak{X}}, \mathcal{L}) \text{ anneau de Gorenstein.}$$

(4.1.1) Rappelons que R Gorenstein équivaut à l'existence de N (ici $\geq O$) tel que $\mathcal{L}^N \cong \Omega$, cf (1.3.2). On a alors $N\delta = 2g-2$ si $\delta = \deg \mathcal{L}$.

(4.1.2) Si $\bar{\mathfrak{X}}$ est elliptique i.e. $g = 1$, R est toujours Gorenstein puisque $\mathcal{L}^O \cong \mathcal{O}_{\bar{\mathfrak{X}}} = \Omega$.

(4.1.3) Si la suite $(F_1, \ldots, F_t)$, définissant (E) comme en (2.1.3), est régulière dans $\mathbb{C}[X_o, \ldots, X_s]$ alors R est Gorenstein et on peut prendre $N = \sum_{j=1}^{t} c_j - \sum_{i=O}^{s} d_i$ où $\deg F_j = c_j$, cf ([**I.1**] § 3).

En particulier, si X, Y, Z sont des indéterminées et si F est un polynôme homogène de degré $n \geq 3$ dans $S = \mathbb{C}[X, Y, Z]$ avec $\deg X = \deg Y = \deg Z = 1$, posons $R = \frac{S}{FS}$.

Supposons que R soit normal (intègre), $\bar{\mathfrak{X}} = \mathrm{Proj}\ R$ étant une courbe de genre $g(\geq 1)$.

On a évidemment $\bar{\mathfrak{X}} \overset{j}{\hookrightarrow} \mathbb{P}^2 = \mathrm{Proj}\ S$, le faisceau $\mathcal{L} = j^*\mathcal{O}_{\mathbb{P}^2}(1)$ est très ample et l'on a

$$R = \bigoplus_{i\geq 0} H°(\mathcal{L}^i)$$

Le genre de $\bar{\mathfrak{X}}$ est $g = \frac{1}{2}(n-1)(n-2)$, on a $\mathcal{L}^{n-3} = \Omega$, $\deg \mathcal{L} = n$; l'anneau $\mathcal{D}$(R) est alors étudié dans [**V**], nous voulons généraliser cette étude à $R(\bar{\mathfrak{X}}, \mathcal{L})$ Gorenstein quelconque.

(4.1.4) Signalons enfin que si $g \geq 1$ et $\mathfrak{L} = \mathcal{O}_{\tilde{\mathfrak{X}}}(P)$ pour P un point de $\bar{\mathfrak{X}}$ la condition $R(\bar{\mathfrak{X}}, \mathfrak{L})$ Gorenstein est caractérisée dans ([G.W.] 5.2.6).

4.2 Nous allons reprendre rapidement la preuve de quelques résultats figurant dans [B.G.G], [V] et [I.1.] ; nous ne rappellerons pas toujours les références précises de ces résultats.

(4.2.1) LEMME : Soit $mN \leq \ell-1$ alors :

(a) $\forall\, O \leq i \leq m-1,\ H^1(\vartheta^i \otimes \mathfrak{L}^{\ell}) = H^1(\sigma_i \otimes \mathfrak{L}^{\ell}) = O$

(b) L'application $\Psi_{i,\ell}$ de $(*)'_{i,\ell}$ en (2.3.5) est surjective pour $O \leq i \leq m$:

$$H^\circ(\vartheta^i \otimes \mathfrak{L}^{\ell}) \cong \frac{H^\circ(\sigma_i \otimes \mathfrak{L}^{\ell})}{I\, H^\circ(\sigma_{i-1} \otimes \mathfrak{L}^{\ell})}$$

__Preuve__ : On vérifie que $mN \leq \ell-1$ implique $H^\circ(\mathfrak{L}^{(i+1)N-\ell}) = O$ pour $O \leq i \leq m-1$. On fait alors une récurrence, grâce à (2.3.5), pour trouver $H^1(\sigma_i \otimes \mathfrak{L}^{\ell}) = O$ pour $O \leq i \leq m-1$. On en déduit la surjectivité de $\Psi_{i,\ell}$ pour $O \leq i \leq m$, et comme $\Phi_{i,\ell}$ est la multiplication par I on a les résultats cherchés.

(4.2.2) LEMME : Soient $\ell \geq q_o$ et $m \geq 1$ tels que $mN \leq \ell-1$. Alors :

(a) $\forall\ 1 \leq i \leq m$, $H^o(\sigma_i \otimes \mathfrak{L}^{\ell}) = \dfrac{Der^{i}_{\ell}(R)}{Der^{i-1}_{\ell}(R)}$

(b) Pour $mN+d \geq q_o$,

$$Der^{m}_{mN+d}(R) \supsetneqq Der^{m-1}_{mN+d}(R) + I\, Der^{m-1}_{mN+d}(R)$$

__Preuve__ : Rappelons que q_o et d sont introduits dans (1.3.1) et (2.1.3) respectivement.

(a) Pour $i = 1$: $H^\circ(\sigma_i \otimes \mathfrak{L}^{\ell}) = H^\circ(\Delta^1_O \otimes \mathfrak{L}^{\ell}) = H^\circ(\Delta^1_{\ell}) = Der^1_{\ell}(R)$ et comme $Der^O_{\ell}(R) = (O)$ le résultat est vrai.

Montrons par récurrence sur i que $H^1(\Delta^{i-1}_{\ell}) = H^1(\sigma_i \otimes \mathfrak{L}^{\ell}) = O$ pour $1 \leq i \leq m-1$. Par (2.3.5) on a :

$$\cdots \to H^1(\Delta^{i-1}_{\ell}) \to H^1(\Delta^{i}_{\ell}) \to H^1(\sigma_i \otimes \mathfrak{L}^{\ell}) \to O$$

Donc si $H^1(\Delta^{i-1}_{\ell}) = O$ il vient $H^1(\Delta^{i}_{\ell}) = H^1(\sigma_i \otimes \mathfrak{L}^{\ell})$ ce qui vaut O par (4.2.1) si $1 \leq i \leq m-1$.

La suite $(**)'_{i,\ell}$ donne ainsi pour $1 \leq i \leq m$:

$$O \to H^o(\Delta^{i-1}_{\ell}) \to H^o(\Delta^{i}_{\ell}) \to H^o(\sigma_i \otimes \mathfrak{L}^{\ell}) \to O.$$

On termine par (2.2.2).

(b) Comme $d \geq 1$, $\ell = mN+d \geq mN+1$ et (a) s'applique :

$$H^\circ(\sigma_i \otimes \mathfrak{L}^{\ell}) = \frac{\mathrm{Der}^{i}_{\ell}(R)}{\mathrm{Der}^{i-1}_{\ell}(R)} \quad \text{pour } i = m-1, m.$$

De (4.2.1) on tire $H^\circ(\vartheta^m \otimes \mathfrak{L}^{\ell}) \cong \dfrac{H^\circ(\sigma_m \otimes \mathfrak{L}^{\ell})}{I\, H^\circ(\sigma_{m-1} \otimes \mathfrak{L}^{\ell})}$, mais $H^\circ(\vartheta^m \otimes \mathfrak{L}^{mN+d}) = H^\circ(\mathfrak{L}^d) \neq O$ ($\mathfrak{L}^N = \Omega = \vartheta^v$)

d'où le résultat.

4.3. Plaçons nous maintenant dans le cas où $\bar{\mathfrak{X}}$ est elliptique, i.e. $N = O$. Alors les énoncés de 4.2 prennent une forme simple, et l'on retrouve les résultats de [B.G.G].

(4.3.1) LEMME : $\forall\ m \geq O,\ \forall \ell \geq 1\ H^\circ(\mathfrak{L}^{\ell}) = \dfrac{H^\circ(\sigma_m \otimes \mathfrak{L}^{\ell})}{I\, H^\circ(\sigma_{m-1} \otimes \mathfrak{L}^{\ell})}$.

Preuve : C'est (4.2.1) en notant que $N = O$ et $\vartheta \cong \mathcal{O}_{\hat{\mathfrak{X}}}$.

(4.3.2.) LEMME : (a) $\forall\ m \geq 1, \forall \ell \geq q_o\ H^\circ(\sigma_m \otimes \mathfrak{L}^{\ell}) = \dfrac{\mathrm{Der}^{m}_{\ell}(R)}{\mathrm{Der}^{m-1}_{\ell}(R)}$

(b) $\forall\ m \geq 1, \forall \ell \geq q_o\ \dim_{\mathbb{C}} \dfrac{\mathrm{Der}^{m}_{\ell}(R)}{\mathrm{Der}^{m-1}_{\ell}(R) + I\, \mathrm{Der}^{m-1}_{\ell}(R)} = \delta\ell > O.$

Preuve (a) est (4.2.2) (a)

(b) résulte de (4.3.1) et (a) ci-dessus puisque $\dfrac{\mathrm{Der}^{m}_{\ell}(R)}{\mathrm{Der}^{m-1}_{\ell}(R) + I\, \mathrm{Der}^{m-1}_{\ell}(R)} \cong H^\circ(\mathfrak{L}^{\ell})$ qui est de dimension $\delta\ell$ car

$\bar{\mathfrak{X}}$ est elliptique et $\deg \mathfrak{L}^{\ell} = \delta\ell > O$.

(4.3.3) THEOREME : Fixons $\ell \geq q_o$ (> O) et pour tout $k \geq O$ posons comme dans [B.G.G.] :

$$\mathcal{J}_k = \sum_{n \geq O} I^n \mathcal{D}^{k}_{\ell}(R) + \sum_{i \geq \ell+1} \mathcal{D}_i(R).$$

Alors $(\mathcal{J}_k)_{k \in \mathbb{N}}$ est une suite strictement croissante d'idéaux bilatères de $\mathcal{D}(R)$, (qui n'est donc pas noethérien).

Preuve : Rappelons que $\mathcal{D}^{s}_{t}(R)\,\mathcal{D}^{k}_{\ell}(R)$ et $\mathcal{D}^{k}_{\ell}(R)\,\mathcal{D}^{s}_{t}(R)$ sont contenus dans $\mathcal{D}^{s+k}_{\ell+t}(R)$.

Il est clair que $\mathcal{J}_k$ est stable par addition. Soit $P \in \mathcal{D}^{s}_{t}(R)$ deux cas sont possibles :

- si $t = O$, $P \in \mathbb{C}\,1 + \dots + \mathbb{C}\, I^s$ par (3.3.5) et alors :

$$P\mathcal{D}^{k}_{\ell}(R) + \mathcal{D}^{k}_{\ell}(R)\,P \subset \sum_{n \geq O} I^n \mathcal{D}^{k}_{\ell}(R)$$

$$P\mathcal{D}_j(R) + \mathcal{D}_j(R)\,P \subset \sum_{i \geq \ell+1} \mathcal{D}_i(R) \text{ si } j \geq \ell+1.$$

- si $t > O$, $P\mathcal{D}^k_j(R)$ et $\mathcal{D}^k_j(R)P$ sont contenus dans $\sum_{i\geq \ell+1} \mathcal{D}_i(R)$, pour tout $j \geq \ell$.

On a donc montré que $\mathcal{J}_k$ est un idéal bilatère de $\mathcal{D}$; il est évident que $\mathcal{J}_k \subset \mathcal{J}_{k+1}$.

Remarquons que :

$$\forall D \in \mathrm{Der}^k(R), \forall n \geq O, I^n D \notin \mathrm{Der}^{k+n-1}(R) \text{ si } \mathrm{ord}\, D = k.$$

En effet gr $\mathcal{D}(R)$ est intègre et si $I^n D \in \mathrm{Der}^{k+n+1}(R)$ avec ord $D = k$, il en découle : $\sigma(I)^n \sigma(D) = O$ dans gr$\mathcal{D}(R)$ avec $\sigma(D) \neq O$. Cela force $\sigma(I) = O$ ce qui est impossible.

De cette remarque on tire les égalités :

$$\begin{cases} \mathcal{J}_k \cap \mathrm{Der}^{k+1}_{\ell}(R) = \mathrm{Der}^{k}_{\ell}(R) + I\, \mathrm{Der}^{k}_{\ell}(R) \\ \mathcal{J}_{k+1} \cap \mathrm{Der}^{k+1}_{\ell}(R) = \mathrm{Der}^{k+1}_{\ell}(R) \end{cases}$$

Grâce à (4.3.2) (b) on conclut $\mathcal{J}_{k+1} \supsetneqq \mathcal{J}_k$.

(4.3.4) THEOREME L'algèbre $\mathcal{D}(R)$ n'est pas de type fini sur $\mathbb{C}$.

__Preuve__ : Par (3.3.5) et (2.1.3) si $\mathcal{D}(R)$ est engendré par $P_o,\ldots, P_\upsilon$ sur $\mathbb{C}$ on peut toujours supposer :

$P_o = x_o,\ldots, P_s = x_s$, deg $P_i = d_i > O$, ord $P_i = O = m_i$, $O \leq i \leq s$,

$P_{s+1} = I$, deg $P_{s+1} = O$, ord $P_{s+1} = 1$.

deg $P_j = d_j > O$, ord $P_j = m_j > O$ si $s+2 \leq j \leq \upsilon$.

Remarquons que par (4.2.2) (b) il existe un P_j tel que deg $P_j = d_j > O$ et ord $P_j = m_j > O$.

Pour $Q \in \mathrm{Der}^m_{\ell}(R)$ écrivons $Q = \sum a_\alpha P_\alpha$, $a_\alpha \in \mathbb{C}$, $O \neq P_\alpha$ produit de P_j que l'on peut supposer par homogénéité de degré ℓ. Si P_j apparaît ℓ_j fois, $\ell_j \geq O$, dans P_α on a donc :

(*) $\deg P_\alpha = \ell = d_o \ell_o + \ldots\ldots + d_s \ell_s + d_{s+2} \ell_{s+2} + \ldots\ldots + d_\upsilon \ell_\upsilon$.

Prenons $Q \in \mathrm{Der}^m_{\ell}(R) \setminus (I\, \mathrm{Der}^m_{\ell}(R) + \mathrm{Der}^{m-1}_{\ell}(R))$ avec $\ell \geq q_o$ et $m > \ell(m_o + \ldots + m_\upsilon) \geq 1$, c'est possible par (4.2.2) (b). Remarquons que si $P_{s+1} = I$ apparaît dans un P_α, on peut toujours soustraire ce P_α à Q et finalement avoir $\ell_{s+1} = O$ dans tout P_α.

Par (*) on a $\ell_i \leq \ell$ pour $i \geq s+2$ et il vient :

$$\mathrm{ord}\ P_\alpha \leq \ell_{s+2} m_{s+2} + \ldots. + \ell_\upsilon m_\upsilon \leq \ell\, (m_o + \ldots + m_\upsilon)$$

Donc $m = \mathrm{ord}\, Q \leq \ell\, (m_o + \ldots + m_\upsilon)$ et une contradiction.

__4.4__ Nous allons maintenant supposer que $\overline{\mathfrak{X}}$ n'est pas elliptique c'est à dire $N > O$. Cette hypothèse sera en vigueur dans toute cette section.

Nous allons énoncer un résultat technique énoncé en ([I.1] Lemma 9), tiré de [V]. Rappelons que l'on a posé $d = \min\{d_i, 0 \leq i \leq s\} \geq 1$.

(4.4.1) LEMME Soit $\ell \leq mN+d-1$ avec $m \geq 1$. Avec les notations de (2.3.5) on a $\Psi_{m,\ell} = 0$. Donc $\Phi_{m,\ell}$: $H°(\sigma_{m-1} \otimes \mathfrak{X}^{\ell}) \to H°(\sigma_m \otimes \mathfrak{X}^{\ell})$ est un isomorphisme donné par la multiplication par I.

__Preuve__ C'est celle de ([I.1.] Lemma 9) que l'on écrit avec les notations précédentes. Il faut remarquer qu'à la fin de cette démonstration on obtient une flèche $\mathbb{C} = H°(\mathcal{O}_{\tilde{\mathfrak{X}}}) \to H^1(\sigma_{m-1} \otimes \Omega^m) \to 0$; il s'agit de s'assurer que $H^1(\sigma_{m-1} \otimes \Omega^m) \neq 0$ pour conclure à l'isomorphisme (et à $\psi_{m,mN} = 0$). Cela peut se montrer en considérant la longue suite exacte $(*)'_{m-1,mN}$ et en utilisant $H^1(\Omega) \neq 0$. Remarquons que l'on a $H°(\vartheta^m \otimes \mathfrak{X}^{mN+j}) = H°(\mathfrak{X}^j)$ et la nullité de $\psi_{m,mN+j}$ est évidente si $0 < j < d$ ou si $j < 0$.

(4.4.2) PROPOSITION Supposons $m \geq 1$ et $q_o \leq \ell \leq mN+d-1$, alors $\mathrm{Der}^1_{\ell}(R) = R_{\ell} I$ si $m = 1$

et $\mathrm{Der}^m_{\ell}(R) = I\ \mathrm{Der}^{m-1}_{\ell}(R) + \mathrm{Der}^{m-1}_{\ell}(R)$ si $m \geq 2$.

__Preuve__ : Cela résulte de (4.4.1) combiné avec (4.4.2) (a)

Nous voulons nous affranchir de la condition $\ell \geq q_o$ dans (4.4.2) nous commencerons par le résultat élémentaire suivant :

(4.4.3) LEMME : Il existe des éléments $f_1, \ldots, f_v$ de R tels que $\mathfrak{X}_o = \bigcup_{i=1}^{v} U_i$, $U_i = \mathrm{Spec}\, R_{f_i}$ et :

$$\forall\, 1 \leq i \leq v,\ \exists\, \partial_i \in \mathrm{Der}^1(R) \text{ de sorte que } \mathrm{Der}^1(R_{f_i}) = R_{f_i} . I \oplus R_{f_i}\, \partial_i.$$

__Preuve__ : Rappelons que $R = \dfrac{\mathbb{C}[X_o,\ldots,X_s]}{(F)} = \mathbb{C}[x_o,\ldots,x_s]$ $\mathrm{Sing}\, R = \{\mathfrak{M} = (x_o,\ldots,x_s)\}$; on pose $V_i = \mathrm{Spec}\, R_{x_i}$, on a donc V_i ouvert affine et $\mathfrak{X}_o = \bigcup_{i=0}^{s} V_i$.

(1) Soit $P = (x_o - \alpha_o, \ldots, x_s - \alpha_s) \in \mathrm{Max}\, R_{x_o}$, donc $\alpha_o \neq 0$. Alors il existe $j, \ell \in \{0,\ldots,s\}$ tels que

$$\mathcal{P} = PR_P = (x_j - \alpha_j, x_{\ell} - \alpha_{\ell}) \text{ avec } \alpha_j \neq 0.$$

La preuve de cette assertion est facile, esquissons la.

Si $x_o - \alpha_o \notin \mathcal{P}^2$, $j = 0$ convient. Si $x_o - \alpha_o \in \mathcal{P}^2$, supposons que $\mathcal{P} = (x_1, x_2)$ (quitte à renuméroter les x_j) et cherchons une contradiction. Il existe $\varphi \in R \backslash P$ tel que

(*) $\varphi(x_o - \alpha_o) = \sum_{i_1,i_2} b_{i_1,i_2} x_1^{i_1} x_2^{i_2}$ avec $b_{i_1,i_2} \in R$. Posons $\varphi = \sum_j \varphi^{(j)}$ avec $\varphi^{(j)}$ homogène et soit $\varphi^{(h)}$ la première composante n'appartenant pas à P, c'est à dire $\deg \varphi^{(j)} \geq \deg \varphi^{(j-1)}$, $\varphi^{(h+j)} \in P$ si $j \geq 1$, $\varphi^{(h)} \notin P$. Considérons la composante homogène de degré $d_o + \deg \varphi^{(h)}$ du membre de gauche de (*) ; comme $\deg x_o\ \varphi^{(j)} < \deg x_o\ \varphi^{(h)}$ si $j < h$ il vient $x_o\ \varphi^{(h)} - \alpha_o\ \varphi^{(q)} = \sum c_{j_1,j_2} x_1^{j1} x_2^{j2}$ (avec $q > h$).

Evaluant en P on trouve $x_o \varphi^{(h)} \in P$ et $x_o \notin P$, $\varphi^{(h)} \notin P$ donne la contradiction.

(2) Soit P comme en (1) ; alors I fait partie d'une base de $Der^1(R_P)$. En effet $E = Der^1(R_P)$ est libre de rang 2 sur R_P et $E/\mathcal{P}E$ contient la classe de I qui est non nulle par $I_*(x_j-\alpha_j) = d_j x_j$ et $x_j \notin \mathcal{P}$. Par le lemme de Nakayama il existe donc $\partial \in Der^1(R_P)$ telle que $Der^1(R_P) = R_P I \oplus R_P.\partial$.

(3) De (2) on déduit l'existence de $f \in R\backslash P$ tel que $Der^1(R_f)$ soit libre de base $\{I,\partial\}$ avec $\partial \in Der^1(R)$.

(4) Par (3) on peut recouvrir Spec R_{x_o} par des ouverts Spec R_{f_j}, $j \in \{1,\ldots,\upsilon_o\}$ tels que $Der^1(R_{f_j})$ soit libre de base $\{I,\partial_j\}$. On fait cette construction pour tous les Spec R_{x_i}, $O \le i \le s$; la collection des Spec $R_{f_j} = U_j$, $1 \le j \le \upsilon_i$, $O \le i \le s$, convient.

(4.4.4) LEMME : On conserve les notations de (4.4.3). Dans l'anneau gr $\mathcal{D}(R_{f_j})$ l'idéal engendré par le symbole de I est premier.

__Preuve__ : Posons $f = f_j$; puisque R_f est régulier on a gr $\mathcal{D}(R_f) = S_{R_f}(Der^1(R_f))$ (l'algèbre symétrique du module $Der^1(R_f)$).

Donc de $Der^1(R_f) = R_f . I \oplus R_f.\partial.$, $\partial = \partial_j$, on déduit le résultat.

Au vu de (4.4.4) il est raisonnable de supposer que $\sigma(I)$ engendre un idéal radiciel de gr $\mathcal{D}(R)$, ce que nous ferons dans la proposition qui suit..

(4.4.5) PROPOSITION : Supposons d = 1, i.e. $H^\circ(\mathcal{L}) \neq O$; alors si $m \ge 1$ et $\ell \le mN$ on a $Der^1_\ell(R) = R_\ell I$ si $m = 1$ et $Der^m_\ell(R) = I\,Der^{m-1}_\ell(R) + Der^{m-1}_\ell(R)$ si $m \ge 2$.

__Preuve__ : Soit $D \in Der^m_\ell(R)$. Prenons $m \ge 2$, le cas m = 1 étant analogue. Si $\ell = O$ on a le résultat par (3.3.5). Si $\ell \ge 1$,

$D^{q_o} \in Der^{mq_o}_{\ell q_o}(R)$ et de (4.4.2), puisque $q_o \le \ell q_o \le (mq_o)N$ si $\ell \le mN$, on tire :

$$D^{q_o} \in I\,Der^{mq_o-1}_{\ell q_o}(R) + Der^{mq_o-1}_{\ell q_o}(R).$$

En passant dans gr$\mathcal{D}(R)$ on trouve $\sigma(D)^{q_o} \in \sigma(I)$ gr $\mathcal{D}(R)$. On en déduit $\sigma(D) \in \sigma(I)$ gr $\mathcal{D}(R)$ et le résultat.

Ainsi pour tout $j \in \{1,\ldots,\upsilon\}$ il existe $P_j, T_j \in Der^{m-1}_\ell(R_{f_j})$ tels que $D = I P_j - T_j$; alors $I(P_j - P_k) = T_j - T_k$ pour $1 \le j,k \le \upsilon$ sur $U_j \cap U_k$. En raisonnant comme dans la preuve de (4.4.2) on trouve $D \in I\,Der^{m-1}_\ell(R) + Der^{m-1}_\ell(R)$.

(4.4.6) La condition d = 1 est vérifiée dans les exemples suivants :

(a) $\bar{\mathfrak{X}}$ quelconque, $\mathcal{L} = \Omega$. (b) $\bar{\mathfrak{X}} = \mathrm{Proj}\,\frac{S}{FS}$, $\mathcal{L} = \mathcal{O}_{\bar{\mathfrak{X}}}(1)$ comme en (4.1.3)

(et bien sûr pour $\bar{\mathfrak{X}}$ elliptique, $\mathcal{L}$ quelconque mais ceci a été exclu de cette section).

(4.4.7) Pour simplifier et compte tenu de (4.4.5) nous supposerons désormais que :

$$\forall\ O \le \ell \le mN+d-1,\ Der^m_\ell(R) = I\ \ Der^{m-1}_\ell(R) + Der^{m-1}_\ell(R) \text{ si } m \ge 2 \text{ et } Der^1_\ell(R) = R_\ell I \text{ si } m = 1.$$

(cela est vrai pour $\ell = O$ par (3.3.3)).

(4.4.8) COROLLAIRE : Fixons $k > 0$ et soit $D \in Der^m_{Nk+d}(R)$. Supposons que $D = f D_1 D_2 \dots D_r$ avec $f \in R$ et $D_i \in Der(R)$ homogène et d'ordre $< k$, pour tout i. Alors il existe P d'ordre $< k$ tel que $D = I\Delta + P$, $\Delta \in Der^{m-1}(R)$.

Preuve : Ce résultat est celui de ([V] Lemme 3), la preuve est identique, rappelons la. Si $m < k$ il n'y a rien à montrer ; supposons donc $m = \text{ord}(D) \geq k$ (donc $r \geq 2$). Il existe j tel que $\deg D_j \leq N \text{ ord } D_j + d-1$ sinon $\deg D = \sum_i \deg D_i \geq mN + rd$ et alors $Nm + rd \leq Nk + d$ ce qui est contradictoire.

Par (4.4.7) $D_j \in I\ Der^{p-1}_{\ell}(R) + Der^{p-1}_{\ell}(R)$, où $\ell = \deg D_j$, $p = \text{ord } D_j$. On en déduit $D = I\,D' + D''$ avec D'' somme d'opérateurs du même type que D mais d'un ordre strictement inférieur. Une récurrence termine la preuve.

On peut maintenant démontrer l'analogue de ([V] Proposition 5).

(4.4.9) PROPOSITION : Fixons $k > 0$, notons H^k_i le sous-espace vectoriel de $Der^k_i(R)$ formé des sommes de produits d'opérateurs d'ordre $< k$. Alors on a : $Der^k_{Nk+d}(R) \underset{\neq}{\supset} H^k_{Nk+d}$.

Preuve : Un élément D de H^k_{Nk+d} s'écrit $D = \sum f D_1 D_2 \dots D_r$ avec $f D_1 D_2 \dots D_r \in Der^m(R)$ vérifiant les hypothèses de (4.4.8). Donc $D = \sum (I\Delta + P)$ avec $\text{ord } P < k$, après une mise en facteur $D = I\,D' + D''$ avec D'' d'ordre $< k$. Comme on a nécessairement $\text{ord } D' < k$, il vient :

$$H^k_{Nk+d} = I\ Der^{k-1}_{Nk+d}(R) + Der^{k-1}_{Nk+d}(R).$$

L'assertion découle donc de (4.2.2) (b).

Comme dans ([V] Théorème) il vient :

(4.4.10) THEOREME : L'algèbre $\mathcal{D}(R)$ n'est pas de type fini sur $\mathbb{C}$.

Preuve : Si $\mathcal{D}(R)$ était engendrée par un nombre fini d'éléments, il existerait $k > 0$ tel que : $Der^k_{Nk+d}(R) = H^k_{Nk+d}$, (par homogénéité), avec H^k_{Nk+d} comme dans (4.4.9). D'où une contradiction.

(4.4.11) THEOREME : Pour $k \geq 0$ on pose : $C_k = \{i \in \mathbb{N}^* \,/\, i = pN+d,\ p \geq k\}$,

$$\mathcal{J}_k = \sum_{C_k} \sum_{n \geq 0} I^n \mathcal{D}^{p-1}_{pN+d}(R) + \sum_{\mathbb{N}^* \setminus C_k} \mathcal{D}_i(R)$$

Alors $(\mathcal{J}_k)$ est une suite croissante d'idéaux bilatères de $\mathcal{D}(R)$ et $\mathcal{J}_{k+1} \underset{\neq}{\supset} \mathcal{J}_k$ si $kN + d \geq q_0$. En particulier $\mathcal{D}(R)$ n'est pas noethérien.

Preuve : De (4.4.7) il découle facilement que :

$\forall\ m \geq 1, \forall\ O \leq \ell \leq mN+d-1,\ \mathcal{D}_{\ell}(R) = \sum_{n\geq O} I^n \mathcal{D}_{\ell}^{m-1}(R)$. Ceci implique en particulier $\mathcal{D}_{\ell}(R) = O$ pour $O < \ell < d$. Montrons que $\mathcal{J}_k \subseteq \mathcal{J}_{k+1}$. Par définition $C_k = C_{k+1} \cup \{kN+d\}$ et $\mathcal{J}_k$ est égal à $\sum_{C_{k+1}} \sum_{n\geq O} I^n \mathcal{D}^{p-1}_{pN+d}(R) + \sum_{\mathbb{N}^*\setminus C_k} \mathcal{D}_i(R) + \sum_{n\geq O} I^n \mathcal{D}^{k-1}_{kN+d}(R)$ qui est contenu dans $\sum_{C_{k+1}} \sum_{n\geq O} I^n \mathcal{D}^{p-1}_{pN+d}(R)$ $+ \sum_{\mathbb{N}^*\setminus C_k} \mathcal{D}_i(R) + \mathcal{D}_{kN+d}(R)$ ce qui n'est autre que $\mathcal{J}_{k+1}$.

Montrons que $\mathcal{J}_k$ est un idéal de $\mathcal{D}(R)$. Par (3.3.4) $\mathcal{D}_o(R) = \sum_{n\geq O} \mathbb{C}\, I^n$ et il est clair que $\mathcal{D}_o(R)\, \mathcal{J}_k + \mathcal{J}_k \mathcal{D}_o(R) \subseteq \mathcal{J}_k$.

Par définition

$$\mathcal{J}_k \cap \mathcal{D}_i(R) = \begin{cases} \mathcal{D}_i(R),\ i \in \mathbb{N}^*\setminus C_k \\ \sum_{n\geq O} I^n \mathcal{D}^{p-1}_{pN+d}(R),\ i = pN+d \in C_k \end{cases}$$

Il suffit donc d'établir, pour $i > O$ et $t > O$, l'assertion suivante : (#) $\mathcal{D}_t(R)\ \mathcal{D}_i(R) \subseteq \mathcal{J}_o$.

Remarquons que (#) est triviale si $O < i < d$, on supposera donc $d \leq i$.

Soient $m \geq 1$ et $\alpha \geq 1$ tels que $(m-1)N < t \leq mN$ et $(\alpha-1)N+d-1 < i \leq \alpha N+d-1$. De $t \leq mN$ et $i \leq \alpha N+d-1$ on tire $\mathcal{D}_t(R) = \sum_{n\geq O} I^n \mathcal{D}_t^{m-1}(R)$ et $\mathcal{D}_i(R) = \sum_{n\geq O} I^n \mathcal{D}_t^{\alpha-1}(R)$. Comme $\mathcal{D}_t^{m-1}(R)\ .\ \mathcal{D}_i^{\alpha-1}(R) \subseteq \mathcal{D}_{i+t}^{\alpha+m-2}(R)$, il suffit pour obtenir (#) de prouver que $\mathcal{D}_{i+t}^{\alpha+m-2}(R) \subseteq \mathcal{J}_o \cap \mathcal{D}_{i+t}(R)$. Si $i+t \in \mathbb{N}^*\setminus C_o$ on a $\mathcal{D}_{i+t}(R) \subseteq \mathcal{J}_o$ et (#) est vraie.

Supposons $i + t = pN+d$ et montrons que $\alpha+m-2 \leq p-1$, ce qui prouvera (#). Sinon $p \leq \alpha + m-2$ et de $-t < -(m-1)N$ il vient $i = pN+d-t < (\alpha+m-2)N + d-(m-1)N = (\alpha-1)N+d$, contredisant le choix de α.

Comme en (4.3.3) on remarque que :

$$\mathcal{J}_k \cap \mathrm{Der}^k_{kN+d}(R) = I\, \mathrm{Der}^{k-1}_{kN+d}(R) + \mathrm{Der}^{k-1}_{kN+d}(R) \subseteq \mathcal{J}_{k+1} \cap \mathrm{Der}^k_{kN+d}(R) = \mathrm{Der}^k_{kN+d}(R).$$

Par (4.2.2) (b) cette inclusion est stricte dès que $kN+d \geq q_o$. Ceci achève la preuve du théorème

(4.4.12) Rappelons que (4.4.10) et (4.4.11) sont vrais sous l' hypothèse (4.4.7).

D'autre part si $N = O$ les mêmes résultats ont été démontrés en (4.3.3 et 4.3.4)

4.5 Ici, comme dans 3.4 on peut obtenir les mêmes résultats en remplaçant $\mathcal{D}(R)$ par l'anneau des germes d'opérateurs différentiels analytiques au voisinage de O.

5. OPERATEURS DIFFERENTIELS SUR LES SURFACES NORMALES MUNIES D'UNE BONNE C*-ACTION

5.1 Afin d'étudier $\mathcal{D}(A)$ pour une $\mathbb{C}$-algèbre de type fini A, graduée, normale, de dimension 2, on va rappeler quelques résultats de [O.W] et [P.2]).

(5.1.1.) Soit $A = \bigoplus_{k\geq 0} A_k$ un anneau normal gradué de dimension 2 avec $A_o = \mathbb{C}$ et A noethérien. Posons $\mathcal{Y} = \operatorname{Spec} A$ et désignons par $\{O\}$ le point fermé correspondant à l'idéal maximal $\bigoplus_{k>0} A_k$, on a donc $\operatorname{Sing} \mathcal{Y} \subset \{O\}$.

Si $\operatorname{Sing} \mathcal{Y} = \emptyset$, A est un anneau de polynômes et l'anneau $\mathcal{D}(A)$ est une algèbre de Weyl, pour cette raison nous ne nous intéresserons qu'au cas $\operatorname{Sing} \mathcal{Y} = \{O\}$.Pour abréger l'écriture nous dirons que $\mathcal{Y}$ est une $\underline{\mathbb{C}^*\text{-surface normale}}$, cf [P.2].

Un exemple est évidemment fourni par les $R(\overline{\mathfrak{X}}, \mathfrak{L})$ du § 1.

(5.1.2) Comme pour toute surface il existe une bonne résolution minimale $\pi : \widetilde{\mathcal{Y}} \to \mathcal{Y}$ de la singularité isolée en O. Si $\pi^{-1}(O) = E_1 \cup \ldots\ldots \cup E_t$, on a :

(i) Les E_j sont des courbes projectives lisses à croisement normaux.

(ii) Le cardinal de $E_i \cap E_j$, est ≤ 1 si $i \neq j$.

(iii) $E_i \cap E_j \cap E_k = \emptyset$ si i, j, k 2 à 2 distincts.

A cette résolution on associe un graphe (pondéré) de la manière suivante :

- les sommets sont les E_j affectés du poids $-b_i = (E_i . E_i)$ (on note comme d'habitude $(E_i . E_j)$ le nombre d'intersection de E_i et E_j).
- Une arête $\underset{i}{\bullet}\!-\!\!\!-\!\underset{j}{\bullet}$ existant si $E_i \cap E_j \neq \emptyset$.

(5.1.3) Supposons que $\mathcal{Y}$ soit une $\mathbb{C}^*$-surface normale, deux cas sont alors possibles pour la résolution (5.1.2).:

(a) On a $E_j \cong \mathbb{P}^1$ pour tout j et le graphe de la résolution est

$$\underset{-b_1}{\circ}\text{———}\underset{-b_2}{\circ}\text{———}\circ\text{- - - - - -}\underset{-b_r}{\circ}$$

Alors la singularité de $\mathcal{Y}$ est analytiquement isomorphe à une singularité cyclique quotient $\mathbb{C}^2/\mathbb{Z}_m$ de type (m,e). C'est à dire $\mathbb{C}^2/\mathbb{Z}_m = \operatorname{Spec} \mathbb{C}[X,Y]^{\mathbb{Z}_m}$, $\mathbb{Z}_m$ opérant par $(X,Y) \mapsto (\xi X, \xi^e Y)$ où ξ racine primitive m-ième de l'unité et $1 \leq e < m$, e et m premiers entre eux. Le type (m,e) étant déterminé par le développement en fractions continues : $[b_1,\ldots, b_r] = \frac{m}{e} = b_1 - \cfrac{1}{b_2 - \cfrac{1}{b_3 \ldots}}$. On dira ici aussi que $\mathcal{Y}$ est une singularité cyclique quotient de type (m,e).

(b) Il existe une unique courbe E parmi les E_j ayant l'une des deux propriétés suivantes : E intersecte plus de deux autres E_j, ou le genre de E est ≥ 1. On dit alors que E est la courbe centrale.

Si $E_j \neq E$ on a $E_j = \mathbb{P}^1$ et le graphe de la résolution a la forme d'une étoile à n branches :

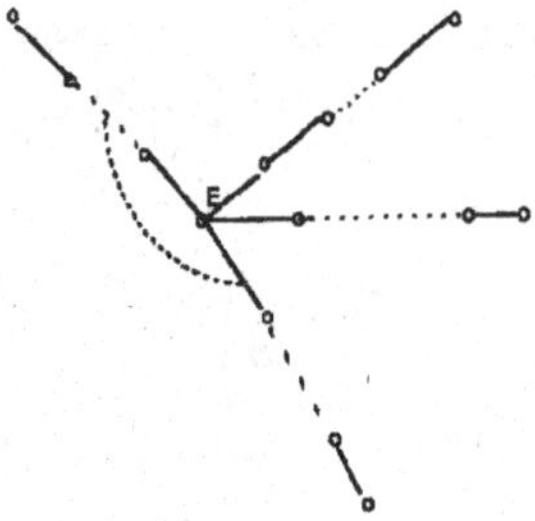

chaque branche étant : $\underset{-b}{\overset{E}{o}}$ ——— $\underset{-b_{i,1}}{\overset{E_{i,1}}{o}}$ - - - - - o ——— $\underset{-b_{i,r_i}}{\overset{E_{i,r_i}}{o}}$ avec $b_{i,j} = -(E_{i,j}.E_{i,j}) \geq 2$ et $b = -(E.E) \geq 1$.

On détermine des entiers e_i, d_i premiers entre eux tels que $1 \leq e_i < d_i$ par le développement $\frac{d_i}{e_i} = [b_{i,1}, b_{i,2}, \ldots, b_{i,r_i}]$.

On sait par **[O.W]** que le graphe "détermine" (à isomorphisme analytique près) la singularité de $\mathcal{Y}$, cf. **[P.2]** pour un énoncé précis.

(5.1.4) Soit $\mathcal{Y} = \mathfrak{X} = \operatorname{Spec} R(\overline{\mathfrak{X}}, \mathfrak{L})$ comme au § 1. Alors $\tilde{\mathcal{Y}} = \mathbb{V}(\overline{\mathfrak{X}}, \mathfrak{L}) \xrightarrow{\pi} \mathfrak{X}$, comme en (1.1.5), donne le graphe :

$$\underset{-b=-\deg \mathfrak{L}}{o}, \text{ avec } E = \pi^{-1}(O) \cong \overline{\mathfrak{X}}.$$

(5.1.5) L'importance des surfaces $\mathfrak{X} = \operatorname{Spec} R(\overline{\mathfrak{X}}, \mathfrak{L})$ est illustrée par le théorème suivant de **[P.2]**.

THEOREME : Soit $\mathcal{Y}$ une $\mathbb{C}^*$-surface normale. Alors il existe une courbe projective lisse $\overline{\mathfrak{X}}$, un groupe fini G d'automorphismes de $\overline{\mathfrak{X}}$, un faisceau inversible ample et G-invariant $\mathfrak{L}$ tels que :

(a) G opère librement sur $\mathfrak{X} = \operatorname{Spec} R(\overline{\mathfrak{X}}, \mathfrak{L})$ en dehors de {O}.

(b) La singularité de $\mathcal{Y}$ est (analytiquement) isomorphe à celle de $\mathfrak{X}/G = \operatorname{Spec} R(\overline{\mathfrak{X}}, \mathfrak{L})^G$.

<u>Remarque</u> : Il est démontré dans **[R]**, qu'avec les notations du théorème précédent, la surface $\mathcal{Y}$ est algébriquement isomorphe à un quotient $\mathfrak{X}/G = \operatorname{Spec} R(\overline{\mathfrak{X}}, \mathfrak{L})^G$ et que l'on peut prendre G abélien (fini).

(5.1.6) Nous voulons donner des exemples de $\mathbb{C}^*$-surfaces normales $\mathcal{Y}$ à singularité rationnelle telles que $\mathfrak{D}(\mathcal{Y})$ ne soit pas simple et pas noethérien. Avant cela nous devons rappeler la définition d'une singularité rationnelle.

Soit $\mathcal{Y}$ une variété irréductible affine et normale. On dit que $\mathcal{Y}$ est à singularités rationnelles ($\mathcal{Y}$ est R.S. pour simplifier) si pour une, et alors toute, résolution des singularités $\pi : \tilde{\mathcal{Y}} \to \mathcal{Y}$ on a $R^j\pi_* \mathcal{O}_{\tilde{\mathcal{Y}}} = O$ pour $j \geq 1$. C'est une condition locale et analytique.

Lorsque $\mathcal{Y}$ est une surface cette condition s'écrit $\dim H^1(\tilde{\mathcal{Y}}, \mathcal{O}_{\tilde{\mathcal{Y}}}) = O$.

(5.1.7) Exemple. Considérons $\mathcal{X} = \operatorname{Spec} R(\bar{\mathcal{X}}, \mathcal{L})$ comme au § 1. On a la résolution : $\pi : \mathbb{V} \to \mathcal{X}$ de (1.1.5). Utilisant la projection $p : \mathbb{V} \to \bar{\mathcal{X}}$, on obtient :

$$H^1(\mathbb{V}, \mathcal{O}_{\mathbb{V}}) = \bigoplus_{j \geq 0} H^1(\bar{\mathcal{X}}, \mathcal{L}^j).$$

Ainsi $\mathcal{X}$ est R.S si et seulement si $\bar{\mathcal{X}} = \mathbb{P}^1$.

(5.1.8) Soit $\mathcal{Y}$ une $\mathbb{C}^*$-surface normale, on peut décider à l'aide de (5.1.3) si $\mathcal{Y}$ est R.S. En effet par **[P.2]** :

(a) Si $\mathcal{Y}$ est une singularité cyclique quotient $\mathcal{Y}$ est R.S.

(b) Sinon, $\mathcal{Y}$ est R.S. si et seulement si la courbe centrale est rationnelle et $\forall\, k > O$, $kb - \sum_{i=1}^{n} \{\frac{ke_i}{d_i}\} > -2$ (où $\{a\}$ est le plus petit entier $\geq a$).

On pourra également consulter [Wa] pour une autre approche de ce critère.

(5.1.9) Grâce à **[P.2]** on dispose d'une méthode pour construire les singularités de $\mathbb{C}^*$-surfaces normales, dont le graphe de la résolution est comme en (5.1.3). Résumons cette construction. Soient E une courbe projective lisse, D un diviseur sur E, $P_1, \ldots, P_n$ des points de E, $(e_i, d_i)_{1 \leq i \leq n}$ des couples d'entiers premiers entre eux avec $1 \leq e_i < d_i$. Formons les diviseurs :

$$D^{(k)} = kD - \sum_{i=1}^{n} \{\frac{ke_i}{d_i}\} P_i, \ k \in \mathbb{N}.$$

Notons $b = \deg D = \deg \mathcal{O}_E(D)$, alors si $b - \sum_{i=1}^{n} \frac{e_i}{d_i} > O$, l'anneau $A = \bigoplus_{k \geq 0} H^\circ(E, \mathcal{O}_E(D^{(k)}))$ est tel que $\mathcal{Y} = \operatorname{Spec} A$ et est une $\mathbb{C}^*$-surface normale ayant le graphe demandé. Une description analogue dûe à M. Demazure est donnée dans [Wa].

5.2. Compte tenu de (5.1.5), l'étude des anneaux d'opérateurs différentiels $\mathcal{D}(\mathcal{Y})$ pour $\mathcal{Y}$ $\mathbb{C}^*$-surface normale (tout au moins des germes d'opérateurs différentiels en $\{O\}$) se décompose en l'étude de $\mathcal{D}(\mathcal{X})$, $\mathcal{X} = \operatorname{Spec} R(\bar{\mathcal{X}}, \mathcal{L})$, suivie de l'action d'un groupe fini opérant librement en dehors de $\{O\}$. Nous allons détailler ce dernier point, l'étude de $\mathcal{D}(\mathcal{X})$ ayant été (partiellement) faite aux § 3 et § 4.

Les résultats qui vont suivre apparaîssent dans [L], [I.2] et [I.3] ; toutes les notions utilisées y sont définies.

(5.2.1) PROPOSITION : Soit B une $\mathbb{C}$-algèbre, G un sous-groupe fini de $\operatorname{Aut}_{\mathbb{C}} B$. On pose $A = B^G$, l'anneau des invariants, et on suppose qu'il existe $Z \subset \operatorname{Spec} B$ tel que :

$$\begin{cases} \text{(i)} \ \forall\, p \in Z \ , \operatorname{prof} B_p \geq 2 \\ \text{(ii) Le morphisme naturel } \operatorname{Spec} B \to \operatorname{Spec} A \text{ est étale en dehors de } Z . \end{cases}$$

Alors pour tout $m \in \mathbb{N}$ on a des isomorphismes naturels (induits par la restriction) :

$\mathrm{Der}^m(A) \cong (\mathrm{Der}^m(B))^G$

$\mathcal{D}^m(A) \cong \mathcal{D}^m(B)^G$.

Et ainsi $\mathcal{D}(A) \cong \mathcal{D}(B)^G$.

On rappelle que G opère sur $\mathcal{D}(B)$ par :

$$\forall\ g \in G, \forall\ D \in \mathcal{D}(B),\ \forall\ \varphi \in B,\ (gD) * \varphi = g[D * (g^{-1}\varphi)].$$

(5.2.2) COROLLAIRE Soient B, G comme en (5.2.1) ; on suppose de plus que B est une $\mathbb{C}$-algèbre de type fini ayant la propriété (S_2) : $\forall\ p \in \mathrm{Spec}\ B$, $\mathrm{prof}\ B_p \geq \inf(2, \mathrm{htp})$. Pour $p \in \mathrm{Spec}\ B$ on pose : $G_i(p) = \{g \in G/g$ induit l'identité sur $B/p\}$.

On suppose que $Z = \{p \in \mathrm{Spec}\ B/\ G_i(p) \neq 1\}$ vérifie $\mathrm{codim}_{\mathrm{Spec}\ B}\ Z \geq 2$. Alors :

$$\forall\ m \in \mathbb{N},\ \mathrm{Der}^m(B^G) \cong [\mathrm{Der}^m(B)]^G \text{ et donc}$$

$$\mathcal{D}(B)^G \cong \mathcal{D}(B^G).$$

__Preuve__ : L'hypothèse faite sur Z assure que Spec B $\to$ Spec A est étale en dehors de Z ; comme B vérifie (S_2) on a $\mathrm{prof}\ B_p \geq 2$ pour $p \in Z$ et on applique (5.2.1).

(5.2.3) Rappelons quelques résultats bien connus sur l'action d'un groupe fini sur une $\mathbb{C}$-algèbre, cf **[Mo]**. Nous les écrirons dans le cadre de ce qui précède. Soient B une $\mathbb{C}$-algèbre de type fini et $G \subset \mathrm{Aut}_{\mathbb{C}}\ B$. On suppose B intègre. Alors :

(a) $\mathcal{D}(B)$ est un anneau simple si et seulement si $\mathcal{D}(B)^G$ l'est.

(b) Si $\mathcal{D}(B)^G$ est noethérien alors $\mathcal{D}(B)$ est un $\mathcal{D}(B)^G$-module de type fini (à droite et à gauche). En particulier $\mathcal{D}(B)$ est alors noethérien et si $\mathcal{D}(B)^G$ est de plus une $\mathbb{C}$-algèbre de type fini c'est aussi le cas pour $\mathcal{D}(B)$.

(c) Inversement si $\mathcal{D}(B)$ est noethérien $\mathcal{D}(B)^G$ l'est aussi.

(5.2.4) On peut maintenant donner une description des opérateurs différentiels sur une $\mathbb{C}^*$-surface normale, ceci grâce à (5.1.5).

THEOREME : Soient $\bar{\mathfrak{X}}$, $\mathfrak{D}$, G comme en (5.1.5). On pose $\mathfrak{X} = \mathrm{Spec}\ R(\bar{\mathfrak{X}}, \mathfrak{D})$, alors :

(a) $\mathcal{D}(\mathfrak{X})^G \cong \mathcal{D}(\mathfrak{X}/G)$

(b) $\mathcal{D}(\mathfrak{X}/G)$ est un anneau noethérien si et seulement si $\mathcal{D}(\mathfrak{X})$ l'est.

(c) $\mathcal{D}(\mathfrak{X}/G)$ est un anneau simple si et seulement si $\mathcal{D}(\mathfrak{X})$ l'est.

__Preuve__ (a) On vérifie aisément que (5.2.2) s'applique à $B = R(\bar{\mathfrak{X}}, \mathfrak{D})$.

(b) et (c) sont conséquences de (5.2.3).

(5.2.5) On conserve les notations de (5.2.4). De (3.2.2) on déduit :

COROLLAIRE : Si le genre de $\bar{\mathfrak{X}}$ est ≥ 1, $\mathcal{D}(\mathfrak{X})$, et donc $\mathcal{D}(\mathfrak{X}/G)$, n'est pas simple.

(5.2.6) Si $\overline{\mathfrak{X}} = \mathbb{P}^1$ et $\mathfrak{L} = \mathcal{O}_{\mathbb{P}^1}(n)$ on a vu en (1.4.2) que $R(\overline{\mathfrak{X}}, \mathfrak{L}) = \mathbb{C}[X,Y]^{\mathbb{Z}_n}$, la description de l'action de $\mathbb{Z}_n$ donnée alors permet d'utiliser (5.2.2) et (5.2.3) (ou même (5.2.4)), on en déduit que $\mathcal{D}(\mathfrak{X})$ est une $\mathbb{C}$-algèbre noethérienne qui est simple ; on peut voir facilement que c'est une $\mathbb{C}$-algèbre de type fini. Il en découle que $\mathcal{D}(\mathfrak{X}/G)$ est un anneau simple noethérien (et même une $\mathbb{C}$-algèbre de type fini).

Notons que ceci traite le cas des singularités cycliques quotients et justifie le fait que l'étude de $\mathcal{D}(\mathfrak{X})$ ait été menée sous l'hypothèse $g \geq 1$ aux § 3 et § 4.

Il faut noter que l'anneau des opérateurs différentiels sur $\mathbb{C}[X,Y]^G$, G groupe fini, a été traité pour la première fois par J.M. Kantor (cf. [L]).

(5.2.7) Lorsque $R(\overline{\mathfrak{X}}, \mathfrak{L})$ est un anneau de Gorenstein on a :

COROLLAIRE : On suppose $g \geq 1$ et qu'il existe N tel que $\mathfrak{L}^N \cong \Omega$. Si $g = 1$ ou sous l'hypothèse (4.4.7), on a :

(a) $\mathcal{D}(\mathfrak{X})$ et $\mathcal{D}(\mathfrak{X}/_G)$ ne sont pas noethériens.

(b) $\mathcal{D}(\mathfrak{X})$ n'est pas une $\mathbb{C}$-algèbre de type fini.

<u>Preuve</u> : (a) résulte de (4.3.3) et (4.4.11) suivis de (5.2.4).

(b) résulte de (4.3.4) et (4.4.10).

<u>Remarque</u> : Il est probable que $\mathcal{D}(\mathfrak{X}/_G)$ n'est pas une $\mathbb{C}$-algèbre de type fini.

(5.2.8) Rappelons que le corollaire (5.2.7) s'applique dans les cas suivants :

(a) $\overline{\mathfrak{X}}$ elliptique $\mathfrak{L}$ quelconque.

(b) $\overline{\mathfrak{X}}$ quelconque de genre ≥ 1, $\mathfrak{L} = \Omega$.

(c) $\overline{\mathfrak{X}} = \operatorname{Proj} \dfrac{\mathbb{C}[X,Y,Z]}{(F)}$, deg X = deg Y = deg Z = 1, F polynôme homogène et

$$\mathfrak{L} = \mathcal{O}_{\mathbb{P}^2}(1)|_{\overline{\mathfrak{X}}} \quad \text{de sorte que} \quad R = \frac{\mathbb{C}[X,Y,Z]}{(F)}.$$

Le cas (a) généralise donc [B.G.G], le cas (c) avait déjà été traité dans [V].

<u>5.3</u> Nous allons appliquer les résultats de <u>5.2</u> pour donner des exemples de $\mathbb{C}^*$-surfaces à singularités rationnelles telles que l'anneau des opérateurs différentiels associé ne soit ni simple ni noethérien. Avant de passer aux exemples expliquons rapidement la raison de cette question : comme nous l'avons noté dans l'introduction tous les exemples connus de variétés $\mathcal{Y}$ telles que $\mathcal{D}(\mathcal{Y})$ ait un "bon" comportement sont R.S., de plus la discussion de <u>5.2</u> a montré que si $\mathfrak{X} = \operatorname{Spec} R(\overline{\mathfrak{X}}, \mathfrak{L})$ $\mathcal{D}(\mathfrak{X})$ est simple si et seulement si $\overline{\mathfrak{X}} = \mathbb{P}^1$ c'est à dire si et seulement si $\mathfrak{X}$ est R.S. (cf. (5.1.7)). Il est donc tentant de voir si cette condition géométrique est suffisante pour assurer un "bon" anneau d'opérateurs différentiels.

(5.3.1) Nous allons reprendre un des exemples de [P.2].

On part de $E = \mathbb{P}^1$, $n = 4$, $e_i = 1$, $d_i = 2$, $1 \le i \le 4$, $D = 3P_4$ et $b = 3$, dans les notations de (5.1.9) ; on fixe $a \in \mathbb{C} \setminus \{O,1\}$, $P_1 = O$, $P_2 = 1$, $P_3 = a$, $P_4 = \infty$. On trouve un anneau $A = \bigoplus_{k\ge 0} H^\circ(E,\mathcal{O}_E(D^{(k)}))$ ayant la présentation $A \cong \frac{\mathbb{C}[X_1, X_2, X_3, Y_1, Y_2]}{J}$, $\deg X_j = 2$, $\operatorname{Deg} Y_i = 3$, J étant l'idéal engendré par les mineurs 2×2 de la matrice

$$\begin{pmatrix} X_1 & X_2 & Y_1 \\ X_2 & X_3 & Y_2 \\ Y_1 & Y_2 & (X_3-X_2)(X_2-aX_1) \end{pmatrix}$$

Le critère de (5.1.8) montre que Spec A est R.S. Pour appliquer les résultats de 5.2 il nous faut montrer que $A \cong R^G$ avec $R = R(\bar{\mathfrak{X}}, \mathfrak{L})$ tel que $\mathcal{D}(R)$ ne soit ni simple ni noethérien. En examinant la démonstration de [P.2] cela n'est pas difficile et on obtient la description suivante.

Soient : $\bar{\mathfrak{X}}$ la courbe elliptique d'équation dans $\mathbb{P}^2$: $Y^2 Z = X(X-Z)(X-aZ)$.

$Q_1 = (O,O,1)$, $Q_2 = (1,O,1)$, $Q_3 = (a,O,1)$, $Q_4 = (O,1,O)$, ce sont des points sur $\bar{\mathfrak{X}}$.

Δ le diviseur $5Q_4 - Q_1 - Q_2 - Q_3$, $\mathfrak{L} = \mathcal{O}_{\bar{\mathfrak{X}}}(\Delta)$.

G le sous-groupe de Aut $\bar{\mathfrak{X}}$ défini par $(\alpha, \beta, \gamma) \mapsto (\alpha, -\beta, \gamma)$, donc $G \cong \mathbb{Z}_2$.

On a : $R(\bar{\mathfrak{X}}, \mathfrak{L}) = R = \dfrac{\mathbb{C}[U, Z_1, Z_2]}{(U^2 - Z_1 Z_2^3 + (a+1) Z_1^2 Z_2^2 - a Z_1^3 Z_2)}$

avec $\deg U = 2$, $\deg Z_i = 1$. Le groupe G opère sur R par $(U, Z_1, Z_2) \mapsto (-U, -Z_1, -Z_2)$. On vérifie que $A \cong R^G$.

Comme $\bar{\mathfrak{X}}$ est elliptique $\mathcal{D}(R)$ et $\mathcal{D}(A)$ ne sont ni simples ni noethériens ; on a vu que Spec A est R.S.

(5.3.2) L'exemple suivant est tiré de ([W.1] § 4.8).

Soient : $E = \mathbb{P}^1$, $n = 3$, $e_i = 1$, $d_i = 3$, $1 \le i \le 3$, $P_1 = O$, $P_2 = 1$, $P_3 = \infty$, $D = 2P_3$, notations (5.1.9).

L'algèbre $A = \bigoplus_{k\ge 0} H^\circ(E,(D^{(k)})$ est donnée par :

$$A \simeq \frac{\mathbb{C}[U_o, U_1, V_1, V_2, V_3, V_4]}{J}, \quad \deg U_i = 2, \deg V_j = 3$$

où J est l'idéal engendré par les mineurs 2×2 de

$$\begin{pmatrix} V_1 & V_2 & V_3 & U_o \\ V_2 & V_3 & V_4 & U_1 \\ U_o^2 & U_o U_1 & U_1^2 & V_3 - V_2 \end{pmatrix}$$

Par (5.1.8) Spec A est R.S.

Soit maintenant $\bar{\mathfrak{X}}$ la courbe elliptique d'équation $X^3 + Y^3 + Z^3 = O$ dans $\mathbb{P}^2$; on considère les points de $\bar{\mathfrak{X}}$: $Q_1 = (O, -1, 1)$ $Q_2 = (O, -\omega, 1)$, $Q_3 = (O, -\omega^2, 1)$ ω étant une racine primitive 3-ième de l'unité.

Si $\mathfrak{L} = \mathcal{O}_{\bar{\mathfrak{X}}}(\Delta)$ et $\Delta = 5Q_3 - Q_1 - Q_2$ on a :

$$R(\bar{\mathfrak{X}}, \mathfrak{L}) = R \cong \frac{\mathbb{C}[X,Y,Z]}{(X^3 + Y^3 + Z^3)}, \deg X = \deg Y = \deg Z = 1.$$

Le groupe $G = \mathbb{Z}_3$ opère sur $\mathfrak{X}$ par $(\alpha, \beta, \gamma) \mapsto (\omega\alpha, \beta, \gamma)$, et donne l'action sur R : $(X,Y,Z) \mapsto (\omega^2 X, \omega Y, \omega Z)$.

On montre que $R^G \cong A$, comme $\mathcal{D}(R)$ n'est ni simple ni noethérien ; il en est de même pour $\mathcal{D}(A)$ et on a vu que Spec A est R.S.

REFERENCES

[B.G.G] J.M. BERNSTEIN, I.M. GELFAND and S.I. GELFAND :*Differential operators on the cubic cone.* Russian Math. Surveys, 27 (1972) 466-488

[E.G.A.] A. GROTHENDIECK : *Eléments de Géométrie Algébrique. Chapitres II et IV.* Publ. I.H.E.S. n° 8 et 32 (1961 et 1967)

[G.W.] S. GOTO and K. WATANABE : *On graded Rings I.* J. Math. Soc. Japan. Vol 3O n° 2 (1978) 179-213.

[H] R. HARTSHORNE : *Algebraic Geometry*. Graduate Texts in Math. Vol 52 (1983) Springer Verlag

[I.1.] Y. ISHIBASHI : *Remarks on a Conjecture of Nakai.* J. of Algebra 95 (1985) 31-45.

[I.2.] Y. ISHIBASHI : *Nakai's Conjecture for invariant subrings.* Hiroshima Math. J. 15 (1985) 429-436.

[I.3.] Y. ISHIBASHI : *A Note on Nakai's Conjecture for invariant subrings.* Comm. in Algebra 15 (1987) 1349-135

[L] T. LEVASSEUR : *Anneaux d'opérateurs différentiels.* Séminaire d'Algèbre M.P. Malliavin L.N. in Math. Vol 867 Springer Verlag (1981) 157-173.

[L.Sm] T. LEVASSEUR and S.P. SMITH : *Primitive ideals and nilpotent orbits in type G_2.* J. of Algebra 114 n° 1 (1988) 81-1O5.

[L.S.S] T. LEVASSEUR, S.P. SMITH and J.T. STAFFORD :*The minimal nilpotent orbit,the Joseph Ideal and differential operators.* J. of Algebra, to appear

[L.St] T. LEVASSEUR and J.T. STAFFORD : *Rings of differential operators on classical rings of invariants.* Preprint, (1988)

[Mo] S. MONTGOMERY : *Fixed rings of Finite Automorphism groups of Associative Rings .* L.N. in Maths. Vol 818. Springer Verlag (198O)

[Mu] I.M. MUSSON : *Rings of Differential operators on Invariant Rings of Tori.* Trans. of the Amer. Math. Soc. Vol 3O3, n° 2. (1987), 8O5-827.

[O.W] P. ORLIK and P. WAGREICH : *Isolated singularities of algebraic surfaces with $\mathbb{C}^*$-action.* Ann. of Math. 93 (1971) 2O5-228.

[P.1.] H.C. PINKHAM : *Deformations of Cones with negative Grading.* J. of Algebra 30 (1974) 92-102.

[P.2.] H.C. PINKHAM : *Normal surface singularities with* $\mathbb{C}^*$*-action.* Math. Annalen 227 (1977) 183-193.

[R] B. RUNGE : *Quasihomogeneous Singularities.* Math. Annalen. 281 (1988), 295-313.

[S.S] S.P. SMITH and J.T. STAFFORD : *Differential operators on an affine curve.* Proc. London Math. Soc., to appear

[V] J.P. VIGUE : *Opérateurs différentiels sur les cônes normaux de dimension 2.*

C.R. Acad. Sci. Paris tome 278 (1974) 1047-1050.

[W.1] J. WAHL : *Equations defining Rational Singularities.* Ann. Sc. Ec. Normale Sup. 10 (1977) 231-264.

[W.2] J. WAHL : *Derivations of Negative Weight and Non-Smoothability of Certain Singularities.* Math. Ann. 258 (1982) 383-398.

[W.3] J. WAHL : *A cohomological characterization of* $\mathbb{P}^n$. Inv. Math. 72 (1983) 315-322.

[Wa] K. WATANABE : *Some remarks concerning Demazure's Construction of normal graded rings.*

Nagoya Math. J. vol 83 (1981) 203-211.

Strongly Filtered Rings Applied to Gabber's Integrability Theorem and Modules with Regular Singularities

Li Huishi*
Dept. of Mathematics
Shaanxi Normal University, Xian, P. R. China.

Freddy Van Oystaeyen
Dept. of Mathematics
University of Antwerp, UIA, Antwerp, Belgium.

[*This paper is in final form, and no version of it will be submitted for publication elsewhere.*] .

0. Introduction.

In the theory of rings of differential operators, $\mathcal{D}$ and $\mathcal{E}$-modules, several authors have tried to provide purely algebraic techniques in order to deal with some of the less technical problems. On one hand O. Gabber's theorem was already purely algebraic in nature and the methods appearing in this proof have been pushed further e.g. by J. E. Björk cf. [3] so they can be applied to more general filtered rings. On the other hand part of the micro-local analysis of [6] has been replaced by algebraic methods by A. Van den Essen in [11] using the forementioned theorem of Gabber and an algebraic treatment of microlocalization as developed in [10]. In [0] a completely algebraic theory of microlocalization was provided by making use of the Rees ring of the filtration in a more elaborate way; it turned out that the theory works very well under the assumption that the Rees ring of the filtration is Noetherian, a condition that is weaker than the Σ-Noetherian condition appearing in [10],

---

* This author is a research fellow at UIA, supported by a grant from the Province of Antwerp, jumelated to Shaanxi province.

[11] or the Zariskian condition appearing in [3]. As a consequence of this we can avoid certain completeness conditions or certain closure properties in filtered modules and this in turn makes it possible to present the theory in a very unified way by using strongly filtered rings. The latter class of rings includes the E-rings used in [11], in fact "microlocally" the rings we consider will be E-rings but whereas E-rings only occur "microlocally" in practice, the strongly filtered ones appear more often (and naturally) and they also constitute a general class of filtered rings having an interest in its own right. We have tried to make this paper as self-contained as possible so that it is semi-expository in nature. After the prelininaries, and the general theory of strongly filtered rings in Section 2, we revisite Gabber's theorem in Section 3; modules with regular singularities over strongly filtered rings are the topic of Section 4. After linking arbitrarily filtered rings to certain strongly filtered rings in Section 5, we can turn to the study of modules with regular singularities over more general filtered rings in the final Section 6.

1. Preliminaries.

Let R be a filtered ring with filtration $FR = \{F_nR, n \in \mathbb{Z}\}$, i.e. an ascending chain : $\ldots \subset F_nR \subset F_{n+1}R \subset \ldots$, with $F_nR.F_mR \subset F_{n+m}R$. A filtered R-module M is an R-module M together with a filtration $FM = \{F_nM, n \in \mathbb{Z}\}$ i.e. an ascending chain : $\ldots F_nM \subset F_{n+1}M \subset \ldots$, with $F_nRF_mM \subset F_{n+m}M$ for all $n, m \in \mathbb{Z}$. In this paper we only consider exhaustive filtrations i.e. we always assume that $R = \cup_n F_nR$, $M = \cup_n F_nM$. As in [7] we denote the category of filtered left R-modules by R-filt. The associated graded ring $G(R) = \oplus_{n\in\mathbb{Z}} F_nR/F_{n-1}R$ is graded by putting $G(R)_n = F_nR/F_{n-1}R$ and the associated graded module $G(M)$ defined by $G(M) = \oplus_{n\in\mathbb{Z}} F_nM/F_{n-1}M$ is a graded $G(R)$-module. On a filtered R-module M we may define an order function v_F associated to the filtration F as follows : $v_F(m) = -\infty$ whenever $m \in \cap_{n\in\mathbb{Z}} F_nM$, $v_F(m) = n$ if $m \in F_nM - F_{n-1}M$. The symbol-map σ is defined by putting $\sigma(m) = 0$ if $v(m) = -\infty$ and $\sigma(m) = m + F_{n-1}M$ in $G(M)_n$ if $\sigma(m) = n \in \mathbb{Z}$. To the filtration FM we can associate the pseudo-norm $|m|_{FM} = 2^{v_F(m)}$ on M and similarly we define $|\ \ |_{RF}$ on R. A filtration FM is separated if $\cap_n F_nM = 0$ and if FM is separated then $|\ \ |_{FM}$ is a norm on M. To a filtered ring R we associate the Rees ring $\widetilde{R} = \oplus_{n\in\mathbb{Z}} F_nR$ with obvious addition and multiplication and it is clear that $\widetilde{R}$ is a $\mathbb{Z}$-graded ring. The element $1 \in F_1R$ viewed as an element of $\widetilde{R}_1$ is denoted by X; then we may identify $\widetilde{R}$ and the subring $\sum_{n\in\mathbb{Z}} F_nRX^n$ in $R[X,X^{-1}]$. If $M \in R$-filt then $\widetilde{M} = \oplus_{n\in\mathbb{Z}} F_nM$ is in the obvious way a graded $\widetilde{R}$-module and it may be viewed as $\sum_{n\in\mathbb{Z}} F_nM.X^n$ in $M[X,X^{-1}] = R[X,X^{-1}]\otimes_R M$. The following easy lemmas contain some very basic observations :

1.1. Lemma. (cf. [0]) With notations as introduced above :

1. $\widetilde{R}/\widetilde{R}X \cong G(R)$, $\widetilde{M}/X\widetilde{M} \cong G(M)$.

2. $\widetilde{R}/\widetilde{R}(1-X) \cong R$, $\widetilde{M}/(1-X)\widetilde{M} \cong M$.

3. If $\widetilde{R}_X$ denotes the localization of $\widetilde{R}$ at the multiplicatively closed central set $\{1, X, X^2, \ldots, X^n, \ldots\}$ then $\widetilde{R}_X \cong R[X,X^{-1}]$; also $\widetilde{M}_X \cong M[M,X^{-1}]$.

4. Let $\mathcal{F}_X$ be the full subcategory of $\widetilde{R}$-gr consisting of the X-torsion free graded $\widetilde{R}$-(left) modules then $\sim$: R-filt $\to \widetilde{R}$-gr, $M \to \widetilde{M}$, defines an equivalence of categories R-filt $\xrightarrow{\sim} \mathcal{F}_X$.

5. The filtration FM is separated if and only if the $\widetilde{R}X$-adic filtration on $\widetilde{M}$ is separated.

6. An $M \in R$-filt has a good filtration FM if and only if $\widetilde{M}$ is finitely generated in $\widetilde{R}$-gr. Recall that FM is said to be good if M is filt-finitely generated in the sense of [7], i.e. if there exist R-generators $m_1, \ldots, m_d$ such that for all $n \in \mathbb{Z}$, $F_nM = F_{n-n_1}R.m_1 + \ldots + F_{n-n_d}R.m_d$ for some $n_1, \ldots, n_d \in \mathbb{Z}$.

1.2. Lemma. (cf. [0]) For a filtered ring R, the following statements are equivalent :

1. $F_{-1}R \subset J(F_0R)$, where $J(F_0R)$ is the Jacobson radical of F_0R.

2. $X \in J^g(\widetilde{R})$, where $J^g(\widetilde{R})$ is the gr-Jacobson radical of $\widetilde{R}$ in the sense of [7].

3. If $\widetilde{M} \in \widetilde{R}$-gr is finitely generated then $X\widetilde{M} = \widetilde{M}$ if and only if $\widetilde{M} = 0$.

4. If $\widetilde{M} \in \mathcal{F}_X$ is finitely generated then $X\widetilde{M} = \widetilde{M}$ if and only if $\widetilde{M} = 0$.

5. If $M \in R$-filt has good filtration then $G(M) = 0$ if and only if $M = 0$.

6. FR is a faithful filtration (one can use 5. for a definition of this notion).

Now let FR be the filtration on the filtered ring R. Let $\sigma(S) \subset (G(R))$ be a multiplicative closed subset in $G(R)$ consisting of homogeneous elements and such that $0 \notin \sigma(S)$, then the set $S = \{s \in R, \sigma(s) \in \sigma(S)\}$ is multiplicatively closed in R! If $\sigma(S)$ is a (left) Ore set of $G(R)$ then S need not be a (left) Ore set of R; in [10] it has been shown that S satisfies some kind of approximative Ore conditions and one can define microlocalization from this point of view. Another, even more purely algebraic approach has been provided in [0]. The microlocalization $Q_S^\mu(R)$ of R, resp. $Q_S^\mu(M)$ of $M \in R$-filt, can be defined such that it has the following properties.

ML1. $Q_S^\mu(R)$ resp. $Q_S^\mu(M)$, is a complete separated filtered ring, resp. filtered $Q_S^\mu(R)$-module, and there is a canonical morphism of filtered rings $\phi : R \to Q_S^\mu(R)$, resp. morphism of filtered modules $\phi_M : M \to Q_S^\mu(M)$

ML2. For every filtered morphism $h : R \to R'$, resp. for every filtered morphism of filtered R-modules $h_M : M \to M'$, such that FR', resp. FM', is complete and separated and such that for every $s \in S$, $h(s)$ is invertible in R' with $h(s)^{-1} \in F_{-n}R'$ if $\sigma(s) \in G(R)_n$, (this condition may be expressed in pseudo-norms as follows : $|h(s)^{-1}|_{R'} \leq |s|_R^{-1}$) there is a unique morphism of filtered rings $\chi : Q_S^\mu(R) \to R'$ satisfying $\chi\phi = h$, resp. $\chi_M : Q_S^\mu(M) \to M'$ satisfying $\chi_M \phi_M = h_M$.

The foregoing properties characterize microlocalization at S. A further property (it follows from the others) is that $\sigma(S)^{-1}G(R) \cong G(Q_S^\mu(R))$ and in fact the isomorphism may be given by $\Psi_R : \sigma(S)^{-1}G(R) \to G(Q_S^\mu(R))$, $\Psi_R(\sigma(s)^{-1}\sigma(a)) = \phi(s)^{-1}\phi(a) + F_{n-1}Q_S^\mu(R)$ if $\sigma(s)^{-1}\sigma(a) \in (\sigma(S)^{-1}G(R))_n$. For a filtered R-module M we have $\Psi_M : \sigma(S)^{-1}G(M) \to G(Q_S^\mu(M))$ defined by $\Psi_M(\sigma(s)^{-1}\sigma(m) = \phi(s)^{-1}\phi(m) + F_{n-1}Q_S^\mu(M)$ for all $\sigma(s)^{-1}\sigma(m) \in (\sigma(S)^{-1}G(M))_n, n \in \mathbb{Z}$, and Ψ_M is an isomorphism of graded $\sigma(S)^{-1}G(R)$-modules. For further properties and detail concerning the functor $Q_S^\mu(-)$ we refer to [0] where the theory is expounded under the weakest Noetherien hypotheses.

2. Strongly Filtered Rings

The properties of strongly graded rings, cf. [7], may be compared to those of strongly filtered rings defined below. It turns out that these classes of rings have very similar properties and up to a few subtle points where the difference between filtrations and gradations really plays, this section may be considered as a rather straightforward adaption of the corresponding graded theory. Including it here, makes the paper more self-contained also for those readers not so familiar with the standard tricks concerning strongly graded rings.

2.1. Definition. Let R be a filtered ring, then R is said to be strongly filtered (or FR is said to be strong) if $F_nRF_mR = F_{n+m}R$ for $n, m \in \mathbb{Z}$.

2.2. Examples. 1. Let R be a subring of R', I an ideal of R. We say that I is **invertible** in R' if there is an R-bimodule J in R' such that $JI = IJ = R$; we put $J = I^{-1}$. We now define FR' on R' by taking $F_nR' = I^n, n \in \mathbb{Z}$, where the negative powers of I correspond to the suitable powers of J, and $F_0R' = R$. In particular, if R' is a field and R a discrete valuation ring of R' with maximal ideal I then FR' is an exhaustive and strong filtration on the field R' !

2. A filtered ring R is said to be an ***E*-ring** in the sense of [10] if it satisfies the following conditions :

E.1. There is an element $s \in F_1R - F_0R$ invertible in R with $s^{-1} \in F_{-1}R$.

E.2. $G(R)$ is a commutative $\mathbb{Q}$-algebra

E.3. FR is Σ-Noetherian (see [10] for this definition).

The condition **E.1.** alone makes an E-ring into a strongly filtered ring !

3. Filtered rings with superradical filtration in the sense of [1], we refer to loc. cit. for some detail on the so-called radical filtrations (this is not a necessity for the sequel).

One easily verifies the following properties :

2.3. Some Properties. Let R be a strongly filtered ring.

1. For all $n \in \mathbb{Z}$, F_nR is a finitely generated projective F_0R-module.

2. For every $M \in R$-filt and all $n, m \in \mathbb{Z}$: $F_nRF_mM = F_{n+m}M$, i.e. every $M \in R$-filt is strongly filtered (definition is exactly given by the foregoing property).

3. If H is any nonzero R-submodule of a filtered R-module M then $F_mM \cap H \neq 0$ for all $m \in \mathbb{Z}$. If we consider the induced filtration FH given by $F_nH = H \cap F_nM$ then $F_nH = F_nRF_0H$ hence $H = RF_0H$.

4. Let $M \in R$-filt. Clearly F_0M is filtered by $\{F_nM, n \leq 0\} = F^0M$. Define a functor

$$R \bigotimes_{F_0R} - : F_0R\text{-filt} \to R\text{-filt}, M_0 \mapsto R \bigotimes_{F_0R} M_0$$

where $F_n(R \otimes_{F_0R} M_0) = F_nR \otimes M_0$ for all $n \in \mathbb{Z}$. A morphism $g : M_0 \to M_0'$ in F_0R-filt yields $1 \otimes g : R \otimes_{F_0R} M_0 \to R \otimes_{F_0R} M_0'$ in R-filt.
(Note : $F_nR \otimes_{F_0R} M_0$ may be considered as an F_0R-submodule of $R \otimes_{F_0R} M_0$, indeed). On the other hand, the restriction functor $(-)_0$ may be defined by,

$(-)_0 : R$-filt $\to F_0R$-filt, $M \mapsto F_0M$ where a filtered morphism (of degree zero, of course) $f : M \to M'$ restricts to $f_0 : F_0M \to F_0M'$ in F_0R-filt.

2.4. Theorem. For a filtered ring R the following properties are equivalent :

1. R is strongly filtered.

2. $G(R)$ is strongly graded

3. For all $n, m \in \mathbb{Z}$, $F_nRF_mM = F_{n+m}M$ for every $M \in R$-filt.

4. The functors $R \otimes_{F_0R} -$ and $(-)_0$ define an equivalence between the categories R-filt and F_0R-filt.

Proof. 1. ⇒ 2. If R is strongly filtered then $\widetilde{R}$ is strongly graded and thus $\widetilde{R}/X\widetilde{R} \cong G(R)$ is strongly graded too (because $X\widetilde{R}$ is graded).
2. ⇒ 1. From $\widetilde{R}_1\widetilde{R}_{-1} + X\widetilde{R} \supset F_0R = \widetilde{R}_0$ it follows that $\widetilde{R}_1\widetilde{R}_{-1} + X\widetilde{R}_{-1} = \widetilde{R}_0$ hence $\widetilde{R}_1\widetilde{R}_{-1} = \widetilde{R}_0$ because $X \in \widetilde{R}_1$. Thus $F_1RF_{-1}R = F_0R$, and similarly $F_{-1}RF_1R = F_0R$ holds. That $F_nRF_{-n}R = F_0R$ for all $n \in \mathbb{Z}$ is then also easily derived and then R is clearly strongly filtered.
1. ⇔ 3. is easy, and in fact it follows from 2.3.2. In order to establish the equivalence 3 ⇔ 4 it suffices to establish 3 ⇒ 4. Take $M \in R$-filt, clearly the canonical R-linear morphism

$$\delta_M : R \bigotimes_{F_0R} F_0M \to M, \sum_i r^{(i)} \otimes m^{(i)} \mapsto \sum_i r^{(i)} m^{(i)}$$

is filtered of degree zero and it is strict and surjective (recall that a filtered morphism $f(M) \cap \to N$ is strict if $f(F_nM) = f(M) \cap F_nN$ for $n \in \mathbb{Z}$). Restricting δ_M to $F_0R \otimes_{F_0R} F_0M \to F_0M$ we learn that $\mathrm{Ker}(\delta_M) \cap F_0(R \otimes_{F_0R} F_0M) = 0$ and then it follows again from the property 2.3.2. that δ_M is injective. It is easy to check that δ_M is a natural transform of the composed functor $(R \otimes_{F_0R} -) \circ (-)_0$ to the identity functor on R-filt. Conversely, if we take F_0R-filt and $\alpha_{M_0} : M_0 \to F_0(R \otimes_{F_0R} M_0) = F_0R \otimes_{F_0R} M_0$; $m_0 \mapsto 1 \otimes m_0$ the canonical isomorphism then it is clear that α_{M_0} defines a natural transform between $(-)_0 \circ (R \otimes_{F_0R} -)$ and the identity functor on F_0R-filt.

2.5. Proposition. Let R be a strongly filtered ring, let M, M' be in R-filt and $f' : M \to M'$ a morphism in R-filt.

1. The filtered morphism f is monomorphic if and only if the restriction $f_n : F_nM \longrightarrow F_nM'$ is monomorphic for some (hence all) $n \in \mathbb{Z}$.

2. If f is strict then f is an epimorphism if and only if f_n is an epimorphism for some (hence all) $n \in \mathbb{Z}$.

Proof. 1. It suffices to show $\mathrm{Ker}(f) = 0$ when $\mathrm{Ker}(f_n) = 0$ for some $n \in \mathbb{Z}$ but this follows 2.3.3.

2. Since f is strict $f(F_n M) = f_n(F_n M) = \mathrm{Imf} \cap F_n M'$. If f is surjective then $f_n(F_n M) = F_n M'$ follows. If f_n is surjective then we have $F_n M' \subset \mathrm{Imf}$ hence $F_0 M' = F_{-n} R F_n M' \subset \mathrm{Imf}$ hence $M' = RF_0 M' \subset \mathrm{Imf}$ or f is surjective. □

2.6. Proposition. Let R be a strongly filtered ring, $M \in R$-filt and $T_n(M) \in R$-filt defined by $F_m T_n(M) = F_{n+m} M$.

1. The canonical morphism $R \otimes_{F_0 R} F_n M \longrightarrow T_n(M), r \otimes x \mapsto rx$, is an isomorphism in R-filt for every $n \in \mathbb{Z}$.

2. For every $n, m \in \mathbb{Z}$, $F_n R \otimes_{F_0 R} F_m R \to F_{n+m} R$ is an isomorphism of $F_0 R$-bimodules in $F_0 R$-filt.

Proof. Easy. □

2.7. Remark. If $M \in R$-filt and R is strongly filtered then we let $L(M)$ be the lattice of R-submodules of M, $L(F_0 M)$ the lattice of $F_0 R$-submodules of $F_0 M$. Then $L(M) \to L(F_0 M), H \to H \cap F_0 M$ is an injective lattice morphism. Hence, if $F_0 M$ is a Noetherian (Artinian) $F_0 R$-module then M is a Noetherian (Artinian) R-module.

The following results may be compared to results of [1], resp. Corollary 3.5., Proposition 3.9. and Proposition 3.10. we leave the easy proofs as exercises.

2.8. Lemma. Let R be strongly filtered, $M \in R$-filt. The following statements are equivalent :

1. FM is a good filtration.

2. $F_0 M$ is a finitely generated $F_0 R$-module.

3. $\widetilde{M}_0$ is a finitely generated $\widetilde{R}_0$-module.

2.9. Lemma. Let R be strongly filtered and $M \in R$-filt, then FM is faithful if and only if $F_0 M \in F_0 R$-filt is faithful.

2.10. Proposition. Le R be strongly filtered and assume that $FR \subset J(F_0 R)$ and good induces good filtrations on finitely generated left ideals.

1. If M is finitely generated and $G(M)$ is generated by n elements (resp. Artinian) then M can be generated by n elements (resp. Artinian).

2. If $G(M)$ has Krull dimension then M has Krull dimension and moreover $\mathrm{Kdim}_R M \leq \mathrm{Kdim}_{G(R)} G(M)$.

3. If $\mathrm{Kdim}_R M = \alpha$ and $\mathrm{Kdim}_{G(R)} G(M) = \alpha$ then M is α-critical if $G(M)$ is α-critical.

Proof. Under the condition assumed in the statement of the proposition one easily proves that all good filtered modules M are separated hence FR is faithful, then one can apply [7] Theorem 4.2.6.(4). If S is a multiplicatively closed subset of R, a strongly filtered ring, then $Q_S^{\mu}(R)$ is also strongly filtered (since $GQ_S^{\mu}(R)$ is obviously strongly graded) in particular the completion $F\hat{R}$ of FR is strongly filtered (by taking $S = 1$). Moreover if $F_{-1}R \subset J(F_0R)$ then $F_{-1}Q_S^{\mu}(R) \subset J(F_0Q_S^{\mu}(R))$ and $F_{-1}\hat{R} \subset J(F_0\hat{R})$.

2.11. Corollary. Let R satisfy the condition of Proposition 2.10.

1. If $M \in R$-filt is such that $G(M)$ is Noetherian (Artinian) in $G(R)$-mod then $M, \hat{M}$ and $Q_S^{\mu}(M)$ are all Noetherian (Artinian).

2 If $G(R)$ is gr Artinian gr-simple then R and $\hat{R}$ are Artinian simple. If $\sigma(S)^{-1}G(R)$ is gr-Artinian gr-simple then $Q_S^{\mu}(R)$ is Artinian simple.

3. Let $S = \{s \in R, \sigma(s)$ is homogeneous regular in $G(R)\}$. If $G(R)_0$ is a prime Goldie ring then the microlocalization $Q_R^{\mu}(R)$ of R at S exists and it is an Artinian simple ring.

Proof. 1. and **2.** follow from Proposition 2.10. and [7, Theorem 4.2.6.4.]. For 3 we note that $G(R)$ is stronlgy graded hence it satisfies condition (E) as required in [8], so we may conclude that $\sigma(S)^{-1}G(R)$ is gr-simple gr-Artinian and apply **2.** □

3. A Variation on Gabber's Theorem.

Recall that a **biderivation** on a commutative ring R is a $\mathbb{Z}$-bilinear map $\delta : R \times R \longrightarrow R$ satisfying $\delta(a_1a_2, b) = a_1\delta(a_2, b) + \delta(a_1, b)a_2$, $\delta(a, b_1b_2) = b_1\delta(a, b_2) + \delta(a, b_1)b_2$, for all a_1, a_2, b_1, b_2 in R. An ideal I in R is **δ-stable** if $\delta(a, b) \in I$ for all $a, b \in I$. If R is a filtered ring such that $G(R)$ is commutative then we can define a $\mathbb{Z}$-bilinear map on $G(R)$ as follows (cf. [4]) $\{\ ,\ \} : G(R) \times G(R) \to G(R), (f, g) \mapsto \{f, g\}$, where $f = \sigma(a), g = \sigma(b)$ for some $a \in F_nR, b \in F_mR, \{f, g\} = [a, b] + F_{m+n-2}R$ ($[a, b]$ being the Lie commutator). Clearly $\{\ ,\ \}$ is a biderivation on $G(R)$ and it is called the Poisson product. An ideal I of $G(R)$ is said to be **involutive** if I is $\{\ ,\ \}$-stable. While introducing some more notation we also give a short review of the approach used in [11] but soon we will turn to a more general setting.

Let R be an E-ring with an invertible element $s \in F_1R - F_0R$, $s^{-1} \in F_{-1}R - F_{-2}R$. In [11] an ideal J of $G(R)_0$ is said to be **involutive** if $\sigma(s)\{f,g\} \in J$ for all $f,g \in J$. For an arbitrary F_0R-submodule of M, $M \in R$-filt having a good filtration, then we put, for all $n \in \mathbb{Z}$:

$$Q(n,N) = F_nM \cap N/F_{n-1}M \cap N$$
$$I(n) = \mathrm{Ann}_{G(R)_0}Q(n,N), J(n) = \mathrm{rad}(I(n))$$

Left multiplication by s^{-1} yields an ascending sequence of ideals in $G(R)_0$: $I(1) \subset I(r) \subset \dots$. Since R is an E-ring, $G(R)_0$ is Noetherian and thus $I(n) = I(n_0)$ for all $n \geq n_0$ for some $n_0 \in \mathbb{Z}$. Hence $J(n) = J(n_0)$ for all $n \geq n_0$. We put $J = J(N) = J(n_0) = U_{n\in\mathbb{Z}}(J(n))$.

3.1. Gabber's Theorem. The ideal J, as defined above, is involutive.

Let us now extend the theory from E-rings to certain strongly filtered rings. First we need another definition of involutive.

3.2. Definition. Let R be a filtered ring such that $G(R)$ is commutative. We say that an ideal J of $G(R)_0$ is involutive (new sense !) if $G(R)_1\{J,J\} \subset J$, where $\{\ ,\ \}$ is as before and

$$G(R)_1\{J,J\} = \{a\{f,g\}, a \in G(R)_1, f,g \in J\}$$

We leave to the reader the easy verification that for an E-ring the new definition coincides with the earlier one as used in [11].

3.3. Lemma. Let R be filtered such that $G(R)$ is commutative.

1. Let J be an ideal of $G(R)_0$. If $G(R)J$ is involutive in $G(R)$ then J is involutive in $G(R)_0$; if R is strongly filtered then J is involutive if and only if $G(R)J$ is involutive in $G(R)$.

2. Let I be a graded ideal of $G(R)$. If I is involutive in $G(R)$ then $I \cap G(R)_0$ is involtive in $G(R)_0$; if R is strongly filtered then I is involutive in $G(R)$ if and only if $I \cap G(R)_0$ is involutive in $G(R)_0$.

Proof. 1. Suppose that $G(R)J$ is involutive in $G(R)$ then $G(R)_1\{J,J\} \subset G(R)J \cap G(R)_0 = J$, hence J is involutive in $G(R)_0$. If R is strongly filtered then $G(R)$ is strongly graded, hence $G(R)_{-1}G(R)_1 = G(R)_0$. Let J be involutive in $G(R)$ and fix a decomposition $1 = \sum_i u_iv_i$ with $u_i \in G(R)_{-1}$, $v_i \in$

$G(R)_1$. It follows from : $G(R)\{J,J\} = G(R)\sum_i u_i v_i\{J,J\} \subset G(R)J$, that : $\{G(R)J, G(R)J\} = G(R)(G(R)\{J,J\} + \{J, G(R)\}J) + (G(R)\{G(R), J\} + \{G(R), G(R)\}J)J \subset G(R)J$ (since $\{\ ,\ \}$ is a biderivation on $G(R)$). Therefore $G(R)J$ is involutive in $G(R)$.

2. Let I be involutive in $G(R)$, then $G(R)_1\{I \cap G(R)_0, I \cap G(R)_0\} \subset I \cap G(R)_0$, i.e. $I \cap G(R)_0$ is involutive in $G(R)_0$. If R is strongly filtered and $I \cap G(R)_0$ is involutive then it follows from **1.** that $G(R)(I \cap G(R)_0) = I$ is involutive in $G(R)$. □

In the sequel (of this section) we always assume that R satisfies the following conditions :

(1) R strongly filtered

(2) $G(R)$ is a commutative Noetherian $\mathbb{Q}$-algebra.

3.4. Lemma. Let J be an ideal of $G(R)_0$, then $J \in \text{Spec}(G(R)_0)$ if and only if $G(R)J \in \text{Spec}^{(g)}(G(R))$. If I is a graded ideal of $G(R)$ then $I \in \text{Spec}^{(g)}(G(R))$ if and only if $I \cap G(R)_0 = I_0 \in \text{Spec}(G(R)_0)$, where $\text{Spec}^{(g)}(G(R))$ denotes the graded prime spectrum of $G(R)$.

2. Let J be an ideal of $G(R)_0$ and $p \in \text{Spec}(G(R)_0), p \supset J$. Then p is minimal over J if and only if $G(R)p$ is a minimal prime of $G(R)J$. Let I be a graded ideal of $G(R)$ and q a graded prime ideal of $G(R)$ containing I, then q is minimal over I if and only if $q \cap G(R)_0$ is minimal over I_0.

3. If $M \in R$-filt has good filtration and $J(M)$ is its characteristic ideal, i.e. $J(M) = \text{rad}(\text{Ann}_{G(R)} G(M))$, then $J(M) \cap G(R)_0$ is the radical of $\text{Ann}_{G(R)}(G(M)) \cap G(R)_0$ in $G(R)_0$ and it is involutive in $G(R)_0$.

Proof. 1. and **2.** are easy

3. Follows from Lemma 3.3. and Lemma 3.4. □

Now take $p_0 \in \text{Spec}(G(R)_0), p = G(R)p_0 \in \text{Spec}^{(g)}(G(R))$ and put :

$$\sigma(S_{p_0}) = G(R)_0 - p_0, S_{p_0} = \{s_0 \in F_0 R, \sigma(s_0) \in G(R)_0 - p_0\}$$
$$\sigma(S_p) = G(R) - p, S_p = \{s \in R, \sigma(s) \in G(R) - p\}$$

Note that, if $\sigma(s) \in \sigma(S_p) \cap G(R)_n$ then $G(R)_{-n}\sigma(s) \not\subset p_0$ and since $G(R)_{-n} \not\subset p$ we may take some $\sigma(s') \in \sigma(S_p)$ such that $\sigma(s')\sigma(s) \in G(R)_0 - p_0 = \sigma(S_{p_0})$.

3.5. Proposition. With notations and assumptions as before :

1. For any $s \in S_p$ there is an $s' \in S_p$ such that $s's \in S_{p_0}$.

2. The (left) ring of fractions $\sigma(S_p)^{-1}G(R)$ is also the left (left) ring of fractions with respect to $\sigma(S_{p_0})$.

3. The ring $\sigma(S_p)^{-1}G(R)$ is strongly graded and $(\sigma(S_p)^{-1}G(R))_0 = \sigma(S_{p_0})^{-1}G(R)_0, (\sigma(S_p)^{-1}G(R))_k = \sigma(S_{p_0})^{-1}G(R)_k$ for all $k \in \mathbb{Z}$.

4. The microlocalization $Q^{\mu}_{S_p}(R)$ exists and coincides with $Q^{\mu}_{S_{p_0}}(R)$. Moreover $Q^{\mu}_{S_p}(R)$ is an E-ring and $G(Q^{\mu}_{S_p}(R))$ is a graded-local ring isomorphic to $\sigma(S_p)^{-1}G(R)$, moreover $G(Q^{\mu}_{S_p}(R))_0 \cong (\sigma(S_p)^{-1}G(R))_0 = \sigma(S_{p_0})^{-1}G(R)_0$, $\sigma(Sp_0)^{-1}G(R)_k = (\sigma(S_p)^{-1}G(R))_k = G(Q^{\mu}_{S_p}(R))_k$ for all $k \in \mathbb{Z}$.

Proof. 1., 2. and **3.** are easy.

4. Since there are elements in $\sigma(S_p) \cap G(R)_1$ it follows that $G(Q^{\mu}_{S_p}(R))$ contains a unit of degree 1, and also it is clear that $Q^{\mu}_{S_p}(R)$ is an E-ring (see also [11], Proposition 1,10). □

3.6. Remark. 1. The ring $\sigma(S_{p_0})^{-1}G(R)_0$ is local with unique maximal ideal $p_0\sigma(S_{p_0}^{-1}G(R)_0$.

2. The Poisson product on $G(R)$ may be extended to $\sigma(S_p)^{-1}G(R)$ and the extension coincides with the Poisson product defined in the usual way on $G(Q^{\mu}_{S_p}(R))$. Let us write $p_0^e = \sigma(S_{p_0})^{-1}G(R)_0 p_0$ and assume p_0^e is involutive in $G(Q^{\mu}_{S_p}(R))_0$. If $\nu : G(R) \to \sigma(S_p)^{-1}G(R)$ is the canonical morphhism, then $\nu^{-1}(p_0^e) \subset p_0$ and $\nu(G(R)_1\{p_0, p_0\} \subset (\sigma(S_p)^{-1}G(R))_1\{p_0^e, p_0^e\} \subset p_0^e$, therefore $G(R)_1\{p_0, p_0\} \subset \nu^{-1}(p_0^e) \subset p_0$, i.e, p_0 is involutive in $G(R)_0$.

Let the canonical maps from R to $Q^{\mu}_{S_p}(R)$, resp. M to $Q^{\mu}_{S_p}(M)$, both be denoted by ϕ_p (this will not give rise to ambiguity). We recall a few statements from [11] (even if these may be reformulated in the formalism of [0] we stick here to the original formulation). The set $\{\phi_p(s)^{-1}\phi_p(m)$, $(s, m) \in S_p \times M\}$ is dense in $Q^{\mu}_{S_p}(M)$ in the $|\ \ |_{M,S_p}$-topology, where $|\ \ |_{M,S_p}$ is the localized pseudo-norm associated to the filtration $FQ^{\mu}_{S_p}(M)$, i.e. $|m|_{M,S_p} = \inf_{s \in S_p} (s)_R^{-1}|sm|_M$ where $|\ \ |_R$ resp. $|\ \ |_M$ is the pseudo-norm associated to FR, resp. FM.

Observation. 1. If $|m|_{M,S_p} \neq 0$ then $|sm|_M = |sm|_{M,S_p}$ for some $s \in S_p$.
2. If $|m|_M = |m|_{M,S_p}$ then $|tm|_{M,S_p} = |tm|_M$ for all $t \in Sp$.

3.8. Lemma. If $m \in M$ is such that $|m|_{M,S_p} \neq 0$ then we have that $|sm|_{M,S_p} = |sm|_M$ for some $s \in S_{p_0}$.

Proof. The foregoing observation combined with Proposition 3.5(1) yields the proof. □

Now consider an $M \in R$-filt with good filtration FM and let N be an arbitrary F_0R-submodule of M. Define, for all $n \in \mathbb{Z}$, $Q(n,N) = F_nM \cap N/F_{n-1}N \cap N$. Clearly $Q(n,N)$ is a $G(R)_0$-submodule of $G(M)$, $Q(n,N) \subset G(M)_n$. Since $G(R)$ is strongly graded we may fix a decomposition $1 = \sum_{i=1}^{l} u_i v_i$ where $u_i \in G(R)_1$, $v_i \in G(R)_{-1}$, $i = 1,\ldots,l$. The commutativity of $G(R)$ yields : for any $t \in \mathrm{Ann}_{G(R)_0} Q(n,N)$ and for any $\overline{m} \in Q(n+1,N)$ we have : $t\overline{m} = t\sum_{i=1}^{l} u_i v_i \overline{m} = \sum_{i=1}^{l} u_i t v_i \overline{m} = 0$. Consequently : $\mathrm{Ann}_{G(R)_0} Q(n,N) \subset \mathrm{Ann}_{G(R)_0} Q(n+1,N)$ for all $n \in \mathbb{Z}$.

Put $I(n) = \mathrm{Ann}_{G(R)_0} Q(n,N)$, $J(n) = \mathrm{rad}(I(n))$ for all $n \in \mathbb{Z}$. For a certain $n_0 \in \mathbb{N}$ we have $I(n) = I(n_0)$ for all $n \geq n_0$, $J(n) = J(n_0)$ for all $n \geq n_0$, and we put $J(N) = J(n_0), J(N) = \cup_n J(n)$. We define $N(p)$ as the $F_0 Q^{\mu}_{S_p}(R)$-submodule in $Q^{\mu}_{S_p}(M)$ generated by the elements $\phi_p(m)$ with $m \in N$, and we put :

$$Q(n,N(p)) = (F_n Q^{\mu}_{S_p}(M) \cap N(p))/(F_{n-1} Q^{\mu}_{S_p}(M) \cap N(p))$$

for $n \in \mathbb{Z}$. Then $Q(n,N(p))$ is a $G(Q^{\mu}_{S_p}(R))_0$-submodule of $G(Q^{\mu}_{S_p}(M))$. On the other hand, $Q(n,N(p))$ is a $G(R)_0$-module via the canonical morphism $G(R)_0 \longrightarrow \sigma(S_{p0})^{-1} G(R)_0 \cong G(Q^{\mu}_{S_p}(R))_0$.

We may define a $G(R)_0$-linear map as follows :

$$\chi : Q(n,N) \to Q(n,N(p)), m+(F_{n-1}M \cap N) \mapsto \phi_p(m) + F_{n-1} Q^{\mu}_{S_p}(M) \cap N(p)$$

Since $Q(n,N(p))$ is a $\sigma(S_{p0})^{-1} G(R)_0$-module we may extend χ to a $\sigma(S_{p_0})^{-1} G(R)_0$-module homomorphism :
$\widetilde{\chi} : \sigma(S_{p0})^{-1} Q(n,N) \to Q(n,N(p))$, obtained by localization : i.e. for $t = \sigma(s)^{-1}\sigma(a)$, with $s \in Sp_0$, $\widetilde{\chi}(t\overline{m}) = \phi_p(s)^{-1}\phi_p(am) + F_{n-1} Q^{\mu}_{Sp}(M)$.

3.9. Proposition. With notations as above : $\widetilde{\chi}$ is an isomorphism.

Proof. A modification of the proof of lemma 4.10. in [11] that we include for completeness' sake.

a. $\widetilde{\chi}$ **is injective.** Pick $m \in F_nM \cap N$ and suppose that $\phi_p(m) \in F_{n-1} Q^{\mu}_{S_p}(M)$ i.e. $|m|_{M,S_p} \leq 2^{n-1}$. If $|m|_{M,S_p} = 0$ then by definition of the pseudo-norm $|\ \ |_{M,S_p}$ and by the choice of elements $s \in S_p$ it follows that for some $s \in S_p$, $sm \in \cap_{n\in \mathbb{Z}} F_nM$. There is a $s' \in S_p$ such that $t = s's \in S_{p0}$

and $tm \in \cap_{n\in\mathbb{Z}} F_nM$, see Proposition 3.5.1. Hence $tm \in F_{n-1}M \cap N$ and $\sigma(t)\overline{m} = 0$ in $Q(n,N)$, where $\sigma(t) = t + F_{-1}R$ and $\overline{m} = m + F_{n-1}M \cap N$. It follows that $\overline{m} = 0$ in $\sigma(Sp_0)^{-1}Q(n,N)$. In case $|m|_{M,S_p} \neq 0$ then for some $s \in S_{p_0}$ we have $|sm|_{M,S_p} = |sm|_M \leq 2^{n-1}$.

Therefore $sm \in F_{n-1}M \cap N$ and $\sigma(s)\overline{m} = 0$ in $Q(n,N)$; consequently $\overline{m} = 0$ in $\sigma(S_{p_0})^{-1}Q(n,N)$.

b. $\widetilde{\chi}$ **is surjective.** Since every element of $F_nQ^{\mu}_{S_p}(M) \cap N(p)$ is a finite sum of elements of the form $\alpha\phi_p(m)$ with $\alpha \in F_0Q^{\mu}_{S_p}(R)$ and $m \in N$. Therefore it suffices to show that all these elements $\alpha\phi_p(m)$ belong to $Im(\widetilde{\chi})$. Pick a $\alpha\phi_p(m)$; then $\phi_p(m) \in F_{n_0}Q^{\mu}_{S_p}(M)$ for some $n_0 \geq n-1$. The density of $\{\phi_p(s)^{-1}\phi_p(m),\ s \in S_p, m \in M\}$ in $Q^{\mu}_{S_p}(M)$ entails that we may choose $(t,a) \in S_p \times R$ such that $\alpha - \phi_p(t)^{-1}\phi_p(a) \in F_{-n_0+n-1}Q^{\mu}_{S_p}(R)$. Then $\alpha\phi_p(m) = \phi_p(t)^{-1}\phi_p(am) \text{mod } (F_{n-1}Q^{\mu}_{S_p}(R) \cap N(p))$ and $\phi_p(t)^{-1}\phi_p(a) \in F_0Q^{\mu}_{S_p}(R)$. By the lemma, we may assume that $|am|_{M,S_p} = |am|_M$. Similarly we may assume $|a|_{R,S_p} = (a)_R$ and $t \in S_{p_0}$. Hence $\sigma(t) \in G(R)_0$ and $a \in F_0R$; thus $am \in N$ and $|am|_M = |\phi(am)| \in 2^n$, i.e. $am \in F_nM \cap N$.

Finally we arrive at :

$\alpha\phi_p(m) + (F_{n-1}Q^{\mu}_{S_p}(M) \cap N(p)) = \widetilde{\chi}(\sigma(t)^{-1}(am + F_{n-1}M \cap I\!N)) \in Im\widetilde{\chi}$ □

Recall the following :

3.10. Remark : (Corollary 3.20 in [0]) If $\widetilde{R}$ is left Noetherian then the functor $Q^{\mu}_S(R) \otimes_R -$ preserves strict maps and it is exact on R-modules. If $M \in R$-filt has good filtration then $Q^{\mu}_S(R) \otimes_R M \cong Q^{\mu}_S(M)$ as filtered R-modules and thus $Q^{\mu}_S(M)$ is a finitely generated $Q^{\mu}_S(R)$-module with good filtration.

Consider $p_0 \in \text{Spec}(G(R)_\circ)$ with $p_0 \supset I(n)$, minimal as such then the radical of $\text{Ann}_{G(Q^{\mu}_{S_p}(R))_0}Q(n,N(p))$ in $G(Q^{\mu}_{S_p}(R)_0$ equals $J(Q(n,N(p)) = \text{rad}(\text{Ann}_{G(Q^{\mu}_{S_p}(R))_0}Q(n,N(p)) = \sigma(S_{p_0})^{-1}G(R)_0p_0)$.

In view of Proposition 3.5.4. and Gabber's theorem (applied locally here) it follows that $\sigma(S_{p_0})^{-1}G(R)_0p_0$ is involutive in $\sigma(S_{p_0})^{-1}G(R)_0$; further, from Remark 3.6. (2) it then follows that p_0 is involutive in $G(R)_0$. If $\text{Min}(I(n))$ denotes the set if prime ideals of $G(R)_0$ that are minimal over $I(n)$, for all $n \in \mathbb{Z}$, then we obtain :

3.11. Theorem. Let R be a strongly filtered ring such that $G(R)$ is a commutative Noetherian $\mathbb{Q}$-algebra. Let $M \in R$-filt have a good filtration and let N be an arbitrary F_0R-submodule of M. Every $p_0 \in \mathrm{Min}(I(n))$ is involutive and hence $J(n) = \mathrm{rad}(I(n))$ is involutive in the sense of Definition 3.2., for all $n \in \mathbb{Z}$. In particular, $J(N)(= J(n_0)$ as defined before) is involutive.

3.12. Remark. This extends Gabber's theorem to strongly filtered rings under a weaker Noetherian condition. It is possible to give a direct proof, avoiding the use of Gabber's theorem "locally", but this is agian a modification of the proof of Theorem 4.9. on p. 28 in [11].

4. Modules with Regular Singularities over Strongly Filtered Rings.

In this section we extend results concerning modules with regular singularities over an E-ring, as expounded in [11], to more general filtered rings satisfying the following conditions :

a. FR is faithful and strong

b. $G(R)$ is a commutaive $\mathbb{Q}$-algebra

c. F_0R is left Noetherian (equivalently : $\widetilde{R}$ is left Noetherian).

Throughout this section R will always be a filtered ring satisfying the foregoing conditions. In view of Remark 2.7. it follows that R as well as $G(R)$ are left Noetherian in this case. Moreover, it follows from (a) that a nonzero $M \in R$-filt with good filtration FM has $G(M) \neq 0$ and thus $\mathrm{Ann}_{G(R)}G(M) \neq G(R)$. Compared to the foregoing section we just added a stronger Noetherian condition, this condition just reduces to the condition $\widetilde{R}$ is left Noetherian that turned out to be basic in [0] too.

4.1. Definitions. Let I be an ideal of $G(R), M \in R$-filt. A good filtration FM on M is said to be **I-good** when $I \subset \mathrm{Ann}_{G(R)}G(M)$. In case $I = J(M) = \mathrm{rad}(\mathrm{Ann}_{G(R)}G(M))$ then an I-good filtration is said to be **very good.**
Put $\mathcal{I} = \{p \in \mathrm{Spec}^{(g)}(G(R))$ such that p is involutive in $G(R)\}$ and $\mu_R = \sup_{p\in\mathcal{I}}\{ht(p)\}$. If $M \in R$-filt is finitely generated then $\mathrm{Min}(J(M)) \subset \mathcal{I}$, so $\mathrm{ht}p \leq \mu_R$ for all $p \in \mathrm{Min}(J(M))$. A nonzero finitely generated $M \in R$-filt is said to be **holonomic** if $ht(p) = \mu_R$ for all $p \in \mathrm{Min}(J(M))$ or else if $M = 0$. Clearly, if there is a nonzero holonomic module then μ_R is finite since $G(R)$ is Noetherian and $\mu_R = ht(p)$ for some $p \in \mathrm{Spec}^{(g)}(G(R))$.

In this section I will be a graded involutive radical ideal of $G(R)$. Put $\mathcal{G}(I) = \{\tau \in F_1R, \tau + F_0R \in I\}$ and let $\mathcal{R}(I)$ be the subring of R generated by $\mathcal{G}(I)$ over F_0R.

4.2. Lemma. $\mathcal{G}(I)$ is a finitely generated F_0R-module and a Lie algebra.

Proof. That $\mathcal{G}(I)$ in an F_0R-submodule of F_1R is obvious. Since F_1R is a finitely generated F_0R-module (left and right) it follows that $\mathcal{G}(I)$ is a finitely generated F_0R-module too (by the Noetherian assumption on F_0R). If $\tau, \tau' \in \mathcal{G}(I)$, put $[\tau, \tau'] = \tau\tau' - \tau'\tau$. Then $[\tau, \tau'] \in F_1R$ because $G(R)$ is commutative. If $[\tau, \tau'] \in F_0R$ then $[\tau, \tau'] + F_0R = 0$ in $G(R)_1$, if $[\tau, \tau'] \in F_1R - F_0R$ then $\sigma([\tau, \tau']) = \{\sigma(\tau), \sigma(\tau')\}$ and the latter is in I because I is involutive. □

4.3. Definition. Let $M \in R$-filt be finitely generated. We say that M has **regular singularities along** I (M has *R.S.* along I) if M possesses an I-good filtration. We say that M has **regular singularities** (M has *R.S.*) if M possesses a very good filtration.

4.4. Proposition. Let $M \in R$-filt be finitely generated. The following statements are equivalent :

1. M has *R.S.* along I
2. There exists a finitely generated F_0R-submodule M_0 of M such that $RM_0 = M$ and $\mathcal{G}(I)M_0 \subset M_0$
3. If N is a finitely generated F_0R-submodule of M, then $\mathcal{R}(I)N$ is a finitely generated F_0R-module.
4. If N is a finitely generated $\mathcal{R}(I)$-submodule of M then N is a finitely generated F_0R-module
5. For all $m \in M$, $\tau \in G(I)$, the F_0R-module $P_\tau(m) = \sum_{i=0}^{\infty} F_0R\tau^i m$ is finitely generated.

Proof. 1. ⇒ 2. Let FM be an I-good filtration on M. In view of Lemma 2.8 F_0M is finitely generated as an F_0R-module and $RF_0M = M$. If $\tau \in G(I)$ and $\tau \in F_0R$ then certainly $\tau F_0M \subset F_0M$ but if $\tau \in F_nR - F_0R$ then $\sigma(\tau) \in I$ by definition of $\mathcal{G}(I)$, hence $\sigma(\tau) \in \mathrm{Ann}_{G(R)}G(M)$. In particular $\sigma(\tau)(F_0M/F_{-1}M) = 0$ i.e. $\tau F_0M \subset F_0M$ holds in any case.

2. ⇒ 1. Let M_0 be as in 2., define $F_nM = F_nR.M_0$ for $n \in \mathbb{Z}$. By Lemma 2.8. again FM is a good filtration (i.e. M is filt-finitely generated) Take $\tau \in F_kR - F_{k-1}R$ with $\sigma(\tau) \in I$. Then $0 \neq F_{-k+1}R\tau \subset \mathcal{G}(I)$ since R is strongly filtered. Therefore $F_{-k+1}R\tau M_0 \subset M_0$ and $\tau M_0 \subset F_0R\tau M_0 = F_{k-1}RF_{1-k}R\tau M_0 \subset F_{k-1}R.M_0$. Hence $\tau F_nRM_0 \subset F_{n+k-1}R.M_0$ for all

$n \in \mathbb{Z}$ (since $ram = arm + [r, a]m$, all $a \in F_nR$, all $m \in M_0$). Consequently $\sigma(r)G(M)_n = 0$ for all $n \in \mathbb{Z}$ i.e. $I \subset \mathrm{Ann}_{G(R)}G(M)$.

2. $\Rightarrow$ **3.** Since N is a finitely generated F_0R-module in $M = RM_0$ there exists a $k \in \mathbb{N}$ such that $N \subset F_kRM_0$. Pick $\tau \in \mathcal{G}(I)$, $r \in F_kR$, $m \in M_0$; then $\tau rm = r\tau m + [\tau, r]m \in F_kR.M_0$ since $\tau m \in M_0$ and $[\tau, r] \in F_nR$. Hence $\mathcal{R}(I)F_kR.M_0 \subset F_kR.M_0$ and therefore $\mathcal{R}(I)N \subset F_kR.M_0$. But $F_kR.M_0$ is a finitely generated F_0R-module, hence $\mathcal{R}(I)N$ is a finitely generated F_0R-module by the Noetherian assumption on F_0R.

3. $\Rightarrow$ **4.** If $\{n_i, i = 1, \ldots q\}$ generate N over $\mathcal{R}(I)$, put $N_0 = \sum_{i=1}^{q} F_0R.n_i$. Then $N = \mathcal{R}(I)N_0$ so we may apply 3.

4. $\Rightarrow$ **5.** $P_\tau(m) \subset \mathcal{R}(I)m$ and $\mathcal{R}(I)m$ is a finitely generated F_0R-module by 4., hence the statement follows.

5. $\Rightarrow$ **2.** By Lemma 4.2. we may write $\mathcal{G}(I) = \sum_{i=1}^{d} F_0R\tau_i$ for certain $\tau_i \in \mathcal{G}(I)$. Pick $m \in M$, from **5.** it follows that there exists a $k \in \mathbb{N}$ such that $\tau_i^k m \in \sum_{j=0}^{k-1} F_0R.\tau_i^j m$ for all $1 \le i \le d$. Since $\mathcal{G}(I)$ is a Lie-algebra it follows that $\mathcal{R}(I)m$ is generated as an F_0R-module by the elements $\tau_1^{i_1} \ldots \tau_d^{i_d} m$ with $0 \le i_j \le k-1, j \in \{1, \ldots, d\}$. Thus, for all $m \in M$, $\mathcal{R}(I)m$ is a finitely generated F_0R-module. Since M is a finitely generated R-module by assumption, say $M = \sum_{i=1}^{t} Rm_i$, we may put $M_0 = \sum_{i=1}^{t} \mathcal{R}(I)m_i$ and this finitely generated F_0R-module satisfies the statement 2. □

4.5. Remark. The foregoing proposition establishes a one-to-one correspondence between I-good filtrations on M and the finitely generated F_0R-submodules M_0 of M satisfying $RM_0 = M$ and $\mathcal{G}(I)M_0 \subset M_0$. The very good filtrations on M correspond one-to-one to F_0R-modules M_0 as before but also satisfying $I = J(M)$.

Let M have a good filtration FM and let N be an arbitrary F_0R-submodule of M. Clearly N is finitely generated as an F_0R-module if and only if $N \subset F_{n_0}M$ for some $n_0 \in \mathbb{Z}$, if and only if $Q(n, N) = 0$ for all $n \ge n_0$ (notations of Section 3).

4.6. Lemma. If $N \subset M$ is not finitely generated as an F_0R-module then the following properties hold :

1. $\mathrm{Min}(I(n)) \neq \emptyset$ or in other words $Q(n,N) \neq 0$ for all $n \in \mathbb{Z}$

2. Let $p_0 \in \mathrm{Min}(I(n))$ then $p = G(R)p_0$ satisfies $htp \leq \mu_R$.

3. $\mathrm{Min}(J(N)) \neq \emptyset$. If $p_0 \in \mathrm{Min}(J(N))$ then for $p = G(R)p_0$, the $F_0Q^{\mu}_{S_p}(R)$-module $N(p)$ is not finitely generated.

Proof. 1. If $Q(n_1,N) \neq 0$ for some $n_1 \in \mathbb{Z}$ then $\mathrm{Min}(I(n_1)) \neq \emptyset$, say $p_0 \in \mathrm{Min}(I(n_1))$. Because each $Q(n,N)$ is a finitely generated $G(R)_0$-module we see that $\sigma(Sp_0)^{-1}Q(n,N) = 0$ if and only if $\sigma(S_{p_0}) \cap I(n) \neq \emptyset$. Now suppose $\sigma(S_{p_0})^{-1}Q(n,N) = 0$ for some $n < n_1$, hence there is an $s_0 \in S_{p_0}$ such that $\sigma(s_0)Q(n,N) = 0$ and thus there is an $s_{n-n_1} \in F_{n-n_1}R \cap S_p - F_{n-n_1-1}R$ such that $\sigma(s_{n-n_1})Q(n_1,N) \subset Q(n,N)$ since $n < n_1$. But then we obtain $\sigma(s_0)\sigma(s_{n-n_1})Q(n_1,N) \subset \sigma(s_0)Q(n,N) = 0$. On the other hand, in view of Proposition 3.5.1. we may find a $\sigma(s') \in S_p$ such that $\sigma(s')\sigma(s_{n-n_1}) \in \sigma(S_{p_0})$, hence $\sigma(s_0 s' s_{n-n_1}) \in \sigma(S_{p_0}) \cap I(n_1) \subset p_0$, a contradiction. If N is not finitely generated as an F_0R-module then $Q(n,N) \neq 0$ or $\mathrm{Min}(I(n)) \neq \emptyset$ for all $n \in \mathbb{Z}$.

2. A p_0 as in the statement is involutive because of Theorem 3.11., hence $p = G(R)p_0$ is involutive in $G(R)$ and then $htp \leq \mu_R$ follows.

3. Since $\mathrm{Min}(J(N)) = \mathrm{Min}(J(n_0))$ for some $n_0 \in \mathbb{N}$, it follows from 1. that $\mathrm{Min}(J(N)) \neq \emptyset$. Take $p_0 \in \mathrm{Min}(J(N))$ and $p = G(R)p_0$. Suppose that $N(p)$ is a finitely generated $F_0Q^{\mu}_{S_p}(R)$-module, then $Q(n,N(p)) = 0$ for all $n \geq n'$, some $n' \in \mathbb{N}$. Therefore, for $n \geq n'$, $\sigma(S_{p_0})^{-1}Q(n,N) = 0$. However $p_0 \supset J(N) \supset J(n)$ for all $n \in \mathbb{N}$. Hence $p_0 \supset I(n)$ and this yields $\sigma(S_{p_0})^{-1}Q(n,N) \neq 0$, a contradiction. Therefore $N(p)$ is not finitely generated. □

4.7. Corollary. Suppose that M as above is moreover holonomic and N an F_0R-submodule of M that is not finitely generated. If $p_0 \in \mathrm{Min}(J(N))$ then $p = G(R)p_0 \in \mathrm{Min}(J(M))$.

Proof. By the lemma, $Q(n,N(p)) \neq 0$ for all $n \in \mathbb{N}$, hence $Q^{\mu}_{S_p}(M) \neq 0$ and thus $p \supset J(M)$ and $p \supset p'$ for some $p' \in \mathrm{Min}(J(M))$. The assumption $htp' = \mu_R$ then leads to $htp \geq \mu_R$ and therefore (by the lemma (2) again) $p = p'$ follows (because μ_R is finite !). □

4.8. Proposition. Let M be holonomic. The following statements are equivalent :

1. N is a finitely generated F_0R-submodule of M.

2. $N(p)$ is a finitely generated $F_0Q^{\mu}_{S_p}(R)$-submodule of $Q^{\mu}_{S_p}(M)$ for every $p \in \mathrm{Min}(J(M))$.

Proof. 1. $\Rightarrow$ **2.** Obvious because ϕ_p (set of generators of N) generates $N(p)$.

2. $\Rightarrow$ **1.** Suppose N is not finitely generated as an F_0-module, than $\mathrm{Min}(J(N)) \neq \emptyset$. Pick $p_0 \in \mathrm{Min}(J(N))$ and put $p = G(R)p_0$. Then $N(p)$ cannot be finitely generated as an $F_0Q^{\mu}_{S_p}(R)$-module, so we arrive at a contradiction and the conclusion that N must be finitely generated as an F_0R-module.

□

The main result of this section may be phrased as follows :

4.9. Theorem. Suppose M with good filtration and holonomic. The following statements are equivalent :

1. M has $R.S.$ as an R-module.

2. $Q^{\mu}_{S_p}(M)$ has $R.S.$ as a $Q^{\mu}_{S_p}(R)$-module for all $p \in \mathrm{Min}(J(M))$.

3. $Q^{\mu}_{S_p}(M)$ has $R.S.$ as a $Q^{\mu}_{S_p}(R)$-module for all $p \in \mathrm{Spec}^{(g)}(G(R))$.

Proof. 1. $\Rightarrow$ **3.** and **3.** $\Rightarrow$ **2.** are easy. Let us prove **2.** $\Rightarrow$ **1.**.

Take $m \in M$, $\tau \in \mathcal{G}(J(M))$. Clearly we only have to show that $N = \sum_i F_0R\tau^i m$ is a finitely generated F_0R-module. By the foregoing proposition we only have to show that $N(p)$ is a finitely generated $F_0Q^{\mu}_{S_p}(R)$-module for every $p \in \mathrm{Min}(J(M))$. We have

$$\phi_p(\tau) \in \mathcal{G}(J(Q^{\mu}_{S_p}(M)) \text{ and } N(p) = \sum_i F_0Q^{\mu}_{S_p}(R)\phi_p(\tau)^i\phi_p(m)$$

Applying Proposition 4.4. it follows that $N(p)$ is indeed finitely generated $F_0Q^{\mu}_{S_p}(R)$-module. □

In concluding this section we just mention some consequences of the foregoing theorem.

4.10. Proposition. Let $M \in R$-filt gave good filtration FM and assume that M is holomorphic. Suppose that I is a graded involutive radical ideal of $G(R)$. Then M has $R.S.$ along I if and only if M has $R.S.$ and $I \subset J(M)$.

Let $0 \to M' \to M \to M'' \to 0$ be an exact sequence of filt-finitely generated R-modules. Using Proposition 4.4.5. one easily arrives at :

4.11. Lemma. M has $R.S.$ along I if and only if M' and M'' have $R.S.$ along I.

4.12. Corollary. Let $0 \to M' \to M \to M'' \to 0$ be an exact sequence of holonomic R-modules and all modules having good filtrations. Then M has $R.S.$ if and only if M' and M'' has $R.S.$.

Proof. Noting that $J(M) = J(M') \cap J(M'')$, one only has to apply the lemma. □

5. The Strongly Filtered Ring of an Arbitrary Filtered Ring.

Using a technique somewhat similar to external homogenization in the theory of graded rings we extend a filtration FR to a strong filtration on $R[X, X^{-1}]$. In doing so we avoid again the Σ-Noetherian condition, replacing it by the weaker condition. $\widetilde{R}$ is Noetherian when applying it in Section 6. Again the effect of this is that our method can be applied "globally" without having to shift to certain completions. Note that, even if R is Σ-Noetherian, $R[X, X^{-1}]$ will not necessarily be an E-ring (but the completion at the extended filtration is). We identify $R[X, X^{-1}] \otimes_{R[X]} M[X]$ with the localization of $M[X]$ at $\{1, X, X^2, \ldots\}$ i.e. with $M[X, X^{-1}]$.

The ring $R[X, X^{-1}]$ may be made into a filtered ring by putting $F_n(R[X, X^{-1}]) = \{\Sigma r_i X^i, v_X(\Sigma r_i X^i) \leq n\}$ for $n \in \mathbb{Z}$, where the order-function v_X is defined by $v_X(\Sigma r_i X^i) = \sup_i(i + v(r_i))$ where v is the order-function corresponding to FR. Since $X \in F_1(R[X, X^{-1}]) - F_0(R[X, X^{-1}])$ and $X^{-1} \in F_{-1}(R[X, X^{-1}]) - F_{-2}(R[X, X^{-1}])$ it follows that $F(R[X, X^{-1}])$ is a strong filtration (up to assuming that $1 \notin F_{-1}R$ in order to avoid the trivial filtration). We say that $R[X, X^{-1}]$ with $F(R[X, X^{-1}])$ is the strongly filtered ring associated to R and FR. In a similar way one defines $F(M[X, X^{-1}])$ for a filtered R-module M, and it turns out to be a filtered $R[X, X^{-1}]$-module. Note that $F_0(R[X, X^{-1}]) \cong \widetilde{R}$, $F_0(M[X, X^{-1}]) \cong \widetilde{M}$. Hence if $\widetilde{R}$ is left Noetherian then so is $F_0(R[X, X^{-1}])$ and also $R[X, X^{-1}]$. If FM is separated then so is $F(M[X, X^{-1}])$. It is easy checked that $G(R[X, X^{-1}]) \cong G(R)[X, X^{-1}]$ as graded rings, where $G(R)[X, X^{-1}]$ is graded by putting $(G(R)[X, X^{-1}])_n = \sum_{i+j=n} G(R)_i X^j$ for all $n \in \mathbb{Z}$. Similarly $G(M[X, X^{-1}]) \cong G(M)[X, X^{-1}]$.

It is also clear how we define a filtration on $R[X]$ and a gradation on $G(R)[X]$ such that $G(R[X]) \cong G(R)[X]$ as graded rings and $G(M[X]) = G(M)[M]$. Then we may consider $(R[X])^{\sim} = \oplus_{n\in\mathbb{Z}} F_n(R[X]), (M[X])^{\sim} = \oplus_{n\in\mathbb{Z}} F_n(M[X])$. If we localize $(R[X])^{\sim}$, resp.$(M[X])^{\sim}$ at $S = \{1, X, X^2, \ldots\}$ then its can easily be verified that $\tilde{Q}^{\mu}_{\tilde{S}}((R[X])^{\sim}) = (S^{-1}(R[X])^{\sim})^{\wedge g}$ (cf.[0]), where the latter is the graded completion of $S^{-1}(R[X]^{\sim})$. Hence the microlocalization of $R[X]$ at S is given as :

$$Q^{\mu}_S(R[X]) = (S^{-1}(R[X])^{\sim}/(X-1)S^{-1}(R[X])^{\sim})^{\wedge F}$$

in view of [0] Lemma 2.1. and Proposition 2.5., where the completion is the completion with respect to the filtration of $S^{-1}(R[X])^{\sim}/(X-1)S^{-1}(R[X])^{\sim}$. One easily verifies that the map $R[X, X^{-1}] \to S^{-1}(R[X])^{\sim}/(X-1)S^{-1}(R[X])^{\sim}$ (and X is the image of 1 in $((R[X])^{\sim})_1$) given by : $\Sigma' r_i X^i \mapsto \Sigma' r_i X^i + (X-1)S^{-1}(R[X])^{\sim}$, is in fact an isomorphism of filtered rings. Consequently we obtain :

$$Q^{\mu}_S(R[X]) \cong (R[X, X^{-1}])^{\wedge F}, Q^{\mu}_S(M[X]) \cong (M[X, X^{-1}])^{\wedge F}$$

Under the usual conditions we impose on R, $G(R)$, the ring $Q^{\mu}_S(R[X])$ is an E-ring in the sense of [11] but as $F(R[X, X^{-1}])$ is not necessarily Σ-Noetherian (even if FR is) it follows that $R[X, X^{-1}]$ is generally only a strongly filtered ring satisfying the conditions set forth at the beginning of Section 4. In the sequel we again restrict to the case where $G(R)$ is commutative. If I is an ideal of $G(R)$ then we write $I^e = I.G(R)[X, X^{-1}]$ and for an ideal J of $G(R)[X, X^{-1}]$ we put $J^c = J \cap G(R)$. The following results are checked in a straightforward way.

5.1. Lemma 1. If $p \in \mathrm{Spec}(G(R))$ then $p^e \in \mathrm{Spec}(G(R)[X, X^{-1}])$

2. For an ideal I of $G(R)$: $I^{ec} = I$

3. For ideals I, J, K in $G(R)$, in $G(R)$, if $I = J \cap K$ then $I^e = J^e \cap K^e$.

4. If $r(I)$ denotes the radical of I in $G(R)$ then $r(I^e) = (r(I))^e$, where $r(I^e)$ is the radical of I^e in $G(R)[X, X^{-1}]$.

5. If $M \in R$-filt then : $(\mathrm{Ann}_{G(R)} G(M))^e = \mathrm{Ann}_{G(R)[X,X^{-1}]}(G(M)[X, X^{-1}])$ and $J(M[X, X^{-1}]) = (J(M))^e$ (where $J(-)$ is the notation used in Section 3.

5.2. Corollary. Assume that $G(R)$ is Noetherian, let $0 \neq M \in R$-filt and let $J(M) = p_1 \cap \ldots \cap p_r$ be an irredundant minimal prime decomposition,

then $J(M)^e = p_1^e \cap \ldots \cap p_r^e$ is an irredundant minimal prime decomposition of $J(M[X,X^{-1}])$ in $G(R)[X,X^{-1}]$ and $\mathrm{ht}p_i = \mathrm{ht}p_i^e$.
If $G(R)$ is a commutative Noetherian $\mathbb{Q}$-algebra and FR is faithful then for every finitely generated $M \in R$-filt, $M \neq 0$, with good filtration FM we have $G(M) \neq 0$ and hence $\mathrm{Min}(J(M)) \neq \phi$. As a consequence of the foregoing corollary we have : $\mathrm{Min}(J(M)^e) = \{p^e, p \in \mathrm{Min}(J(M))\} \neq \phi$. By Gabber's theorem in the generality of Section 3, all $p_e \in \mathrm{Min}(J(M)^e)$ are involutive. Furthermore, if we assume that M is holonomic then $htp^e = \mu_R$ for all $p^e \in \mathrm{Min}(J(M)^e)$, and vice-versa. Obviously, if M is holomorphic and $\mu_{R[X,X^{-1}]} = \mu_R$ then the filtered $R[X,X^{-1}]$-module $M[X,X^{-1}]$ is holomorphic too. We want to characterize the situation where $\mu_R = \mu_{R[X,X^{-1}]}$.
Put :

$$\mathcal{I} = \{p \in \mathrm{Spec}(G(R)), p \text{ is involutive in } G(R)\}$$

$$\nu_R = \sup_{p\in\mathcal{I}} \mathrm{ht}p$$

5.3. Proposition. With notation as above : $\mu_{R[X,X^{-1}]} = \mu_R$ if and only if $\mu_R = \nu_R$.

Proof. Since one only has to deal with the graded rings $G(R)$ and $G(R)[X,X^{-1}]$, whilst $G(R)[X,X^{-1}]$ is the associated graded ring of $Q_S^\mu(R[X])$ ($= E_X(-)$ in the notation of [10,11]) the proof given in [11,6] applies without modification. □

6. Modules with Regular Singularities over More General Filtered Rings.

This section continues the generalization of the results of [11] over more general filtered rings. The filtered ring R considered in this section will satisfy the following conditions :

(a) FR is faithful

(b) $G(R)$ is a commutative $\mathbb{Q}$-algebra

(c) $\widetilde{R}$ is left Noetherian

(d) $\mu_R = \nu_R$

6.1. Remark. Under these hypotheses the results in Section 5 hold and also it is clear that all results of Section 4 may be applied to $R[X,X^{-1}]$.

6.2. Lemma. With notations as in Lemma 5.1., let $M \in R$-filt and consider an ideal I of $G(R)$. If FM is I-good on M then $F(M[X,X^{-1}])$ is I^e-good on $M[X,X^{-1}]$; if FM is very good then $F(M[X,X^{-1}])$ is very good too.

Proof. If FM is I-good on M then FM is a good filtration, so $F_nM = \sum_i F_{n-d_i}Rm_i$ for some $v_i \in \mathbb{Z}$, $m_i \in M$ and all $n \in \mathbb{Z}$. In order to prove that $F_n(M[X,X^{-1}]) = \sum_i F_{n-d_i}(R[X,X^{-1}])m_i$ it suffices to consider $mX^j \in F_n(M[X,X^{-1}])$ and to show that $mX^j \in \sum_i F_{n-d_i}(R[X,X^{-1}])m_i$ since elements of $F_n(M[X,X^{-1}])$ are finite sums of elements of the form mX^j with $v(m)+j \leq n$. But if $mX^j \in F_n(M[X,X^{-1}])$ then $v(m)+j \leq n$ i.e. $m \in F_{n-j}M$ or $m \in \sum_i F_{n-j-d_i}R.m_i$ and $mX^j \in \sum_i F_{n-d_i}(R[X,X^{-1}])m_i$. We have checked that $F(M[X,X^{-1}])$ is good. From Lemma 5.1.5. we retain that $I^e \subset (\mathrm{Ann}_{G(R)}(G(M)))^e = \mathrm{Ann}_{G(R)[X,X^{-1}]}(G(M)[X,X^{-1}])$ if $I \subset \mathrm{Ann}_{G(R)}(G(M))$.

Hence $F(M[X,X^{-1}])$ is I^e-good on $M[X,X^{-1}]$. Again from Lemma 5.2.5. we have $J(M[X,X^{-1}]) = J(M)^e \subset (\mathrm{Ann}_{G(R)}(G(M)))^e$ if FM is very good and therefore $F(M[X,X^{-1}])$ is very good too. □

Now let $\mathbb{F}(M[X,X^{-1}])$ be an arbitrary filtration on $M[X,X^{-1}]$ with $M \in R$-filt, making $M[X,X^{-1}]$ into a filtered $R[X,X^{-1}]$-module with respect to $F(R[X,X^{-1}])$ as defined before. Let $i : M \to M[X,X^{-1}]$ be the canonical inclusion map. Put $F'_nM = i^{-1}(\mathbb{F}_n(M[X,X^{-1}]))$ for all $n \in \mathbb{Z}$, the filtration induced on M by $\mathbb{Z}$.

6.3. Proposition. If $M \in R$-filt is finitely generated and $\mathbb{F}$ is a J-good filtration on $M[X,X^{-1}]$ where J is a graded ideal of $G(R)[X,X^{-1}]$ then $F'M$ is J^c-good on M. Similar conclusions hold for very good filtrations.

Proof. Since $\mathbb{F}$ is good it follows from Lemma 2.8. that $\mathbb{F}_0(M[X,X^{-1}])$ is a finitely generated $F_0(R[X,X^{-1}])$-module; moreover, $\mathbb{F}_n(M[X,X^{-1}]) = F_n(R[X,X^{-1}])\mathbb{F}_0(M[X,X^{-1}])$. Since M is finitely generated we may choose a good filtration F^*M on M and then $F^*(M[X,X^{-1}])$ is good on $M[X,X^{-1}]$ because of Lemma 6.2., and we have : $F^*_n(M[X,X^{-1}]) = F_n(R[X,X^{-1}])F^*_0(M[X,X^{-1}])$. Consequently there exists a $c \in \mathbb{N}$ such that :

$$F_{-c}(R[X,X^{-1}])F^*_0(M[X,X^{-1}]) \subset \mathbb{F}_0(M[X,X^{-1}]) \subset F_c(R[X,X^{-1}])F^*_0(M[X,X^{-1}])$$

and also, for all $n \in \mathbb{Z}$:

$$i^{-1}(F_{n-c}(R[X,X^{-1}])F^*_0(M[X,X^{-1}])) \subset i^{-1}(\mathbb{F}_n(M[X,X^{-1}])) \subset i^{-1}(F_{n+c}(R[X,X^{-1}])F^*_0(M[X,X^{-1}])$$

The definition of F^* yields $F_n^*M = i^{-1}(F_n^*(M[X,X^{-1}]))$ for all $n \in \mathbb{Z}$, hence we may derive that : $F_{n-c}^*M \subset F_n'M \subset F_{n+c}^*M$, for all $n \in \mathbb{Z}$. It follows (cf. e.g. [AVO, Proposition 1.2.]) that $F'M$ is a good filtration on M too. Pick $\sigma(a) \in J^c$ with $v(a) = k$ and $m \in F_m'M$; then $(i(a) + F_{k-1}(R[X,X^{-1}]))(i(m) + \mathbb{F}_{n-1}(M[X,X^{-1}]) = 0$ since we have assumed that $J \subset \mathrm{Ann}_{G(R)[X,X^{-1}]}(\mathbb{G}(M[X,X^{-1}]))$ where $\mathbb{G}$ is associated to $\mathbb{F}$.

Therefore, $i(a)i(m) \in \mathbb{F}_{k+n-1}(M[X,X^{-1}])$ or in other words $am \in F_{k+n-1}'M$. It follows from this fact that $J^c \subset \mathrm{Ann}_{G(R)}G'(M)$ (here $G'(M)$ depends on $F'M$).

In case $\mathbb{F}(M[X,X^{-1}])$ is very good on $M[X,X^{-1}]$ we have to establish that $F'M$ is very good on M. Noting that $\mathbb{F}$ and $F'(M[X,X^{-1}])$ are equivalent because both are good filtrations on $M[X,X^{-1}]$ we may apply a result of [O.G.] and conclude that $J_{F'}(M[X,X^{-1}]) = J_{\mathbb{F}}(M[X,X^{-1}])$. The final statement follows therefore from the proof of the first part combined with Lemma 5.1. □

For $M \in R$-filt, $q \in \mathrm{Spec}^{(g)}(G(R))$ let us fix notations as follows. $\phi_M : M \to Q^{\mu}_{S_q}(M)$ is the canonical localization morphism, $G(\phi_M) : G(M) \to G(Q^{\mu}_{S_q}(M))$ is the induced graded homomorphism, $j_M : G(M) \to \sigma(S_q)^{-1}G(M)$ is the canonical graded localization map.

$$\psi_M : \sigma(S_q)^{-1}G(M) \xrightarrow{\cong} G(Q^{\mu}_{S_q}(M)), \sigma(s)^{-1}\sigma(m) \mapsto \phi_R(s)^{-1}\phi_M(m) + F_{n-1}Q^{\mu}_{S_q}(M)$$

We know that $G(\phi_M) = \psi_M \circ j_M, G(\phi_R) = \psi_R \circ j_R$. For an arbitrary ideal I of $G(R)$ we put :

$$I^q = G(Q^{\mu}_{S_q}(R))G(\phi_R)(I) = \Psi_R(\sigma(S_q)^{-1}I)$$

If we take an $M \in R$-filt then, in particular :

$$(\mathrm{Ann}_{G(R)}G(M))^q = \Psi_R(\sigma(S_q)^{-1}\mathrm{Ann}_{G(R)}G(M)) = \mathrm{Ann}_{G(Q^{\mu}_{S_q}(R))}(G(Q^{\mu}_{S_q}(M))))$$

Now, put $p = q^e \subset G(R)[X,X^{-1}]$ as defined in Section 5, then $p \in \mathrm{Spec}^{(g)}(G(R[X,X^{-1}])$. It is easily checked one obtains commutative diagrams of filtered morphisms :

$$\begin{array}{ccc} R & \xrightarrow{i} & R[X,X^{-1}] \\ \phi_R \downarrow & & \downarrow \phi_{R[X,X^{-1}]} \\ Q^{\mu}_{S_q}(R) & \xrightarrow[\hat{i}]{} & Q^{\mu}_{S_p}(R[X,X^{-1}]) \end{array}$$

$$\begin{array}{ccc} G(R) & \xrightarrow{G(i)} & G(R[X,X^{-1}]) \\ G(\phi_R) \downarrow & & \downarrow G(\phi_{R[X,X^{-1}]}) \\ G(Q^{\mu}_{S_q}(R)) & \xrightarrow[G(\hat{i})]{} & G(Q^{\mu}_{S_p}(R[X,X^{-1}]) \end{array}$$

and similar diagrams exist if R is replaced by $M \in R$-filt.

The following relations arise almost trivially :

$$\begin{aligned} &\mathrm{Ann}_{G(Q^{\mu}_{S_q}(R))} G(Q^{\mu}_{S_q}(M)) \\ &\qquad = G(Q^{\mu}_{S_q}(R))G(\phi_R)(\mathrm{Ann}_{G(R)}G(M)) \\ &\mathrm{Ann}_{G(Q^{\mu}_{S_p}(R[X,X^{-1}J)]}G(Q^{\mu}_{S_p}(M[X,X^{-1}])) \\ &\qquad = G(Q^{\mu}_{S_p}(R[X,X^{-1}])G(\phi_{R[X,X^{-1}]})(\mathrm{Ann}_{G(R[X,X^{-1}])}G(M[X,X^{-1}])) \\ &G(\hat{i})(\mathrm{Ann}_{G(Q^{\mu}_{S_q}(R))}G(Q^{\mu}_{S_q}(M))) \\ &\qquad = G(\hat{i})(G(Q^{\mu}_{S_q}(R)))G(\hat{i})G(\phi_R)(\mathrm{Ann}_{G(R)}G(M)) \\ &\qquad = G(\hat{i})(G(Q^{\mu}_{S_q}(R)))G(\phi_{R[X,X^{-1}]})G(i)(\mathrm{Ann}_{G(R)}G(M)) \\ &\qquad \subset G(Q^{\mu}_{S_p}(R[X,X^{-1}]))G(\phi_{R[X,X^{-1}]})(\mathrm{Ann}_{G(R[X,X^{-1}])}G(M[X,X^{-1}])) \\ &\qquad = \mathrm{Ann}_{G(Q^{\mu}_{S_p}(R[X,X^{-1}]))}G(Q^{\mu}_{S_p}(M[X,X^{-1}])) \end{aligned}$$

Furthermore, as a consequence of Lemma 5.1. we may add :

$$J(Q^{\mu}_{S_q}(M)) = G(Q^{\mu}_{S_q}(R))G(\phi_R)(J(M))$$

$$\begin{aligned} &J(Q^{\mu}_{S_p}(M[X,X^{-1}])) \\ &\qquad = G(Q^{\mu}_{S_p}(R[X,X^{-1}))G(\phi_{R[X,X^{-1}]})(J(M[X,X^{-1}])) \\ &\qquad = G(Q^{\mu}_{S_p}(R[X,X^{-1}]))G(\phi_{R[X,X^{-1}]})(G(R)[X,X^{-1}]G(i)(J(M))) \\ &\qquad \subset G(Q^{\mu}_{S_p}(R[X,X^{-1}]))G(\phi_{R[X,X^{-1}]})G(i)(J(M)) \\ &\qquad = G(Q^{\mu}_{S_p}(R[X,X^{-1}]))G(\hat{i})G(\phi_R)(J(M))) \end{aligned}$$

Combining the latter with the relations preceding them we arrive at the observation that $G(\phi_R)(J(M)) \subset \mathrm{Ann}_{G(Q^{\mu}_{S_q}(R))}G(Q^{\mu}_{S_q}(M))$ whenever $J(Q^{\mu}_{S_q}(M)) \subset \mathrm{Ann}_{G(Q^{\mu}_{S_q}(R))}G(Q^{\mu}_{S_q}(M))$ and thus :

$$G(\hat{\imath})G(\phi_R)(J(M)) \subset G(\hat{\imath})(\mathrm{Ann}_{G(Q^{\mu}_{S_q}(R))}G(Q^{\mu}_{S_q}(M))) \subset \\ \mathrm{Ann}_{G(Q^{\mu}_{S_p}(R[X,X^{-1}]))}(G(Q^{\mu}_{S_p}(M[X,X^{-1}])))$$

Finally this yields :

$$J(Q^{\mu}_{S_p}(M[X,X^{-1}])) \subset \mathrm{Ann}_{G(Q^{\mu}_{S_p}(R[X,X^{-1}]))}G(Q^{\mu}_{S_p}(M[X,X^{-1}]))$$

and this proves :

6.4. Proposition. If $Q^{\mu}_{S_q}(M)$ has R.S. as a $Q^{\mu}_{S_q}(R)$-module then $Q^{\mu}_{S_p}(M[X,X^{-1}])$ has R.S. as a $Q^{\mu}_{S_p}(R[X,X^{-1}])$-module.

Of course the proof given is only terrible because of notation, in fact it is a rather easy application of the external homogenization techniques for graded rings and modules (cf. [7]).

6.5. Theorem. Let $M \in R$-filt be a holonomic module, then the following statements are equivalent

1. M has R.S.
2. $Q^{\mu}_{S_q}(M)$ has R.S. s a filtered $Q^{\mu}_{S_q}(R)$-module for all $q \in \mathrm{Min}(J(n))$
3. $Q^{\mu}_{S_q}(M)$ has R.S. as a filtered $Q^{\mu}_{S_q}(R)$-module for all $q \in \mathrm{Spec}(G(R))$
4. $Q^{\mu}_{S_p}(M[X,X^{-1}])$ has R.S. as a filtered $Q^{\mu}_{S_p}(R[X,X^{-1}])$-module, for all $p = q^e$, $q \in \mathrm{Min}(J(M))$.
5. $M[X,X^{-1}]$ has R.S. as a filtered $R[X,X^{-1}]$-module.

Proof. 1 $\Rightarrow$ 3 see [11] Proposition 2.9.

3 $\Rightarrow$ 3 Obvious

2 $\Rightarrow$ 4 Follows Proposition 6.4.

4 $\Rightarrow$ 5 Follows from Section 5

5 $\Rightarrow$ 1. Follows from Proposition 6.3. if we replace $\mathbb{F}$ by $F(M[X,X^{-1}J])$. □

References.

[0] M. J. Asensio, M. Van den Bergh, F. Van Oystaeyen, *A New Algebraic Approach to Microlocalization of Filtered Rings*, to appear.

[1] M. Awami, F. Van Oystaeyen, *On Filtered Rings with Noetherian Associated Graded Rings*, Proceedings of Ring Theory Meeting, Granada 1986, LNM, Springer Verlag, Berlin, 1988.

[2] J. E. Björk, *Rings of Differential Operators*, Math. Library Series, Vol 21, North Holland, Amsterdam, 1979.

[3] J. E. Björk, *Unpublished Notes*, 1987.

[4] O. Gabber, *The Integrability of Characteristic Varieties*, Am. Journal of Math. 103, 1981, 445-468.

[5] M. Kashiwara, T. Kawai, *On Holomorphic Systems with Regular Singularities*, III, RIMS Kyoto Univ. Vol. 17, 1981, 813-979.

[6] M. Kashiwara, T. Oshima, *Systems of Differential Equations with Regular Singularities and their Boundary Values Problems*, Ann. of Math. 106, 1977, 145-200.

[7] C. Năstăsescu, F. Van Oystaeyen, *Graded Ring Theory*, Library of Math. Series Vol. 28, North Holland, Amsterdam 1982.

[8] C. Năstăsescu, E. Nauwelaerts, F. Van Oystaeyen, *Arithmetically Graded Rings Revisited*, Comm. in Algebra 12, 1984, 745-775.

[9] T. Springer, *Micro-localisation algébrique, Sém. d'Algèbre Dubreil-Malliavin*, Lect. in Math., Springer Verlag, Berlin.

[10] A. Van den Essen, *Algebraic Micro-Localization*, Comm. in Algebra, 1985.

[11] A. Van den Essen, *Modules with Regular Singularies over Filtered Rings*, Report 8508, 1985, Univ. of Nijmegen.

PRIMITIVE IDEALS OF ALGEBRAS OVER UNCOUNTABLE FIELDS

Louis Halle Rowen
Bar-Ilan University
Ramat-Gan, Israel

This paper is in final form and no version of it will be submitted for publication elsewhere.

In memory of I.N. Herstein

One of the recurrent themes in Yitz' work was the utilization of an explicit subdirect decomposition of a ring without nil ideals into prime homomorphic images. In this paper, which is a study of what makes a prime ideal primitive, we shall see that a careful look into this decomposition provides information about the primitive ideals of algebras over uncountable fields.

We focus on conditions (see §2) introduced by Dixmier to investigate the connection between primitive and rational ideals in enveloping algebras, but following [3] and [6] we use Dixmier's conditions as a framework for the study of primitive ideals of arbitrary algebras over a field.

One of our main concerns will be a condition (called "special" here) which falls between the Dixmier conditions (3) and (3'). Under a cardinality assumption made in much of the paper ($[R:F] < |F|$) we see in theorem 1.2 that every special prime ideal is primitive. After some general considerations in §2, we show in §3 that the Dixmier conditions are equivalent in the presence of "enough" commutativity, e.g. for PI-algebras (under this cardinality assumption).

In §4 we bring in the "Kaplansky ring" approach, cf. [3], which is used there to study primitive ideals in terms of the Baire property of the primitive spectrum. Actually we settle on a slightly weaker version, which we call "weak Kaplansky," which formally provides enough information to link the Dixmier conditions (3') and (1), and which is equivalent to "Kaplansky" (and to the equivalence of the Dixmier conditions) in many rings of interest.

In §5 we address the question of lifting "Kaplansky" and "weak Kaplansky" up a finite ring extension, i.e. finitely generated as left module. Considerable progress has been made with these extensions in recent years, cf. [1],[8],[10],[11],[18],[19], and it turns out that the <u>right-handed</u> version of Kaplansky passes up; thus an interesting left-right interplay develops, which is worthy of further investigation. In §6 we show many of the results from §5 can be modified to cover "special" primes; however we do not yet have analogues of the results in [3]. I would like to thank Ami Braun for several enlightening observations, which prevented several errors from appearing in print.

§1. Special primes and a subdirect decomposition of algebras

Throughout we shall assume R is an algebra over a field F. A theorem of Amitsur shows that if F is uncountable then R is semiprimitive, i.e. a subdirect product of primitive algebras. The object of this section is to describe these primitive algebras more explicitly, in order to determine their structure. One of the key inspirations is in a classical result which will be described now.

Definition 1.1. A prime algebra R is special if there is a non-nilpotent element r, such that each nonzero ideal of R contains a power of r.

In [5, lemma 2.2.3] it is shown that any ring without nonzero nil ideals is a subdirect product of special rings. We are ready for the main result of this section, which is to show that special prime algebras are primitive, under a suitable cardinality hypothesis. Recall the "core" of a left ideal is the largest 2-sided ideal it contains; an ideal is primitive iff it is the core of a maximal left ideal.

Theorem 1.2. If R is a special prime algebra over an uncountable field F, with $[R:F] < |F|$ as infinite cardinals then R is primitive.

Proof. Let r be as in definition 1.1. Note if $[R:F]$ is finite then R is simple and thus primitive, so we may assume $[R:F]$ is infinite.

Case I. r is not algebraic over F. Since $F[r]$ has countably infinite dimension over F we see by [17, proposition 2.5.21] that $r-\mu$ is not invertible for "most" μ in F. Take any such $\mu \neq 0$ and let L be any proper maximal left ideal of R containing $R(r-\mu)$. Then L has core 0. Indeed otherwise, by definition 1.1, L contains some power r^n of r. But $r^n-\mu^n = (r^{n-1}+r^{n-2}\mu+\ldots+\mu^{n-1})(r-\mu) \in L$, so $\mu^n \in L$, and thus $1 \in L$, contradiction. Thus core(L) = 0, implying R is primitive.

Case II. r is algebraic over F. Then $r^m = \Sigma_{i>m} \alpha_i r^i$ for suitable m and suitable α_i in F. We claim this implies that every nonzero ideal A of R contains r^m. Indeed by hypothesis A contains some power r^n of r; we choose n minimal possible. If $n>m$ then $r^{n-1} = \Sigma_{i>m} \alpha_i r^{n+i-m-1} = (\Sigma_{i>m} \alpha_i r^{i-m-1})r^n \in A$, contradiction. Thus $n \leq m$, so obviously $r^m \in A$.

On the other hand Amitsur's theorem [17, theorem 2.5.22] implies Jac(R) = Nil(R) = 0, which implies some primitive ideal of R is 0 (for otherwise each primitive ideal contains r^m, contradiction). Hence R is primitive. Q.E.D.

Braun has pointed out that some hypothesis on $[R:F]$ is needed, since any commutative discrete valuation ring is "special" under this definition, but is not a field (whereas any commutative primitive ring is a field.) In general the proof of theorem 1.2 shows, without any restriction on F, that any special prime algebra R either is subdirectly irreducible, or else r (of definition 1.1) is transcendental over F.

Corollary 1.3. If R is an affine algebra over an uncountable field and Nil(R) = 0 then R is a subdirect product of special algebras, each of which is primitive.

Proof. By [5, lemma 2.2.3] and theorem 1.2. Q.E.D.

Definition 1.4. A prime ideal of R is special if R/P is a special ring, i.e. if P is maximal with respect to not containing any power of a given element r.

Corollary 1.5. Suppose F is an uncountable field, and R is of countable dimension over F. If P is a prime ideal of R such that Nil(R/P) = 0 then P is the intersection of special prime ideals, and each of these is primitive. In particular the set of special prime ideals is dense in the primitive spectrum!

Th

Digression: Algebraic ideals

Let us show now that the hypothesis "R has no nonzero nil ideals" implies R has no algebraic ideals, in the instances of interest to us. The following observation is taken from [5].

Remark. If a left ideal L contains an algebraic element $a \neq 0$ then L contains a nontrivial idempotent. (Indeed suppose $p = \alpha + q\lambda^i$ is the minimal polynomial of a in $F[\lambda]$, where α is the constant term of p, and the constant term of q is nonzero (and thus may be assumed to be 1). Then $0 = p(a) = \alpha + q(a)a^i$ so $\alpha \in F \cap L = 0$, and hence $q(a)a = 0$; it follows q(a) is idempotent, and is nonzero by the minimality of p.)

Proposition 1.6. Suppose R is prime left Noetherian but not simple Artinian. Then any algebraic left ideal L of R is 0.

Proof. If $L \neq 0$ take any submodule N of L and any non-nilpotent element a in N (possible since N is a left ideal of R). N contains a nontrivial idempotent $e_1 = e$; thus N is the direct sum of Ne and N(1-e). Note Ne ⊇ Ree = Re ⊇ Le, so Ne = Le. Repeating this procedure we see N(1-e) (if nonzero) contains a nontrivial idempotent e_2, and iterating the procedure we can build a sequence of orthogonal idempotents $e_1, e_2, \ldots$ in N. Since R is left Noetherian, this sequence must terminate at some e_t, and so N = Lf where $f = e_1 + e_2 + \ldots + e_t$ for some t, implying N is a (direct) summand of L. We have shown that every submodule of L is a summand, and thus L is a completely reducible module.

Thus $soc(R) \neq 0$, and in particular R is primitive (since any minimal left ideal is a faithful simple module). But any primitive Noetherian ring with nonzero socle is simple Artinian, by an easy application of the density theorem). Q.E.D.

Note. The hypothesis "left Noetherian" in the proof was used only in three places, and thus could be weakened to the following conditions:

(i) R does not have an infinite set of orthogonal idempotents;

(ii) Every nil left ideal of R is 0.

Modifying corollary 1.5.

In case R satisfies a mild condition, the ACC on prime ideals (e.g. if R has Krull dimension or if R has finite GK dimension), we can obtain various variants of corollary 1.5. Recall a prime ring R is a G-ring if the intersection of its nonzero prime ideals is nonzero. A prime ideal P of a ring R is a G-ideal if R/P is a G-ring. Obviously any semiprimitive G-ring is primitive. Recall R is Jacobson if every prime image is semiprimitive. Thus the G-ideals of a Jacobson ring are all primitive. The next result must certainly be well-known.

Proposition 1.7. If R is prime Jacobson and satisfies ACC (prime ideals) then R is a subdirect product of G-rings.

Proof. Otherwise take a prime ideal P maximal with respect to R/P being a counterexample. Passing to R/P we may assume each proper prime homomorphic image of R is a subdirect product of G-rings.

For each element r of R, take an ideal P_r maximal with respect to $P_r \cap \{r\} = 0$. Then $\cap P_r$ misses all nonzero elements of R, and so is 0. Hence it suffices to prove that each R/P_r is a subdirect product of primitive G-rings. If $P_r \neq 0$ this is true by the first paragraph; if $P_r = 0$ then R itself is a G-ring. Q.E.D.

Meta-remark 1.8. Suppose C is a class of algebras, closed under homomorphic images, such that for every algebra R in C, every nonzero prime ideal of R contains an element having a certain property π. Then in the proof of proposition 1.7 we may stipulate that each element r possesses this property π, and so $\cap P_r$ does not contain an element having property π and is thus 0.

For example, for F infinite, every ideal of a prime, left Noetherian, non-Artinian algebra contains a regular element transcendental over F, so we could require the elements r to be regular and transcendental over F, yielding the assertion, "Every prime left Noetherian, non-Artinian algebra over an infinite field is a subdirect product of G-rings for which the intersection of the nonzero prime ideals contains an element which is regular and transcendental."

Although this process seems to strengthen corollary 1.5, it is not as explicit as the construction of the special prime ideals, and thus does not give us the same flexibility in passing from a ring R to a closely related ring.

§2. Conditions for primitivity

This section brings in some well-known conditions necessary for a prime ideal P to be primitive, or, passing to R/P, for a prime ring to be primitive. Dixmier initiated the study of the following three conditions for a prime Noetherian F-algebra R:

(1) R is primitive.

(2) The center of the ring of fractions of R is algebraic over F.

(3) The intersection of the prime ideals of R is nonzero

Dixmier conjectured that these conditions are equivalent whenever R is a prime homomorphic image of an enveloping algebra; this was verified by Moeglin [14] in case $F = \mathbb{C}$, and the reduction to arbitrary F was carried out by Irving-Small [7]. This raises the question of their equivalence for other prime rings, where even the "Noetherian" condition on R can be dropped if we replace (2) by the condition:

(2) (generalized) The extended centroid (in Martindale's sense) is algebraic over F.

Clearly (3) $\Rightarrow$ (1) for any semiprimitive prime ideal; (1) $\Rightarrow$ (2) if F is uncountable, cf. [17, example 2.12.28 and remark 8.4.17]. On the other hand, (2) does not necessarily imply (3):

Remark 2.1. A centrally closed prime ring need not be semiprimitive. Indeed take the central closure $\hat{R}$ of a prime ring R which has a nonzero locally nilpotent ideal A; then $AZ(\hat{R})$ is a locally nilpotent ideal of $\hat{R}$.

Of course the objection could be raised that this example fails for the "trivial" reason that $\text{Nil}(R) \neq 0$, which cannot hold in a primitive ring. Thus we shall assume that any prime ideal P we consider satisfies the property $\text{Nil}(R/P) = 0$. This could be done formally, by modifying the definition of "prime ideal;" anyway one should note that this property always holds for prime ideals in many classes of rings (PI-rings, rings with ACC(ideals), etc.) We shall refer repeatedly to an example of Lorenz [12,§4] in which (3) does not imply (1), but his example does satisfy the following weakening of (3), also brought to prominence by Dixmier:

(3') R has a countable subset S which separates the ideals of R in the sense that $A \cap S \neq \emptyset$ for every $0 \neq A \triangleleft R$.

If R has a countable base over F then (2) $\Rightarrow$ (3'), cf. [6, theorem 2.2] or [17, proposition 8.4.20]. Thus (1) $\rightarrow$ (2) $\rightarrow$ (3') for any algebra of countable dimension over an uncountable field. Nonetheless one still can find counterexamples to (2) $\Rightarrow$ (1) (and to

(3') ⇒ (1)) by noting that (2) and (3') are left-right symmetric whereas (1) is not; any right but not left primitive ring is a counterexample. This was noted in [6, p.85], in which an affine, left Ore counterexample was given. Similarly one has

Example 2.2. Jategonkar's example of a principal left ideal domain over an arbitrary field (cf. [17, example 2.1.36]), satisfies (2) and (3') but not (1), when we take α to be the first limit ordinal. (Both assertions follow from (vii) of [17, p.160]). Moreover if in the notation of [17, p.161] we take $D_0 = F$ and let D instead be the division subring of F((X)) generated by F and H (the subgroup of the free group generated by all $X_\nu X_\beta X_\nu^{-1}$ for $\beta < \nu$) then D has a countable base over F. (Indeed let E be a countable subfield of F and let D_0 be the division subring generated by E and H. Then D_0 is countable, and $D \approx D_0 \otimes_E F$ implies $[D:F] = [D_0:E]$ is countable.) Consequently $R = R_\alpha$ has a countable base over F. In particular $[R:F] < |F|$ when F is uncountable.

This example shows that the relationship among Dixmier's conditions is cnosiderably subtler than the Nullstellensatz. Inspired by §1, let us also throw in the condition

(3") R is special.

Remark 2.3. (3) ⇒ (3"); indeed ∩{special prime ideals} = 0, so one of the special prime ideals must be 0. On the other hand (3") ⇒ (3') trivially, as seen by taking $S = \{r^i: i \in \mathbb{N}\}$. Thus (3) ⇒ (3") ⇒ (3'). On the other hand, theorem 1.2 showed (3") ⇒ (1) for affine algebras over an uncountable field.

In summary, we could obtain a circle of equivalence by proving either (2) ⇒ (3") or (3') ⇒ (1). Dixmier's conditions can be tied to height 1 (= minimal nonzero) prime ideals.

Proposition 2.4. Suppose R is prime and satisfies ACC(ideals). If R has a countable set separating its ideals then {height 1 prime ideals of R} is countable. Conversely if every prime ideal contains a height 1 prime ideal and {height 1 prime ideals} is countable then R has a countable set separating its ideals.

Proof. (⇒) Suppose $S = \{s_i: i \in \mathbb{N}\}$ is the separating set. For each s_i in S, let $P_{i1}, \ldots, P_{it}$ be the prime ideals of R minimal over s_i, where $t = t(i)$ depends on i, and let $\mathcal{S} = \{P_{iu}: i \in \mathbb{N}, 1 \le u \le t(i)$, and P_{iu} has height 1$\}$, clearly a countable set of height 1 prime ideals. We conclude by showing $\mathcal{S}$ contains every height 1 prime ideal. Indeed if Q is any height 1 prime ideal then some $s_i \in Q$ so $(P_{i1} \cap \ldots \cap P_{it})^k \subseteq Q$ for suitable k (since R satifies ACC(ideals)) and thus some $P_{iu} \subseteq Q$, implying $P_{iu} = Q$ since height(Q) = 1.

($\Leftarrow$) Letting $\{P_i: i \in \mathbb{N}\}$ be the height 1 prime ideals, let $\mathcal{A} = \{P_{i_1}\dots P_{i_t}: i_1,\dots,i_t, t \in \mathbb{N}\}$, allowing repetitions. $\mathcal{A}$ is a countable set of ideals; writing $\mathcal{A} = \{A_1, A_2, \dots\}$ choose $0 \neq s_j \in A_j$ for each j. Any nonzero ideal contains a finite product of prime ideals and thus a finite product of height 1 prime ideals, and thus some s_j, proving $\{s_1, s_2, \dots\}$ separates the ideals of R. Q.E.D.

Proposition 2.5. Suppose R is prime and satisfies ACC(ideals). If R satisfies (3) then R has only a finite number of height 1 primes. Conversely if every prime ideal contains a height 1 prime and {height 1 prime ideals} is finite then R satisfies (3).

Proof. ($\Rightarrow$) Let $A = \cap$(nonzero primes} and let $P_1,\dots,P_t$ be the prime ideals minimal over A. Then obviously these are of height 1, and are the only height 1 primes since if height$(P) = 1$ then $P_1 \dots P_t \subseteq P$ so some $P_j \subseteq P$.

($\Leftarrow$) $\cap$(nonzero primes} = $\cap$(height 1 primes} $\neq 0$. Q.E.D.

Thus the difference between (3) and (3') is in the number of height 1 primes. The conditions of these two propositions are satisfied by wide classes of rings:

Remark 2.6. If R has finite classical Krull dimension then obviously every nonzero prime ideal contains a height 1 prime ideal. In particular this is true if R has finite GK (Gelfand-Kirillov) dimension or finite K-dim (in the sense of Gordon-Robson [4]).

Proposition 2.7. If T is the Laurent extension $R[\lambda, \lambda^{-1}; \sigma]$ and R satisfies ACC(ideals) and DCC(prime ideals) then T satisfies DCC(prime ideals).

Proof. Recall the result of Goldie-Michler (cf. [3]) that one cannot have a chain of three prime ideals of T each contracting to the same ideal of R. On the other hand, an ideal A of R is said to be σ-prime if A is σ-invariant and, whenever B_1, B_2 are σ-invariant ideals of R with $B_1 B_2 \subseteq A$ then $B_1 \subseteq A$ or $B_2 \subseteq A$. Then one sees that any prime ideal Q of T contracts to a σ-prime ideal $Q \cap R$ of R.

Also note that if $A \triangleleft R$ is σ-prime and P is a prime (of R) minimal over A, then each $\sigma^i P$ is also minimal over $\sigma^i A = A$, so there are only a finite number of distinct $\sigma^i P$, and $\cap \sigma^i P = A$.

Now suppose T failed DCC on prime ideals. Then R would fail DCC on σ-prime ideals, by the first paragraph. Take any such chain $A_1 \supset A_2 \supset \dots$ of σ-prime ideals. If P_j is a prime (of R) minimal over A_j then by the second paragraph $\cap_{\text{finite}} \sigma^i P_2 = A_2 \subset P_1$, so some $\sigma^i P_2 \subseteq P_1$, and we could assume $P_2 \subseteq P_1$; in fact $P_2 \subset P_1$, for otherwise $A_2 = \cap \sigma^i P_2 = \cap \sigma^i P_1 = A_1$, which is false. Continuing in this way we see R fails DCC(prime ideals), contradiction. Q.E.D.

Remark 2.8. An analogous argument holds for differential operator extensions.

Remark 2.8. In view of remark 2.6 and proposition 2.7 and [17, corollary 8.4.14] we see that enveloping algebras of finite dimensional Lie algebras and group algebras of polycyclic groups both satisfy the conditions of proposition 2.4.

§3. General results involving (3') $\Rightarrow$ (1)

In this section we map a strategy for showing (3') $\Rightarrow$ (1), which works for PI-algebras. Throughout assume $S = \{s_1, s_2, \ldots\}$ is the countable set separating the ideals of R, and write A_i for $\langle s_i \rangle$.

Remark 3.1. We may replace s_i by any element in A_i. In particular if R is Goldie we may assume each s_i is regular.

The obvious strategy is to build a maximal left ideal which does not contain any s_i and thus has core 0. Of course it would be enough to find elements $\alpha_i \neq 0$ in F and a proper left ideal containing $\alpha_i - s_i$ for each i; indeed expanding this to a maximal left ideal L, we would see that core(L) = 0 (for if $s_i \in L$ then $\alpha_i \in L$, contradiction). Inspired by §1, we would like to use a counting argument when F is uncountable. But one has to be careful, for, as Amitsur has pointed out, if $s_1 s_2 - s_2 s_1 = 1$ then

$$(\alpha_2 - s_2)(\alpha_1 - s_1) - (\alpha_1 - s_1)(\alpha_2 - s_2) = -1$$

so in this case we have no proper left ideal L containing both $(\alpha_1 - s_1)$ and $(\alpha_2 - s_2)$.

For example, this approach could not work for the Weyl algebra $\mathbb{A}_1$. Note, however, that $\mathbb{A}_1$ is a counterexample only to the method, but not to (3') $\Rightarrow$ (1) since $\mathbb{A}_1$ is simple. Let us introduce a new condition, strengthening (3').

(4) Let $\Lambda = \{\lambda_i, i = 1,2\ldots\}$ be a set of commuting indeterminates over a prime algebra R. There is a countable set S separating the ideals of R, and satisfying the additional restriction that $\sum R[\Lambda](1-\lambda_i s_i) \neq R[\Lambda]$.

Proposition 3.2. (4) implies (1), for any (prime) algebra R over a field F, such that $[R:F] \leq |F|$.

Proof. Take a maximal left ideal L of $R[\Lambda]$ containing $\sum R[\Lambda](1-\lambda_i s_i)$, and let $M = R[\Lambda]/L$, a simple $R[\Lambda]$-module. Multiplication by λ_i defines a map $\sigma_i: M \to M$, which is invertible by Schur's lemma. By a theorem of Amitsur (cf. [17, example 2.12.28]) every endomorphism of M is algebraic over F, and in particular σ_i^{-1} is algebraic. Thus $\sigma_i = p_i(\sigma_i^{-1})$ for a suitable polynomial p_i.

We use a trick of Irving [6]. Let $x = 1+L \in M$. Clearly $(1-\lambda_i s_i)x = 0$, so $x = \lambda_i s_i x$, implying $\lambda_i{}^m s_i{}^m x = (\lambda_i s_i)^m x = x$, and so $s_i{}^m x = \sigma_i{}^{-m} x$ for all m. Thus $p(s_i)x = p(\sigma_i{}^{-1})x = \sigma_i x$ yielding

$$0 = (1-s_i\lambda_i)x = (1-s_i p(s_i))x$$

Now let $L_1 = \sum R(1-s_i p(s_i))$. If $L_1 = R$ then $1 = \sum r_i(1-s_i p(s_i))$ for suitable r_i in R; but then $x = \sum r_i(1-s_i p(s_i))x = \sum r_i((1-s_i p(s_i))x) = \sum r_i 0 = 0$, which is false. Hence $L_1 < R$, and thus is contained in a maximal left ideal L_2 of R. But $\text{core}(L_2) = 0$, since otherwise L_2 contains some s_i, and then $1 = s_i p(s_i) + (1-s_i p(s_i)) \in L$, contradiction. This proves R is primitive. Q.E.D.

<u>Theorem 3.3</u>. A separating set S of R satisfies (4) if the elements of S commute with each other, such that every (finite) product of elements of S is nonzero.

<u>Proof</u>. Suppose on the contrary that $\sum R[\Lambda](1-\lambda_i s_i) = R[\Lambda]$. Then $1 \in \sum_{i=1}^{t} R[\Lambda](1-\lambda_i s_i)$ for some t. Write $1 = \sum_{i=1}^{t} h_i(\lambda_1,\lambda_2,\ldots,\lambda_m)(1-\lambda_i s_i)$ for suitable $h_i(\Lambda) = h_i(\lambda_1,\lambda_2,\ldots,\lambda_m)$ in $R[\Lambda]$. Working in the power series ring $R[[\Lambda]]$ let $q_i = \sum_{u=0}^{\infty} (\lambda_i s_i)^u = (1-\lambda_i s_i)^{-1}$. Noting each q_i centralizes S we see

$$q_t = \sum_{i=1}^{t-1} h_i(\Lambda)(1-\lambda_i s_i)q_t + h_t(\Lambda)(1-\lambda_t s_t)q_t = \sum_{i=1}^{t-1} h_i(\Lambda)q_t(1-\lambda_i s_i) + h_t(\Lambda)$$

Taking n greater than the degree of h_t in λ_t, and matching coefficients of $\lambda_t{}^n$ yields

$$s_t{}^n = \sum_{i=1}^{t-1} g_{in}(\Lambda)(1-\lambda_i s_i)$$

where we write $h_i q_t = \sum_{u=0}^{\infty} g_{iu}(\Lambda)\lambda_t{}^u$. (Note each $g_{iu} \in R[\Lambda]$.) Continuing in this way we get

$$s_t{}^{n(t)} s_{t-1}{}^{n(t-1)} \ldots s_2{}^{n(2)} = f(\Lambda)(1-\lambda_1 s_1)$$

for suitable $n(t) = n, n(t-1), \ldots, n(2)$, and $f(\Lambda) \in R[\Lambda]$. Thus

$$s_t{}^{n(t)} s_{t-1}{}^{n(t-1)} \ldots s_2{}^{n(2)} q_1 = f(\Lambda)$$

and taking $n(1) >$ degree of f in λ_1 we see $s_t{}^{n(t)} s_{t-1}{}^{n(t-1)} \ldots s_1{}^{n(1)} = 0$, contrary to hypothesis. Q.E.D.

<u>Corollary 3.4</u>. (4) holds if R is special, since then we can take $s_i = r^i$ where r is

as in definition 1.1.

Corollary 3.5. (4) holds if R is a prime PI-algebra, since by remark 3.1 we an replace s_i by any element in $A_i = \langle s_i \rangle$, and A_i contains a nonzero element of $Z(R)$ (which thus is regular).

Corollary 3.6. (4) holds in $F[G]$ if F is a perfect field and G is a supersovlable group by [12, proposition 3.2].

Theorem 3.7. Conditions (1), (2), (3), (3'), and (4) are equivalent for PI-algebras which are countably generated over an uncountable field.

Proof. Any primitive PI-ring is simple, so obviously (1) implies (3). The remainder follows from corollary 3.5. Q.E.D.

Remark 3.8. Example 2.2 gives a counterexample to (4), since (1) fails to hold!

Digression 3.9. Despite remark 3.7 it is tempting to see how far one could push the proof of theorem 3.3. Let us consider the simplest case: Are there h_1, h_2 in $R[\Lambda]$ such that $h_1(\Lambda)(1-\lambda_1 s_1) + h_2(\Lambda)(1-\lambda_2 s_2) = 1$?

Specializing $\lambda_i \to 0$ for all $i > 2$ still yields a solution, so we may assume $h_1, h_2 \in R[\lambda_1, \lambda_2]$. As before multiplying through by $q_1 = (1-\lambda_1 s_1)^{-1}$ in $R[[\lambda_1, \lambda_2]]$ yields

$$h_1(\lambda_1, \lambda_2) + h_2(\lambda_1, \lambda_2)(1-\lambda_2 s_2)q_1 = q_1$$

Write $h_i = \sum h_{iu}\lambda_1{}^u$ for h_{iu} in $R[\lambda_2]$. Matching components in $\lambda_1{}^n$ yields

$$(*)\ldots\ h_{1n}(\lambda_2) + \Sigma h_{2j}(\lambda_2)(1-\lambda_2 s_2)s_1{}^{n-j} = s_1{}^n$$

Note we could use (*) to solve for $h_{1n}(\lambda_2)$ when nonzero, and on the other hand for large n we use (*) to get the necessary and sufficient condition

$$h_{2j}(\lambda_2)(1-\lambda_2 s_2)s_1{}^{n-j} = s_1{}^n;$$

and we could solve for the coefficients of the h_{2j} using reverse induction. This would give a "generic" solution, which might be useful in building counterexamples.

§4. Farkas' approach, and related conditions

Unifying various approaches to showing (1), (2), and (3') are equivalent (in particular Dixmier [2]), Farkas [3] defined a Baire ring to be one in which the primitive spectrum

Priv(R) is Baire (i.e. any countable intersection of dense open sets is dense); Farkas showed the Baire property for prime semiprimitive rings is equivalent to the following condition:

Condition F. For any countable set $\{A_i: i \in \mathbb{N}\}$ of nonzero ideals of R there is a primitive ideal of R which does not contain any A_i.

Unfortunately condition F requires R to be prime (for if $A_1A_2 = 0$ then clearly every primitive ideal contains A_1 or A_2).

Note. For R satisfying ACC(ideals) each A_i has only a finite number of primes minimal over it; hence in verifying condition F we could assume each A_i is prime.

Definition 4.1. A ring R is Kaplansky if every homomorphic image of R is Baire.

Farkas shows that it is enough to check the prime images of R when R satisfies ACC(ideals).

It is easy to see (3') ⇒ (1) in any Kaplansky ring. (Indeed, given a separating set $S = \{s_i: i \in \mathbb{N}\}$ for a prime Baire ring R, let $A_i = \langle s_i \rangle$ and take Q a primitive ideal not containing any A_i; then Q does not contain any s_i so is 0, proving R is primitive.) Farkas' strategy is to show this Kaplansky property passes from R to T where T is a "suitable" extension of R. Farkas proved this for skew Laurent extensions in [3], and Irving for differential operator rings in [6]. Nevertheless one comes up against the following difficulty.

Suppose we would like to apply condition F to "lift" Kaplansky up a ring extension from R to T. It is enough to prove any prime homomorphic image T/P of T satisfies condition F, but $P \cap R$ need not be a prime ideal of R, so we cannot apply condition F to $R/P \cap R$. In order to bypass this difficulty we weaken Farkas' definition, and define the weak primitive spectrum WPriv(R) to be the set of finite intersections of primitive ideals. Note that we do not require these ideals to be prime; in fact if $A = P_1 \cap \ldots \cap P_t$ is prime with each P_i primitive then $P_1 \ldots P_t \subseteq A$ so some $P_i = A$ and A is primitive.

Definition 4.2. R is weakly Baire if the weak primitive spectrum is Baire.

This definition leads also to

Condition F'. For any countable set $\{A_i: i \in \mathbb{N}\}$ of ideals not contained in Nil(R) there is some member of WPriv(R) not containing any A_i.

Remark 4.3. If a prime ring R satisfies condition F' then (3') ⇒ (1). (Indeed, there are primitive ideals $P_1, \ldots, P_m$ such that $P_1 \cap \ldots \cap P_m$ fails to contain any of a separating set and is thus 0; hence $P_1 \ldots P_m = 0$ so some $P_j = 0$.)

Thus we are led to mimick Farkas' set-up, using weakly Baire and condition F'.

Lemma 4.4. A prime semiprimitive ring R is weakly Baire iff R satisfies condition F'.

Proof. (⇒) Let $\mathcal{P}_i$ = {Ideals in WPriv(R) not containing A_i}, an open subset of WPriv(R) which is dense since R is prime (so that every nonempty open subset is dense!) $\cap\, \mathcal{P}_i \neq \emptyset$ by hypothesis, so some member of WPriv(R) fails to contain any A_i.

(⇐) Suppose we are given dense open subsets $\mathcal{P}_i$ of WPriv(R); by definition there is $A_i \subset R$ such that $\mathcal{P}_i$ = {Ideals in WPriv(R) not containing A_i}. Replacing A_i by RA_iR we may assume $A_i \triangleleft R$.

By hypothesis there is an ideal of WPriv(R) not containing any A_i; call such a set A_i-admissible, and let B be the intersection of all A_i-admissible ideals. We need to show B = 0. Otherwise replacing A_i by A_iB we could find an A_iB-admissible ideal B'; in particular $B \not\subseteq B'$ and each $A_i \not\subseteq B'$, implying B' is A_i-admissible contrary to $B \not\subseteq B'$. Q.E.D.

Definition 4.5. R is weakly Kaplansky if every homomorphic image of R is weakly Baire.
It turns out that the key to the Kaplansky property is weaker even than condition F':

Condition F" For any countable set $\{A_i: i \in \mathbb{N}\}$ of ideals not contained in Nil(R) there is some semiprime ideal of R not containing any A_i. (Of course in a ring with ACC(ideals), each semiprime ideal is a finite intersection of prime ideals.)

Theorem 4.6. The following conditions are equivalent for a Jacobson ring R satisfying ACC(ideals):

(i) R is weakly Kaplansky;
(ii) Every homomorphic image of R satisfies condition F';
(iii) Every prime homomorphic image of R satisfies condition F';
(iv) Every prime homomorphic image of R either is primitive or satisfies condition F'';
(v) Every prime homomorphic image of R is weakly Baire.

Proof. (i) ⇒ (v) a fortiori.
(v) ⇒ (iii) by lemma 4.4.
(iii) ⇒ (iv) a fortiori.
(iv) ⇒ (iii) By Noetherian induction we may assume every prime homomorphic image of R does satisfy condition F', but R is a prime ring not satisfying condition F', and in particular R is not primitive. Suppose $\{A_i: i \in \mathbb{N}\}$ are nonzero ideals of R. By (iv)

there are nonzero prime ideals $P_1,\dots,P_t$ with $A_i \not\subseteq P_1 \cap\dots\cap P_t$. Applying condition F' to each R/P_j in turn gives us some B_j in WPriv(R) such that $A_i \not\subseteq P_j \Rightarrow A_i \not\subseteq B_j$, and thus each $A_i \not\subseteq B_1 \cap\dots\cap B_t$, proving R indeed satisfies condition F'.

(iii) ⇒ (ii) Suppose $\bar{R}$ is any homomorphic image of R. Condition F' does not recognize $\mathrm{Jac}(\bar{R})$, so we may mod out $\mathrm{Jac}(\bar{R})$ and assume $\mathrm{Jac}(\bar{R}) = 0$. But then 0 is a finite intersection of prime ideals since $\bar{R}$ satisfies ACC(ideals); take prime ideals $P_1,\dots,P_t$ with $P_1 \cap\dots\cap P_t = 0$. Suppose $\{A_i: i \in \mathbb{N}\}$ are nonzero ideals of $\bar{R}$. Then there is B_j/P_j in $\mathrm{WPriv}(R/P_j)$ such that $(A_i+P_j)/P_j \not\subseteq B_j/P_j$; thus $A_i \not\subseteq B_j$ whenever $A_i \not\subseteq P_j$. But $P_1 \cap\dots\cap P_t = 0$, so each $A_i \not\subseteq B_1 \cap\dots\cap B_t$.

(ii) ⇒ (i) We need to show every homomorphic image $\bar{R}$ of R is weakly Baire. But every prime homomorphic image is weakly Baire, by lemma 4.4. We may assume $\mathrm{Jac}(\bar{R}) = 0$; then 0 is a finite intersection of prime ideals, so we are done by a standard topological argument, cf. [3, lemma 2]. Q.E.D.

Corollary 4.7. Every Jacobson Kaplansky ring is weakly Kaplansky. (Proof: Check the prime images, noting condition F implies condition F'.)

In many situations "Kaplansky" and "weakly Kaplansky" are actually the same, as we see in the following result.

Theorem 4.8. Suppose R is Jacobson and satisfies ACC(ideals) and DCC on prime ideals. Then the following conditions are equivalent:

(i) (3') ⇒ (1) in every prime image of R;

(ii) R is Kaplansky;

(iii) R is weakly Kaplansky.

Proof. We have already seen (ii) ⇒ (iii) ⇒ (i), so it remains to show (i) ⇒ (ii). Assuming otherwise take a prime ideal P maximal to R/P not being Kaplansky. Passing to R/P, we may assume each proper prime image of R is Kaplansky. Moreover, we shall prove R is Baire, contrary to assumption. By theorem 4.6 (iv), it is enough to show for any countable set $\{A_i: i \in \mathbb{N}\}$ of nonzero prime ideals of R, there is some prime Q not containing any A_i. By hypothesis each A_i contains a height 1 prime; using this prime instead of A_i we may assume $\mathrm{height}(A_i) = 1$.

{height 1 prime ideals} of R is countable, for otherwise some height 1 prime B would not contain any A. But then R satisfies (3'), by proposition 2.4, so R is primitive by (i). Hence we are done taking Q = 0. Q.E.D.

Remark 4.9. By remarks 2.8 and 2.9 we see, for the cases treated in the literature ([2], [3], and [6]), that the Kaplansky property is necessary as well as sufficient to prove the equivalence of (1) and (3') in prime images of R.

The same philosophy is applicable to condition (3), when we observe that the only property of Priv(R) need in the proof of theorem 4.8 is that R is Jacobson, i.e. for every closed subset C of Spec(R), we have Priv(R)$\cap C$ is dense in C, under the relative topology. Thus we could generalize theorem 4.8 to the following meta-theorem:

Theorem 4.10. Suppose $S \subset$ Spec(R) has the property that $S \cap C$ is dense in C, for every closed subset C of Spec(R). We say R is S-Baire if S satisfies the Baire property; R is weakly S-Baire if {finite intersections of members of S} satisfies the Baire property. R is (weakly) S-Kaplansky if every prime image of R is (weakly) S-Baire. Then the following conditions are equivalent, whenever R satisfies DCC (prime ideals):

(i) For every P in Spec(R), if R/P has a countable separating set then $P \in S$;
(ii) R is S-Kaplansky;
(iii) R is weakly S-Kaplansky.

These results put the counterexample of [12, §4] in context; this is an example of a Kaplansky ring failing (3') ⇒ (3), and thus it is not S-Kaplansky, where S is the set of G-ideals. On the other hand we have

Remark 4.11. By [14] and [7], conditions (1) and (3) are equivalent for homomorphic images of an enveloping algebra of a finite dimensional Lie algebra. By [12], conditions (1) and (3) are equivalent for homomorphic images of group algebras of supersolvable groups. But then theorem 4.10 says these rings are S-Kaplansky, where S = {G-ideals}.

In the next section we shall encounter a problem with left-right symmetry. Let us define R to be resp. right Baire, right weakly Kaplansky, right Kaplansky if the right-handed analogues of the original definitions hold, i.e. if the opposite ring R^{op} is resp. Baire, weakly Kaplansky, Kaplansky.

Remark 4.12. If R is right Noetherian, Jacobson and right Kaplansky over an uncountable field then the skew Laurent extension $R[\lambda,\lambda^{-1};\sigma]$ (with respect to an automorphism) and the differential operator ring $R[\lambda;\delta]$ are each right Kaplansky. (Indeed $(R[\lambda,\lambda^{-1};\sigma])^{op} \approx R^{op}[\lambda,\lambda^{-1};\sigma^{-1}]$ so this construction is left-right symmetric, as is $R[\lambda;\delta]$ Thus we can appeal to [3] and [6].)

Recall an involution is an anti-automorphism of degree 2.

Proposition 4.13. A ring R with involution is Kaplansky (resp. weakly Kaplansky) iff R is right Kaplansky (resp. weakly Kaplansky).

Proof. Let (*) be the involution of R. We need to show that if condition F (resp. condition F') holds for the prime homomorphic images of R then the right-handed version also holds. So suppose P is a prime ideal of R, and $\{A_i: i \in \mathbb{N}\}$ is a countable set of nonzero ideals of R each containing P; we need some right primitive ideal $Q \supseteq P$ not containing any A_i. First note that we may assume each $A_i^* \subseteq A_i$; indeed if $A_i^* \subseteq P$ this is obvious, and if not then replace A_i by $A_i \cap A_i^*$.

Applying condition F to the prime image R/P^* gives a primitive ideal $Q \supseteq P^*$ not containing any A_i; thus Q^* is the desired right primitive ideal. (The argument for condition F' is analogous.) Q.E.D.

§5. Finitely generated extensions

$T \supseteq R$ is called a finite extension if T is finitely generated as left R-module (merely). As noted in §4, Farkas' strategy was to show if R is Kaplansky and $R \subseteq T$ is a particular type of extension then T is Kaplansky; this was worked out in [3] for skew Laurent extensions, and in [6] for differential operator extensions. Here we shall apply results of Letzter and of Warfield to §4, to prove that the "weak Kaplansky" approach yields analogous results for finite extensions. We start with variations on a theme of Lance Small:

Lemma 5.1 (Cortzen-Small [1]). If T is a prime ring and a finite extension of a left Noetherian ring R, then every ideal of T intersects R nontrivially.

Cortzen-Small essentially gave two proofs. The first is elementary, relying on the fact that any nonzero ideal contains a regular element a, and $a^{n+1} = \sum_{j=u}^{n} r_j a^j$ for suitable n, since Ta is Noetherian as R-module; cancelling a^u gives $r_u \in R \cap Ta$. The second proof is actually a modification of an argument Cortzen-Small used in a different theorem. It uses more machinery, but gives somewhat more. Recall any left Noetherian ring R possesses noncommutative Krull dimension, denoted K-dim, in the sense of Gordon-Robson [4].

Proposition 5.2. Suppose T is a prime ring containing a ring R having K-dim, and is f.g. as left R-module. If $A \triangleleft T$ then $R \cap A \not\subseteq \mathrm{Nil}(R)$ (so in particular is nonzero!)

Proof. (following the Cortzen-Small argument) Throughout K-dim is taken with respect to the ring R. Note Nil(R) is nilpotent, by [4, theorem 5.1]. Also T is Goldie by [4, corollary 3.4].

If $R \cap A \subseteq \mathrm{Nil}(R)$ then $\text{K-dim}(R/R\cap A) = \text{K-dim } R = \text{K-dim } T > \text{K-dim } T/A$ (since T is prime). But T/A is finite over $R/R \cap A$, so $\text{K-dim}(R/R\cap A) = \text{K-dim } T/A$, contradiction . Q.E.D.

Aside 5.2'. This proof also works for Gabriel dimension, proving the assertion, "Suppose T is a prime Goldie ring containing a ring R having G-dim, and is f.g. as left R-module. If $A \triangleleft T$ then $R \cap A$ is not nilpotent. (The obvious non-Noetherian example is affine PI-algebras.)

When R is Noetherian the proof becomes quite easy, but in this case Letzter [11, lemma 2.3] has proved the stronger result that $R \cap A$ is not contained in any height 0 prime ideal of R.

Remark 5.3. Suppose $S \subseteq R$ separates the ideals of R, and every ideal of T intersects R nontrivially. Then S separates the ideals of T. (Indeed if $0 \neq A \triangleleft T$ then $A \cap R \neq 0$ so $0 \neq (A \cap R) \cap S = A \cap S$.)

This raises the question of whether R primitive implies T primitive; if not then (3') ⇒ (1) fails for T! On the other hand Letzter [10, theorem 1.3] has proved that if R is primitive Noetherian and T is prime and both left and right finite over R then T indeed is primitive.

If we look a bit closer at Letzter's results we see he proved a bit more. (Please observe that he deals with the right-handed version.) Assume T is a finite extension of a Noetherian ring R, and T is prime. Then T is an R-T bimodule, finitely generated on each side, which is faithful and thus torsion-free on the right, cf. [17, proposition 3.5.77]. By [10, lemma 1.1], for any minimal prime P of R one can find an R-T bimodule M (which is a subfactor of T), which is torsion-free as (R/P)-T bimodule. On the other hand, [10, lemma 1.2] says that if there is a T-R bimodule which is torsion free and f.g. on each side and T is primitive then R is primitive; from this one concludes the following.

Proposition 5.4. (Letzter) Suppose T is a finite extension of a Noetherian ring R, and P is a minimal prime ideal of R.

(i) If T is primitive then R/P is primitive.

(ii) If R/P is right primitive then T is right primitive.

Proof. (i) is immediate; for (ii) note that M (defined above) is a T^{op}-$(R/P)^{op}$ bimodule, and $(R/P)^{op}$ is primitive, so [10, lemma 1.2] shows T^{op} is primitive.

Thus we have returned to the question of left-right symmetry in primitivity. This can be deflected somewhat by an approach of Small-Warfield.

Proposition 5.5. Suppose T is prime and is a finite extension of a left Noetherian ring R. Also suppose T is a left R-submodule of $R^{(n)}$. If some minimal prime ideal P of R is primitive then T is a primitive ring.

Proof. (Following an idea in a lecture of Small) By Lambek [9, p. 55, ex. 3.1.2,

3.1.3], $P = \cap\{$maximal left ideals L (in R) having core P$\}$, so $T \neq TL$ for some such L, implying TL is contained in a maximal ideal L' of T. But $\text{core}_T(L') \cap R \subseteq \text{core}_R(L) = P$, so $\text{core}_T(L') = 0$ by [11, lemma 2.3]. Q.E.D.

The crux of the matter here is whether T is a R-submodule of $R^{(n)}$, for R left Noetherian. This is known by an old result of L. Levy (cf. [17, exercise 3.1.13]) when T is torsion free over R and R is right Goldie. Warfield [19, §1, lemma] has verified Levy's criterion when R has finite GK dimension and T is right Goldie (as well as left Noetherian), thereby yielding

Corollary 5.6. If T is prime, right Goldie, of finite GK dimension, and a finite extension of a left Noetherian ring R, and if some minimal prime ideal of R is primitive, then T is primitive.

Of course a subtler approach would be to see whether or not the Kaplansky property lifts from R to T. Actually we are interested in lifting the weak Kaplansky property from R to T, in view of the discussion preceding definition 4.2, and the remaining obstacle in completing a proof is in lifting primitive ideals from R to T. Specifically we are led to

Definition 5.7. An extension $R \subset T$ is generically admissible if the following two properties hold:

(i) For every $B \triangleleft T$ we have $B \cap R \not\subseteq \text{Nil}(R)$.

(ii) There is a non-nil ideal A_0 of R, such that for every primitive ideal P of R with $A_0 \not\subseteq P$ there is a primitive ideal Q of T with $Q \cap R \subseteq P$.

$R \subseteq T$ is fully admissible if for every prime ideal P of T the extension $R/P\cap R \subseteq T/B$ is generically admissible.

Proposition 5.8. Suppose $R \subseteq T$ is fully admissible and T is Jacobson. If R is weakly Kaplansky then T is weakly Kaplansky.

Proof. Passing to a prime image of T, we may assume T is prime and need only verify property F'. Suppose $\{B_i: i \geq 1\}$ are nonzero ideals of T and let $A_i = B_i \cap R$. Then each $\not\subseteq \text{Nil}(R)$ so by hypothesis (throwing in A_0 of definition 5.7) there is a primitive ideal P of R for which each $A_i \not\subseteq P$. Take a primitive ideal Q of T with $Q \cap R \subseteq P$. Then certainly each $A_i \not\subseteq Q \cap R$, so $B_i \not\subseteq Q$. Q.E.D.

Corollary 5.9. Suppose $R \subseteq T$ is admissible and T satisfies ACC(ideals), DCC(prime ideals), and is Jacobson. If R is Kaplansky then T is Kaplansky. (Proof: Combine proposition 5.8 with theorem 4.8.)

Our main goal is to apply all this to finite extensions. Suppose $R \subseteq T$. We say LO holds from R to T if every prime ideal P of R is the intersection of R with an ideal of T, i.e. $P = R \cap TPT$. There are several situations in which this occurs, one of which is when $R \subseteq T \subseteq M_n(R)$, as pointed out by Robson and Small [16] (since then $R \cap TPT \subseteq R \cap M_n(R)PM_n(R) = R \cap M_n(P) = P$). Using the regular representation one concludes that LO holds from R to T if T is f.g. free as left R-module (since then $T \subseteq M_n(R)$ by the regular representation). Another example where LO holds is "liberal extensions," by a theorem of Bergman.

Let us say LO holds generically if there is a non-nil ideal A_0 of R such that $P = R \cap TPT$ for every prime ideal P not containing A_0. There are several extra examples of this phenomenon, such as when T is a prime PI-ring and $R = Z(T)$, where we take A_0 to be the "central kernel," cf. [17, theorem 6.1.38].

It turns out that a slightly weaker version of LO holds considerably more generally. We say $Q \lhd T$ almost lies over a prime ideal P of R if P is a minimal prime over $Q \cap R$; almost lying over (ALO) holds generically from R to T if there is a non-nil ideal A_0 of R such that for every prime ideal P (of R) not containing A_0 there is an ideal Q of T almost lying over P. (Then by the standard Zorn lemma argument we may assume Q is maximal such and thus is prime.)

Definition 5.10. AGU holds generically from R to T if ALO holds generically from $R/B \cap R$ to T/B, for all prime ideals B of T.

A theorem of Heinecke-Robson (cf. [15] and [17, exercise 2.12.17]) shows normalizing extensions satisfy AGU. Warfield [19, §5, theorem] has proved that AGU holds generically for any finite extension of a Noetherian ring of finite Gelfand-Kirillov dimension. Similarly, Letzter [11, theorem 4.6] shows AGU holds generically for any finite extension $R \subseteq T$ in which R, T both satisfy Jategaonkar's "second layer condition," cf. [8]. (The second layer condition holds for enveloping algebras of finite dimensional solvable Lie algebras, group algebras of polycyclic-by-finite groups, and Noetherian PI-algebras.) Actually we are interested in AGU for primitive ideals:

Question 5.11. If P is a primitive ideal of R is there a primitive ideal of T almost lying over P?

When T is a finite extension of a Noetherian ring R having either finite GK dim or the second layer condition, we "almost" have an affirmative answer, by proposition 5.4(ii) (except that "primitive" must be replaced by "right primitive"). However, using the ideas of [19] one can bypass this whole discussion, "generically".

Proposition 5.12. Suppose T is a finite extension of a prime Goldie ring R. Let $A_0 = \Sigma \{Rf(T) : f \in \mathrm{Hom}_R(T,R)\}$. Then $A_0 \neq 0$; moreover, if P is a right primitive ideal ideal of

R not containing A_0, then there is a right primitive ideal Q of T with $Q \cap R \subseteq P$.

Proof. As explained in [19,§2], $A_0 \neq 0$ since T is not torsion over R (indeed T has regular elements!), so we can tensor up by the simple Artinian ring $\hat{R}$ of fractions of R. $\hat{T} = T \otimes_R \hat{R}$ is a projective module over $\hat{R}$, so there is a nonzero module map $\hat{T} \to \hat{R}$, and we get a nonzero map $T \to R$ since T is finite over R. By hypothesis there is some $f: T \to R$ with $f(T) \not\subseteq P$. Note $f(T)$ is a left ideal of R. Hence $f(T) \not\subseteq I$ for any maximal right ideal I having core P, but $f(IT) = If(T) \subseteq I$, implying $IT \neq T$. Thus IT is contained in a maximal right ideal of T, whose core Q is a primitive ideal of T satisfying $Q \cap R \subseteq P$. Q.E.D.

Theorem 5.13. Suppose T is a finite extension of a Jacobson ring R.

(i) If R is left Goldie having Gabriel dimension and is right weakly Baire, and if T is prime then T is right weakly Baire.

(ii) If R is left Noetherian and right weakly Kaplansky then T is right weakly Kaplansky.

(iii) If R is Noetherian of finite GK dimension and weakly Kaplansky then T is Kaplansky

(iv) If $R \subseteq T$ is normalizing and R is weakly Kaplansky then T is weakly Kaplansky.

Proof. T is Jacobson by [1], so it remains to apply proposition 5.8.

(i) By proposition 5.2 (and aside 5.2') and proposition 5.12.

(ii) Apply (i) to the prime homomorphic images of T.

(iii) $R \subseteq T$ is admissible by proposition 5.2, [19, §5, theorem], and corollary 5.6. T is Kaplansky by corollary 5.9, since T satisfies DCC(ideals). (Indeed T has finite GK dimension.)

(iv) Primitive ideals "lift," by [17, exercise 2.5.12 and 2.12.17]. Q.E.D.

Clearly (ii) and (iii) are the focal parts of the theorem, so the following result is useful.

Proposition 5.14. Suppose T is a prime ring which is a finite extension of a left Noetherian ring R. If R satisfies DCC (prime ideals) then T satisfies DCC (prime ideals).

Proof. Let $Q_1 \supset Q_2 \supset \dots$ be a chain of primes of T. We define a chain $P \supset P_1 \supset \dots$ of primes of R such that P_i is minimal over $Q_i \cap R$. Indeed suppose inductively we have found P_i; we need to find $P_{i+1} \subset P_i$. Passing to T/Q_{i+1} we may assume $Q_{i+1} = 0$, and $Q_i \neq 0$. Thus P_i is not a minimal prime of R, by aside 5.2', and we simply take P_{i+1} to be a minimal prime of R contained in P_i.

But R satisfies DCC(prime ideals), so we have a contradiction. Q.E.D.

Corollary 5.15. In theorem 5.13(ii) if R also satisfies DCC (prime ideals) then T is Kaplansky.

Proof. Apply theorem 4.8.

§6. Special primes and Dixmier's conditions

In theorem 4.8 we noted that Farkas' theory could also be invoked to study whether (3') implies (3''), or whether (3') implies (3). In this section we look at these conditions briefly, but without conclusive results. In line with the terminology of theorem 4.10 we write "special Kaplansky" to denote the Kaplansky property with regard to {special prime ideals}, and "G-Kaplansky" to denote the Kaplansky property with respect to {G-ideals}. Often we shall be interested in the property "special weakly Kaplansky." Although this is a mouthful, it should be viewed merely as the most natural class of rings in which (3') implies (3''). If P is a special prime ideal of R and r is as in definition 1.1 we shall say P is r-special.

Proposition 6.1. Suppose every ideal of T intersects R nontrivially. If P is an r-special prime of R and there is an ideal Q of T such that $Q \cap R = P$, then P is the intersection of R with an r-special prime ideal of T.

Proof. By hypothesis Q misses all powers of r, and so can be enlarged to an r-special prime ideal Q' of T. But $Q' \cap R$ misses all powers of r and contains P, implying $Q' \cap R = P$. Q.E.D.

Thus LO for prime ideals implies LO for special prime ideals. There is a rather easy consequence for
finite extensions, i.e. free as left module.

Corollary 6.2. Suppose T is a free finite extension over a left Noetherian ring R. Any r-special prime ideal of R lifts to a r-special prime ideal of T.

Proof. Viewing $T \subseteq M_n(R)$, we are done by [16, lemma 2.1] and proposition 6.1.

On the other hand the general finite extension poses a problem for special primes:

Question 6.3. Suppose $R \subseteq T$ and P is a special prime ideal of R. If $Q \triangleleft T$ is maximal such that $Q \cap R \subseteq P$ then standard arguments show Q is prime, but is Q special prime?

We shall not go into detail here, but question 6.3 (for skew Laurent extensions, resp. for differential operator extensions) is the only serious obstruction in obtaining the analog of [3] and [6] for "special Kaplansky". It seems to me that the skew Laurent extension constructed in [12,§4] is a counterexample to this question. Let us hazard the conjecture that question 6.3 holds in these cases whenever R is Noetherian and Jacobson over an uncountable field and T has finite Gelfand-Kirillov dimension.

Of course we want to be rid of the hypothesis that T is free over R. Towards this end we could consider r-special prime ideals P for which (classical) Krull dim R/P is minimal possible (as a possibly infinite ordinal). Utilizing [8, theorem 8.2.8], Letzter [11] shows ALO holds for a finite extension T of R, if R,T both Noetherian satisfy the second layer condition; then by considering Krull dimension one can show that any prime almost lying over P indeed is r-special. A similar result can be had for finite extensions over Noetherian rings with finite GK dimension, using [19], but I do not yet see any useful applications for these results. Instead we return to examine the implications of question 6.3 for finite extensions.

<u>Proposition 6.4</u>. Suppose T is a finite extension of a left Noetherian ring R; also assume every semiprime ideal of R is the intersection of R with a semiprime ideal of T. (This would be implied by LO.) If R is special weakly Kaplansky then T is special weakly Kaplansky.

<u>Proof</u>. Otherwise take a prime ideal Q of T maximal with respect to $\mathrm{Spec}(T/Q)$ <u>not</u> satisfying the special version of condition F'. In particular we may asume Q is not special prime. Given $S = \{A_i: i \in \mathbb{N}\}$ with each $A_i \supset Q$, we wish to find prime ideals $Q_1,\dots,Q_t \supset Q$ with each $A_i \not\subseteq Q_1 \cap \dots \cap Q_t$, for then we could use the special version of theorem 4.6(iv) to obtain a contradiction. Note we can replace S by $\{A_i^k: i,k \in \mathbb{N}\}$ since this set also is countable, and thereby assume that if $A \in S$ then $A^k \in S$ for each $k>0$.

Let $\bar{T} = T/Q$ and $\bar{R} = R/Q\cap R$. Then $\bar{A}_i \cap \bar{R} \not\subseteq \mathrm{Nil}(\bar{R})$ by proposition 5.2, so there is a semiprime ideal B of R such that each $\bar{A}_i \cap \bar{R} \not\subseteq \bar{B}$. Thus $A_i \cap R \not\subseteq B$, and by hypothesis we have a semiprime ideal $\tilde{B}$ of T lying over B. Clearly $A_i \not\subseteq \tilde{B}+Q$ since $Q \subset A_i$. Let $Q_1,\dots,Q_t$ be the prime ideals of T minimal over $\tilde{B}+Q$. Then $(Q_1 \cap \dots \cap Q_t)^k \subseteq \tilde{B}+Q$ for some k; if $A_i \subseteq Q_1 \cap \dots \cap Q_t$ we conclude $A_i^k \subseteq \tilde{B}+Q$, contrary to $A_i^k \in S$. Q.E.D.

<u>Corollary 6.5</u>. If T is a finite free extension of a left Noetherian ring R then "special weakly Kaplansky" lifts from R to T. (Proof: Apply corollary 6.2 to proposition 6.4.)

On the other hand one can dispense of LO altogether, and employ instead an interesting connection between (3) and (3").

<u>Proposition 6.6</u>. Suppose T is prime and is a finite extension of a left Noetherian ring R, and (3") fails for T. Then (3) fails for the minimal prime ideals of R.

Proof. Suppose P is a minimal prime ideal of R. For each non-nilpotent a in R take an a-special prime ideal Q_a of T. Note $Q_a \neq 0$ by hypothesis, and $\cap Q_a = 0$ by §1. Let $P_{a1},\ldots,P_{at}$ be the prime ideals of R minimal over $Q_a \cap R$. Note by aside 5.2' that none of the P_{aj} are minimal primes of R, so each $P_{aj} \neq P$. $Q_a \cap R$ contains some product of the P_{aj}, implying some P_{aj} does not contain any power of a. Letting P_a denote this particular P_{aj} we see $\cap_a P_a$ misses all non-nilpotent elements of R and so is nil. Let $\mathcal{P} = \{P_a: P \subseteq P_a\}$, and $\mathcal{P}' = \{P_a: P \not\subseteq P_a\}$, and let $B = \cap\{P \in \mathcal{P}\}$ and $B' = \cap\{P \in \mathcal{P}'\}$. Then $BB' = 0$ but clearly $B' \not\subseteq P$ (since B' contains the intersection of the other minimal prime ideals of R). Thus $B \subseteq P$, so $B = P$, proving P fails (3).

Q.E.D.

Corollary 6.7. If T is a finite extension of a weakly G-Kaplanksy ring R then T is weakly special Kaplansky, so in particular (3') ⇒ (3") for T.

Remark 4.11 shows there are several well-known classes of G-Kaplansky rings, including enveloping algebras of finite dimensional Lie algebras, and group algebras of supersolvable groups. I had hoped to use the methods of this section to obtain a simpler proof of [14], but thus far this has not come about. Of course we would like the conclusion of corollary 6.7 to state that T is weakly G-Kaplansky; Letzter [19, theorem 2.6] concludes this in the rather special situation where T is finite free of finite GK dimension.

Proposition 6.8. Suppose T is a finite extension of a left Noetherian ring R, and, for every prime ideal B of T, $B \cap R$ is a semiprime ideal of R. If R is weakly G-Kaplansky then T is weakly G-Kaplansky.

Proof. We need to show if (3) fails for T then T fails for the minimal prime ideals of R, and we proceed as in the proof of proposition 6.9, this time taking Q_a to be a prime ideal not containing a. (Then some prime ideal of R minimal over $Q_a \cap R$ also fails to contain a.) Q.E.D.

Examples of this condition include normalizing extensions, and enveloping algebras over enveloping algebras of ideals. Of course any result about G-Kaplansky must contend with Lorenz' counterexample ([12, §4], and we conclude by showing it is a "natural" obstruction.

Proposition 6.9.. Suppose $T = R[\lambda,\lambda^{-1};\sigma]$ with R a left Noetherian, Jacobson, G-Kaplansky algebra over an uncountable field. If T is not G-Kaplansky then there is an example of a non-G-Kaplansky ring $\bar{T}$ which is a skew polynomial ring over a homomorphic image $\bar{R}$ of R, satisfying the further extra properties:

(i) $\bar{T}$ is primitive

(ii) $\bar{R}$ is prime

(iii) Every nonzero ideal of $\bar{T}$ intersects $\bar{R}$ nontrivially

Proof. By Farkas' theorem [3] T is Kaplansky, so it suffices to prove that every primitive image of T satisfies (3). In particular, it is enough to show T is weakly G-Kaplansky; furthermore, by Noetherian induction we may assume T is primitive and not weakly G-Baire, and may assume every proper prime homomorphic image of T is weakly G-Kaplanksy. Thus we have a set $\{A_i: i \in \mathbb{N}\}$ of prime ideals of R, such that every G-ideal of T contains some A_i.

Let P be a minimal prime ideal of R. Then $\sigma P, \sigma^2 P, \ldots$ are minimal prime ideals o, and so are finite in number, implying $\sigma^t P = P$ for some t. Let $T_0 = R[\lambda^t, \lambda^{-t}; \sigma^t] \subset T$. T is a finite extension of T_0 satisfying the conditions of proposition 6.8, so T_0 is not weakly G-Baire. Thus we may replace T by T_0 and thereby assume $\sigma P = P$. But then $TP \triangleleft T$ and $(TPT\sigma P \ldots T\sigma^{t-1}P)^k \subseteq T(P\sigma P \ldots \sigma^{t-1}P)^k = 0$ for some k, implying $TP = 0$ and so $P = 0$. Thus R is prime. Furthermore if there were a prime ideal Q of T with $Q \cap R = 0$ then by [3] there would be an uncountable number of such prime ideals, and hence we could assume each $A_i \not\subseteq Q$; passing to T/Q we conclude by Noetherian induction that there are G-ideals $Q_1, \ldots, Q_t$ each containing Q, with each $A_i \not\subseteq Q_1 \cap \ldots \cap Q_t$, contradiction. Thus every prime ideal of T intersects R nontrivially, implying easily that every nonzero ideal of T intersects R nontrivially. Q.E.D.

REFERENCES

[1] Cortzen B. and Small L., preprint.

[2] Dixmier J., "Idéaux primitifs dans les algèbres enveloppantes," J. Algebra **48** (1977), 96-112.

[3] Farkas, D., "Baire categories and Laurent Extensions," Can. J. Math. **31** (1979), 824-830.

[4] Gordon R. and Robson C., Krull Dimension, Mem. Amer. Math. Soc. 133 (1973).

[5] Herstein, I.N., Noncommutative Rings, Carus Mathematical Monographs **15** (1968), Math. Assoc. Amer., Providence, R.I.

[6] Irving R. "Prime ideals of certain Noetherian algebras," Math. Z. **169** (1979), 77-92.

[7] Irving, R. and Small, L., "On the characterization of primitive ideals in enveloping algebras," Math. Z. **173** (1980), 217-221.

[8] Jategaonkar A.V., Localization in Noetherian Rings, London Mathematical Lecture Note Series 98 (1986), Cambridge Univ. Press.

[9] Lambek J., Lecture Notes on Rings and Modules (1966), Blaisdell.

[10] Letzter E., "Primitive ideals in finite extensions of Noetherian rings," preprint.

[11] _________, "Prime ideals in finite extensions of Noetherian rings," preprint.

[12] Lorenz, M., "Primitive ideals in group algebras of supersoluble groups, Math. Ann. **225**, 115-122.

[13] _________, "Completely prime primitive ideals in group algebras of finitely

generated nilpotent-by-finite groups," Comm. in Algebra **6**, 717-734.

[14] Moeglin C., "Idéaux primitifs des algèbres enveloppantes," J. Math. Pures et Appl. 59 (1980), 265-336.

[15] Passman D., "Prime ideals in normalizing extensions," J. Algebra 73, pp. 556-572.

[16] Robson C. and Small L., "Liberal extensions," Proc. London Math. Soc. (3) **42** (1981), 87-103.

[17] Rowen, L.H., Ring Theory, 2 volumes in Pure and Applied Mathematics Series (127 and 128), Academic Press (1988).

[18] Warfield R., "The trace ideal and lying over for links," preprint (1987-88).

[19] ————, "Miscellaneous remarks," preprint (1987).

"FORMES REDUITES DES AUTOMORPHISMES ANALYTIQUES DE $\mathbb{C}^n$ A VARIETE LINEAIRE FIXE ET REPULSIVE"

Danielle COUTY - 22 rue de Béarn - 65310 - ODOS -

[Cet article apparaît dans sa forme définitive et aucune autre version de ce travail ne sera soumise ailleurs pour publication].

INTRODUCTION

Soit $\mathcal{F}$ un germe d'application analytique au voisinage d'une variété linéaire Ω de $\mathbb{C}^n$ tel que $\mathcal{F}(z) = z$ pour tout $z \in \Omega$. On s'intéresse ici à la recherche d'une "forme réduite" $\mathcal{N}$ de $\mathcal{F}$. Une forme réduite $\mathcal{N}$ de $\mathcal{F}$ est un automorphisme polynomial de $\mathbb{C}^n$ tel que $\mathcal{N}(z) = z$, pour tout $z \in \Omega$ et vérifiant entre autres propriétés l'équation fonctionnelle $\mathcal{F} \circ \mathcal{P} = \mathcal{P} \circ \mathcal{N}$ sur un ouvert contenant Ω, $\mathcal{P}$ étant le germe d'une application analytique au voisinage de Ω.

Ce problème n'est pas soluble en général [18] même dans le cas où Ω se réduit à un point.

On se place ici dans le cas particulier où la variété Ω est répulsive pour $\mathcal{F}$. (Pour une définition précise d'une variété répulsive, voir au chapitre III la définition 3.1.1).

Le but de ce travail est de prouver que $\mathcal{F}$ admet alors une forme réduite $\mathcal{N}$, Ω étant fixe et répulsive pour $\mathcal{N}$. S'il existe un ouvert $\mathcal{V}$ contenant Ω tel que $\mathcal{F}(\mathcal{V}) \subset \mathcal{V}$, l'application $\mathcal{P}$ satisfaisant $\mathcal{F} \circ \mathcal{P} = \mathcal{P} \circ \mathcal{N}$ peut en fait s'étendre en une application entière. De plus, si $\mathcal{F}$ est injective sur $\mathcal{V}$, $\mathcal{P}$ est injective sur $\mathbb{C}^n$ et $\mathcal{P}(\mathbb{C}^n)$ est l'ensemble des $z \in \bigcap_{m \geq 1} \mathcal{F}^m(\mathcal{V})$ tels que la suite $(\mathcal{F}^{-m}(z))_{m\geq 1}$ converge vers un élément de Ω.

Dans certains cas on peut obtenir une forme réduite plus simple, appelée "forme normale" de $\mathcal{F}$. L'étude d'une forme normale pour des automorphismes analytiques admettant une sous variété de points fixes a été par exemple entreprise par Nishimura [19] dans le cas d'une sous-variété Y d'une variété analytique complexe X. Nishimura se limite au cas particulier où les valeurs propres de l'automorphisme dans la direction transversale à Y sont sans relation. (Pour une définition des relations entre les valeurs propres voir chapitre I, définition 1.3.3. Si $\mathcal{F} : \mathcal{U} \subset \mathbb{C}^{\ell} \times \mathbb{C}^p \to \mathbb{C}^{\ell} \times \mathbb{C}^p$, $(x,y) \to (\varphi(x,y), \Psi(x,y))$ où $\mathcal{U}$ est un ouvert de $\mathbb{C}^n$ contenant la sous-variété $Y = \mathbb{C}^{\ell} \times 0_{\mathbb{C}^p}$, les valeurs propres dans la direction transversale à Y sont les valeurs propres de $\frac{\partial \Psi}{\partial y}(x,0)$).

On montre au chapitre III que dès que la variété linéaire de points fixes est répulsive ou attractive pour un germe d'application analytique au voisinage de Ω, ses valeurs propres sont constantes : cette remarque simplifie beaucoup l'étude des formes réduites pour les applications analytiques admettant une variété linéaire de points fixes répulsifs ou attractifs.

Parallèlement à l'étude de telles applications, il existe de nombreux travaux sur les germes de fonctions analytiques au voisinage de l'origine, 0 étant un point fixe attractif ou répulsif. On trouve par exemple dans Stehle [25] que, si $\mathcal{F}$ est une application holomorphe injective au voisinage de 0 et si $\mathcal{F}(0)=0$, alors l'équation fonctionnelle $\mathcal{F} \circ \mathcal{P} = \mathcal{P} \circ \mathcal{F}'(0)$ admet une solution si $\mathcal{F}'(0)= kI$ avec $|k|>1$. Si de plus, $\mathcal{F}$ est un automorphisme de $\mathbb{C}^n$, $\mathcal{P}$ s'étend en une fonction entière injective telle que $\mathcal{P}(\mathbb{C}^n) = \{z \in \mathbb{C}^n \,/\, \lim_{m \to +\infty} \mathcal{F}^{-m}(z) = 0\}$ (voir aussi Kodaira [15], Lattes [16], Sadullaev [24]).

Plus généralement, il existe un automorphisme polynomial $\mathcal{N}$ tel que l'équation fonctionnelle $\mathcal{F} \circ \mathcal{P} = \mathcal{P} \circ \mathcal{N}$ admette une solution locale $\mathcal{P}$, définie au voisinage de 0, même dans le cas où les valeurs propres de $\mathcal{F}'(0)$ ont des relations (Reich [22] et [23]). Nous renvoyons à [8] pour un aperçu historique de ces questions déjà étudiées par exemple par Poincaré dans [21] et par Picard dans [20].

J. Esterle et PG. Dixon se sont intéressés [8,9] à l'étude d'équations fonctionnelles du type $\mathcal{F} \circ \mathcal{P} = \mathcal{P} \circ \mathcal{N}$, comme moyen de construction de "fonctions de Bieberbach" (pour une définition voir [8] p. 154). Ces fonctions sont des fonctions entières injectives, dont le jacobien est constamment égal à 1 et d'image non dense. On trouve les premiers exemples de fonctions de Bieberbach dans [2] et dans [11]. Nishimura donne dans [19] un exemple de fonction de Bieberbach dont l'image évite le voisinage d'une droite complexe, ce voisinage étant décrit dans [6].

Dans un manuscrit non publié, [10], J. Esterle aborde l'étude des formes normales d'automorphismes analytiques admettant l'origine comme point fixe répulsif ou attractif en présence de relations. La méthode utilisée ici reprend la démarche de [10], en l' appliquant au cas d'une variété linéaire répulsive de points fixes. On peut assez facilement montrer que pour tout $x \in \Omega$, il existe effectivement un germe d'application $\mathcal{P}(x)$ défini au voisinage de x vérifiant $\mathcal{F} \circ \mathcal{P}(x) = \mathcal{P}(x) \circ \mathcal{N}$. La difficulté essentielle de ce travail est d'obtenir une application $\mathcal{P}$ définie sur tout un voisinage de Ω coïncidant avec $\mathcal{P}(x)$ au voisinage de chaque x tout en imposant à $\mathcal{N}$ des conditions garantissant que $\mathcal{P}$ sera en fait entière si $\mathcal{F}$ laisse stable un voisinage de Ω. (Il ne suffit pas pour cela que $\mathcal{N}$ soit polynomial). Cette approche du problème de la normalisation est à notre connaissance nouvelle en présence de relations. (En l'absence de relations des propriétés d'unicité des applications $\mathcal{P}(x)$ permettent d'effectuer facilement des recollements le long de Ω).

Au chapitre I, on trouve une présentation des résultats formels. Au §3 on rappelle le théorème de Jordan - Chevalley, qui est un des théorèmes-clés de la méthode de [10].

Aux chapitres II et III, on prouve que l'on peut remplacer l'étude générale d'un germe d'application analytique au voisinage de Ω par l'étude d'un germe d'application analytique ayant une forme plus simple. C'est-à-dire, si $\Omega = \{(x,y) \in \mathbb{C}^{\ell} \times \mathbb{C}^{p} \ / \ y=0\}$ toute application analytique admettant Ω comme variété répulsive de points fixes peut se mettre sous la forme $\mathfrak{F} = (\varphi,\Psi)$ ($\varphi : \mathbb{C}^{\ell+p} \to \mathbb{C}^{\ell}$ et ψ $\mathbb{C}^{\ell+p} \to \mathbb{C}^{p}$) : avec $\varphi(x,y) = x$ pour tout $x \in \mathbb{C}^{\ell}$. Pour cela, on étudie au chapitre II un changement de variable formel, alors que l'objet du chapitre III est une étude analytique du même problème.

Notons alors $\mathfrak{F}(x,y) = (x,F(x,y))$. On peut s'inspirer des méthodes utilisées dans [8], section 6, en les appliquant à l'application F_x (où F_x désigne, pour $x \in \mathbb{C}^{\ell}$, l'application $y \to F(x,y)$). C'est ce qui est fait au chapitre IV, où on obtient ainsi la forme réduite cherchée, d'abord dans le cas particulier où $\mathfrak{F}(x,y) = (x,F(x,y))$, puis dans le cas général à l'aide des résultats du chapitre III.

Additif Mars 1989 : Le cours de 3° cycle de M. CHAPERON sur les systèmes dynamiques différentiels [31], publié dans Astérisque, contient une étude des actions formelles et des germes d'actions différentiables de certains groupes qui recoupe largement les trois premières parties de ce travail (on peut certainement déduire les résultats de la 3° partie de son "théorème du point fixe à tiroirs" [31, p. 89], ou de sa Note de 1975 ([28]) . Compte tenu de l'extrême diversité des points de vue sous lesquels la question des formes normales a été abordée dans la littérature il m'a paru souhaitable de publier ce travail dans sa forme actuelle. Aussi curieux que cela puisse paraître l'étude de l'aspect formel au Chapitre I, dont tous les résultats se trouvent dans la thèse de LEAU publiée aux Annales de Toulouse en 1897 [17], présente un aspect original. Aucun des nombreux auteurs qui se sont penchés sur la question ne semble avoir remarqué que la clef des formes normales formelles est donnée par le théorème de Jordan-Chevalley sur les groupes algébriques.

Je remercie M. CHAPERON d'avoir attiré mon attention sur son cours de 3° cycle [31] et sur d'autres références concernant les formes normales pour les germes d'automorphismes ou les germes de champs de vecteurs ([28], [29], [30], [32], [33], [34], [35], [36]).

CHAPITRE I

Nous nous intéressons dans ce chapitre au problème de la diagonalisation "d'applications formelles" définies par une famille de séries formelles et plus généralement au problème de l'existence de "formes normales" pour ces "applications formelles" dans le cas non diagonalisable. Ce sujet est très classique (voir Poincaré [21], Leau [17], Sternberg [26], Reich [22], etc...).

§ 1 - Problème formel de diagonalisation

Soit Δ l'ensemble des séries formelles à coefficients dans $\mathbb{C}$ en p variables commutatives $(X_1,...,X_p)$, sans terme constant. (p étant fixé une fois pour toutes et étant quelconque).

Soit Δ_k l'ensemble des polynômes en p variables commutatives $(X_1,...,X_p)$ sans terme constant et de degré inférieur ou égal à k (pour $k \geq 1$).

Posons $\mathcal{Q} = \Delta^p$. Soit $F \in \mathcal{Q}$ et $G \in \mathcal{Q}$. Définissons $G \circ F$:

- Si $g \in \Delta$ et $F = (f_1,...,f_p) \in \mathcal{Q}$, on définit $g \circ F$ comme l'élément de Δ obtenu en remplaçant (X_i) par (f_i) dans le développement de g.

Si $G = (g_1,...,g_p) \in \mathcal{Q}$ alors on pose $G \circ F = (g_1 \circ F,..., g_p \circ F)$.

La loi o est associative sur $\mathcal{Q}$ et admet $I=(X_1,...,X_p)$ comme élément neutre. $\mathcal{Q}$ est donc ainsi muni d'une structure de monoïde, non abélien.

Si $f \in \Delta$, on note $\Pi_k(f) \in \Delta_k$ le polynôme obtenu en tronquant la série f à l'ordre k. De même, si $F =(f_1,...,f_p) \in \mathcal{Q}$, on pose

$$\Pi_k(F) = (\Pi_k(f_1),..., \Pi_k(f_p)).$$

On vérifie la relation suivante pour tout F et G éléments de $\mathcal{Q}$

(α) $\Pi_k [\Pi_k (F) \circ (\Pi_k (G))] = \Pi_k (F \circ G)$.

Soit alors dans $\mathcal{Q}$ la relation d'équivalence $\mathcal{R}_k$ définie par

$$F \, \mathcal{R}_k \, G \Leftrightarrow \Pi_k (F) = \Pi_k (G).$$

Par (α), $\mathcal{R}_k$ est compatible avec la loi o. Considérons alors $\mathcal{Q}_k = \mathcal{Q} / \mathcal{R}_k$.

On peut identifier $\mathcal{Q}_k$ avec $\Pi_k (\mathcal{Q}) = [\Delta_k]^p$ et $\mathcal{Q}_k$ a une structure de monoïde pour la loi Θ définie pour tout F et G $\in \mathcal{Q}_k$ par $F \Theta G = \Pi_k (F \Theta G)$. Pour tout $f \in \Delta$ et tout $F \in \mathcal{Q}$, on vérifie la relation

(β) $\Pi_k (f) \Theta \Pi_k (F) = \Pi_k (f \circ F)$.

Si $F \in \mathcal{Q}$, $\Pi_1(F)$ est la "différentielle formelle" de F que l'on note F'(0). Il est clair que si F est définie par des séries convergentes sur un voisinage de l'origine, F'(0) est la différentielle de F au sens usuel. On sait que F est inversible pour la loi o si et seulement si $F'(0) \in GL(\mathbb{C}^p)$. Soit $\mathcal{G}$ le groupe des éléments de $\mathcal{Q}$ inversibles pour la loi o. Soit $\mathcal{G}_k$ le groupe quotient $\mathcal{G} / \mathcal{R}_k$. $\mathcal{G}_k$ est le groupe des éléments de $\mathcal{Q}_k$ inversible pour la loi Θ.

On vérifie alors facilement la proposition suivante :

PROPOSITION 1.1.1 -

(i) *Soit* $F \in \mathcal{Q}$. *Soit* $\hat{F} : \Delta \to \Delta$

$$f \to \hat{F}(f) = f \circ F.$$

Alors $\hat{F} \in \mathrm{End}(\Delta)$ *et pour tout* $F, G \in \mathcal{Q}$, *on a* $\widehat{F \circ G} = \hat{G} \circ \hat{F}$.

(ii) *Soit* $F \in \mathcal{Q}_k$. *Soit* $\hat{F}^{(k)} : \Delta_k \to \Delta_k$

$$f \to \hat{F}^{(k)}(f) = f \Theta F.$$

Alors $\hat{F}^{(k)} \in \mathrm{End}\, \Delta_k$ *et pour tout* $F, G \in \mathcal{Q}_k$, *on a* $\widehat{F \Theta G}^{(k)} = \hat{G}^{(k)} \circ \hat{F}^{(k)}$.

Définition 1.1.2. - Soit $D \in \mathcal{Q}$. On dit que D est diagonal si $D = (\lambda_1 X_1, \ldots, \lambda_p X_p)$ avec $(\lambda_1, \ldots, \lambda_p) \in \mathbb{C}^p$.

Définition 1.1.3. - Soit $F \in \mathcal{A}$. On dit que F est diagonalisable dans $\mathcal{A}$ s'il existe $P \in \mathcal{G}$ et E diagonal tels que $PoFoP^{-1} = D$.

Définition 1.1.4. - Soit $F \in \mathcal{A}_k$. On dit que F est diagonalisable dans $\mathcal{A}_k$ s'il existe $P \in \mathcal{G}_k$ et D diagonal tels que $P \ominus F \ominus P^{-1} = D$.

Définition 1.1.5. - Soit $F \in \mathcal{A}$ ou $F \in \mathcal{A}_k$. On dit que F est unipotent si toutes les valeurs propres de la différentielle à l'origine F'(0) sont égales à 1.

THEOREME 1.1.6 -

(i) *Un élément* F *de* $\mathcal{A}_k$ *est diagonalisable si et seulement si* $\hat{F}$ *est diagonalisable dans* $\mathrm{End}(\Delta_k)$.

(ii) *Si* $F \in \mathcal{A}_k$, *la famille des valeurs propres de* $\hat{F}$ *est* $\{\lambda_1^{i_1} \dots \lambda_p^{i_p} \, / \, 1 \leq i_1 + \dots + i_p \leq k\}$ *où* $(\lambda_1, \dots, \lambda_p)$ *est la famille des valeurs propres de* F'(0).

Démonstration - Montrons (i). Supposons $\hat{F}$ diagonalisable. Alors $\hat{F}$ admet un système de vecteurs propres qui est une base de Δ_k. On en déduit qu'on peut en extraire un système libre de p vecteurs, soit $(g_1, \dots, g_p)$ tel que $(\Pi_1(g_1), \dots, \Pi_p(g_p))$ engendre Δ_1 et soit un système libre. Posons alors $P = (g_1, \dots, g_p)$. D'après le choix des $(g_i)_{1 \leq i \leq p}$, on a $P'(0) \in GL(\mathbb{C}^p)$ donc $P \in \mathcal{G}_k$. Montrons que $P \ominus F \ominus P^{-1}$ est diagonal :

$$P \ominus F \ominus P^{-1} = P \ominus [F \ominus P^{-1}] = (g_1 \ominus [F \ominus P^{-1}], \dots, g_p \ominus [F o P^{-1}])$$

g_i est un vecteur propre de $\hat{F}$. Notons λ_i la valeur propre à laquelle il est associé. Alors on a :

$$\begin{aligned} g_i \ominus [F \ominus P^{-1}] = (g_i \ominus F] \ominus P^{-1} = [\hat{F}(g_i)] \ominus P^{-1} &= \lambda_i g_i \ominus P^{-1} \\ &= \lambda_i [g_i \ominus P^{-1}] \\ &= \lambda_i X_i \end{aligned}$$

donc $P \ominus F \ominus P^{-1} = (\lambda_1 X_1, \dots, \lambda_p X_p)$.

Réciproquement, supposons F diagonalisable ; alors il existe $P \in \mathcal{G}_k$ tel que $P \ominus F \ominus P^{-1} = D$ avec D diagonal.

Pour tout F_1 et $F_2 \in \Delta_k$, on a $F_1 \ominus F_2 = \hat{F}_2 \, o \, \hat{F}_1$ donc on a $\hat{P}^{-1} o \hat{F} o \hat{P} = \hat{D}$. Il suffit donc de montrer que si D est diagonal, alors $\hat{D}$ est diagonalisable.

Soit $D = (\lambda_1 X_1,\dots,\lambda_p X_p)$ avec $\lambda_i \in \mathbb{C}$ pour tout $i, 1 \le i \le p$. Alors tout monôme $X_1^{i_1}\dots X_p^{i_p}$ ($1 \le i_1 + \dots i_p \le k$) est un vecteur propre de $\hat{D}$. En effet

$$\hat{D}(X_1^{i_1}\dots X_p^{i_p}) = \lambda_1^{i_1} X_1^{i_1} \dots \lambda_p^{i_p} X_p^{i_p}$$
$$= \lambda_1^{i_1} \dots \lambda_p^{i_p} X_1^{i_1} \dots X_p^{i_p}.$$

Les monômes formant une base de Δ_k, $\hat{D}$ est donc diagonalisable.

Montrons (ii). Soit $P[X] = (X-\lambda_1)\dots(X-\lambda_p)$ .Posons $\hat{P}[X] = \Pi(X-\lambda_1^{i_1}\dots\lambda_p^{i_p})_{1 \le i_1+\dots+i_p \le k}$.

Pour $F \in \mathcal{G}_k$, notons $P_F[X]$ le polynôme caractéristique de $F'(0)$ et $P_{\hat{F}}[X]$ le polynôme caractéristique de $\hat{F}$. Si D est diagonal, l'argument ci-dessus prouve que $P_{\hat{D}}[X] = \hat{P}_D[X]$. Si F est diagonalisable, il existe $P \in \mathcal{G}_k$ et D diagonal tels que $F = P^{-1} \ominus D \ominus P$, donc $F'(0)$ est semblable à D et $\hat{F}$ est semblable à $\hat{D}$. On a donc

$$P_{\hat{F}}[X] = P_{\hat{D}}[X] = \hat{P}_D[X] = \hat{P}_F[X].$$

On veut montrer que l'égalité (1) $P_{\hat{F}}[X] = \hat{P}_F[X]$ est vraie pour tout $F \in \mathcal{G}_k$.

Choisissons dans Δ_k, la base $B = (e_{i_1,\dots,i_p})_{1 \le i1+\dots+ ip \le k}$ avec $e_{i_1,\dots,i_p} = X_1^{i_1}\dots X_p^{i_p}$.

Soit $F \in \mathcal{G}_k$ avec $F = (f_1,\dots,f_p)$. Pour tout i, $1 \le i \le p$ $f_i \in \Delta_k$ et s'écrit dans la base B . Notons q la dimension de $\mathcal{G}_k$ et $(\alpha_1(F),\dots,\alpha_q(F))$ les coordonnées de F dans $\mathcal{G}_k$, obtenues en regroupant les coordonnées des $(f_i)_{1 \le i \le p} \in \Delta_k$. Soit maintenant $\hat{F} \in End(\Delta_k)$. les vecteurs colonnes de la matrice de $\hat{F}$ dans la base B sont les vecteurs $\hat{F}(X_1^{i_1}\dots X_p^{i_p}) = \Pi_k(f_1^{i_1}\dots f_p^{i_p})$. Les coefficients de la matrice de $\hat{F}$ sont donc des fonctions polynômes des coefficients $\alpha_1(F),\dots,\alpha_q(F)$. Notons r la dimension de Δ_k et $P_{\hat{F}}[X] = X^r + a_{r-1} X^{r-1} + \dots a_o$. Alors chaque a_i, $1 \le i \le r$ est fonction polynôme des coefficients $\alpha_1(F),\dots,\alpha_q(F)$.

D'autre part, $P_F[X] = (X-\lambda_1)\dots(X-\lambda_p)$ est un polynôme dont les coefficients sont des fonctions polynômes des coordonnées de la matrice associée à $F'(0)$ donc fonctions polynômes de $\alpha_1(F),\dots,\alpha_q(F)$. Les coefficients de $\hat{P}_F[X]$ sont donnés par des expressions symétriques de

$\lambda_1,...,\lambda_p$, donc peuvent s'exprimer comme fonction polynôme des coefficients de $P_F[X]$. Finalement, les coefficients de $\hat{P}_F[X]$ - $P_{\hat{F}}[X]$ sont donc des fonctions polynômes de $\alpha_1(F),...,\alpha_q(F)$.

Par ailleurs, notons $R(P_{\hat{F}})$ le résultant du polynôme $P_{\hat{F}}[X]$ et de son polynôme dérivé. Alors on sait que $\hat{F}$ a ses valeurs propres distinctes si et seulement si $R[P_{\hat{F}}] \neq 0$. Finalement, si $R(P_{\hat{F}}) \neq 0$, $\hat{F}$ est diagonalisable, donc F est diagonalisable (d'après (i)) et l'égalité (1) est vérifiée.

Comme $R(P_{\hat{F}}) = S(a_{r-1},..., a_0)$ où S est un polynôme, et où les $(a_I)_{0 \leq i \leq r}$ sont les coefficients de $P_{\hat{F}}[X]$, $R(P_{\hat{F}}) = Q[\alpha_1(F),...,\alpha_q(F)]$ où Q est un polynôme. On peut choisir $D = (\lambda_1 X_1,...,\lambda_p X_p)$ telle que $(\lambda_1^{i_1},...,\lambda_p^{i_p})$ $1 \leq i_1+...i_p \leq k$ soient distincts, et alors $Q[\alpha_1(D),...,\alpha_q(D)] \neq 0$. Q n'est pas identiquement nul, donc Q est non nul sur un ensemble dense de $\mathbb{C}^q$.

Les coefficients de $\hat{P}_F[X]$ - $P_{\hat{F}}[X]$ sont donc des fonctions polynômes de $\alpha_1(F)...\alpha_q(F)$ nulles sur un ensemble dense de $\mathbb{C}^q$, donc nulles partout.

Donc $\hat{P}_F[X] = P_{\hat{F}}[X]$ pour tout $F \in \mathcal{Q}_k$.

COROLLAIRE 1.1.7 - *Un élément* F *de* $\mathcal{Q}_k$ *est unipotent si et seulement si* $\hat{F}$ *est unipotent dans* $\mathrm{End}(\Delta_k)$.

COROLLAIRE 1.1.8 - *Un élément* F *de* $\mathcal{Q}_k$ *est inversible si et seulement si* $\hat{F}$ *est inversible dans* $\mathrm{End}(\Delta_k)$.

PROPOSITION 1.1.9 - *Un élément* F *de* $\mathcal{Q}$ *est diagonalisable si et seulement si* $\Pi_k(F)$ *est diagonalisable dans* $\mathcal{Q}_k$ *pour tout* $k \geq 1$.

Démonstration - Notons $\Pi_k(F) = F_k$ pour tout $k, k \geq 1$. S'il existe $P \in \mathcal{G}$ et $(\lambda_1,...,\lambda_p) \in \mathbb{C}^p$ tels que $PoFoP^{-1} = (\lambda_1 X_1,...,\lambda_p X_p)$ alors on a pour tout $k \geq 1$:

$$P_k \ominus F_k \ominus P_k^{-1} = \Pi_k(\lambda_1 X_1,...,\lambda_p X_p) = (\lambda_1 X_1,...,\lambda_p X_p).$$

Donc F_k est diagonalisable dans $\mathcal{Q}_k$ pour tout $k \geq 1$. Réciproquement, supposons que F_k est diagonalisable dans $\mathcal{Q}_k$ pour tout $k \geq 1$. Notons V_k l'ensemble des valeurs propres de F_k et, pour tout $\lambda \in V_k$, notons $E_{\lambda,k}$ le sous-espace propre de $\hat{F}_k$ associé à la valeur propre λ. Pour $\lambda \notin V_k$ on pose $E_{\lambda,k} = \{0\}$. D'après le théorème 1.1.6 on sait que $V_k \subset V_{k+1}$ pour tout $k \geq 1$. D'autre part, si $\lambda \in V_k$ on a $\Pi_k(E_{\lambda,k+1}) \subset E_{\lambda,k}$ et si $\lambda \in \mathbb{C} \setminus V_k$ on a $\Pi_k(E_{\lambda,k+1}) = 0$. Comme F_k est diagonalisable dans $\mathcal{A}_k$, on a finalement $\Delta_k = \bigoplus_{\lambda \in \mathbb{C}} E_{\lambda,k}$ et $\Delta_k = \Pi_k(\Delta_{k+1}) = \bigoplus_{\lambda \in \mathbb{C}} \Pi_k(E_{\lambda,k+1})$ pour tout $k \geq 1$. Donc $E_{\lambda,k} = \Pi_k(E_{\lambda,k+1})$ pour tout $\lambda \in \mathbb{C}$, pour tout $k \geq 1$. Notons $(\lambda_1,\ldots,\lambda_p)$ les valeurs propres de $\hat{F}_1$ (c'est-à-dire de F'(0)) et soit dans Δ_1 une base $(e_1,\ldots,e_p)$ telle que $\hat{F}_1(e_i) = \lambda_i e_i$ pour tout i, $1 \leq i \leq p$. D'après ce qui précède, on peut construire pour tout i, une famille $(e_i^{(n)})_{n\geq 1}$ vérifiant

$$e_i^{(n)} \in E_{\lambda_i,n} \quad \text{et} \quad \Pi_i(e_i^{(n+1)}) = e_i^{(n)} \quad (n \geq 1).$$

Soit $g_i \in \Delta$ tel que $\Pi_k(g_i) = e_i^{(k)}$ $(k\geq 1)$. En particulier

$$\Pi_1(g_i) = e_i \text{ et } \Pi_k(g_i o F) = e_i^{(k)} \odot \Pi_k(F) = \hat{F}_k(e_i^{(k)}) = \lambda_i e_i^{(k)} = \Pi_k(\lambda_i g_i) \text{ pour tout } k \geq 1$$

donc $g_i \circ F = \lambda_i g_i$.

Finalement en posant $G = (g_1,\ldots,g_p)$ et $D = (\lambda_1 X_1,\ldots,\lambda_p X_p)$ on a GoF = DoG. Comme $(e_1,\ldots,e_p)$ est une base de Δ_1, G est un élément de $\mathcal{G}$, donc F est diagonalisable et ceci achève la démonstration.

On considère $\mathcal{G}_k$ et $GL(\Delta_k)$ munies de leurs topologies naturelles, c'est-à-dire que $\mathcal{G}_k$ est muni de la topologie de la convergence coordonnée par coordonnée, qui fait de $\mathcal{G}_k$ un groupe topologique et $GL(\Delta_k)$ est muni de la topologie induite par la topologie usuelle de $End(\Delta_k)$. On peut alors vérifier la proposition suivante.

PROPOSITION 1.1.10 - *Soit* $\psi : \mathcal{G}_k \to GL(\Delta_k)$

$F \to \psi(F) = \hat{F}$.

Alors ψ *est un antihomomorphisme de groupe injectif et continu pour les topologies naturelles.*

§ 2 - Théorème de Jordan-Chevalley

Nous avons besoin des notions classiques de variété affine et de topologie de Zariski, qui sont exposées, entre autres, dans [14].

On dit que V est une variété affine de $\mathbb{C}^n$ s'il existe une famille finie $(P_i)_{1\leq i\leq k}$ de polynômes de $\mathbb{C}[X_1,...,X_n]$ tels que

$$V = \{x \in \mathbb{C}^n / P_i(x) = 0 \quad \forall_i,\ 1 \leq i \leq k\}.$$

Toute variété affine est un fermé pour la topologie usuelle de $\mathbb{C}^n$. Les variétés affines de $\mathbb{C}^n$ vérifient les axiomes des fermés d'une topologie. Cette topologie, définie par ses fermés, est appelée topologie de Zariski sur $\mathbb{C}^n$. Elle est moins fine que la topologie usuelle et elle n'est pas séparée.

Notons $\mathbb{C}[X_{i,j}, X_{n^2+1}]$ l'algèbre des polynômes sur $\mathbb{C}$ en n^2+1 variables $X_{11},.. X_{1n}, X_{21}... X_{nn}, X_{n^2+1}$.

Soit $d = \det (X_{ij})_{\substack{1\leq i\leq n\\ 1\leq j\leq n}}$

Posons $G_n = \{(x, \frac{1}{d(x)}),\ x \in GL(n,\mathbb{C})\}$.

On peut identifier $GL(n,\mathbb{C})$ à G_n et G_n est une variété affine de $\mathbb{C}^{n^2+1}$. En effet $G_n = \{ X \in \mathbb{C}^{n^2+1} / P(X) = 0\}$ où $P = 1 - X_{n^2+1}$ où $P \in \mathbb{C}[X_{ij}, X_{n^2+1}]$.

PROPOSITION 1.2.1 - *Soit*

$$\psi : \mathcal{G}_k \longrightarrow GL(\Delta_k)$$
$$F \longrightarrow \hat{F}.$$

Alors $\psi(\mathcal{G}_k) = \hat{\mathcal{G}}_k$ *est un fermé pour la topologie de Zariski de* $GL(n,\mathbb{C})$ *où* $n = \dim \Delta_k$.

Démonstration - Soit $u \in GL(\Delta_k)$. Si $u \in \hat{\mathcal{G}}_k$, il existe $v = (v_1,..., v_p) \in \mathcal{G}_k$ tel que $u = \hat{v}$. Alors nécessairement, pour tout ℓ, $1 \leq \ell \leq p$, $v_\ell = \hat{v}(X_\ell) = u(X_\ell)$. Donc si $u \in \hat{\mathcal{G}}_k$, on a :

$$u = \widehat{(u(X_1),..., u(X_p))}$$

et on voit aisément que la réciproque est vraie. Or

$$u = \widehat{(u(X_1),..., u(X_p))}$$

si et seulement si, pour tout $(i_1,...,i_p) \in \mathbb{N}^p$ tel que $1 \leq i_1 + ... + i_p \leq p$ on a :

$$u(X_1^{i_1}\dots X_p^{i_p}) = X^{i_1}\dots X_p^{i_p} \ominus (u(X_1),\dots,u(X_p))$$

ou encore

$$u(X_1^{i_1}\dots X_p^{i_p}) = \Pi_k\,[u(X_1)^{i_1})\dots u(X_p)^{i_p}) \qquad (1).$$

En effet la famille $(X_1^{i_1}\dots X_p^{i_p})_{1\leq i_1+\dots+i_p\leq k}$ est une base de Δ_k.

Chaque égalité du type (1) entre deux éléments de Δ_k traduit la nullité de n relations polynomiales entre les coefficients de la matrice de u dans cette base. On peut donc considérer $\hat{\mathcal{G}}_k$ comme un fermé de Zariski de $GL(n,\mathbb{C})$.

Rappelons le théorèmede Jordan-Chevalley dont on peut trouver une démonstration par exemple dans [14].

THEOREME 1.2.2. - *Soit* H *un sous-groupe de* $GL(n,\mathbb{C})$ *fermé pour la topologie de Zariski. Si* $x \in H$, *il existe un couple unique* (s,u) *d'éléments de* H *tels que* : x=su=us *avec* s *diagonalisable et* u *unipotent.*

§ 3 - Forme réduite et forme normale formelle

On va démontrer ici que tout élément de $\mathcal{G}_k$ (respectivement de $\mathcal{G}$) admet une décomposition formelle de "type Jordan" en un élément diagonalisable de $\mathcal{G}_k$ (respectivement de $\mathcal{G}$) et en un élément unipotent de $\mathcal{G}_k$ (respectivement de $\mathcal{G}$). La démonstration repose d'une part sur le théorèmè de Jordan-Chevalley, d'autre part sur les résultats formels obtenus dans le paragraphe 1.

THEOREME 1.3.1 - *Pour tout élément* Q *de* $\mathcal{G}_k$ *, il existe un couple unique* (G,H) *d'éléments de* $\mathcal{G}_k$ *, tels que* g *soit diagonalisable dans* $\mathcal{Q}_k$, H *unipotent et* $Q=G \ominus H=H \ominus G$. *De plus* G *et* H *commutent avec tous les éléments de* $\mathcal{Q}_k$ *qui commutent avec* Q.

Démonstration : $\hat{\mathcal{G}}_k$ est un sous-groupe de $GL(n,\mathbb{C})$, fermé pour la topologie de Zariski (Proposition 1.2.1) . D'après le théorème de Jordan-Chevalley, si $Q \in \mathcal{G}_k$, il existe un couple unique $(\hat{G},\hat{H})$ d'éléments de $\hat{\mathcal{G}}_k$ tels que $\hat{Q} = \hat{G}o\hat{H} = \hat{H}o\hat{G}$ avec $\hat{G}$ diagonalisable et $\hat{H}$ unipotent. Comme $\psi : \mathcal{G}_k \longrightarrow \hat{\mathcal{G}}_k$, $F \longrightarrow \hat{F}$, est un antihomomorphisme de groupe injectif, on a $Q = G\Theta H = H\Theta G$ (1), avec G et H dans $\mathcal{G}_k$. D'après le théorème 1.1.6 et le Corollaire 1.1.7, G est diagonalisable dans $\mathcal{G}_k$ et H est unipotent. La décomposition (1) est unique à cause de l'unicité de la décomposition dans $\hat{\mathcal{G}}_k$. De plus, d'après les propriétés de la décomposition de Jordan dans $GL(n,\mathbb{C})$, $\hat{G}$ et $\hat{H}$ commutent avec tout élément de $End(\mathbb{C}^n)$ commutant avec $\hat{Q}$. Donc si $L \in \mathcal{Q}_k$ et si L commute avec F, L commute avec G et H.

THEOREME 1.3.2. - *Pour tout élément* Q *de* $\mathcal{G}$ *, il existe un couple unique* (G,H) *d'éléments de* $\mathcal{G}$ *tels que* G *soit diagonalisable dans* $\mathcal{Q}$, H *unipotent et* $Q = GoH = HoG$. *De plus* G *et* H *commutent avec tous les éléments de* $\mathcal{Q}$ *qui commutent avec* Q.

Démonstration - Soit $Q \in \mathcal{G}$. Notons $\Pi_k(Q) = Q_k$ pour tout $k \geq 1$. D'après le théorème précédent, pour tout $k \geq 1$, il existe un couple unique (G_k , H_k) d'éléments de $\mathcal{G}_k$ avec G_k diagonalisable et H_k unipotent tel que $Q_k = G_k \ominus H_k = H_k \ominus G_k$. Le couple (G_k , H_k) étant unique pour tout k, on a $\Pi_k(G_{k+1}) = G_k$ et $\Pi_k(H_{k+1}) = H_k$. Soit alors $G \in \mathcal{G}$ et $H \in \mathcal{G}$ tels que, pour tout $k \geq 1$: $\Pi_k(G) = G_k$ et $\Pi_k(H) = H_k$. D'après la proposition 1.1.9 G est diagonalisable ; il est clair que H est unipotent, car H_1 est unipotent.

De plus, pour tout $k \geq 1$ on a :

$\Pi_k(GoH) = \Pi_k(HoG) = \Pi_k(Q)$ donc $GoH = HoG = Q$.

D'autre part, si $L \in \mathcal{Q}$ et si L commute avec Q, alors pour tout $k \geq 1$, $L_k = \Pi_k(L)$ commute avec Q_k, donc avec G_k et H_k. Finalement L commute avec G et H.

Définition 1.3.3 - Soit $A \in \mathcal{G}_1$. Notons $\lambda_1,...,\lambda_p$ les valeurs propres de A. On dit que les valeurs propres de A sont en relation s'il existe un multi-indice $i=(i_1,...,i_p) \in \mathbb{N}^p$ avec $\sum_{j=1}^{p} i_j \geq 2$ et un indice m, $1 \leq m \leq p$ tel que $\lambda_m = \lambda_1^{i_1} \ldots \lambda_p^{i_p}$. L'ordre de la relation $\lambda_m = \lambda_1^{i_1} \ldots \lambda_p^{i_p}$ est par définition $i_1 + ... + i_p$.

On peut facilement vérifier la remarque suivante.

Remarque 1.3.4. - Soit $A \in \mathcal{G}_1$. Soient $(\lambda_1,...,\lambda_p)$ les valeurs propres de A. Si $0<|\lambda_p|\leq ...\leq |\lambda_1|<1$ (respectivement $1<|\lambda_1|\leq ...\leq |\lambda_p|$) il n'y a qu'un nombre fini de relations entre les valeurs propres de A. Plus précisément, les valeurs propres de A n'ont aucune relation d'ordre strictement supérieur à q si $|\lambda_1|^{q+1}<|\lambda_p|$ (respectivement $|\lambda_p|<|\lambda_1|^{q+1}$).

PROPOSITION 1.3.5. - *Soient* $G \in \mathcal{G}$. *Supposons que* G *commute avec un élément diagonal* $D=(\lambda_1 X_1,...,\lambda_p X_p)$ *de* $\mathcal{G}_1$. *Si les valeurs propres de* D *n'ont aucune relation d'ordre strictement supérieur à* q, *alors* $G \in \mathcal{G}_q$.

Démonstration - soit $G=(g_1,..,g_p)$ et $D=(d_1,..,d_p)$. Comme DoG = GoD, on a, pour tout m, $1\leq m\leq p$ $d_m oG = g_m oD$. Posons

$$g_m = \sum_{(i_1,...,i_p)\in \mathbb{N}^p} \alpha_{i_1,...,i_p,m} X_1^{i_1}...X_p^{i_p}$$

on a alors

$$\lambda_m \sum_{(i_1,...,i_p)\in \mathbb{N}^p} \alpha_{i_1,...,i_p,m} X_1^{i_1}...X_p^{i_p} = \sum_{(i_1,...,i_p)\in \mathbb{N}^p} \alpha_{i_1,...,i_p,m} \lambda_1^{i_1}...\lambda_p^{i_p} X_1^{i_1}...X_p^{i_p}.$$

Donc on a soit :

$$\alpha_{i_1,...,i_p,m} = 0.$$

Soit :

$$\lambda_m = \lambda_1^{i_1} ... \lambda_p^{i_p}. \qquad (2)$$

Comme $i_1+...+i_p\leq q$ dans la relation (2), si $i_1+...+i_p>q$, alors $\alpha_{i_1,...,i_p,m}=0$. Donc $g_m \in \Delta_q$ et finalement $G \in \mathcal{G}_q$.

PROPOSITION 1.3.6. - *Soit* $F \in \mathcal{G}$, $N \in \mathcal{G}$ *et* $R \in \mathcal{G}_q$. *Si les valeurs propres de* F'(0) *n'ont aucune relation d'ordre strictrement supérieur à* q, *alors il existe au plus un élément* S *de* $\mathcal{G}$ *vérifiant :*

(i) FoS = SoN

(ii) $\Pi_q(S) = R$.

Démonstration - Supposons qu'il existe un autre élément de $\mathcal{G}$, soit P tel que :
PoN = FoP avec Π_q (P) = R. Montrons que P=S.

Posons $Q=PoS^{-1}$ .On remarque que Π_q (Q) = I. D'autre part, d'après le théorème 1.3.2, il existe un couple unique (G,H) d'éléments de $\mathcal{G}$, tels que G soit diagonalisable, H unipotent et tels que F =GoH=HoG.

G est diagonalisable, donc il existe $V \in \mathcal{G}$ tel que $VoGoV^{-1} = D$, avec D diagonal. Posons $M = VoHoV^{-1}$; alors M est unipotent et $VoFoV^{-1} = DoM = MoD$. De plus, d'après le théorème 1.3.2, D et M commutent avec tout élément de $\mathcal{Q}$ qui commute avec $VoFoV^{-1}$. Or $T = VoQoV^{-1}$ commute avec $VoFoV^{-1}$. Donc D commute avec T ; comme les valeurs propres de D sont celles de F'(0), $T \in \mathcal{G}_q$, d'après la proposition 1.3.5. Donc

$$T = \Pi_q (T) = \Pi_q [\Pi_q (V) \circ \Pi_q (Q) \circ \Pi_q (V^{-1})]$$

$$= I \text{ car } \Pi_q (Q) = I.$$

Donc $Q = V^{-1} \circ T \circ V = V^{-1} \circ V = I$, d'où l'unicité de S.

Définition 1.3.7. - Soit $F \in \mathcal{G}$. On appelle forme normale de F tout élément N de $\mathcal{G}$ tel que N=DoU=UoD, avec U unipotent et D diagonal, pour lequel il existe $P \in \mathcal{G}$ tel que $P^{-1}oFoP=N$.

Définition 1.3.8. - Soit $F \in \mathcal{G}$. On appelle forme réduite de F tout élément N de $\mathcal{G}$ tel que N = LoU = UoL, avec U unipotent , $L \in \mathcal{G}_1$ et L diagonalisable pour lequel il existe $P \in \mathcal{G}$ tel que $P^{-1}oFoP=N$.

Il résulte du théoèrme de décomposition formelle dans $\mathcal{G}$ (théorème 1.3.2) que tout élément de $\mathcal{G}$ admet une forme normale.

Définition 1.3.9. - On dit qu'une forme normale N de F est une forme normale de Jordan si N'(0) est une réduite de Jordan de F'(0).

Tout élément F de $\mathcal{G}$ admet une forme normale de Jordan. En effet, connaissant une forme normale N de F, on obtient une forme normale de Jordan par un simple changement linéaire. En effet, soit N=DoU=UoD (avec D diagonal et U unipotent). Alors N'(0) = DoU'(0)=U'(0)oD avec U'(0) unipotente. Il existe $P \in \mathcal{G}_1$ tel que $PoN'(0) \circ P^{-1}$ soit une réduite de Jordan. On a alors

$$PoN'(0) \circ P^{-1} = (PoDoP^{-1}) \circ (PoU'(0)o\, P^{-1})$$
$$= (PoU'(0)o\, P^{-1}) \circ (PoDoP^{-1}).$$

$PoU'(0)oP^{-1}$ est unipotente, $PoDoP^{-1}$ est diagonalisable, donc $PoDoP^{-1}$ est la partie diagonalisable de la matrice de Jordan $PoN'(0)oP^{-1}$. Par conséquent $PoDoP^{-1}$ est diagonale,

$$PoNoP^{-1} = (PoDoP^{-1}) \circ (PoU(0)o\, P^{-1}) = (PoU(0)o\, P^{-1}) \circ (PoDoP^{-1})$$

est une forme normale de F, et cette forme normale est évidemment une forme de Jordan.

De plus, on peut obtenir par le procédé ci-dessus que la partie linéaire de la forme normale de Jordan de F ait ses termes de la diagonale rangés par ordre de module croissants (ou décroissants).

THEOREME 1.3.10 - *Soit* $F \in \mathcal{G}$ *tel que les valeurs propres de* $F'(0)$ *n'aient qu'un nombre fini de relations. Soit* q *un entier supérieur ou égal à l'ordre de toutes les relations entre les valeurs propres de* $F'(0)$. *Alors toute forme réduite* N *de* F *appartient à* $\mathcal{G}_q$, *ainsi que son inverse* N^{-1}.

Démonstration - Posons $N = LoU = UoL$. $L \in \mathcal{G}_1$ et est diagonalisable, donc il existe $T \in \mathcal{G}_1$ tel que $ToLoT^{-1} = D$, avec D diagonal.

Posons $H = ToUoT^{-1}$. Alors $DoH = HoD$, donc d'après la proposition 1.3.5, $H \in \mathcal{G}_q$. Notons $D = (\lambda_1 X_1,...,\lambda_p X_p)$. Alors $(DoH)^{-1} = H^{-1} \circ D^{-1} = D^{-1} \circ H^{-1}$ avec $D^{-1} = (\frac{1}{\lambda_1} X_1,..., \frac{1}{\lambda_p} X_p)$. Les valeurs propres de D^{-1} s'obtenant comme inverses des valeurs propres de D,q est aussi le plus grand ordre possible pour les relations entre les valeurs prores de D^{-1}. Donc $H^{-1} \in \mathcal{G}_q$. Comme $T \in \mathcal{G}_1$, on a $U \in \mathcal{G}_q$ et $U^{-1} \in \mathcal{G}_q$. Donc $N \in \mathcal{G}_q$ et $N^{-1} \in \mathcal{G}_q$.

THEOREME 1.3.11 - *Soit* $F \in \mathcal{G}$. *Supposons que les valeurs propres* $(\mu_1,...,\mu_p)$ *de* $F'(0)$ *vérifient* $1 < |\mu_1| \le ... \le |\mu_p|$ *et soit* q *un entier tel que* $|\mu_1|^{q+1} > |\mu_p|$. *Alors toute forme réduite de* F *appartient à* $\mathcal{G}_q$ *et il existe une forme normale* $N = (N_1,...,N_p)$ *ayant la forme suivante :*

$$\begin{cases} N_1 = \lambda_1 X_1 \\ N_2 = \lambda_1 X_2 + X_1 \\ \vdots \\ N_{r_1} = \lambda_1 X_{r_1} + X_{r_1 - 1} \end{cases}$$

$$\begin{cases} N_{r_1+1} = \lambda_2 X_{r_1+1} + P_{r_1+1}(X_1, \dots, X_{r_1}) \\ N_{r_1+2} = \lambda_2 X_{r_1+2} + X_{r_1+1} + P_{r_1+2}(X_1, \dots, X_{r_1}) \\ \vdots \\ N_{r_1+r_2} = \lambda_2 X_{r_1+r_2} + X_{r_1+r_2-1} + P_{r_1+r_2}(X_1, \dots, X_{r_1}) \end{cases}$$

$$\begin{cases} \vdots \\ N_{r_1+\dots+r_{s-1}} = \lambda_{s-1} X_{r_1+\dots+r_{s-1}} + X_{r_1+\dots+r_{s-1}-1} + P_{r_1+\dots+r_{s-1}}(X_1, \dots, X_{r_1+\dots+r_{s-2}}) \end{cases}$$

$$\begin{cases} N_{r_1+\dots+r_{s-1}+1} = \lambda_s X_{r_1+\dots+r_{s-1}+1} + P_{r_1+\dots+r_{s-1}+1}(X_1, \dots, X_{r_1+\dots+r_{s-1}}) \\ \\ N_{r_1+\dots+r_s} = \lambda_s X_{r_1+\dots+r_s} + X_{r_1+\dots+r_s-1} + P_{r_1+\dots+r_s}(X_1, \dots, X_{r_1+\dots+r_{s-1}}) \end{cases}$$

où

$$1 < |\lambda_1| \leq \ldots \leq |\lambda_s| \quad \text{et où } P_m(X_1,\ldots,X_{r_1+\ldots r_k})$$

est une somme de mônomes de la forme

$$X_1^{i_1} X_2^{i_2} \ldots X_{r_1+\ldots+r_k}^{i_{r_1+\ldots+r_k}} \quad \text{avec} \quad i_1, \ldots, i_{r_1+\ldots+r_k} \quad \text{tels que}$$

$$\lambda_{k+1} = \lambda_1^{i_1+\ldots+i_{r_1}} \lambda_2^{i_{r_1+1}+\ldots+i_{r_1+r_2}} \ldots \lambda_k^{i_{r_1+\ldots+r_{k-1}+1}+\ldots+i_{r_1+\ldots+r_k}}$$

pour $1 \leq k \leq s-1$ et $r_1 + \ldots + r_k < m \leq r_1 + .. + r_{k+1}$ $(r_1 + .. + r_s = p)$.

De plus si $PoFoP^{-1} = QoFoQ^{-1} = N$, $\Pi_q(P) = \Pi_q(Q)$ avec $P,Q \in \mathcal{G}$, alors $P = Q$.

Démonstration : Soit N une forme normale de Jordan de F où les termes de la diagonale de N'(0) sont rangés par ordre de modules croissants et soit $P \in \mathcal{G}$ tel que $P^{-1}oFoP=N$. D'après la proposition 1.3.6, P est unique si $\Pi_q(P)$ est connu.

On a N = DoU = UoD avec D diagonal et U unipotent.

Notons $\lambda_1,\ldots,\lambda_s$ les valeurs propres de F'(0) correspondant aux différents "blocs de Jordan" de N'(0).

Si $1 \leq k \leq s$, notons r_k l'ordre du bloc correspondant à λ_1 et posons $v_0=0,\ldots,v_k=r_1+\ldots+r_k$ .Si $1 \leq t \leq r_{k+1} - r_k$ alors $\mu_{v_k+t} = \lambda_{k+1}$. Soit $m \in \{1\ldots p\}$. Alors il existe k tel que $m=v_k+t$ d'où $\mu_m = \lambda_{k+1}$.

Posons $U = (u_1,\ldots,u_p)$ et $D = (d_1,\ldots,d_p)$. Comme DoU = UoD, on a, pour tout $m \in \{1\ldots p\}$, $d_m \circ U = u_m \circ D$ (1). Si on pose $u_m = \sum_{(i_1,\ldots,i_p)\in \mathbb{N}^p} \alpha_{i_1,\ldots,i_p,m} X_1^{i_1}\ldots X_p^{i_p}$ la relation (1) s'écrit :

$$\mu_m \, \Sigma \alpha_{i_1,\ldots,i_p,m} X_1^{i_1} \ldots X_p^{i_p} = \Sigma \alpha_{i_1,\ldots,i_p,m} \mu_1^{i_1} \ldots \mu_p^{i_p} X_1^{i_1} \ldots X_p^{i_p} .$$

Donc si $\alpha_{i_1,\ldots,i_p,m} \neq 0$, on a : $\mu_m = \mu_1^{i_1} \ldots \mu_p^{i_p}$ (2).

On en déduit $\lambda_{k+1} = \lambda_1^{\beta_1} \ldots \lambda_s^{\beta_s}$ où $\beta_j = i_{v_{j-1}+1} + \ldots i_{v_j}$ (3).

S'il existe $j \geq k+1$ tel que $\lambda_{k+1} = \lambda_1^{\beta_1} \ldots \lambda_j^{\beta_j} \ldots \lambda_s^{\beta_s}$ avec $\beta_j \neq 0$ et $i_1 + \ldots + i_p \geq 2$ alors

$$\left| \frac{\lambda_{k+1}}{\lambda_j} \right| = |\lambda_1|^{\beta_1} \ldots |\lambda_j|^{\beta_j - 1} \ldots |\lambda_s|^{\beta_s}$$

est strictement plus grand que 1. Par ailleurs

$$\left| \frac{\lambda_{k+1}}{\lambda_j} \right| \leq 1 \quad \text{car} \quad j \geq k+1.$$

Donc (3) est impossible dès qu'il existe $j \geq k+1$ tel que $\beta_j \neq 0$. On en déduit, que quand $i_1 + \ldots + i_p \geq 2$, la relation (2) est impossible s'il existe $1 \geq r_1 + \ldots + r_k + 1$ tel que $i_1 \neq 0$.

On peut donc écrire $u_m = \Pi_1(u_m) + \mathcal{P}_m(X_1 \ldots X_{v_k})$ où $\mathcal{P}_m$ est un polynôme ne contenant que les variables $X_1 \ldots X_{v_k}$. De plus, d'après la relation (3), $\mathcal{P}_m$ ne contient que les monômes $X_1^{i_1} \ldots X_p^{i_{v_k}}$ avec $i_1, \ldots, i_{v_k}$ tels que

$$\lambda_1^{i_1+\ldots+i_{v_1}} \lambda_2^{i_{v_1}+\ldots+i_{v_2}} \ldots \lambda_k^{i_{v_{k-1}}+\ldots+i_{v_k}} = \lambda_{k+1} .$$

Comme D est diagonale et N = DoU, on peut écrire de même $N_m = \Pi_1(N_m) + P_m(X_1 \ldots X_{v_k})$ où P_m est un polynôme en $X_1 \ldots X_{v_k}$ ne contenant que des monômes du type précédent.

Il résulte du théorème 1.3.10 et de la remarque 1.3.4, que si $F \in \mathcal{G}$ et si toutes les valeurs propres de F'(0) sont de module strictement inférieur à 1 (ou de module strictement supérieur à 1) alors toute forme normale N de F est un automorphisme polynomial de $\mathbb{C}^p$ (ainsi que N^{-1}, mais il est de toutes façons connu que l'inverse d'un automorphisme polynomial est un automorphisme polynomial). On peut montrer [10] que si les séries définissant F convergent sur un voisinage de l'origine, il en est de même des séries définissant P où P est un élément de $\mathcal{G}$ quelconque vérifiant P^{-1} of F o P = N, N désignant une forme normale quelconque de F.

CHAPITRE II

§ 1 - Les sous-groupes $\mathcal{H}$, $\mathcal{U}$ et $\mathcal{L}$ du groupe $\mathcal{G}$

On fixe $p \geq 1$ et $\ell < p$. L'application formelle $(X_1,...,X_\ell,0,...,0)$ est notée Π.

Définition 2.1.1 - On pose

$$\mathcal{H} = \{F \in \mathcal{G} : F \circ \Pi = \Pi\} \quad \mathcal{U} = \{F \in \mathcal{G} : F \circ \Pi = \Pi \circ F\}$$
$$\mathcal{L} = \{F \in \mathcal{G} : F \circ \Pi = \Pi \circ F \circ \Pi\}.$$

Remarque 2.1.2

(α) Soit $f = (f_1,...,f_p) \in \mathcal{G}$. Notons $f_j = \Sigma\, \alpha_{i_1,...,i_p,j} X_1^{i_1}...X_p^{i_p} (i_1,...,i_p) \in \mathbb{N}^p$. $\mathcal{L}$ est l'ensemble des éléments de $\mathcal{G}$ tels que

$$f_j = \Sigma\, \alpha_{i_1,...,i_p,j}\, X_1^{i_1}...X_p^{i_p} \quad \text{pour } j \geq \ell + 1$$

$$i_{\ell+1} + ... + i_p > 0$$

$\mathcal{H}$ est l'ensemble des éléments de $\mathcal{L}$ tels que

$$f_j = X_j + \Sigma\, \alpha_{i_1,...,i_p,j}\, X_1^{i_1}...X_p^{i_p} \quad \text{pour } j \geq \ell$$

$$i_{\ell+1} + ... + i_p > 0$$

(β) Soit Ω la variété linéaire suivante :

$$\Omega = \{(x_1,...,x_p) \in \mathbb{C}^p : x_{\ell+1} = ... = x_p = 0\}.$$

Un automorphisme analytique de $\mathbb{C}^p$ appartient à $\mathcal{L}$ si et seulement si, $F(\Omega) \subset \Omega$, et un automorphisme analytique de $\mathbb{C}^p$ appartient à $\mathcal{H}$ si, et seulement si, $F(z) = z$ pour tout z de Ω.

(γ) $\mathcal{U} \subset \mathcal{L}$ et $\mathcal{U} \cap \mathcal{H}$ est l'ensemble des éléments de $\mathcal{L}$ tels que $f_j = X_j$ pour $j \leq \ell$.

(δ) Un élément F de g appartient à $\mathcal{H}$ (respectivement à $\mathcal{U}$, à $\mathcal{L}$) si et seulement si, $\Pi_k(F)$ appartient à $\mathcal{H}$ (respectivement à $\mathcal{U}$, à $\mathcal{L}$) pour tout $k \geq 1$.

(ε) Si $F \in \mathcal{H}$, alors $F'(0) \circ \Pi = \Pi$, donc 1 est valeur propre de $F'(0)$ avec un ordre de multiplicité supérieur ou égal à ℓ.

PROPOSITION 2.1.3. - *$\mathcal{H}, \mathcal{U}$ et $\mathcal{L}$ sont des sous-groupes de $\mathcal{G}$. De plus $\mathcal{H}$ est un sous-groupe distingué de $\mathcal{L}$.*

Démonstration - Soit $G \in \mathcal{L}$; démontrons que $G^{-1} \in \mathcal{L}$. Posons, $\Pi_k(G) = G_k$ pour $k \geq 1$.

On a $G \in \mathcal{L}$, donc pour tout $k \geq 1$, $G_k \in \mathcal{L}$. Comme Π est linéaire on a $\hat{\Pi} \circ \hat{G}_k \circ \hat{\Pi} = \hat{\Pi} \circ \hat{G}_k$ (dans cette relation, on considère $\Pi \in \mathrm{End}\ (\Delta_k)$) et on a (1) ${}^t\hat{\Pi} \circ {}^t\hat{G}_k \circ {}^t\hat{\Pi} = {}^t\hat{G}_k \circ {}^t\hat{\Pi}$ où t désigne la transposée. Comme ${}^t\hat{\Pi}$ est une projection, d'après la relation (1), on a : ${}^t\hat{G}_k(U) \subset U$ où $U = \mathrm{Im}({}^t\hat{\Pi})$. Comme $G \in \mathcal{G}$, ${}^t\hat{G}_k$ est une application linéaire inversible, donc ${}^t\hat{G}_k \mid U$ est injective et par conséquent surjective car U est de dimension finie. Donc ${}^t\hat{G}_k(U) = U$. On en déduit que ${}^t(\widehat{G^{-1}})_k(U) = {}^t(\hat{G}_k)^{-1}(U)$ et donc

$$\bullet \qquad {}^t\hat{\Pi} \circ {}^t(\hat{G}^{-1})_k \circ {}^t\hat{\Pi} = {}^t(\hat{G}^{-1})_k \circ {}^t\hat{\Pi}.$$

Finalement on a $\Pi \ominus (G^{-1})_k \ominus \Pi = (G^{-1})_k \ominus \Pi$ pour tout $k \geq 1$. Donc $\Pi \ominus G^{-1} \ominus \Pi = G^{-1} \ominus \Pi$, c'est-à-dire que $G^{-1} \in \mathcal{L}$. Soient $F, G \in \mathcal{L}$. On a

$$\begin{aligned} \Pi \circ F \circ G^{-1} \circ \Pi &= \Pi \circ F \circ \Pi \circ G^{-1} \circ \Pi \\ &= F \circ \Pi \circ G^{-1} \circ \Pi \\ &= F \circ G^{-1} \circ \Pi. \end{aligned}$$

Donc $F \circ G^{-1} \in \mathcal{L}$ et $\mathcal{L}$ est un sous-groupe de $\mathcal{G}$.

Soient $F, G \in \mathcal{H}$. Alors $G \circ \Pi = \Pi$ donc $\Pi = G^{-1} \circ \Pi$ et $F \circ G^{-1} \circ \Pi = F \circ \Pi = \Pi$. Donc $\mathcal{H}$ est un sous-groupe de $\mathcal{G}$. Il est clair que $\mathcal{H} \subset \mathcal{L}$.

De plus si $F \in \mathcal{H}$ et $P \in \mathcal{L}$ alors $P^{-1} \in \mathcal{L}$.

Donc

$$\begin{aligned} P \circ F \circ P^{-1} \circ \Pi = P \circ F \circ \Pi \circ P^{-1} \circ \Pi &= P \circ \Pi \circ P^{-1} \circ \Pi \\ &= P \circ P^{-1} \circ \Pi \\ &= \Pi. \end{aligned}$$

Donc $PoFoP^{-1} \in \mathcal{H}$ et $\mathcal{H}$ est un sous groupe distingué de $\mathcal{B}$.

Soient $F,G \in \mathcal{U}$. Alors $G^{-1} \in \mathcal{U}$ et $FoG^{-1}o\Pi = Fo\Pi = F \circ \Pi \circ G^{-1} = \Pi \circ F \circ G^{-1}$. Donc $\mathcal{U}$ est un sous-groupe de $\mathcal{G}$.

THEOREME 2.1.4 - Si $F \in \mathcal{H}$, *alors la partie diagonalisable* G *de* F *et la partie unipotente* H *de* F *appartiennent à* $\mathcal{H}$.

Démonstration - Notons $\Pi_k(F) = F_k$ et $\Pi_k(G) = G_k$ pour $k \geq 1$. Pour tout $k \geq 1$, on sait qu'il existe un polynôme S_k tel que $\hat{G}_k = S_k(\hat{F}_k)$. Comme 1 est valeur propre de $F'(0)$ (remarque 2.1.2) et donc de $\hat{F}_k$ (théorème 1.1.6) le polynôme S_k est tel que S_k-1 soit un multiple de $(X-1)^{\alpha_k}$ avec $\alpha_k \geq 1$. Comme $F \in \mathcal{H}$, on a $\hat{\Pi}o\hat{F}_k = \hat{\Pi}$ donc $\hat{\Pi}o\hat{G}_k = \hat{\Pi} \circ S_k(\hat{F}_k) = S_k(1)\,\hat{\Pi}$, donc $\hat{\Pi}o\hat{G}_k = \Pi$. Finalement, pour tout $k \geq 1$, $G_k \ominus \Pi = \Pi$ et par conséquent $G \circ \Pi = \Pi$. Donc $G \in \mathcal{H}$ et $H = FoG^{-1} \in \mathcal{H}$.

Remarque - Comme pour le groupe $\mathcal{G}$, on peut introduire pour $\mathcal{H}$ les groupes $\mathcal{H}_k$ et $\hat{\mathcal{H}}_k$. On peut démontrer facilement que $\hat{\mathcal{H}}_k$ est un sous-groupe algébrique de $\hat{\mathcal{G}}_k$ pour tout $k \geq 1$ et obtenir le théorème précédent comme application du théorème de Jordan-Chevalley.

§ 2 - Normalisation des éléments de $\mathcal{H}$

LEMME 2.2.1 - *Soient* $D = (\lambda_1 X_1,\dots,\lambda_p X_p)$ *avec* $\lambda_1 = \dots = \lambda_\ell = 1$ *et soit* $W = (W_1,\dots,W_p)$ *un élément de* $\mathcal{G}$. *Si* $\lambda_{\ell+1}^{i_{\ell+1}} \dots \lambda_p^{i_p} \neq 1$ *pour* $i_{\ell+1} \geq 0 \dots i_p \geq 0$ *et* $i_{\ell+1} + \dots + i_p > 0$ *et si*

$WoD = DoW$ *alors* $W \in \mathcal{U}$.

Démonstration - Soit $W = (W_1,\dots,W_p)$ tel que $W \circ D = D \circ W$. Posons $W_j = \Sigma\, \alpha_{i_1,\dots,i_p,j}\, X_1^{i_1} \dots X_p^{i_p}$. Puisque $DoW = WoD$ et $\lambda_1 = \dots = \lambda_\ell = 1$ on a

$$\lambda_j\, \alpha_{i_1,\dots,i_p,j} = \lambda_{\ell+1}^{i_{\ell+1}} \dots \lambda_p^{i_p}\, \alpha_{i_1,\dots,i_p,j}\ .$$

Donc pour $j \leq \ell$ on a

(1) si $i_{\ell+1} + \dots + i_p > 0$ alors $\alpha_{i_1,\dots,i_p,j} = 0$ (car $\lambda_{\ell+1}^{i_{\ell+1}} \dots \lambda_p^{i_p} \neq 1$).

Pour $j > \ell$ on a

(2) si $i_{\ell+1} + \ldots + i_p = 0$ alors $\alpha_{i_1,\ldots,i_p,j} = 0$ (car $\lambda_j \neq 1$).

D'après les conditions (1) et (2), $Wo\Pi = \Pi oW$, c'est-à-dire que $W \in \mathcal{U}$. Notons que si $W \in \mathcal{H}$, alors $W = (X_1,\ldots,X_\ell, W_{\ell+1},\ldots W_p)$.

THEOREME 2.2.2 - Soit $F \in \mathcal{H}$ *et soient* $(\lambda_1,\ldots,\lambda_p)$ *les valeurs propres de* $F'(0)$ *avec* $\lambda_1=..=\lambda_\ell=1$. *On suppose que* $\lambda_{\ell+1}^{i_{\ell+1}} \ldots \lambda_p^{i_p} \neq 1$ *pour* $i_{\ell+1} \geq 0 \ldots i_p \geq 0$ *et* $i_{\ell+1} + \ldots + i_p > 0$.

Soit $D = (\lambda_1 X_1,\ldots,\lambda_p X_p)$ *et soit* G *la partie diagonalisable de* F.

Alors tout élément Q de $\mathcal{G}$ tel que $QoGoQ^{-1}=D$ appartient à $\mathcal{L}$ et vérifie $QoFoQ^{-1}=U$ avec $\Pi oU = Uo\Pi = \Pi$ (c'est-à-dire que $U \in \mathcal{U} \cap \mathcal{H}$)

Démonstration - Soit $Q \in \mathcal{G}$ tel que $QoGoQ^{-1} = D$. Démontrons que $Q \in \mathcal{L}$.

On a $QoG=DoQ$, donc $QoGo\Pi=DoQo\Pi$. Or $Go\Pi=\Pi$ car $G\in\mathcal{H}$. Donc on a $Qo\Pi= DoQo\Pi$. D'après cette égalité, en posant $Qo\Pi=(\theta_1,\ldots,\theta_p)$ on a $\theta_{\ell+1}= \ldots = \theta_p=0$. Donc $\Pi oQo\Pi = Qo\Pi$ et $Q \in \mathcal{L}$.

Soit maintenant H la partie unipotente de F. Posons $QoHoQ^{-1} = W$ et $U = DoW = WoD$ avec $W=(W_1,\ldots,W_p)$. Comme $H \in \mathcal{H}$ et que $\mathcal{H}$ est sous-groupe distingué de $\mathcal{L}$, on a $W \in \mathcal{H}$ et donc $U \in \mathcal{H}$. De plus, d'après le lemme 2.2.1, $W = (X_1 \ldots X_\ell, W_{\ell+1} \ldots W_p)$ donc $U=(X_1 \ldots X_\ell, W_{\ell+1} \ldots W_p)$.

Ce théorème prouve donc l'existence d'éléments $Q \in \mathcal{L}$ tels que $QoFoQ^{-1} \in \mathcal{U} \cap \mathcal{H}$. A la fin de ce chapitre, on aboutit à l'existence de $P \in \mathcal{H}$ tel que $PoFoP^{-1} \in \mathcal{U} \cap \mathcal{H}$.

Le lemme suivant est bien connu.

LEMME 2.2.3 - Soit E *un espace vectoriel de dimension finie. Soit* $T \in \mathcal{L}(E)$ *et* λ *une valeur propre de* T. *Si la dimension de* $Ker(T-\lambda I)$ *est égale à l'ordre de multiplicité de la valeur propre* λ, *alors* $T-\lambda I$ *est une bijection de* $(T-\lambda I)(E)$ *sur lui-même.*

Pour le lemme 2.2.4, le corollaire 2.2.5 et le théorème 2.2.6, on a les hypothèses suivantes sur $G \in \mathcal{G}$:

Soit λ une valeur propre de $G'(0)$. On suppose que, pour tout $k>0$, la dimension de $\mathrm{Ker}(\Pi_k(G)-\lambda I)$ est égale à l'ordre de multiplicité de λ en tant que valeur propre de $\Pi_k(G)$.

Notons $\mathcal{P}_k$ l'ensemble des polynômes homogènes de degré k.

LEMME 2.2.4 - *Pour tout* $k > 0$

(1) $\widehat{\Pi_k(G)-\lambda I}$ *est une bijection de* $(\widehat{\Pi_k(G)-\lambda I})(\Delta_k)$ *sur lui-même.*

(2) $\widehat{\Pi_k(G)-\lambda I}$ *est une bijection de* $(\widehat{\Pi_k(G)-\lambda I})(\Delta_k) \cap \mathcal{P}_k$ *sur lui-même.*

Démonstration - D'après le lemme précédent, il est clair que $\widehat{\Pi_k(G)-\lambda I}$ est une bijection de $(\widehat{\Pi_k(G)-\lambda I})(\Delta_k)$ sur lui-même. D'autre part, si $g \in \mathcal{P}_k$

$$(\widehat{\Pi_k(G)-\lambda I})(g) = g \,\Theta [\Pi_k(G)] - \lambda g = g \,\Theta\, G'(0) - \lambda g .$$

Donc $(\widehat{\Pi_k(G)-\lambda I})(g) \in \mathcal{P}_k$.

D'après ce qui précède, la restriction de $\widehat{\Pi_k(G)-\lambda I}$ à $(\widehat{\Pi_k(G)-\lambda I})(\Delta_k) \cap \mathcal{P}_k$ est injective. Donc $\widehat{\Pi_k(G)-\lambda I}$ est une bijection de $(\widehat{\Pi_k(G)-\lambda I})(\Delta_k) \cap \mathcal{P}_k$ sur lui-même.

COROLLAIRE 2.2.5 - $\hat{G}-\lambda I$ *est une bijection de* $(\hat{G}-\lambda I)(\Delta)$ *sur lui-même.*

Démonstration - Soit $\varphi \in (\hat{G}-\lambda I)(\Delta)$; alors $\Pi_k(\varphi) \in (\widehat{\Pi_k(G)-\lambda I})(\Delta_k)$ pour $k > 0$.
En appliquant deux fois le lemme précédent, on trouve, pour tout $k>0$, ψ_k et θ_k tels que :

(α) $\Pi_k(\varphi) = \widehat{\Pi_k(G)}(\psi_k) - \lambda\psi_k$

(β) $\psi_k = \widehat{\Pi_k(G)}(\theta_k) - \lambda\theta_k$ avec $\theta_k \in (\widehat{\Pi_k(G)-\lambda I})(\Delta_k)$.

Soient alors ψ_{k+1} et θ_{k+1} tels que

$$\Pi_{k+1}(\varphi) = \widehat{\Pi_{k+1}(G)}(\psi_{k+1}) - \lambda\psi_{k+1}$$

$$\psi_{k+1} = \widehat{\Pi_{k+1}(G)}(\theta_{k+1}) - \lambda\theta_{k+1} \quad \text{avec} \quad \theta_{k+1} \in (\widehat{\Pi_{k+1}(G)-\lambda I})(\Delta_{k+1}).$$

On a :

(γ) $\Pi_k(\varphi) = \widehat{\Pi_k(G)}[\Pi_k(\psi_{k+1})] - \lambda\Pi_k(\psi_{k+1})$

(δ) $\Pi_k(\psi_{k+1}) = \widehat{\Pi_k(G)}[\Pi_k(\theta_{k+1})] - \lambda\Pi_k(\theta_{k+1})$

$(\Pi_k(G)- \lambda I)$ est une bijection de $(\Pi_k(G)- \lambda I)(\Delta_k)$ sur lui-même. On déduit donc des relations (α) et (γ) que $\Pi_k(\psi_{k+1}) = \psi_k$ pour tout $k>0$ et des relations (β) et (δ) que $\Pi_k(\theta_{k+1}) = \theta_k$.

Soit ψ tel que $\Pi_k(\psi) = \psi_k$ pour tout $k > 0$ et θ tel que $\Pi_k(\theta) = \theta_k$ pour tout $k>0$. Alors $\Pi_k(\varphi) = \Pi_k[\widehat{G(\psi)} - \lambda\psi]$ pour tout $k > 0$ et $\Pi_k(\psi)=\Pi_k[\hat{G}(\theta)-\lambda\theta]$. On en déduit que $\varphi=\hat{G}(\psi)-\lambda\psi$ avec $\psi \in (\hat{G} - \lambda I)(\Delta)$.

Donc $\hat{G} - \lambda I\,|_{(\hat{G}-\lambda I)(\Delta)}$ est surjective . D'autre part, si $\varphi \in (\hat{G} - \lambda I)(\Delta) \cap \mathrm{Ker}(\hat{G}-\lambda I)$ alors $\Pi_k(\varphi) \in (\widehat{\Pi_k(G)}- \lambda I)(\Delta_k) \cap \mathrm{Ker}(\widehat{\Pi_k(G)}- \lambda I$. D'après la propriété (1) du lemme précédent, $\Pi_k(\varphi) = 0$ pour tout $k > 0$. Donc $\varphi = 0$ et $(\hat{G} - \lambda I)\,|_{(\hat{G}-\lambda I)(\Delta)}$ est bijective.

THEOREME 2.2.6 - *Soit* $h \in \Delta_1$ *tel que* $hoG'(0) = \lambda h$. *Il existe un unique* $\theta \in \Delta$ *tel que l'on ait les propriétés suivantes :*

(1) $\theta oG = \lambda\theta$

(2) $\theta\text{-}h \in (\hat{G} - \lambda I)(\Delta)$.

De plus $\theta'(0) = h$.

Démonstration - Comme $\hat{G} - \lambda I$ est une bijection de $(\hat{G} - \lambda I)(\Delta)$ sur lui-même (corollaire 2.2.5), il existe $\psi \in (\hat{G} - \lambda I)(\Delta)$ tel que

(α) $\hat{G}(h) - \lambda h = \hat{G}(\psi) - \lambda\psi$.

Posons $\theta=h-\psi$, alors $\lambda\theta = \hat{G}(\theta) = \theta oG$ et $\theta\text{-}h \in (\hat{G} - \lambda I)(\Delta)$. Démontrons que $\theta'(0) = h$.

Notons $\delta = \Pi_2[\hat{G}(h) - \lambda h]$; $\delta \in (\widehat{\Pi_2(G)} -\lambda I)(\Delta_2)$. D'après ($\alpha$) on a :

$$\delta = \Pi_2[\hat{G}(\psi) - \lambda\psi] = \widehat{\Pi_2(G)}[\Pi_2(\psi)] - \lambda\Pi_2(\psi) \quad \text{avec } \Pi_2(\psi) \in (\widehat{\Pi_2(G)} -\lambda I)(\Delta_2).$$

Comme $\widehat{\Pi_2(G)} -\lambda I$ est une bijection de $(\widehat{\Pi_2(G)} -\lambda I)(\Delta_2)$ sur lui-même, $\Pi_2(\psi)$ est l'unique antécédent de δ par $\widehat{\Pi_2(G)} -\lambda I$. Or $\delta \in \mathcal{P}_2$ car $\Pi_1(\delta) = hoG'(0) - \lambda h = 0$. Comme $\widehat{\Pi_2(G)} -\lambda I$ est une bijection de $(\widehat{\Pi_2(G)} -\lambda I)(\Delta_2) \cap \mathcal{P}_2$ sur lui-même, $\Pi_2(\psi) \in \mathcal{P}_2$. Donc $\theta'(0) = h$.

De plus θ est unique : en effet, soit θ_1 vérifiant également les propriétés (1) et (2). Alors $\theta_1 - \theta \in (\hat{G} - \lambda I)(\Delta) \cap \mathrm{Ker}(\hat{G} - \lambda I)$. D'après le corollaire 2.2.5 $\theta_1 = \theta$.

A l'aide du théorème 1.1.6 et de la proposition 1.1.9, on en déduit le résultat suivant :

COROLLAIRE 2.2.7 - *Soit* $G \in \mathcal{G}$ *diagonalisable, soient* $(\lambda_1,...,\lambda_p)$ *les valeurs propres de* $G'(0)$ *et soit* $D = (\lambda_1 X_1,...,\lambda_p X_p)$.

Il existe $L \in \mathcal{G}_1$ tel que $L \circ G'(0) \circ L^{-1} = D$ et pour tout $U=(U_1,...,U_p) \in \mathcal{G}_1$ tel que $U \circ G'(0) \circ U^{-1} = D$, il existe un unique $Q = (Q_1,...,Q_p) \in \mathcal{G}$ tel que

(1) $Q'(0) = U$
(2) $Q \circ G \circ Q^{-1} = D$
(3) $Q_i - U_i \in (\hat{G} - \lambda_i I)(\Delta) \quad (1 \le i \le p)$.

LEMME 2.2.8 - *Soit* $F \in \mathcal{H}$ *et soient* $(\lambda_1,...,\lambda_p)$ *les valeurs propres de* $F'(0)$ *avec* $\lambda_1 = ... = \lambda_\ell = 1$. *On suppose que* $\lambda_{\ell+1}^{i_{\ell+1}} \cdots \lambda_p^{i_p} \neq 1$ *pour* $i_{\ell+1} \ge 0 \ldots i_p \ge 0$, $i_{\ell+1} + \ldots + i_p > 0$.

Notons $\mathcal{E}_k$ l'ensemble des polynômes homogènes de degré k dont tous les monômes contiennent $X_{\ell+1}, X_{\ell+2}, \ldots$ ou X_p et posons $V_k = X_{\ell+1}(\mathbb{C} \oplus \Delta_{k-1}) + \ldots + X_p(\mathbb{C} \oplus \Delta_{k-1})$. Alors on a :

(1) La dimension de $\mathrm{Ker}(\widehat{\Pi_k(F)} - I)$ est égale à l'ordre de multiplicité de 1 en tant que valeur propre de $\widehat{\Pi_k(F)}$.

(2) $\widehat{\Pi_k(F)} - I_{|\mathcal{E}_k} = \widehat{\Pi_1(F)} - I_{|\mathcal{E}_k}$ est une bijection

(3) $V_k = (\widehat{\Pi_k(F)} - I)(\Delta_k)$

(4) $\widehat{\Pi_k(F)} - I_{|V_k}$ est une bijection.

Démonstration - Posons, pour tout $k > 0$, $A_k = \widehat{\Pi_1(F)}_{|\Delta_k}$ et $F_k = \Pi_k(F)$.

Démontrons tout d'abord que $A_k - I$ est une bijection de $(A_k - I)(\Delta_k)$ sur lui-même. Comme $\Pi_1(F) \in \mathcal{H}$, la matrice de $A_1 - I$ dans la base $(X_1,...,X_p)$ a la forme suivante :

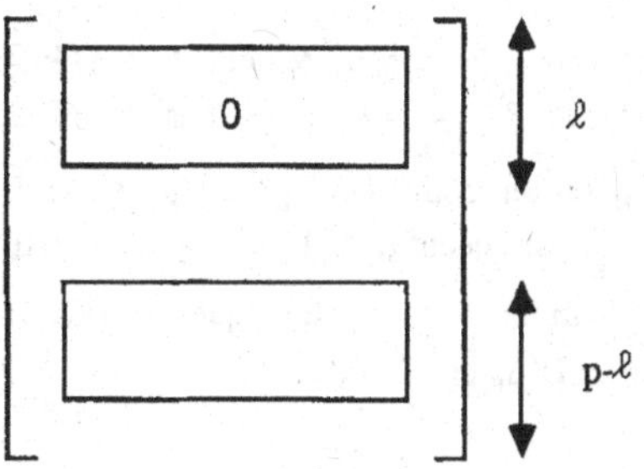

Donc dim Ker $(A_1\text{-}I) \geq \ell$ et $\mathrm{Im}(A_1\text{-}I) \subset \{a_{\ell+1}X_{\ell+1}+...+a_p X_p, a_j \in \mathbb{C}$ pour $\ell+1\leq j\leq p\}$.

Or 1 est valeur propre d'ordre ℓ de A_1 donc dim Ker $(A_1\text{-}I) = \ell$ et $\mathrm{Im}(A_1\text{-}I) = \{a_{\ell+1}X_{\ell+1}+...+a_p X_p, a_j \in \mathbb{C}$ pour $\ell+1\leq j\leq p\}$.

D'après le lemme 2.2.3, on sait en particulier que A_1-I est une bijection de $(A_1\text{-}I)(\Delta_1)$ sur lui-même. Donc, pour $i\leq\ell$, il existe $Y_i \in (A_1\text{-}I)(\Delta_1)$ tel que $(A_1\text{-}I)(Y_i) = (A_1\text{-}I)(-X_i)$. Posons pour $i \leq \ell$, $v_i = X_i+Y_i$. On a alors $A_i(v_i) = v_i$ avec $v_i \in \Delta_1$.

Donc pour $i \leq \ell$, $v_i \circ \Pi_1(F) = v_i$ et pour $i_1+...+i_\ell \leq k$ $\; v_1^{i_1}... v_\ell^{i_\ell} \circ \Pi_1(F) = v_1^{i_1}... v_\ell^{i_\ell}$. De plus, comme $v_i = X_i+Y_i$, pour $i \leq \ell$, avec $Y_i \in \{a_{\ell+1}X_{\ell+1}+...+a_p X_p, a_j \in \mathbb{C}$ pour $\ell+1\leq j\leq p\}$, $(v_1^{i_1}... v_\ell^{i_\ell})_{i_1+...+i_\ell\leq k}$ est un système libre de Δ_k. Or $v_1^{i_1}... v_\ell^{i_\ell}$ est un vecteur propre de A_k pour la valeur propre 1 si $i_1+...+i_\ell\leq k$.

Donc dim Ker $(A_k\text{-}I) \geq r$ où $r = \mathrm{card}\,\{(i_1+...+i_\ell)\}_{i_1+...+i_\ell\leq k}$. Comme $\lambda_{\ell+1}^{i_{\ell+1}}... \lambda_p^{i_p} \neq 1$ pour $i_{\ell+1} \geq 0 ... i_p \geq 0$, $i_{\ell+1}+ ... i_p>0$, 1 est valeur propre d'ordre r de A_k, donc dim Ker$(A_k\text{-}I)=r$ et d'après le lemme 2.2.3, A_k-I est une bijection de $(A_k\text{-}I)(\Delta_k)$ sur lui-même.

Démontrons maintenant la propriété (3) pour $\Pi_1(F)$. Comme $F = (f_1,...,f_p) \in \mathcal{H}$, tous les monômes de $\Pi_1(f_i)\text{-}X_i$ contiennent $X_{\ell+1}$ ou $X_{\ell+2}$ ... ou X_p (si $i \leq \ell$). On en déduit que tous les monômes de $g \,\Theta\, (\Pi_1(F)\text{-}1)$ contiennent de même $X_{\ell+1}$ ou $X_{\ell+2}$ ... ou X_p , si $g \in \Delta_k$. Donc $(A_k\text{-}I)(\Delta_k) \subset V_k$. Or dim V_k=dim Δ_{k-r}, donc dim V_k=dim $(A_k\text{-}I)(\Delta_k)$ et $V_k=(A_k\text{-}I)(\Delta_k)$.

D'autre part, si v est un polynôme homogène de degré k et si $G \in \mathcal{G}_k$, alors $\hat{G}(v) = \widehat{\Pi_1(G)}(v)$ et $\hat{G}(v)$ est homogène de degré k. Donc $F_{k|\mathcal{E}_k} = A_{k|\mathcal{E}_k}$. Pour obtenir (2), il suffit donc de démontrer que A_k-I est une bijection de $\mathcal{E}_k$ sur lui-même. Or $\mathcal{E}_k \subset V_k$ et $(A_k\text{-}I)(\mathcal{E}_k) \subset \mathcal{E}_k$. Comme $(A_k\text{-}I)_{|V_k}$ est injectif, $(A_k\text{-}I)_{|\mathcal{E}_k}$ est injectif donc bijectif et on a : $(A_k\text{-}I)(\mathcal{E}_k)=\mathcal{E}_k$. Ceci établit (2). Démontrons la propriété (3) par récurrence sur k. On a vu que $(A_1\text{-}I)(\Delta_1) = V_1$. Supposons maintenant que $(\hat{F}_k\text{-}I)(\Delta_k) = V_k$. On a $(\hat{F}_{k+1}\text{-}I)(\Delta_{k+1}) \subset V_{k+1}$.

Soit $f \in V_{k+1}$, alors d'après l'hypothèse de récurrence, $\Pi_k(f) \in (\hat{F}_k\text{-}I)(\Delta_k)$. Donc il existe $g \in \Delta_k$ tel que $\Pi_k[f\text{-} (\hat{F}_{k+1}\text{-}I)(g)]=0$. On a $f\text{-}(\hat{F}_{k+1}\text{-}I)(g) = (\hat{F}_{k+1}\text{-}I)(h)$. Donc $f = (\hat{F}_{k+1}\text{-}I)(h\text{-}g)$ et $f \in (\hat{F}_{k+1}\text{-}I)(\Delta_{k+1})$. Donc $(\hat{F}_{k+1}\text{-}I)(\Delta_{k+1}) = V_{k+1}$.

Démontrons maintenant la propriété (1).

D'après le théorème 1.1.6, on sait que 1 est valeur propre de $\hat{F}_k$ d'ordre card $\{(i_1,\dots,i_p) : (i_1+\dots+i_p \le k, \lambda_1^{i_1}\dots\lambda_p^{i_p}=1\}$. Comme $\lambda_{\ell+1}^{i_{\ell+1}}\dots\lambda_p^{i_p} \neq 1$ pour $i_{\ell+1} \geq 0 \dots i_p \geq 0$ et $i_{\ell+1}+\dots i_p>0$, card $\{(i_1,\dots,i_p) : (i_1+\dots+i_p \le k, \lambda_1^{i_1}\dots\lambda_p^{i_p}=1\} = r$. Or $r = \dim \Delta_k - \dim V_k$. Donc $r = \dim \Delta_k - \dim(\hat{F}_k\text{-}I)(\Delta_k) = \dim \operatorname{Ker}(\hat{F}_k\text{-}I)$.

Donc, d'après le lemme 2.2.3 $\hat{F}_k$-I est une bijection de $(\hat{F}_k\text{-}I)(\Delta_k)$ sur lui-même, ce qui démontre (4) et achève la démonstration.

On garde les mêmes hypothèses sur F. Comme la dimension de Ker($\widehat{\Pi_k(F)}$-I) est égale à l'ordre de multiplicité de 1 en tant que valeur propre de $\widehat{\Pi_k(F)}$, on peut appliquer le théorème 2.2.6 à F. On peut aussi, dans ce cas, préciser $(\hat{F}\text{-}I)(\Delta)$.

COROLLAIRE 2.2.9 -

$$(\hat{F}\text{-}I)(\Delta) = X_{\ell+1}(\mathbb{C} \oplus \Delta) + \dots + X_p(\mathbb{C} \oplus \Delta).$$

Démonstration - Posons de nouveau, pour tout k>0, $F_k = \Pi_k(F)$

$$V_k = X_{\ell+1}(\mathbb{C} \oplus \Delta_{k-1}) + \ldots + X_p(\mathbb{C} \oplus \Delta_{k-1})$$

et posons

$$V = X_{\ell+1}(\mathbb{C} \oplus \Delta) + \ldots + X_p(\mathbb{C} \oplus \Delta).$$

D'après le lemme précédent $V_k\ (\hat{F}_k - I)(\Delta_k)$. Donc, si $\varphi \in (\hat{F}-I)(\Delta)$, $\Pi_k(\varphi) \in V_k$ pour tout $k > 0$. Donc $\varphi \in V$. Réciproquement, si $\varphi \in V$, alors $\Pi_k(\varphi) \in V_k$ pour k>0, donc $\Pi_k(\varphi) \in (\hat{F}_k - I)(\Delta_k)$. Donc, il existe, pour tout k>0, $\psi \in (\hat{F}_k - I)(\Delta_k)$ tel que

$$(\alpha)\quad \Pi_k(\varphi) = \hat{F}_k(\psi_k) - \psi_k$$

Soit alors ψ_{k+1} tel que $\Pi_{k+1}(\varphi) = \hat{F}_{k+1}(\psi_{k+1}) - \psi_{k+1}$. On a

$$(\beta)\quad \Pi_k(\varphi) = \hat{F}_k(\Pi_k(\psi_{k+1}) - \Pi_k(\psi_{k+1})$$ avec $\Pi_k(\psi_{k+1}) \in (\hat{F}_k - I)(\Delta_k)$ car $\psi_{k+1} \in (\hat{F}_{k+1} - I)(\Delta_{k+1})$. $\hat{\Pi}_k(F) - I$ est une bijection de $(\widehat{\Pi_k(F)} - I)(\Delta_k)$ sur lui-même. On déduit donc des relations (α) et (β) que $\Pi_k(\psi_{k+1}) = \psi_k$ pour tout k>0.

Soit ψ tel que $\Pi_k(\psi) = \psi_k$ pour tout k>0. On a $\Pi_k(\varphi) = \Pi_k\ [\hat{F}(\psi) - \psi]$ pour tout k>0 ; donc $\varphi = \hat{F}(\psi) - \psi$ et $\varphi \in (\hat{F}-I)(\Delta)$. Donc $(\hat{F}-I)(\Delta) = V$.

THEOREME 2.2.10 - *Soit* $F \in \mathcal{H}$ *et soient* $(\lambda_1, \ldots, \lambda_p)$ *les valeurs propres de* F'(0) *avec* $\lambda_1 = \ldots = \lambda_\ell = 1$. *On suppose que* $\lambda_{\ell+1}^{i_{\ell+1}} \ldots \lambda_p^{i_p} \neq 1$ *pour* $i_{\ell+1} \geq 0 \ldots i_p \geq 0$ *et* $i_{\ell+1} + \ldots i_p > 0$.

Il existe $L \in \mathcal{H} \cap \mathcal{G}_1$ *tel que* $\Pi o L o F'(0) = \Pi o L$, *et pour tout* $Q \in \mathcal{H}$ *tel que* $\Pi o Q'(0) o F'(0) = \Pi o Q'(0)$, *il existe un unique* $P \in \mathcal{H}$ *tel que*

(1) $P'(0) = Q'(0)$

(2) $(I-\Pi) o P = (I-\Pi) o Q$

(3) $\Pi o (P o F o P^{-1}) = (P o F o P^{-1}) o \Pi = \Pi$ *(c'est-à-dire que* $P o F o P^{-1} \in \mathcal{U} \cap \mathcal{H}$).

Démonstration - Comme $F'(0) o \Pi = \Pi$, $X_1, \ldots, X_\ell$ sont les vecteurs propres de F'(0). Comme $\lambda_i \neq 1$ pour $i > \ell$, on a $\dim \mathrm{Ker}(F'(0)-I) = \ell$. Donc $\dim \mathrm{Ker}\ (\hat{F}'(0)-I) = \ell$, ℓ étant l'ordre de multiplicité de 1 comme valeur porpre de $\hat{F}'(0)$ (on fait agir ici $\hat{F}'(0)$ sur Δ_1). Donc, d'après le lemme 2.2.3, $\hat{F}'(0) - I \;|_{(\hat{F}'(0)-I)(\Delta_1)}$ est bijective . D'autre part , d'après le lemme 2.2.8,

$\widehat{(F'(0)}$-I$(\Delta_1) = X_{\ell+1}\mathbb{C} + ..+X_p\mathbb{C}$. Donc, pour tout i, $(1 \le i \le \ell)$, il existe $Y_i \in X_{\ell+1}\mathbb{C} + ..+X_p\mathbb{C}$. tel que Y_i o $F'(0) - Y_i = X_i$ o $F'(0) - X_i$. On a alors $(X_i - Y_i)$ o $F'(0) = X_i - Y_i$. Posons, pour tout i,$(1 \le i \le \ell)$: $\nu_i = X_i - Y_i$ et $L = \nu_1 ,...,\nu_\ell , X_{\ell+1}, X_p)$. Alors $L \in \mathcal{H} \cap \mathcal{G}_1$ et

$$\Pi oLoF'(0)=(\nu_1 ,...,\nu_\ell ,0...0) = \Pi oL.$$

Posons $\Pi oQ'(0) = (p_1,...,p_\ell ,0...0)$. Comme $\Pi oQ'(0)oF'(0) = \Pi oQ'(0)$, d'après le théorème 2.2.6, pour tout $i \le 1$, il existe $P_i \in \Delta$ tel que P_i o $F = P_i$ et $P_i - p_i \in (\hat{F}\text{-}I)(\Delta)$.

Comme $P'_i(0) = p_i$ et que $(F\text{-}I)(\Delta) = X_{\ell+1}(\mathbb{C} \oplus \Delta)+ ..+X_p(\mathbb{C} \oplus \Delta)$ on a finalement $P_i - p_i \in X_{\ell+1} \Delta+ ..+X_p\Delta$.

Posons $P =(P_1,...P_\ell, Q_{\ell+1}...Q_p$ où $Q = (Q_1,...Q_p)$. $P \in \mathcal{H}$ car $Q \in \mathcal{H}$ car

$$\begin{cases} P_i - Q_i \in X_{\ell+1}\Delta + \ldots + X_p\Delta \quad \text{pour} \quad i \le \ell \\ P_i - Q_i = 0 \quad \text{pour} \quad \ell+1 \le i \le p. \end{cases}$$

Avec ce choix de P, on a $P'(0) = Q'(0)$ et $(I\text{-}\Pi)$ o $P = (I\text{-}\Pi)$ o Q. De plus, comme $PoFoP^{-1}o\Pi=\Pi$ et comme P_i o $F = P_i$ pour $i \le \ell$ on a $\Pi oPoF = \Pi oP$. Donc $\Pi oPoFoP^{-1}= \Pi$. Soit $R \in \mathcal{H}$. Supposons que R vérifie les propriétés (1) (2) et (3). On a $R=(R_1,...,R_\ell, Q_{\ell+1}...Q_p)$.

De plus, comme $R \in \mathcal{H}$ et $Q \in \mathcal{H}$ on a : $R_i\text{-}Q_i \in X_{\ell+1}(\mathbb{C} \oplus \Delta)+ ..+X_p(\mathbb{C} \oplus \Delta)$, pour $i \le \ell$ et $Q_i\text{-}X_i \in X_{\ell+1}(\mathbb{C} \oplus \Delta)+ ..+X_p(\mathbb{C} \oplus \Delta)$, pour $i \le \ell$ .Donc $R_i - Q_i \in X_{\ell+1}\Delta+ ..+X_p\Delta$ pour $i \le \ell$, car $R'_i(o) = Q'_i(o)$ pour $i \le \ell$. D'autre part $\Pi oRoFoR^{-1} = \Pi$, donc R_i o $F = R_i$ pour $i \le \ell$.

D'après le théorème 2.2.6, on a $R_i = P_i$ pour $i \le \ell$. Donc R=P.

CHAPITRE III

§ 1 - Variétés attractives et répulsives

Définition 3.1.1. - Soit $\Omega = \{x_1,...,x_p) \in \mathbb{C}^p / x_{\ell+1} =...= x_{\ell+m} = 0\}$ (avec $\ell+m=p$) une variété linéaire de $\mathbb{C}^p$. Soit $\mathcal{U}$ un ouvert de $\mathbb{C}^p$ contenant Ω et soit $F:\mathcal{U}\to\mathbb{C}^p$, $(x,y)\mapsto(\varphi(x,y),\psi(x,y))$ une application analytique telle que $F(z) = z$ pour tout $z \in \Omega$. On dit que Ω est attractive (respectivement répulsive) pour F si toutes les valeurs propres de $\frac{\partial\psi}{\partial y}(x,0)$ sont de module strictement inférieur à 1 pour tout $x \in \mathbb{C}^\ell$ (respectivement supérieur à 1).

Remarque 3.1.2. - Si F est définie sur un ouvert $\mathcal{U}$ contenant Ω avec $F(z) = z$ pour tout $z \in \Omega$ et si $\frac{\partial\psi}{\partial y}(x,0) \in GL(\mathbb{C}^m)$ pour tout $x \in \mathbb{C}^\ell$, on déduit du théorème d'inversion locale qu'il existe un ouvert $\mathcal{V}\subset\mathcal{U}$ et un ouvert $\mathcal{W}$ contenant Ω tel que F est une bijection de $\mathcal{V}$ sur $\mathcal{W}$.

Dans ce cas $F^{-1} : \mathcal{W} \to \mathcal{V}$ est analytique et $F^{-1}(z) = z$ pour tout $z \in \Omega$. On voit alors aisément que Ω est attractive (respectivement répulsive) pour F si et seulement si Ω est répulsive (respectivement attractive) pour F^{-1}.

PROPOSITION 3.1.3. - *Si Ω est attractive ou répulsive pour* F, *les valeurs propres de* $\frac{\partial\psi}{\partial y}(x,0)$ *sont constantes.*

Démonstration : Les coefficients de la matrice de $\frac{\partial\psi}{\partial y}(x,0)$ sont des fonctions entières de x, donc le polynôme caractéristique de $\frac{\partial\psi}{\partial y}(x,0)$ est de la forme

$$P_x(X) = X^m + a_{m-1}(x)X^{m-1} + ... + a_{m-k}(x), X^{m-k} + ... + a_0(x),$$

où $a_0,..., a_{m-1}$ sont des fonctions entières de x.

Comme $a_0,..., a_{m-1}$ sont des fonctions symétriques des racines de $P_x(X)$, si Ω est attractive, $a_0,..., a_{m-1}$ sont bornées sur C^ℓ, donc constantes d'après le théorème de Liouville ([3] p. 138). Donc $P_x(X)$ est indépendant de x et les valeurs propres de $\frac{\partial\psi}{\partial y}(x,0)$ sont constantes. On a évidemment un résultat analogue quand Ω est répulsive.

Le résultat suivant justifie le terme "attractif" dans la définition 3.1.1.

THEOREME 3.1.4. - *Soit* $\Omega = \mathbb{C}^{\ell} \times 0_{\mathbb{C}^m}$ *une variété linéaire de* $\mathbb{C}^p$. *Soit* $\mathcal{U}$ *un ouvert de* $\mathbb{C}^p$ *contenant* Ω *et soit* $F : \mathcal{U} \to \mathbb{C}^p$ *une application analytique telle que* $F(z) = z$ *pour tout* $z \in \Omega$. *Alors* Ω *est attractive pour* F *si et seulement si il existe un ouvert* $\mathcal{V} \subset \mathcal{U}$ *contenant* Ω *et une application analytique* $\theta : \mathcal{V} \to \Omega$ *tels que* $F^n(\mathcal{V}) \subset \mathcal{U}$ *pour tout* n *et tels que la suite* F^n *converge vers* θ *uniformément sur tout compact de* $\mathcal{V}$. *On a alors* $\theta(z) = z$ *pour tout* $z \in \Omega$.

Démonstration - Supposons qu'il existe θ telle que la suite F^n converge vers θ uniformément sur tout compact de $\mathcal{V}$.

Posons

$$F(x,y) = (\varphi(x,y), \psi(x,y)), \quad F^n(x,y) = (\varphi_n(x,y), \psi_n(x,y)).$$

Pour tout $x \in \mathbb{C}^{\ell}$, on a $\varphi_n(x,0)=x$ et $\psi_n(x,0)=0$ donc $(F^n)'(x,0)$ est définie par la matrice

$$\begin{bmatrix} I_{\mathbb{C}^{\ell}} & \dfrac{\partial \varphi_n}{\partial y}(x,0) \\ 0 & \dfrac{\partial \Psi_n}{\partial y}(x,0) \end{bmatrix}$$

On en déduit immédiatement en utilisant le théorème de différentiation des applications composées que

$$\frac{\partial \Psi_n}{\partial y}(x,0) = \left[\frac{\partial \Psi}{\partial y}(x,0) \right]^n$$

D'autre part $(F^n)'(z)$ converge uniformément sur tout compact de Ω vers $\theta'(z)$ pour la topologie de $\mathrm{End}(\mathbb{C}^p)$. On a $\theta(x,y) = (\omega(x,y),0)$ où ω est une application analytique de $\mathbb{C}^p$ dans $\mathbb{C}^{\ell}$. Comme

$$\left[\frac{\partial \Psi}{\partial y}(x,0) \right]^n = \frac{\partial \Psi_n}{\partial y}(x,0)$$

on a

$$\lim_{n\to+\infty} \left\| \left[\frac{\partial\Psi}{\partial y}(x,0) \right]^n \right\| = 0$$

et le rayon spectral de $\frac{\partial\psi}{\partial y}(x,0)$ est strictement inférieur à 1. Donc Ω est attractive.

Supposons maintenant que Ω est attractive pour F. Soit x_0 fixé dans $\mathbb{C}^\ell$. En utilisant la théorie des formes de Jordan, on voit qu'il existe $P \in GL(\mathbb{C}^m)$ telle que $\| P \circ \frac{\partial\psi}{\partial y}(x_0,0) \circ P^{-1} \| < 1$. Choisissons comme norme sur $\mathbb{C}^m$ la norme p définie par $p(x) = \|P(x)\|$ et posons $G = P\circ\psi_{x_0}\circ P^{-1}$. Alors $G'(0) = P \circ \frac{\partial\psi}{\partial y}(x_0,0)\circ P^{-1}$. Or

$$p\left(\frac{\partial\Psi}{\partial y}(x_0,0)\right) = \sup_{p(x)<1} \|G'(0) \circ P(x)\| = \|G'(0)\| < 1.$$

Soit $\mu \in \left] \|P \circ \frac{\partial\psi}{\partial y}(x_0,0)\circ P^{-1}\|, 1 \right[$, soit B_1 la boule unité fermée de $\mathbb{C}^\ell$ et B_2 la boule unité fermée de $(\mathbb{C}^m,p)$. Comme $\frac{\partial\psi}{\partial y}$ est continue en $(x_0,0)$ il existe un réel $r>0$ tel que $(x_0+rB_1)\times rB_2 \subset \mathcal{U}$ et tel que $p(\frac{\partial\psi}{\partial y}(x,y)) \leq \mu$ pour $(x,y) \in (x_0+rB_1)\times rB_2$.

Il existe également $M > 0$ tel que $\| \frac{\partial\varphi}{\partial y}(x,y)\| \leq M$ pour $(x,y) \in (x_0+rB_1)\times rB_2$. On a alors, pour $(x,y) \in (x_0+rB_1)\times rB_2$

(1) $\|\varphi(x,y)-x\| = \|\varphi(x,y)-\varphi(x,0)\| \leq M\, p(y)$

(2) $p(\psi(x,y)) = p(\psi(x,y) - \psi(x,0)) \leq \mu p(y)$.

Soit $\rho<1$ tel que $\rho + \frac{M\rho}{1-\rho} < r$. Soit $U_{x_0} = \{x \in \mathbb{C}^\ell / \|x-x_0\| < \rho\}$ $V_{x_0} = \{y \in \mathbb{C}^m / p(y) < \rho\}$ $W_{x_0} = U_{x_0} \times V_{x_0}$ et $W'_{x_0} = (x_0+rB_1)\times rB_2$.

Posons, pour $n \geq 0$, $F^n(x,y) = (x_n,y_n)$.

Montrons, par récurrence sur n, que $F^n(W_{x_0}) \subset W'_{x_0} \subset \mathcal{U}$ et que, pour $(x,y) \in W_{x_0}$, on a :

(3) $p(y_n) < \mu^n \rho \quad (n \geq 1)$

(4) $\|x_n - x_0\| < \rho + M\rho\,(1+\ldots+\mu^{n-1}) \quad (n \geq 1).$

D'après (1) et (2) on a :

$$p(y_1) < \mu\rho \quad \text{et} \quad \|x_1 - x\| < M\rho\,, \ \text{donc} \quad \|x_1 - x_0\| < \rho + M\rho\,.$$

On voit que (3) et (4) sont vérifiées pour n=1. De plus, comme $\rho+M\rho < r$, on a $(x_1,y_1) \in W'_{x_0}$. Supposont que $(x_n,y_n) \in W'_{x_0}$ et que x_n et y_n vérifient (3) et (4). D'après (1) et (2) on a :

$$p(y_{n+1}) \leq \mu p\,(y_n) < \mu^{n+1}\rho \quad \text{et} \quad \|x_{n+1} - x_n\| \leq M\,p(y_n) < M\mu^n\rho\,.$$

Donc $\|x_{n+1} - x_0\| \leq \|x_{n+1} - x_n\| + \|x_n - x_0\| < \rho + M\rho(1+\ldots+\mu^n)$.

Donc les propriétés (3) et (4) sont vérifiées à l'ordre n+1, et comme $\mu^{n+1}\rho < r$ et $\rho+M\rho(1+\ldots+\mu^n) < r$, on voit que $(x_{n+1},y_{n+1}) \in W'_{x_0}$. Donc, si $(x,y) \in W_{x_0}$, la suite $(x_n,y_n)_{n\geq 1} = (F^n(x,y))_{n\geq 1}$ est bien définie pour tout $n \geq 1$ et est contenue dans W'_{x_0}.

Choisissons comme norme sur $\mathbb{C}^p$ la norme q définie par $q(x,y) = \|x\| + p(y)$ pour $(x,y) \in \mathbb{C}^\ell \times \mathbb{C}^m$. Alors $q(x_{n+1}-x_n, y_{n+1}-y_n) = \|x_{n+1}-x_n\| + p(y_{n+1}-y_n) < M\,\mu^n\rho + \mu^{n+1}\rho + \mu^n\rho$. Donc $q(F^{n+1}(z) - F^n(z)) < \mu^n\rho(M+\mu+1)$ pour tout $z \in W_{x_0}$ et la suite $F^n(z)$ converge uniformément sur W_{x_0}. Posons $\mathcal{V} = \bigcup_{x \in \mathbb{C}^\ell} W_x$. Alors $\mathcal{V} \subset \mathcal{U}$, $F^n(\mathcal{V}) \subset \mathcal{U}$ pour tout $n \geq 1$ et $\mathcal{V}$ est ouvert. Comme tout compact de $\mathcal{V}$ peut être recouvert par un nombre fini d'ouverts W_{x_i}, la suite (F^n) converge uniformément sur tout compact de $\mathcal{V}$ vers une application analytique $\theta: \mathcal{V} \to \mathbb{C}^p$. Soit $z \in \mathcal{V}$. Posons $\theta(z) = (\omega_1(z) \ldots \omega_p(z))$. D'après la relation (3), on a $\omega_{\ell+1}(z) = \ldots = \omega_p(z) = 0$, donc $\theta(z) \in \Omega$ pour tout $z \in \mathcal{V}$. Comme $F^n(z) = z$ pour tout $z \in \Omega$, on a $\theta(z) = z$ pour tout $z \in \Omega$. Ceci achève la démonstration du théorème.

Remarque 3.1.5 - Soit $\Omega = \mathbb{C}^\ell \times 0_{\mathbb{C}^m}$ une variété linéaire de $\mathbb{C}^p$. Soit $\mathcal{U}$ un ouvert de $\mathbb{C}^p$ contenant Ω et soit $F: \mathcal{U} \to \mathbb{C}^p$, $(x,y) \mapsto (\varphi(x,y), \psi(x,y))$ une application analytique telle que $F(z) = z$ pour tout $z \in \Omega$. Supposons Ω attractive pour F. Alors, pour tout $x_0 \in \mathbb{C}^\ell$, il existe un voisinage V de x_0 dans $\mathbb{C}^\ell$ et un voisinage V' de 0 dans $\mathbb{C}^m$ tels que $\psi_x(V') \subset V'$ pour tout $x \in V$.

De plus, si $F(x,y) = (x, \psi(x,y))$ pour tout $(x,y) \in \mathcal{U} \subset \mathbb{C}^\ell \times \mathbb{C}^m$, il existe un ouvert $\mathcal{V} \subset \mathcal{U}$, contenant, Ω tel que $F^k(\mathcal{V}) \subset \mathcal{V}$ pour tout $k \geq 1$.

Démonstration - Soit x_0 fixé dans $\mathbb{C}^\ell$. En reprenant les notations de la démonstration précédente, on choisit p comme norme sur $\mathbb{C}^m$ et on note B_1 et B_2 les boules unités fermées sur $\mathbb{C}^\ell$ et sur $\mathbb{C}^m$.

Alors on a $p\left(\frac{\partial\psi}{\partial y}(x_0,0)\right) < 1$. Soit $\mu \in \left] p\left(\frac{\partial\psi}{\partial y}(x_0,0)\right), 1\right[$. Comme $\frac{\partial\psi}{\partial y}$ est continue en $(x_0,0)$, il existe un réel r $(0 < r < 1)$ tel que $V \times V' \subset \mathcal{U}$, avec $V = x_0+rB_1$ et $V' = rB_2$, tel que $p\left(\frac{\partial\psi}{\partial y}(x,y)\right) \leq \mu$ pour $(x,y) \in V \times V'$. Donc $p(\psi(x,y)) \leq \mu p(y)$ pour $(x,y) \in V \times V'$. finalement $\psi_x(V') \subset V'$ pour tout $x \in V$.

Supposons maintenant que $F(x,y)=(x,\psi(x,y))$. Posons $V\times V'=V_{x_0}$ et pour $n\geq 0, F^n(x,y)=(x_n,y_n)$. On a immédiatement $x_n = x$. Montrons par récurrence sur n, que $F^n(V_{x_0}) \subset V_{x_0}$ et que pour $(x,y)\in V_{x_0}$, on a :

(1) $p(y_n) < \mu^n r$.

Comme $p(\psi(x,y)) \leq \mu p(y)$ pour $(x,y) \in V \times V'$, on a $p(y_1) \leq \mu p(y) < \mu r$ pour $(x,y) \in V_{x_0}$. (1) est donc vérifiée pour n=1 et $(x_1, y_1) \in V_{x_0}$.

Supposons que $(x_n, y_n) \in V_{x_0}$ et que y_n vérifie (1). Alors $p(\psi(x_n, y_n)) \leq \mu p(y_n) < \mu^{n+1} r$. Donc la propriété (1) est vérifiée à l'ordre n+1 et $(x_{n+1}, y_{n+1}) \in V_{x_0}$. Posons alors $\mathcal{V} = \bigcup_{x \in \mathbb{C}^\ell} V_x$.

Alors $F^n(\mathcal{V}) \subset \mathcal{V}$.

§ 2 - Forme canonique des germes d'applications analytiques possédant une variété linéaire attractive ou répulsive

On a vu au chapitre 2 que si $F \in \mathcal{H}$, $Q \in \mathcal{H}$ avec $\Pi o Q'(0)\ o\ F'(0) = \Pi o Q'(0)$ et si les valeurs propres $(\lambda_1,\dots,\lambda_p)$ de F'(0) vérifient $\lambda_1=\dots=\lambda_\ell=1$ et $\lambda_{\ell+1}^{i_{\ell+1}} \dots \lambda_p^{i_p} \neq 1$ pour $i_{\ell+1} \geq 0 \dots i_p \geq 0$, $i_{\ell+1}+\dots+i_p>0$, il existe un unique $P\in \mathcal{H}$ tel que

(1) $P'(0) = Q'(0)$

(2) $(I-\Pi)oP = (I-\Pi)oQ$

(3) $\Pi o(PoFoP^{-1}) = (PoFoP^{-1})o\Pi$; ce qui signifie que $PoFoP^{-1}=(X_1,...,X_\ell, g_{\ell+1}... g_p)$ où $g_i \in X_{\ell+1}(\mathbb{C}\oplus\Delta) + ... + X_p(\mathbb{C}\oplus\Delta)$ pour $i \geq \ell + 1$.

Il est évidemment possible de choisir $Q \in \mathcal{H}$ définissant une fonction analytique injective au voisinage $\Omega = \mathbb{C}^\ell \times 0_{\mathbb{C}^m}$. Pour ceci, il suffit de prendre Q linéaire et inversible. Le théorème suivant montre que si F est analytique au voisinage de Ω et si Ω est attractive ou répulsive pour F, alors P définit automatiquement une fonction analytique au voisinage de Ω s'il en est de même pour Q.

Un germe Q d'application analytique au voisinage d'une variété linéaire Ω est dit inversible si, pour tout $z \in \Omega$, $Q'(z) \in GL(\mathbb{C}^p)$. Il résulte alors du théorème d'inversion locale qu'il existe un voisinage ouvert V de Ω tel que $Q_{|V}$ est injectif. Posons $\mathcal{W} = Q(\mathcal{V})$. Alors, $\mathcal{W}$ est ouvert d'après le théorème de l'application ouverte. Si $Q(z) = z$ pour tout $z \in \Omega$, on a $\Omega \subset \mathcal{W}$ et $Q^{-1} : \mathcal{W} \to \mathcal{V}$ est analytique et vérifie $Q^{-1}(z) = z$ pour tout $z \in \Omega$. Donc, si Q est un germe inversible d'application analytique au voisinage de Ω tel que $Q(z) = z$ pour tout $z \in \Omega$, il en est de même pour Q^{-1}.

D'autre part si F et G sont des germes d'applications analytiques au voisinage de Ω tels que $F(z) = z$ et $G(z) = z$ pour tout $z \in \Omega$, alors FoG et GoF sont aussi des germes d'applications analytiques au voisinage de Ω tels que $(FoG)(z) = z$ et $(GoF)(z) = z$ pour tout $z \in \Omega$.

THEOREME 3.2.1.- *Soit* $\Omega = \mathbb{C}^\ell \times 0_{\mathbb{C}^m}$ *une variété linéaire de* $\mathbb{C}^p$ *et soit* F *un germe d'application analytique au voisiange de* Ω *vérifiant* $F(z)=z$ *pour tout* $z \in \Omega$. *Supposons* Ω *attractive ou répulsive pour* F. *Soit* Q *un germe inversible d'application analytique au voisinage de* Ω *vérifiant* $Q(z)=z$ *pour tout* $z\in\Omega$ *et* $\Pi oQ'(0)oF'(0) = \Pi oQ'(0)$. *Alors il existe un unique germe inersible* P *d'application analytique au voisinage de* Ω *vérifiant les propriétés suivantes :*

(1) $P'(0) = Q'(0)$

(2) $(I-\Pi)oP = (I-\Pi)oQ$

(3) *Il existe un voisinage* $\mathcal{V}$ *de* Ω *tel que*
$(PoFoP^{-1})(z_1,...,z_p)=(z_1,...,z_\ell,t_{\ell+1}(z_1,...,z_p)...t_p(z_1,...,z_p)$ pour $(z_1,...,z_p) \in \mathcal{V}$

où les t_i *sont des fonctions analytiques sur* $\mathcal{V}$ *vérifiant* $t_i(z)=0$ *pour tout* $z \in \Omega$.

Démonstration - Supposons Ω attractive pour F. Il résulte du théorème 3.1.4 qu'il existe un germe d'application analytique θ au voisinage de Ω tel que $\theta oF=\theta$ et tel que $\theta(z)=z$ pour tout $z\in \Omega$. Posons $Q(z) = (\varphi(z), \psi(z))$ et $\theta(z) = (\omega(z),0)$.

Soit alors $P(z) = (\omega(z), \psi(z))$. P ainsi défini vérifie (2). P est, comme Q, un germe inversible d'application analytique au voisinage de Ω. D'autre part, comme $F(z) = z$ pour tout $z \in \Omega$, la matrice de $F'(0)$ a la forme suivante :

$$\begin{bmatrix} I_{\mathbb{C}^\ell} & A \\ 0 & B \end{bmatrix}$$

Posons $\frac{\partial \omega}{\partial y}(0,0) = C$, $\frac{\partial \varphi}{\partial y}(0,0) = U$ et $\frac{\partial \psi}{\partial y}(0,0) = V$. Comme $\Pi o Q'(0) o F'(0) = \Pi o Q'(0)$, on a $A+UB=U$. Comme $\theta'(0)\, oF'(0) = \theta'(0)$ on a $A+CB=C$. Donc $U(I-B)=C(I-B)$. Comme Ω est attractive pour F, I-B est inversible donc U=C, c'est-à-dire $\frac{\partial \omega}{\partial y}(0,0) = \frac{\partial \varphi}{\partial y}(0,0)$. Donc $P'(0)=Q'(0)$, ce qui démontre (1). De plus, $PoF=(\omega oF, \psi oF)=(\omega, \psi oF)$. Donc $\Pi oPoF=\Pi oP$ et $\Pi oPoFoP^{-1} = \Pi$. Posons

$$PoFoP^{-1}(z_1,\dots,z_p)=(z_1,\dots,z_\ell,t_{\ell+1}(z_1,\dots,z_p),\dots,t_p\,(z_1,\dots,z_p)) .$$

Comme P est un germe inversible d'application analytique au voisinage de Ω et que F est un germe d'application analytique au voisinage de Ω , il existe un voisinage $\mathcal{V}$ de Ω tel que t_i soit une fonction analytique sur $\mathcal{V}$ pour tout i $(1+\ell \leq i \leq p)$. Comme de plus $F(z)=z$ et $P(z)=z$ pour tout $z \in \Omega$, on a $t_i(z) = 0$ pour tout $z \in \Omega$. L'unicité de P est assurée par le théorème formel (théorème 2.2.10).

Supposons maintenant Ω répulsive pour F. ALors, pour tout $z \in \Omega$, $F'(z) \in GL(\mathbb{C}^p)$; F^{-1} est alors un germe inversible d'application analytique au voisinage de Ω, vérifiant $F^{-1}(z)=z$ pour tout $z \in \Omega$, et Ω est attractive pour F^{-1}. D'autre part si $\Pi oQ'(0)oF'(0)=\Pi oQ'(0)$, on a

$$\Pi oQ'(0) = \Pi oQ'(0)o[F'(0)]^{-1} = \Pi oQ'(0) \; o \; (F^{-1})'(0).$$

Donc il existe un unique germe P d'application analytique au voisinage de Ω vérifiant les propriétés (1) et (2) et tel que, sur un voisinage $\mathcal{U}$ de Ω on ait :

$$(PoF^{-1}oP^{-1})(z_1,\dots,z_p)=(z_1,\dots,z_\ell,s_{\ell+1}(z_1,\dots,z_p),\dots s_p\,(z_1,\dots,z_p))$$

où s_i est une fonction analytique sur $\mathcal{U}$ vérifiant $s_i(z)=0$ pour tout $z \in \Omega$. De plus, si on pose $S(z) = (z_1,\dots,z_\ell,s_{\ell+1}(z),\dots,s_p(z))$ on a $S'(z) \in GL(\mathbb{C}^p)$ et $S(z)=z$ pour tout $z \in \Omega$ et on peut supposer S injective sur $\mathcal{U}$.

Posons $S(\mathcal{U}) = \mathcal{V}$ et $S^{-1} : \mathcal{V} \to \mathcal{U}$

$$z \to (z_1,\dots,z_\ell,t_{\ell+1}(z)\dots,t_p\,(z)).$$

Alors pour tout $z \in \mathcal{V}$ on a : $(PoFoP^{-1})(z)=(z_1,...,z_\ell,t_{\ell+1}(z),...t_p(z))$ où t_i est une fonction analytique sur $\mathcal{V}$.

De plus, comme $S(z)=z$ pour tout $z \in \Omega$, on a $t_i(z)=0$ pour tout $z \in \Omega$. La propriété (3) est donc vérifiée, et le théorème est démontré.

Nous utiliserons plus loin la remarque suivante.

Remarque 3.2.2. - Soit F et P vérifiant les conditions du théorème 3.2.1. Posons $T = PoFoP^{-1}$ sur $\mathcal{V}$. Alors Ω est attractive (respectivement répulsive) pour T, si Ω est attractive (respectivement répulsive) pour F . De plus P laisse fixe Ω point par point.

CHAPITRE IV

§ 1 - Linéarisation analytique des applications analytiques à valeurs dans $\mathcal{G}$

Soit F un espace vectoriel de dimension finie sur $\mathbb{C}$ et $\mathcal{O}$ un ouvert de $\mathbb{C}^{\ell}$. On dit que $M : \mathcal{O} \to F \quad x \to M(x)$ est analytique si les coefficients de $M(x)$ dans une base de F sont analytiques sur $\mathcal{O}$.

LEMME 4.1.1 - *Soit* E *un espace vectoriel de dimension finie* n *sur* $\mathbb{C}$. *Soit*

$$M : \mathcal{O} \subset \mathbb{C}^{\ell} \longrightarrow \mathrm{End}(E) \qquad et \qquad a : \mathcal{O} \subset \mathbb{C}^{\ell} \longrightarrow E$$
$$x \longrightarrow M(x) \qquad\qquad\qquad x \longrightarrow a(x)$$

avec M *et* a *analytiques sur* $\mathcal{O}$. *On suppose que, pour tout* $x \in \mathcal{O}$, $M(x)$ *est bijective de* $M(x).E$ *sur lui-même et que* $\dim M(x).E = K$. *Alors, il existe une unique application* χ :

$$\chi : \mathcal{O} \to E$$
$$x \to \chi(x)$$

telle que $M(x)[\chi(x)] = M(x)\,[a(x)]$ *et que* $\chi(x) \in M(x).E \; (x \in \mathcal{O})$.

De plus χ est analytique sur $\mathcal{O}$.

Démonstration - Posons $S(x)=M(x)_{|M(x).E}$. Pour tout $x \in \mathbb{C}^{\ell}$, il existe un unique $\chi(x) \in M(x).E$ tel que $M(x)[\chi(x)] = M(x)[a(x)]$; $\chi(x)$ est défini de la façon suivante : $\chi(x) = [S^{-1}(x) o M(x)][a(x)]$.

Montrons que χ est analytique sur $\mathcal{O}$. Soit x_0 fixé dans $\mathcal{O}$. Comme la matrice de M(x) dans une base de E est à coefficients analytiques sur $\mathcal{O}$ et est de rang k, il existe un déterminant $k \times k$ extrait, soit d(x), non nul sur un voisinage V de x_0. Soient $r_1(x),\dots, r_k(x)$ les vecteurs colonnes qui correspondent au déterminant extrait d(x), et R(x) la matrice associée aux vecteurs colonnes $r_1(x),\dots,r_k(x)$.

Posons $\varphi(x) = M(x)[a(x)]$. Soit alors le système (1) $RX = \varphi$. R est une matrice $n \times k$ à coefficients analytiques sur V.

De plus, pour tout $x \in V$, $\varphi(x)$ appartient à l'espace vectoriel engendré par $r_1(x),\dots, r_k(x)$ et s'écrit de manière unique comme combinaison linéaire de $r_1(x),\dots, r_k(x)$. Le système (1) est donc de rang k et compatible, et d(x) est un déterminant principal du système pour tout $x_0 \in V$. On résout alors sur V le système de Cramer des équations principales. Il existe donc, pour $1 \leq i \leq k$,

$$\alpha_i : V \longrightarrow \mathbb{C} \quad , \quad x \longrightarrow \alpha_i (x),$$

les α_i étant analytiques sur V et telles que :

$$\varphi(x) = \sum_{i=1}^{k} \alpha_i (x) r_i (x) \quad \text{soit} \quad M(x) [a(x)] = \sum_{i=1}^{k} \alpha_i (x) r_i (x) \quad .$$

De même, il existe, pour $1 \leq i \leq k$, $1 \leq j \leq k$, $\mu_{i,j} : V \longrightarrow \mathbb{C}$, $x \longrightarrow \mu_{i,j} (x)$,

les $\mu_{i,j}$ étant analytiques sur V et telles que

$$M(x) [r_j (x)] = \sum_{i=1}^{k} \mu_{i,j} (x) r_i (x) .$$

$(\mu_{i,j} (x))_{\substack{1 \leq i \leq k \\ 1 \leq j \leq k}}$ est la matrice de la bijection S(x) dans la base $(r_1 (x),..., r_k (x))$ sur V. L'inverse de cette matrice est notée $(v_{i,j} (x))_{\substack{1 \leq i \leq k \\ 1 \leq j \leq k}}$.

Sur V , $[S^{-1}(x) \text{ o } M(x)] [a(x)]$ a pour coefficients $(\beta_i(x))_{1 \leq i \leq k}$, avec $\beta_i (x) = \sum_{j=1}^{k} v_{i,j}(x) \alpha_j(x)$, dans la base $(r_1 (x),..., r_k (x))$. Posons $\chi(x) = \sum_{i=1}^{k} \beta_i(x) r_i(x)$, ; alors χ est analytique sur V et $\chi(x)=[S^{-1}(x)oM(x)][a(x)]$, avec $\chi(x) \in M(x).E$.

Comme χ est défini de manière unique sur $\mathcal{O}$ par $\chi(x) = S^{-1}(x)oM(x) [a(x)]$.et que χ est localement analytique, on en déduit que χ est analytique sur $\mathcal{O}$.

LEMME 4.1.2. *- Soit* $G : \mathcal{O} \subset \mathbb{C}^{\ell} \rightarrow \mathcal{G}$ $x \rightarrow G(x)$ *une application analytique sur* $\mathbb{C}^{\ell}$, *telle que, pour tout* $x \in \mathbb{C}^{\ell}$, G(x) *soit diagonalisable. Soient* $(\lambda_1 ,...,\lambda_p)$ *les valeurs propres de* Π_1 [G(x)] *supposées constantes, et soit* $D = (\lambda_1 X_1,...,\lambda_p X_p)$. *Pour toute application*

$T : \mathcal{O} \subset \mathbb{C}^{\ell} \rightarrow \mathcal{G}_1$,$x \rightarrow (T_1(x),...,T_p (x))$ analytique sur l'ouvert $\mathcal{O}$ et telle que $To\Pi_1 [G(x)]oT^{-1} = D$ il existe une unique application $Q=(Q_1,...,Q_p)$ analytique sur $\mathcal{O}$, telle que :

(1) $\Pi_1[Q] = T$

(2) $QoGoQ^{-1} = D$

(3) $Q_i(x) - T_i(x) \in (\hat{G}(x) - \lambda_i I)(\Delta)$ pour tout $1 \leq i \leq p$ et pour tout $x \in \mathcal{O}$.

Démonstration - Posons

$$A_{k,i} : \mathcal{O} \longrightarrow \text{End}(\Delta_k),\ x \longrightarrow \widehat{\Pi_k[G(x)]} - \lambda_i I.$$

D'après le lemme 2.2.4, $A_{k,i}(x)$ est une bijection de $A_{k,i}(x)(\Delta_k)$ sur lui-même. De plus, la dimension de $A_{k,i}(x)(\Delta_k)$ est constante et $A_{k,i}(x)$ est analytique sur $\mathcal{O}$, car G est analytique sur $\mathcal{O}$. Donc, d'après le lemme précédent, il existe une unique application $\chi_{k,i} : \mathcal{O} \longrightarrow \Delta_k$, $x \longrightarrow \chi_{k,i}(x)$ telle que .

(α) $A_{k,i}(x)[\chi_{k,i}(x)] = A_{k,i}(x)[\Pi_k(T_i(x))]$ où $\chi_{k,i}(x) \in A_{k,i}(x)(\Delta_k)$

De même il existe une unique application $\chi_{k+1,i}$ telle que

$$A_{k+1,i}(x)[\chi_{k+1,i}(x)] = A_{k+1,i}(x)[\Pi_{k+1}(T_i(x))]$$

où $\chi_{k+1,i}(x) \in A_{k+1,i}(x)(\Delta_{k+1})$. De cette dernière relation, on déduit :

(β) $\quad A_{k,i}(x)[\Pi_k(\chi_{k+1,i}(x)] = A_{k,i}(x)[\Pi_k(T_i(x))]$ où $\Pi_k(\chi_{k+1,i}(x) \in A_{k,i}(x)(\Delta_k)$.

De (α) et de (β) on déduit que $\Pi_k(\chi_{k+1,i}) = \chi_{k,i}$. De plus $\chi_{k,i}$ et $\chi_{k+1,i}$ sont analytiques sur $\mathcal{O}$.

Soit $\chi_i : \mathcal{O} \longrightarrow \Delta$, $x \longrightarrow \chi_i(x)$ définie par : $\Pi_k[\chi_i(x)] = \chi_{k,i}(x)$ pour tout $x \in \mathcal{O}$.

Alors χ_i est analytique sur $\mathcal{O}$, $\chi_i o G - \lambda_i \chi_i = \hat{G}(T_i) - \lambda_i T_i$ et $\chi_i(x) \in (\hat{G}(x) - \lambda_i I)(\Delta)$. Posons $Q_i(x) = T_i(x) - \chi_i(x)$ pour $x \in \mathcal{O}$. Q_i est analytique sur $\mathcal{O}$ et $\hat{G}(Q_i) = \lambda_i Q_i$. De plus $Q_i(x) - T_i(x) = -\chi_i(x)$ donc $Q_i(x) - T_i(x) \in (\hat{G}(x) - \lambda_i I)(\Delta)$. Comme $T_i(x) \text{ o } \Pi_1[G(x)] = \lambda_i T_i(x)$, on a , d'après le théorème 2.2.6, $\Pi_1[Q_i(x)] = T_i(x)$ pour tout $x \in \mathbb{C}^\ell$. De plus, d'après le corollaire 2.2.7, Q est unique.

LEMME 4.1.3 - *Soit* E *un espace vectoriel de dimension finie* p *sur* $\mathbb{C}$ *et* $\mathcal{B}$ *une base de* E. *Soit* $L : \mathcal{O} \subset \mathbb{C}^\ell \longrightarrow \text{End}(E)$, $x \longrightarrow L(x)$ *une application analytique sur* $\mathcal{O}$. *On suppose que, pour tout* $x \in \mathcal{O}$, $L(x)$ *est diagonalisable et semblable à* D *dont la matrice est diagonale et constante dans la base* $\mathcal{B}$. *Alors, pour tout* $x_0 \in \mathcal{O}$, *il existe un ouvert* $\mathcal{O}_{x_0}$ *contenant* x_0 *et une application* T, *analytique sur* $\mathcal{O}_{x_0}$, $T : \mathcal{O}_{x_0} \longrightarrow \text{End}(E)$, $x \longrightarrow T(x)$ *telle que, sur* $\mathcal{O}_{x_0}$, $T(x) \text{ o } L(x) \text{ o } [T(x)]^{-1} = D$.

Démonstration - Notons $\gamma_1,\ldots,\gamma_m$ les valeurs propres distinctes de L(x) et $k_1,\ldots,k_m$ leurs ordres respectifs. Notons $\Gamma_i(x)$ la matrice de $L(x)-\gamma_i I$ dans la base $\mathcal{B}(1 \le i \le m)$. Γ_i est une matrice $p \times p$ à coefficients analytiques sur $\mathcal{O}$, et pour tout $x \in \mathbb{C}^\ell$, $\Gamma_i(x)$ est de rang $p-k_i$.

Soit le système $\Gamma_i(x).X=0$. Soit x_0 fixé dans $\mathcal{O}$. Comme la matrice $\Gamma_i(x)$ est à coefficients analytiques sur $\mathcal{O}$, et est de rang constant, il existe un voisinage $\mathcal{O}_i$ de x_0 dans $\mathbb{C}^\ell$ et un déterminant extrait $D_i(x)$ tels que D_i ne s'annule pas sur $\mathcal{O}_i$, $D_i(x)$ étant un déterminant principal du système $\Gamma_i(x).X=0$.

Sur $\mathcal{O}_i$, on résout le système de Cramer des équations principales correspondant aux lignes de D_i et on peut trouver k_i vecteurs propres indépendants à coordonnées holomorphes sur $\mathcal{O}_i$ associés à la valeur propre γ_i. Posons $\mathcal{O}_{x_0} = \bigcap_{1 \le i \le m} \mathcal{O}_i$. Il existe donc sur $\mathcal{O}_{x_0}$ une base de vecteurs propres de L(x) à coordonnées holomorphes. Notons $(T_i(x))_{1 \le i \le p}$ cette base. Soit $T : \mathcal{O}_{x_0} \longrightarrow End(E)$ l'application telle que T(x) ait pour matrice $(T_1(x),\ldots,T_p(x))$ dans la base $\mathcal{B}$. Alors pour tout $x \in \mathcal{O}_{x_0}$, on a :

$$T(x) \circ L(x) \circ [T(x)]^{-1} = D.$$

THEOREME 4.1.4 - *Soit* $G : \mathcal{O} \longrightarrow \mathcal{G}, x \longrightarrow G(x)$ *une application analytique sur* $\mathcal{O}$ *telle que, pour tout* $x \in \mathcal{O}$, $G(x)$ *soit diagonalisable et telle que les valeurs propres de* $\Pi_1[G(x)]$ *soient constantes. Alors il existe une application* P, *analytique sur* $\mathcal{O}$, $P : \mathcal{O} \longrightarrow \mathcal{G}, x \longrightarrow P(x)$ *telle que* $PoGoP^{-1} = \Pi_1(G)$ *et* $\Pi_1(P) = I$,.

Démonstration - Soit $\Pi_1(G) : \mathcal{O} \longrightarrow \mathcal{G}_1, x \longrightarrow \Pi_1[G(x)]$. On peut appliquer le lemme précédent à $\Pi_1[G(x)]$ en identifiant $\mathcal{G}_1$ et $GL(\Delta_1)$. Notons $(\lambda_1,\ldots,\lambda_p)$ les valeurs propres de G. Alors il existe un recouvrement de $\mathcal{O}$ par des ovuerts $\mathcal{O}_v$, tels que sur $\mathcal{O}_v$ il existe une application T_v analytique sur $\mathcal{O}_v$, à valeurs dans $\mathcal{G}_1$, vérifiant $T_v \circ \Pi_1(G) \circ T_v^{-1} = D$ où $D = (\lambda_1 X_1,\ldots,\lambda_p X_p)$.

D'après le lemme 4.1.2, il existe Q_v, analytique sur $\mathcal{O}_v$ telle que

$$\Pi_1[Q_v] = T_v$$
$$Q_v \circ G \circ Q_v^{-1} = D$$
$$[Q_v(x) - T_v(x)]_i \in (\hat{G}(x) - \lambda_i I)(\Delta)$$

Soit $x \in \mathcal{O}_\nu \cap \mathcal{O}_\mu$. Notons $Q_\nu(x) = R$, $\Pi_1[Q_\nu(x)] = r$ et $Q_\mu(x) = S$, $\Pi_1[Q_\mu(x)] = s$. On a

(1) $\Pi_1(R) = r$

(2) $RoG(x)oR^{-1} = D$

(3) $R_i - r_i \in (\hat{G}(x) - \lambda_i I)(\Delta)$.

De plus on a :

(α) $r \circ \Pi_1[G(x)] \circ r^{-1} = D$.

On a des relations analogues pour S et s. On en déduit

(1') $\Pi_1[ros^{-1}oS] = r$.

D'autre part $s^{-1}oSoG(x)oS^{-1}o\, s = s^{-1} \circ Dos = \Pi_1[G(x)]$.

Donc

(2') $r \circ s^{-1}oSoG(x)\, oS^{-1} o\, s \circ r^{-1} = D$.

Soit i, $1 \leq i \leq p$, comme $S_i - s_i \in (\hat{G}(x) - \lambda_i I)(\Delta)$, il existe $\varphi_i \in \Delta$ tel que $S_i - s_i = (\hat{G}(x)(\varphi_i) - \lambda_i\varphi_i$, c'est-à-dire $S_i - s_i = \varphi_i \circ G(x) = \lambda_i\varphi_i$. En posant $\varphi = (\varphi_1, ..., \varphi_p)$ on obtient $S\text{-}s = \varphi \circ G(x) - Do\varphi$. Comme $s^{-1} \in \mathcal{G}_1$ et $r \in \mathcal{G}_1$ on a

$$\begin{aligned} r \circ s^{-1}oS - r &= r \circ s^{-1} o(S-s) \\ &= r \circ s^{-1} \circ \varphi \circ G(x) - r \circ s^{-1} \circ D \circ \varphi \\ &= r \circ s^{-1} \circ \varphi \circ G(x) - D \circ r \circ s^{-1} \circ \varphi \end{aligned}$$

Soit i, $1 \leq i\, p$, alors $r_i \circ s^{-1} o\, S - r_i = \hat{G}(x)[\psi_i] - \lambda_i\psi_i$ en posant $\psi = r \circ s^{-1} \circ \varphi$. Donc

(3') $[r \circ s^{-1}oS - r]_i \in (\hat{G}(x) - \lambda_i I)(\Delta)$.

Des relations (α), (1), (2), (3), (1'), (2'), (3') et du corollaire 2.2.7, on déduit que $r \circ s^{-1} \circ S = R$ ou encore $s^{-1} \circ S = r^{-1} \circ R$.

Donc $[\Pi_1(Q_\mu)]^{-1} o\, Q_\mu = [\Pi_1(Q_\nu)]^{-1} o\, Q_\nu$ sur $\mathcal{O}_\nu \cap \mathcal{O}_\mu$. On peut donc définir $P: \mathcal{O} \to \mathcal{G}$ $x \mapsto P(x)$ par $P = \Pi_1(Q_\nu)^{-1} o\, Q_\nu$ sur $\mathcal{O}_\nu$ P est analytique sur $\mathcal{O}$, $PoGoP^{-1} = \Pi_1(G)$ et

$\Pi_1[P]=\Pi_1(Q_v)^{-1} o \Pi_1(Q_v)=I$. Ceci achève la démonstration.

§ 2 - Forme réduite des applications analytiques à valeurs dans $\mathcal{G}$

LEMME 4.2.1 *- Soit* $q : \mathcal{O} \longrightarrow GL(\mathbb{C}^n), x \longrightarrow q(x)$ *une application analytique sur* $\mathcal{O}$. *Pour* $x \in \mathcal{O}$, *soit* $q(x) = g(x) \circ h(x) = h(x) \circ g(x)$ *la décomposition de* $q(x)$, *où* $g(x)$ *est diagonalisable et* $h(x)$ *unipotente. Si les valeurs propres de* $q(x)$ *sont constantes, alors* g *et* h *sont analytiques sur* $\mathcal{O}$.

Démonstration - D'après les résultats sur la décomposition de Jordan d'un automorphisme, il existe un polynôme S de $\mathbb{C}[X]$ tel que, pour tout $x \in \mathcal{O}$, $S[q(x)] = g(x)$. Rappelons une construction possible du polynôme S. Soient (λ_i) $1 \le i \le k$ les valeurs propres de q(x), chacune étant d'ordre α_i. Posons $P_i = \prod_{i=1}^{k} (X-\lambda_i)^{\alpha_i}$ avec $\sum_{i=1}^{k} \alpha_i = n$. Les polynômes P_i sont premiers entre eux deux à deux, donc d'après le théorème chinois, il existe un polynôme S tel que, pour tout i, $1 \le i \le k$, $S-\lambda_i$ soit un multiple de P_i. On peut vérifier que $S[q(x)] = g(x)$. Comme q est analytique sur $\mathcal{O}$, g est analytique sur $\mathcal{O}$. Il en est de même pour h car $h(x) = g(x)^{-1} \circ q(x)$ pour tout $x \in \mathcal{O}$.

THEOREME 4.2.2. - Soit $Q : \mathcal{O} \longrightarrow \mathcal{G}, x \longrightarrow Q(x)$ une application analytique sur $\mathcal{O}$. Pour $x \in \mathcal{O}$, soit $Q(x) = G(x) \circ H(x) = H(x) \circ G(x)$ la décomposition de Q, où G(x) est diagonalisable et H(x) unipotente. Si les valeurs propres de $\Pi_1[Q(x)]$ sont constantes, alors G et H sont analytiques sur $\mathcal{O}$.

Démonstration - Posons $\Pi_k[Q(x)] = Q_k(x)$, $\Pi_k[G(x)] = G_k(x)$ et $\Pi_k[H(x)] = H_k(x)$ pour tout $k, k \ge 1$. Alors on a : $Q_k(x) = G_k(x) \ominus H_k(x) = H_k(x) \ominus G_k(x)$ avec $G_k(x)$ diagonalisable et $H_k(x)$ unipotente. Soit alors

$$\hat{Q}_k : \mathcal{O} \longrightarrow GL(\Delta_k) \quad x \longrightarrow \hat{Q}_k(x) \qquad \hat{G}_k : \mathcal{O} \longrightarrow GL(\Delta_k) \quad x \longrightarrow \hat{G}_k(x)$$

et

$$\hat{H}_k : \mathcal{O} \longrightarrow GL(\Delta_k) \qquad x \longrightarrow H_k(x).$$

Choisissons dans Δ_k la base $(e_{i_1..i_p})_{i_1+...+i_p \le k}$ avec $e_{i_1..i_p} = X_1^{i_1} ... X_p^{i_p}$. Notons n la dimension de Δ_k, on peut alors identifier $GL(\Delta_k)$ et $GL(\mathbb{C}^n)$.

D'après la proposition 1.1.1, on a

$\hat{Q}_k(x) = \hat{G}_k(x) \circ \hat{H}_k(x) = \hat{H}_k(x) \circ \hat{G}_k(x)$.

De plus, d'après le corollaire 1.1.7, $\hat{H}_k(x)$ est unipotente et d'après le théorème 1.1.6, $\hat{G}_k(x)$ est diagonalisable. Comme $G_k(x)$ et $\Pi_1[Q(x)]$ ont les mêmes valeurs propres : on déduit du théorème 1.1.6 que les valeurs propres de $\hat{G}_k(x)$ sont constantes si celles de $\Pi_1[Q(x)]$ le sont.

Donc, dans ce cas, d'après le lemme précédent $\hat{G}_k$ et $\hat{H}_k$ sont analytiques sur $\mathcal{O}$. On en déduit que G_k et H_k sont analytiques sur $\mathcal{O}$, pour tout $k, k \geq 1$, donc que G et H sont analytiques sur $\mathcal{O}$.

THEOREME 4.2.3. - Soit $F : \mathcal{O} \longrightarrow \mathcal{G}$, $x \longrightarrow F(x)$ *une application analytique sur* $\mathcal{O}$. *Soient* $(\mu_1, \ldots, \mu_p)$ *les valeurs propres de* Π_1 (F) *supposées constantes. Alors il existe une application* P *et une application* N *analytiques sur* $\mathcal{O}$:

$$P : \mathcal{O} \longrightarrow \mathcal{G} \qquad N : \mathcal{O} \longrightarrow \mathcal{G}$$
$$x \longrightarrow P(x) \qquad x \longrightarrow N(x).$$

telles que, pour tout $x \in \mathcal{O}$:

a) $F(x) \circ P(x) = P(x) \circ N(x)$ et $\Pi_1[P(x)] = I$
b) $N(x) = L(x) \circ U(x) = U(x) \circ L(x)$
c) $L(x) \in \mathcal{G}_1$ *et est diagonalisable*
d) $U(x)$ *est unipotente.*

Si de plus, $1 \leq |\mu_1| \leq \ldots \leq |\mu_p|$ et $|\mu_p| < |\mu_1|^{q+1}$, alors $U(x) \in \mathcal{G}_q$ et $[U(x)]^{-1} \in \mathcal{G}_q$ pour tout $x \in \mathcal{O}$.

Démonstration - Pour $x \in \mathcal{O}$, soit $F(x) = G(x) \circ H(x) = H(x) \circ G(x)$ la décomposition de F où G(x) est diagonalisable et H(x) unipotente. D'après le théorème précédent, G et H sont analytiques sur $\mathcal{O}$ car $\Pi_1(F)$ a des valeurs propres constantes . Comme $\Pi_1[F(x)]$ et $\Pi_1[G(x)]$ ont les mêmes valeurs propres d'après le théorème 4.1.4, il existe une application P, analytique sur $\mathcal{O}$, $P : \mathcal{O} \longrightarrow \mathcal{G}$, $x \longrightarrow P(x)$, telle que $P^{-1} \circ G \circ P = \Pi_1(G)$ avec $\Pi_1(P) = I$.

Posons $U = P^{-1} \circ H \circ P$ .U est unipotente et analytique sur $\mathcal{O}$. Notons $\Pi_1(G) = L$. Alors

$$\begin{aligned} L \circ U &= \Pi_1(G) \circ U \\ &= P^{-1} \circ G \circ P \circ P^{-1} \circ H \circ P \\ &= P^{-1} \circ G \circ H \circ P \\ &= P^{-1} \circ F \circ P \end{aligned}$$

de même

$$\begin{aligned} U o L &= P^{-1} o H o P o P^{-1} o G o P \\ &= P^{-1} o H o G o P \\ &= P^{-1} o F \; o P. \end{aligned}$$

En posant $N(x) = U(x) \text{ o } L(x)$ pour tout $x \in \mathcal{O}$, on a les propriétés b) c) et d). Comme $G(x)$ est diagonalisable, il existe, pour tout $x \in \mathcal{O}$, $T(x) \in \mathcal{G}_1$ tel que

$$T(x) \text{ o } L(x) \text{ o } [T(x)]^{-1} = D \quad \text{où} \quad D = (\mu_1 X_1, \ldots, \mu_p X_p).$$

Posons $V(x) = T(x) \text{ o } U(x) \text{ o } [T(x)]^{-1}$. Alors $V(x)$ est unipotente et

$$V(x) \text{ o } D = T(x) \text{ o } U(x) \text{ o } L(x) \text{ o } [T(x)]^{-1} = T(x) \text{ o } L(x) \text{ o } U(x) \text{ o } [T(x)]^{-1} = D \text{ o } V(x).$$

D'après la remarque 1.3.4 et la proposition 1.3.5, si $1 < |\mu_1| \leq \ldots \leq |\mu_p|$ et $|\mu_p| < |\mu_1|^{q+1}$, alors $V(x) \in \mathcal{G}_q$ et $[V(x)]^{-1} \in \mathcal{G}_q$. Donc $U(x) = [T(x)]^{-1} \text{o} V(x) \text{o} T(x) \mathcal{G}_q$ et $[U(x)]^{-1} = [T(x)]^{-1} \text{o} [V(x)]^{-1} \text{o} T(x) \in \mathcal{G}_q$.

§ 3 - Equivalence locale de germes d'automorphismes à points fixes répulsifs dépendant analytiquement d'un paramètre.

THEOREME 4.3.1 - *Soit* $x_0 \in \mathbb{C}^\ell$ *et soit* $\mathcal{V} = V \times V'$ *un voisinage ouvert de* $(x_0, 0)$ *dans* $\mathbb{C}^\ell \times \mathbb{C}^p$. *Soient* $G : \mathcal{V} \longrightarrow \mathbb{C}^\ell$ *et* $H : \mathcal{V} \longrightarrow \mathbb{C}^p$ *deux applications analytiques. Notons* $G_x : V' \to \mathbb{C}^p$ *(respectivement* $H_x : V' \to \mathbb{C}^p$*) l'application* $y \longrightarrow G(x,y)$ *(respectivement* $y \longrightarrow H(x,y)$*). On suppose que* $G_x(0) = 0$ *et que* $\Pi_q(G_x) = \Pi_q(H_x)$ $(q \in \mathbb{N})$ *pour tout* $x \in V$ *et que l'on a les deux conditions suivantes :*

(i) $G'_{x0}(0) \in GL(\mathbb{C}^p)$

(ii) $\sigma_0 \sigma_1^{q+1} < 1$ *et* $\sigma_1 < 1$ *où* $q \in \mathbb{N}$ *et où* σ_0 *(respectivement* σ_1*) désigne le rayon spectral de* $G'_{x0}(0)$ *(respectivement de* $[G'_{x0}(0)]^{-1}$*). Alors, il existe un voisinage ouvert* W *de* x_0 *dans* $\mathbb{C}^\ell$, *un voisinage* W' *de* 0 *dans* $\mathbb{C}^p$ *et une application* Q *analytique sur* $\mathcal{W}_{x_0} = W \times W'$ *tels que :*

(1) $(x,y) \longrightarrow H_x^{-1}(y)$ *est définie et analytique sur* $W \times W'$.

(2) Pour tout $x \in W$, *pour tout* $n \geq 1$, *la série définissant le germe* $(Q_n)_x = (G_x)^n \circ (H_x)^{-n}$ *est analytique sur W' et l'application* $Q_n : (x,y) \longrightarrow (Q_n)_x\ (y)$ *est analytique sur* $\mathcal{W}_{x_0}$.

(3) Q_n *converge uniformément sur* $\mathcal{W}_{x_0}$ *vers une application analytique* Q.

(4) $\Pi_q(Q_x) = I_w$, *pour tout* $x \in W$ *où* $Q_x : W' \longrightarrow \mathbb{C}^p$ *est l'application* $y \longrightarrow Q(x,y)$.

(5) La série définissant le germe $G_x \circ Q_x \circ H_x^{-1}$ *est convergente sur* W', *pour tout* $x \in W$ *et :*

$(G_x \circ Q_x \circ H_x^{-1})(y) = Q_x(y)$ *pour tout* $(x,y) \in \mathcal{W}_{x_0}$.

Démonstration - Par hypothèse $\sigma_0 \sigma_1^{q+1} < 1$, c'est-à-dire $\sigma_0 < (\frac{1}{\sigma_1})^{q+1}$. De plus $\sigma_1 < 1$, donc il existe $\nu \in]0,1[$ et $\lambda > 1$ tels que :

$$\sigma_0 < \lambda < (\frac{1}{\nu})^{q+1} .< (\frac{1}{\sigma_1})^{q+1} .$$

Comme $\sigma_0 < \lambda$ et $\sigma_1 < \nu$ il résulte de la théorie des formes de Jordan que, pour x_0 donné dans $\mathbb{C}^\ell$, il existe $\chi \in GL(\mathbb{C}^p)$ telle que :

$$\|\chi^{-1} \circ G'_{x_0}(0) \circ \chi\| < \lambda$$

et $\|\chi^{-1} \circ [G'_{x_0}(0)]^{-1} \circ \chi\| < \nu$ avec $\nu < 1$ et $\lambda \nu^{q+1} < 1$. Considérons alors sur $\mathbb{C}^p$ la norme $\| \ \|_1$, définie par $\|x\|_1 = \|\chi^{-1}(x)\|$. On a alors :

$$\| G'_{x_0}(0)\|_1 < \lambda \quad \text{et} \quad \| [G'_{x_0}(0)]^{-1}\|_1 < \nu \quad .$$

Notons par ailleurs, B la boule unité de $\mathbb{C}^p$ pour la norme $\| \ \|_1$. Si F est analytique sur $\mathcal{V}$, on note $\frac{\partial F}{\partial y}$ la deuxième différentielle partielle de F. Par définition de $G'_{x_0}(0)$, on a $\frac{\partial G}{\partial y}(x_0, 0) = G'_{x_0}(0)$. On a $H(x_0, 0) = 0$ et $H'_{x_0}(0) \in GL(\mathbb{C}^p)$. Donc, d'après le théorème des fonctions implicites (voir par exemple [3] p. 138) pour $x_0 \in \mathbb{C}^\ell$, il existe une application analytique définie sur un voisinage de $(x_0,0)$, dont on note $H_x^{-1}(y)$ la valeur en (x,y) et telle que sur un voisinage de 0,

$(H_x o H_x^{-1})(y) = (H_x^{-1} o H_x)(y) = y.$

De plus H_x^{-1} est l'inverse de H_x dans $\mathcal{G}$. Comme $\Pi_q(G_x) = \Pi_q(H_x)$, on a en particulier $H_x^{-1}(0) = H_x(0) = G_x(0)$ et $H'_x(0) = G'_x(0)$ pour tout $x \in V$. Comme $\frac{\partial H}{\partial y}, \frac{\partial G}{\partial y}$ et $J : GL(\mathbb{C}^p) \longrightarrow GL(\mathbb{C}^p)$, $u \longrightarrow u^{-1}$ sont continues, il existe une boule ouverte W de $\mathbb{C}^\ell$, centrée en x_0 et un réel $r > 0$ tel que $\overline{W} \subset V$, $r\overline{B} \subset V'$ et tels que pour $(x,y) \in W \times rB$, on ait $H'_x(y) \in GL(\mathbb{C}^p)$,

$$\| G'_x(y) \|_1 < \lambda \quad \text{et} \quad \| [H'_x]^{-1}(y) \|_1 < \nu \quad .$$

On a, pour $(x,y) \in W \times rB$

(a) $\| H_x^{-1}(y) \|_1 \leq \nu \, \|y\|_1$.

(b) $\| G_x(y_1) - G_x(y_2) \|_1 \leq \lambda \| y_1 - y_2 \|_1$ $(y_1 \in rB \, ; \, y_2 \in rB)$.

De plus, l'application $\frac{\partial^m (G-H)}{(\partial y)^m}$ est continue pour tout $m \geq 1$. Comme $\overline{W} \subset V$, $r\overline{B} \subset V'$ et comme $\Pi_q(G_x) = \Pi_q(H_x)$, il existe $M > 0$ tel que l'on ait :

(c) $\| G_x(y) - H_x(y) \|_1 \leq M \| y \|_1^{q+1}$ pour tout $(x,y) \in W \times rB$. Posons

$$k = \inf \left[\frac{1}{2\nu} \, , \, \left[\frac{1 - \nu^{q+1} \lambda}{2M\nu^{q+1} r^q} \right]^{\frac{1}{q+1}} \right] \quad (k<1)$$

$k \, r \, B = W'$

$U = W \times W'$.

Notons $\mathcal{H}(U,(\mathbb{C}^p)$ l'ensemble des fonctions holomorphes de U dans $\mathbb{C}^p$, et I l'identité sur W'. Soit $\mathcal{E} = \{ \varphi \in \mathcal{H}(U,\mathbb{C}^p) \, / \, \forall x \in W, \forall y \in W', \ \| \varphi(x,y) - y \|_1 \leq \frac{r}{2}$ et $\Pi_q[\varphi_x] = I \}$

Soit $\mathcal{H}_0$ l'ensemble des applications analytiques bornées de U dans $\mathbb{C}^p$. $\mathcal{H}_0$ est une algèbre de Banach pour la norme

$$\|\varphi\|_U = \sup_{(x,y)\in U} \|\varphi(x,y)\|_1 = \sup_{x\in W} (\|\varphi_x\|_{W'}) \text{ où } \|\varphi_x\|_{W'} = \sup_{y\in W'} \|\varphi(x,y)\|_1$$

$\mathcal{E}$ est un fermé de $\mathcal{H}_0$ donc $\mathcal{E}$ est métrique complet pour la distance

$$d(\varphi,\Psi) = \sup_{(x,y)\in U} \|\varphi(x,y) - \Psi(x,y)\|_1 = \|\varphi-\psi\|_U$$

Soient $\varphi \in \mathcal{E}$, $\psi \in \mathcal{E}$ et x fixé dans W. Montrons d'abord que pour tout $y \in W'$, on a :

$$\|\varphi_x(y) - \Psi_x(y)\|_1 \leq \|\varphi_x - \psi_x\|_{W'} \frac{\|y\|_1^{q+1}}{k^{q+1}\, r^{q+1}} .$$

On fixe y_0 tel que $\|y_0\|_1 = kr$. Soit ℓ une forme linéaire continue sur $\mathbb{C}^p$ telle que $\|\ell\|_1 = 1$. Alors l'application :

$$D(0,1) \to \mathbb{C}$$

$$\xi \to \ell\left[\frac{\varphi_x(\xi y_0) - \Psi_x(\xi y_0)}{\xi^{q+1}}\right]$$

est analytique sur D(0,1) (disque unité ouvert de $\mathbb{C}$). Cela résulte du fait que $\varphi_x(\xi y_0) - \psi_x(\xi y_0) = \xi^{q+1}R(\xi)$ avec R analytique sur D(0,1) (car $\Pi_q(\varphi_x) = \Pi_q(\psi_x)$).

Soit ξ un élément donné de D(0,1), alors pour tout ρ tel que $|\xi| < \rho < 1$ on a :

$$\left|\ell\left[\frac{\varphi(\xi y_0) - \Psi(\xi y_0)}{\xi^{q+1}}\right]\right| \leq \frac{\|\varphi_x - \Psi_x\|_W}{\rho^{q+1}}$$

d'où $|\ell(\varphi_x(\xi y_0) - \psi_x(\xi y_0))| \leq |\xi|^{q+1}\|\varphi_x - \psi_x\|_{W'}$ pour tout y tel que $\|y\|_1 < kr$ il existe y_0

dans $\mathbb{C}^p$ et ξ dans D(0,1) tels que $\|y_0\|_1 = kr$ et $y = \xi y_0$. Donc, pourtout $y \in W'$, on a :

$$|\ell(\varphi_x(y) - \Psi_x(y))| \leq \|\varphi_x - \Psi_x\|_{W'} \frac{\|y\|_1^{q+1}}{k^{q+1} r^{q+1}}$$

si ℓ est une forme linéaire continue de norme 1 .

D'après Hahn-Banach, il existe une forme linéaire ℓ_y continue sur $\mathbb{C}^p$, de norme 1 telle que :

$$|\ell_y(\varphi_x(y) - \psi_x(y))| = \|\varphi_x(y) - \psi_x(y)\|_1 .$$

Donc, pour tout $(\varphi,\psi) \in \mathcal{E}^2$, pour tout $x \in W$ et pour tout $y \in W'$, on a :

$$\text{(d)} \quad \|\varphi_x(y) - \Psi_x(y)\|_1 \leq \|\varphi_x - \Psi_x\|_{W'} \frac{\|y\|_1^{q+1}}{k^{q+1} r^{q+1}} .$$

Il résulte de (a) que W' est stable par $(H_x)^{-1}$, pour tout $x \in W$. On déduit alors de (d) que, pour tout $y \in W'$, on a

$$\text{(e)} \quad \|(\varphi_x \circ H_x^{-1})(y) - (\psi_x \circ H_x^{-1})(y)\|_1 \leq \|(\varphi_x - \psi_x\|_{W'} \, v^{q+1}$$

Appliquons alors (e) en choisissant ψ telle que $\psi(x,y) = y$ sur U ; alors pour tout $y \in W'$ on a :

$$\|(\varphi_x \circ H_x^{-1})(y) - H_x^{-1}(y)\|_1 \leq \|\varphi_x - I\|_{W'} \, v^{q+1} \leq v^{q+1} \frac{r}{2}$$

donc

$$\|(\varphi_x \circ H_x^{-1})(y)\|_1 \leq v\|y\|_1 + \frac{r}{2} v^{q+1} .$$

Donc pour tout $\varphi \in \mathcal{E}$ et tout $y \in W'$ on a :

$$\text{(f)} \quad \|(\varphi_x \circ H_x^{-1})(y)\|_1 \leq r \quad (\text{car} \quad k < \frac{1}{2v} \text{ et } v<1).$$

Soit maintenant $\theta : \mathcal{E} \longrightarrow \mathcal{H}_0$, $\varphi \longrightarrow \theta(\varphi)$ où $\theta(\varphi)(x,y) = (G_x \circ \varphi_x \circ H_x^{-1})(y)$. Il résulte de (f) que θ est bien définie.

Soit $x \in W$. On a

$$\begin{aligned}\Pi_q(G_x \circ \varphi_x \circ H_x^{-1}) &= \Pi_q(G_x) \ominus \Pi_q(\varphi_x) \ominus \Pi_q(H_x^{-1}) \\ &= \Pi_q(H_x) \ominus \Pi_q(H_x^{-1}) \\ &= \Pi_q(H_x \circ H_x^{-1}) \\ &= I.\end{aligned}$$

Donc $\Pi_q[\theta(\varphi)_x] = I$ pour tout $\varphi \in \mathcal{E}$.

Soit $(x,y) \in W \times W'$. Appliquons l'inégalité (b) avec $y_1 = (\varphi_x \circ H_x^{-1})(y)$ et $y_2 = (\psi_x \circ H_x^{-1})(y)$ où $\varphi \in \mathcal{E}$, $\psi \in \mathcal{E}$. Donc, pour tout $(x,y) \in U$ on a :

(g) $$\|(G_x \circ \varphi_x \circ H_x^{-1})(y) - (G_x \circ \psi_x \circ H_x^{-1})(y)\|_1 \leq \lambda \|(\varphi_x \circ H_x^{-1})(y) - (\psi_x \circ H_x^{-1})(y)\|_1$$

Donc

$$\|[\theta(\varphi) - \theta(\psi)](x,y)\|_1 \leq \lambda v^{q+1} \|\varphi_x - \psi_x\|_{W'} \leq \lambda v^{q+1} \|\varphi - \psi\|_U, \text{ (d'après (e))}.$$

Finalement $\|\theta(\varphi) - \theta(\psi)\|_U \leq \lambda v^{q+1} \|\varphi - \psi\|_U$. θ est donc contractante. Montrons maintenant que $\mathcal{E}$ est stable par θ. Soit $(x,y) \in U$.

$$\|(G_x \circ \varphi_x \circ H_x^{-1})(y) - y\|_1 \leq \|(G_x \circ \varphi_x \circ H_x^{-1})(y) - (G_x \circ H_x^{-1})(y)\|_1 + \|(G_x \circ H_x^{-1})(y) - (H_x \circ H_x^{-1})(y)\|_1$$

Appliquons (g) en choisissant ψ telle que $\psi(x,y) = y$ sur U ; alors on a

$$\|(G_x \circ \varphi_x \circ H_x^{-1})(y) - (G_x \circ H_x^{-1})(y)\|_1 \leq \lambda \|(\varphi_x \circ H_x^{-1})(y) - H_x^{-1}(y)\|_1 \leq \lambda v^{q+1} \frac{r}{2}$$

De plus

$$\|(G_x \circ H_x^{-1})(y) - (H_x \circ H_x^{-1})(y)\|_1 \le M \| H_x^{-1}(y)\|_1^{q+1} \le M v^{q+1} \|y\|_1^{q+1} .$$

D'où

$$\|G_x \circ \varphi_x \circ H_x^{-1})(y) - y\|_1 \le \lambda v^{q+1} \frac{r}{2} + M v^{q+1} k^{q+1} r^{q+1} \le \lambda v^{q+1} \frac{r}{2} + \frac{M v^{q+1} r}{2 M v^{q+1}} (1 - v^{q+1} \lambda) \le \frac{r}{2}.$$

Donc, pour tout $(x,y) \in U$ on a : $\|\theta(\varphi)(x,y) - y\|_1 \le \frac{r}{2}$

D'après le théorème du point fixe, il existe un unique $Q \in \mathcal{E}$ telle que $\theta(Q)=Q$ et $\lim_{n \to +\infty} \|Q - \theta^n(\mathcal{J})\|_U = 0$

(où $\mathcal{J} : U \longrightarrow \mathbb{C}^p$ est l'application $(x,y) \longrightarrow y$).

C'est-à-dire, pour tout $(x,y) \in W \times W'$ on a : $(G_x \circ Q_x \circ H_x^{-1})(y) = Q(x,y)$, avec Q analytique sur $W \times W'$, et pour tout $x \in W$, $\Pi_q(Q_x) = I$. De plus $\theta^n(\mathcal{J})$ est bien définie et analytique sur U, et on démontre par récurrence sur n que

$[\theta^n(\mathcal{J})]_x = G_x^n \circ H_x^{-n}$. Donc, si on pose $Q_n(x,y) = (G_x^n \circ H_x^{-n})(y)$, la suite Q_n converge uniformément vers Q sur U. En posant $\mathcal{W}_{x_0} = U$, on obtient toutes les conditions requises.

§ 4 - Forme réduite des applications analytiques définies au voisinage d'une variété linéaire fixe répulsive

Définition 4.4.1 - Soit $\Omega = \mathbb{C}^\ell \times O_{\mathbb{C}^p}$ une variété linéaire de $\mathbb{C}^n$ $(n=\ell+p)$ et soit $\mathcal{F}$ un germe d'application analytique au voisinage de Ω vérifiant $\mathcal{F}(z)=z$ pour tout $z \in \Omega$.

On appelle forme réduite de $\mathcal{F}$ tout automorphisme polynomial $\mathcal{N} : (x,y) \longrightarrow (x,N(x,y))$ de $\mathbb{C}^{\ell+p}$ tel que $\mathcal{N}(z)=z$ pour tout $z \in \Omega$, pour lequel il existe un germe $\mathcal{P}$ d'application analytique au voisinage de Ω et deux applications entières $(x,y) \longrightarrow L(x,y)$ et $(x,y) \longrightarrow U(x,y)$ de $\mathbb{C}^{\ell+p}$ dans $\mathbb{C}^p$ nulles sur Ω et vérifiant :

(1) $\mathcal{P}(z)=z$ pour tout $z \in \Omega$.

(2) $\mathcal{P}'(z) \in GL(\mathbb{C}^n)$ pour tout $z \in \Omega$.

(3) $\mathcal{F} \circ \mathcal{P} = \mathcal{P} \circ \mathcal{N}$ (au sens des germes)

(4) $N_x = L_x \circ U_x = U_x \circ L_x$ pour tout $x \in \mathbb{C}^\ell$

(5) L_x est linéaire et diagonalisable pour tout $x \in \mathbb{C}^\ell$.

(6) U_x est unipotente pour tout $x \in \mathbb{C}^\ell$ où $N_x : \mathbb{C}^p \longrightarrow \mathbb{C}^p$; $L_x : \mathbb{C}^p \longrightarrow \mathbb{C}^p$; $U_x : \mathbb{C}^p \longrightarrow \mathbb{C}^p$ désignent respectivement, pour $x \in \mathbb{C}^\ell$, les applications $y \longrightarrow N(x,y)$; $y \longrightarrow L(x,y)$ et $y \longrightarrow U(x,y)$.

Remarque - On note qu'il n'y a pas unicité de la forme réduite.

Définition 4.4.2 - Soit $\Omega = \mathbb{C}^\ell \times O_{\mathbb{C}^p}$ une variété linéaire de $\mathbb{C}^n$ et soit $\mathcal{F} : (x,y) \longmapsto (x,F(x,y))$ un germe d'application analytique au voisinage de Ω tel que $\mathcal{F}(z)=z$ pour tout $z \in \Omega$ et telle que Ω soit répulsive pour $\mathcal{F}$.

Soit $\mu_1,\ldots,\mu_p$ avec $1 < |\mu_1| \leq \ldots \leq |\mu_p|$ les valeurs propres de $\dfrac{\partial F}{\partial y}(0,0)$ on note q(F) le plus petit entier positif q tel que $|\mu_1|^{q+1} > |\mu_p|$. Il résulte de la proposition 3.1.3 que $(\mu_1,\ldots,\mu_p)$ est la famille des valeurs propres de $\dfrac{\partial F}{\partial y}(x,0)$ pour tout $x \in \mathbb{C}^\ell$.

LEMME 4.4.3 - Soit $\Omega = \mathbb{C}^\ell \times O_{\mathbb{C}^p}$ une variété linéaire de $\mathbb{C}^n$ et soit $\mathcal{F}$ un germe d'application analytique au voisinage de Ω vérifiant $\mathcal{F}(z) = z$ pour tout $z \in \Omega$. Supposons qu'il existe un ouvert $\mathcal{V}$ de $\mathbb{C}^\ell \times \mathbb{C}^p$ contenant Ω tel que $\mathcal{F}(x,y) = (x,F(x,y))$ sur $\mathcal{V}$. Alors, si Ω est répulsive pour $\mathcal{F}$, $\mathcal{F}$ admet une forme réduite $\mathcal{N}$, et il existe un germe $\mathcal{R} : (x,y) \longrightarrow (x,R(x,y))$ d'application analytique au voisinage de Ω, une application entière $T : x \longrightarrow T_x$ de $\mathbb{C}^p$ dans $\mathcal{G}_{q(F)}$ et un voisinage $\mathcal{U}$ de Ω tels que

(1) $\mathcal{F} \circ \mathcal{R} = \mathcal{R} \circ \mathcal{N}$ (au sens des germes)

(2) $\mathcal{R}(z) = z$ pour tout $z \in \Omega$

(3) Le germe d'application $R_n : (x,y) \longrightarrow (F_x^n \circ T_x \circ N_x^{-n})(y)$ est défini et analytique sur $\mathcal{U}$ pour tout n et la suite R_n converge uniformément vers R sur tout compact de $\mathcal{U}$.

(4) $\Pi_1[R_x] = I$ et $\Pi_q[R_x] = T_x$ pour tout $x \in \mathbb{C}^\ell$ où F_x , N_x et R_x désignent respectivement pour $x \in \mathbb{C}^\ell$, les applications $y \longrightarrow F(x,y)$, $y \longrightarrow N(x,y)$ et $y \longrightarrow R(x,y)$.

Démonstration - Posons $q = q(F)$. $F = (f_1 ,\ldots,f_p)$. Soit $j \leq p$ et notons $f = f_j$. Pour $x \in \mathbb{C}^\ell$, soit $f_x : \mathbb{C}^p \longrightarrow \mathbb{C}$ l'application $y \longrightarrow f(x,y)$. Soit $\mathcal{V}$ un ouvert de $\mathbb{C}^n$ contenant Ω sur lequel $\mathcal{F}(x,y)=(x,F(x,y))$. Pour tout x_0 fixé dans $\mathbb{C}^\ell$, il existe une boule V_x de centre x_0 et une boule W_x

de centre 0 dans $\mathbb{C}^p$ telles que $V_{x_0} \times W_{x_0} \subset \mathcal{V}$. Sur $V_{x_0} \times W_{x_0}$, f est développable en série entière, donc :

$$f_x(y) = \sum_{|i| \geq 1} f_{i_1,\dots,i_p,x_0}(x)\, y_1^{i_1} \dots y_p^{i_p}$$

où les $f_{i_1,\dots,i_p,x_0}$ sont analytiques sur V_{x_0} et où $|i| = i_p + \dots + i_1$. De même, si $V_{x_0} \cap V_{x_1} \neq \emptyset$, f est développable en série entière sur $V_{x_1} \times W_{x_1} \subset \mathcal{V}$ et

$$f_x(y) = \sum_{|i| \geq 1} f_{i_1,\dots,i_p,x_1}(x)\, y_1^{i_1} \dots y_p^{i_p}$$

où les $f_{i_1,\dots,i_p,x_1}$ sont analytiques sur V_{x_1}. f est analytique sur $(V_{x_0} \cap V_{x_1}) \times (W_{x_0} \cap W_{x_1})$ et $f_{i_1,\dots,i_p,x_0}(x) = f_{i_1,\dots,i_p,x_1}(x)$ sur $V_{x_0} \cap V_{x_1}$ car le développement de Taylor de f_x sur $W_{x_0} \cap W_{x_1}$ est unique pour $x \in V_{x_0} \cap V_{x_1}$. Il existe donc des fonctions entières sur $\mathbb{C}^\ell$, notées $f_{i_1,\dots,i_p}$ telles que :

$$f_x(y) = \sum_{|i| \geq 1} f_{i_1,\dots,i_p}(x)\, y_1^{i_1} \dots y_p^{i_p} \quad \text{pour} \quad (x,y) \in \bigcup_{x_0 \in \mathbb{C}^\ell} V_{x_0} \times W_{x_0} .$$

Notons $F_x : \mathbb{C}^p \longrightarrow \mathbb{C}^p$, $y \longrightarrow F(x,y)$; l'application $x \longrightarrow F_x$ est une application analytique de $\mathbb{C}^\ell$ dans $\mathcal{G}$.

D'après le théorème 4.2.3 il existe deux applications entières $x \longrightarrow P_x$ et $x \longrightarrow N_x$ de $\mathbb{C}^\ell$ dans $\mathcal{G}$ telles, pour tout $x \in \mathbb{C}^\ell$, on ait :

(a) $F_x \circ P_x = P_x \circ N_x$ où $\Pi_1 [P_x] = I$
(b) $N_x = L_x \circ U_x = U_x \circ L_x$
(c) $L_x \in \mathcal{G}_1$ et est diagonalisable
(d) U_x est unipotente, $U_x \in \mathcal{G}_q$ et $(U_x)^{-1} \in \mathcal{G}_q$.

Posons $T_x = \Pi_q [P_x]$. Alors on a : $\Pi_q (F_x) \Theta\, T_x = T_x \Theta\, N_x$ pour tout $x \in \mathbb{C}^1$. Pour tout $x \in \mathbb{C}^\ell$, T_x et N_x sont des applications polynomiales. On peut donc définir, pour tout $y \in \mathbb{C}^p$ la valeur pour y de T_x et de N_x : notons les respectivement $T_x(y) = T(x,y)$ et $N_x(y) = N(x,y)$. T et N sont deux fonctions analytiques sur $\mathbb{C}^{\ell+p}$ (d'après [13] p. 28, théorème 2.2.8). On a $T(x_0, 0) = 0$ et

$T'_{x_0}(0) \in GL(\mathbb{C}^p)$. Donc, d'après le théorème des fonctions implicites, pour tout $x_0 \in \mathbb{C}^\ell$ il existe une application analytique définie sur un voisinage de $(x_0,0)$, dont on note $T_x^{-1}(y)$ la valeur en (x,y) et telle que sur un voisinage de 0 $(T_x \circ T_x^{-1})(y) = (T_x^{-1} \circ T_x)(y) = y$. De plus, T_x^{-1} est l'inverse de T_x dans $\mathcal{G}$.

Posons $M(x,y) = (T_x \circ N_x \circ T_x^{-1})(y)$; pour tout $(x_0,0) \in \mathbb{C}^n$, il existe un voisinage de $(x_0,0)$ dans $\mathbb{C}^n$ sur lequel M est holomorphe. Soient x_0 fixé dans $\mathbb{C}^\ell$, U un voisinage de x_0 dans $\mathbb{C}^\ell$ et U' un voisinage de 0 dans $\mathbb{C}^p$ tels que F et M soient analytiques sur $\mathcal{U} = U \times U'$.

Comme $\Pi_q(F_x) \ominus T_x = T_x \ominus N_x$, on a, pour tout $x \in U$, $\Pi_q(F_x) = \Pi_q(T_x \circ N_x \circ T_x^{-1})$ c'est-à-dire : $\Pi_q(F_x) = \Pi_q(M_x)$. De plus $F_x(0)=0$ et $F'_x(0) \in GL(\mathbb{C}^p)$ pour tout $x \in \mathbb{C}^\ell$. Notons σ_0 le rayon spectral de $F'_{x_0}(0)$ et σ_1 le rayon spectral de $[F'_{x_0}(0)]^{-1}$. On a, d'après la définition 4.4.2, $\sigma_0 = |\mu_p|$ et $\sigma_1 = \dfrac{1}{|\mu_1|}$ avec $\sigma_0 \sigma_1^{q+1} < 1$.

Il existe donc, d'après le théorème 4.3.1, un voisinage ouvert W de x_0 dans $\mathbb{C}^\ell$, un voisinage W' de 0 dans $\mathbb{C}^p$ et une application Q analytique sur $\mathcal{W}_{x_0} = W \times W'$ tels que :

(1) $(x,y) \longrightarrow M_x^{-1}(y)$ est définie et analytique sur $W \times W'$

(2) Pour tout $x \in W$, pour tout $n \geq 1$, la série définissant le germe $(Q_n)_x = (F_x^n \circ M_x^{-n})$ est analytique sur W' et l'application $Q_n : (x,y) \longrightarrow (Q_n)_x(y)$ est analytique sur $\mathcal{W}_{x_0}$

(3) Q_n converge uniformément sur $\mathcal{W}_{x_0}$ vers une application analytique Q.

(4) $\Pi_q(Q_x) = I_W$, pour tout $x \in W$.

(5) La série définissant le germe $F_x \circ Q_x \circ M_x^{-1}$ est analytique sur W', pour tout $x \in W$, et $(F_x \circ Q_x \circ M_x^{-1})(y) = Q_x(y)$ pour tout $(x,y) \in \mathcal{W}_{x_0}$

Soit maintenant A un voisinage de x_0 dans $\mathbb{C}^\ell$ et rB une boule ouverte dans $\mathbb{C}^p$ de cente 0 et de rayon r tels que $A \times rB \subset \mathcal{W}_{x_0}$. Comme T est continue en $(x_0, 0)$, il existe un ouvert $\mathcal{V}_{x_0} \subset \mathbb{C}^n$ tel que, pour tout $(x,y) \in \mathcal{V}_{x_0}$, $\|T_x(y)\| < r$ (car $(T_x(0)=0)$. Posons $\mathcal{O}_{x_0} = (A \times rB) \cap \mathcal{V}_{x_0}$. Alors, pour tout $(x,y) \in \mathcal{O}_{x_0}$ on a $(x, T_x(y)) \in \mathcal{W}_{x_0}$. On a donc les propriétés suivantes :

(2)' Pour tout $x \in W$, pour tout $n \geq 1$, la série définissant le germe
$(R_n)_x = F_x^n \circ M_x^{-n} \circ T_x = F_x^n \circ T_x \circ N_x^{-n}$

est analytique sur $T_x^{-1}(W')$ et l'application $R_n : (x,y) \longrightarrow (R_n)_x(y)$ est analytique sur $\mathcal{O}_{x_0}$.

(3)' R_n converge uniformément sur $\mathcal{O}_{x_0}$ vers une application analytique R, où $R(x,y) = (Q_x \circ T_x)(y)$ pour tout $(x,y) \in \mathcal{O}_{x_0}$.

(4)' $\Pi_q(R_x) = T_x$ et donc $\Pi_1(R_x) = I$ pour tout $x \in W$.

(5)' La série définissant le germe $F_x \circ Q_x \circ M_x^{-1} \circ T_x = F_x \circ Q_x \circ T_x \circ N_x^{-1}$ est analytique sur

$T_x^{-1}(W')$ pour tout $x \in W$, et : $(F_x \circ Q_x \circ M_x^{-1} \circ T_x)(y) = (Q_x \circ T_x)(y)$ pour tout $(x,y) \in \mathcal{O}_{x_0}$ ou

$(F_x \circ Q_x \circ N_x^{-1})(y) = R_x(y)$.

Comme précédemment, N étant continue en $(x_0,0)$, il existe un ouvert $\mathcal{U}_{x_0} \subset \mathcal{O}_{x_0}$, contenant $(x_0,0)$, tel que, pour tout $(x,y) \in \mathcal{U}_{x_0}$, on ait $(x, N_x(y)) \in \mathcal{O}_{x_0}$.

Donc, pour tout $(x,y) \in \mathcal{U}_{x_0}$, on a :

(6) $(F_x \circ R_x)(y) = (R_x \circ N_x)(y)$.

En raisonnant ainsi pour tout $x_0 \in \mathbb{C}^\ell$, on peut définir R sur $\mathcal{U} = \bigcup_{x_0 \in \mathbb{C}^\ell} \mathcal{U}_{x_0}$, en posant, pour

tout $(x,y) \in \mathcal{U}$: $R(x,y) = \lim_{n \to +\infty} F_x^n \circ T_x \circ N_x^{-n}(y)$. On a, en particulier, $R(x,0)=0$, $\Pi_q(R_x) = T_x$ et $\Pi_q(R_x) = I$

pour tout $x \in \mathbb{C}^\ell$. L'égalité (6) est alors vraie pour tout $(x,y) \in \mathcal{U}$. De plus, la suite R_n où $R_n(x,y) = F_x^n \circ T_x \circ N_x^{-n}(y)$ converge uniformément vers R sur tout compact de $\mathcal{U}$.

Posons, pour tout $(x,y) \in \mathcal{U}$: $\mathcal{R}(x,y) = (x, R(x,y))$ et pour tout $(x,y) \in \mathbb{C}^n$. $\mathcal{N}(x,y) = (x, N(x,y))$ On a $\mathcal{R}(x,0) = (x,0)$ et $\mathcal{N}(x,0) = (x,0)$ pour tout $x \in \mathbb{C}^\ell$. De plus $\mathcal{R}'(x,0) \in GL(\mathbb{C}^n)$. Enfin $(\mathcal{F} \circ \mathcal{R})(z) = (\mathcal{R} \circ \mathcal{N})(z)$ pour tout $z \in \mathcal{U}$ et $\mathcal{N}$ vérifie toutes les propriétés qui font de $\mathcal{N}$ une forme réduite de $\mathcal{F}$.

PROPOSITION 4.4.4 - *Soit* $\Omega = \mathbb{C}^{\ell} \times O_{\mathbb{C}^p}$ *une variété linéaire de* $\mathbb{C}^n$ *et soit* $\mathfrak{F} : (x,y) \longrightarrow (x,F(x,y))$ *un germe d'application analytique au voisinage de* Ω, *telle que* $\mathfrak{F}(z) = z$ *pour tout* $z \in \Omega$ *et telle que* Ω *soit répulsive pour* $\mathfrak{F}$. *Soit* $S : x \to S_x$ *une application entière de* $\mathbb{C}^{\ell}$ *dans* $\mathcal{G}_{q(F)}$.

Si $\mathfrak{N}$ *est une forme réduite de* $\mathfrak{F}$, *il existe au plus un germe d'application analytique* $\mathfrak{P} : (x,y) \longrightarrow (x,P(x,y))$ *au voisinage de* Ω *tel que* :

(1) $\mathfrak{F} \circ \mathfrak{P} = \mathfrak{P} \circ \mathfrak{N}$

(2) $\Pi_{q(F)}(P_x) = S_x$ *pour tout* $x \in \mathbb{C}^1$.

Démonstration - Soit $\mathfrak{P}$ un germe d'application analytique $(x,y) \longrightarrow (x,P(x,y))$ au voisinage de Ω vérifiant les conditions (1) et (2). Il existe un ouvert $\mathcal{U}$ contenant Ω tel que $F_x o P_x (y) = P_x o N_x (y)$ pour tout $(x,y) \in \mathcal{U}$. Donc , pour tout $x \in \mathbb{C}^{\ell}$, on a (au sens des germes)

(i) $F_x o P_x = P_x o N_x$

(ii) $\Pi_q(P_x) = S_x$ (en posant $q(F) = q$).

D'autre part, les valeurs propres de $F'_x(0)$ sont constantes et vérifient $1 < |\mu_1| \leq ... \leq |\mu_p|$ et $|\mu_1|^{q+1} > |\mu_p|$, donc elles n'ont aucune relation d'ordre strictement supérieur à q (Remarque 1.3.4). Donc d'après la proposition 1.3.6, il existe au plus un élément P_x de $\mathcal{G}$ vérifiant (i) et (ii).

THEOREME 4.4.5 - *Soit* $\Omega = \mathbb{C}^{\ell} \times O_{\mathbb{C}^p}$ *une variété linéaire de* $\mathbb{C}^n$ *et soit* $\mathfrak{F}$ *un germe d'application analytique au voisinage de* Ω, *vérifiant* $\mathfrak{F}(z)=z$ *pour tout* $z \in \Omega$. *Alors si* Ω *est répulsive pour* $\mathfrak{F}$, $\mathfrak{F}$ *admet une forme réduite* $\mathfrak{N}$. *(Et* Ω *est répulsive pour* $\mathfrak{N}$*).*

Démonstration - D'après le théorème 3.2.1, il existe un germe inversible $\mathfrak{P}$ d'application analytique au voisinage de Ω, vérifiant la propriété suivante :

(1) $\mathfrak{P} \circ \mathfrak{F} \circ \mathfrak{P}^{-1} = \mathfrak{T}$, où $\mathfrak{T} : (x,y) \longrightarrow (x,T(x,y))$ est un germe de fonction analytique au voisinage de Ω, tel que $\mathfrak{T}(z)=z$ pour tout $z \in \Omega$.

De plus, d'après la remarque 3.2.2, Ω est répulsive pour $\mathfrak{T}$. D'après (1), on a $\mathfrak{F} \circ \mathfrak{P}^{-1} = \mathfrak{P}^{-1} \circ \mathfrak{T}$. D'autre part, d'après le lemme 4.4.3, $\mathfrak{T}$ admet une forme réduite $\mathfrak{N}$. Donc, il existe un germe $\mathfrak{R} : (x,y) \longrightarrow (x,R(x,y))$ d'application analytique au voisinage de Ω vérifiant :

(2) $\mathfrak{T} \circ \mathfrak{R} = \mathfrak{R} \circ \mathfrak{N}$

On a donc $\mathfrak{F} \circ \mathcal{P}^{-1} \circ \mathcal{R} = \mathcal{P}^{-1} \circ \mathcal{T} \circ \mathcal{R} = \mathcal{P}^{-1} \circ \mathcal{R} \circ \mathcal{N}$. Soit $Q = \mathcal{P}^{-1} \circ \mathcal{R}$. Q est un germe d'application analytique au voisinage de Ω, tel que $\mathfrak{F} \circ Q = Q \circ \mathcal{N}$ et tel que Q(z)=z pour tout $z \in \Omega$, et $Q'(z) \in GL(\mathbb{C}^n)$ pour tout $z \in \Omega$. Donc $\mathfrak{F}$ admet, comme $\mathcal{T}$, $\mathcal{N}$ comme forme réduite. Comme $\mathcal{N}$ est une forme réduite, on a $\mathcal{N}(z)=z$ pour tout $z \in \Omega$; montrons que Ω est de plus une variété répulsive pour $\mathcal{N}$.

D'après (2), on a : $T_x o R_x (y) = R_x o N_x (y)$ sur un ouvert contenant Ω. Donc $T_x o R_x = R_x o N_x$ pour tout $x \in \mathbb{C}^l$. On a donc

$$T'_x(0) o R'_x (0) = R'_x (0)\ o N'_x (0) \quad \text{où} \quad R'_x (0) \in GL(\mathbb{C}^p)$$

c'est-à-dire $\frac{\partial T}{\partial y}(x,0) \circ \frac{\partial R}{\partial y}(x,0) = \frac{\partial R}{\partial y}(x,0) \circ \frac{\partial N}{\partial y}(x,0)$. Donc $\frac{\partial N}{\partial y}(x,0)$ et $\frac{\partial T}{\partial y}(x,0)$ ont les mêmes valeurs propres et Ω est bien répulsive pour $\mathcal{N}$.

§ 5 - Applications de Bieberbach associées aux automorphismes de $\mathbb{C}^n$ à variétés linéaires fixes et répulsives

LEMME 4.5.1 - *Soit* E *un espace vectoriel de dimension finie sur* $\mathbb{C}$. *Soit* $G : \mathbb{C}^\ell \longrightarrow \mathrm{End}(E)$, $x \longrightarrow G(x)$ *une application continue sur* $\mathbb{C}^l$.

Supposons que les valeurs propres de G(x) soient constantes et de module strictement inférieur à 1. Soit ν le rayon spectral de G(x) et soit μ tel que $\nu < \mu < 1$. Alors pour tout compact K de $\mathbb{C}^\ell$, il existe $M_K > 0$ tel que, pour tout $x \in K$, $\|(G(x))^n\| \leq M_K\, \mu^n (n \in \mathbb{N})$.

Démonstration - Soit $x \in \mathbb{C}^\ell$; on a $\lim\limits_{p \to +\infty} \|(G(x))^p\|^{1/p} = \nu$, donc il existe un entier n tel que $\|(G(x))^n\| < \mu^n$. Soit K un compact de $\mathbb{C}^\ell$; recouvrons K par les ouverts $\mathcal{O}_n = \{x \in K, \|(G(x))^n\| < \mu^n\}$.

Soit $\mathcal{O}_{n_1} \ldots \mathcal{O}_{n_k}$ un recouvrement fini extrait du recouvrement précédent et soit $n_0 = \sup(n_1, \ldots, n_k)$. Soit $x \in K$ et soit $n > n_0$; il existe $i (1 \leq i \leq k)$ tel que $\|(G(x))^{n_i}\| < \mu^{n_i}$. On a $n = n_i q_i + r_i$ avec $0 \leq r_i < n_i$. Donc

$$\|(G(x))^n\| \leq \|(G(x))^{n_i}\|^{q_i} . \|G(x)\|^{r_i}$$

$$\leq \mu^{n_i q_i} \|G(x)\|^{r_i}$$

$$\leq \mu^{n-r_i} \|G(x)\|^{r_i}$$

$$\leq C\mu^n \text{ où } C = (\max_{i \leq k} \mu^{-r_i}) . \max_{\substack{r_i < n_0 \\ x \in K}} \|G(x)\|^{r_i}$$

Comme les fonctions $\|(G(x))^n\| \mu^{-n}$ sont bornées sur K pour $n \leq n_0$, il existe bien M_k tel que $\|(G(x))^n\| \leq M_K \mu^{-n}$ $(n \in \mathbb{N})$.

LEMME 4.5.2 - *Soit* $\Omega = \mathbb{C}^\ell \times O_{\mathbb{C}^p}$ *une variété linéaire de* $\mathbb{C}^n$. *Soit* $\mathcal{N}$ *une forme réduite d'une application* $\mathcal{F}$ *analytique au voisinage de* Ω, Ω *étant une variété fixe et répulsive pour* $\mathcal{F}$. *Alors la suite* $\mathcal{N}^{-m}$ *converge uniformément sur tout compact de* $\mathbb{C}^{\ell+p}$ *vers* $\Pi : \mathbb{C}^\ell \times \mathbb{C}^p \longrightarrow \mathbb{C}^p$, $(x,y) \longrightarrow (x,0)$.

Démonstration - Reprenons les notations de la définition 4.4.1, c'est-à-dire, notons $\mathcal{N}(x,y)=(x,N(x,y))$ avec $N_x = L_x \circ U_x = U_x \circ L_x$.

Notons, d'autre part, N_x^{-1} l'inverse de N_x dans $\mathcal{G}$, V_x l'inverse de U_x dans $\mathcal{G}$, et Λ_x l'inverse de L_x dans $\mathcal{G}$. Alors on a : $N_x^{-1} = V_x \circ \Lambda_x = \Lambda_x \circ V_x$. Les valeurs propres $\lambda_1, ..., \lambda_p$, de Λ_x vérifient $0 < |\lambda_p| \leq ... \leq |\lambda_1| < 1$ (d'après la définition 4.4.2 en posant $\lambda_i = \frac{1}{\mu_i}$ pour $1 \leq i \leq p$) et V_x est unipotent. Comme les valeurs propres de $V'_x(0)$ sont égales à 1, l'application $(\Lambda_x \circ V_x^3)'(0)$ admet également $\lambda_1, ..., \lambda_p$, pour valeurs propres.

Soit maintenant (a,b) un point quelconque de $\mathbb{C}^\ell \times \mathbb{C}^p$. D'après la remarque 3.1.5 il existe un voisinage borné A de a dans $\mathbb{C}^\ell$ et un voisinage A' de 0 dans $\mathbb{C}^p$ tel que pour tout $x \in A$, on ait

(1) $(\Lambda_x \circ V_x^3)(A') \subset A'$.

Soit B un voisinage borné de b dans $\mathbb{C}^p$,

$B' = \{ (\Lambda_x \circ V_x)(y) ; (x,y) \in A \times B\}$ et $B'' = \{(\Lambda_x \circ V_x)^2 (y) ; (x,y) \in A \times B\}$.

Posons $H = \overline{B} \cup \overline{B}' \cup \overline{B}''$; H est un compact de $\mathbb{C}^p$ et pour tout r tel que $0 \leq r \leq 2$ et pour tout $x \in A$, on a : $(\Lambda_x \circ V_x)^r (B) \subset H$. Soit W un voisinage de l'origine dans $\mathbb{C}^p$. D'après le lemme 4.5.1, il existe un entier n_0 tel que, pour tout $n > n_0$, on ait :

(2) $\Lambda_x^n (H) \subset A'$ et

(3) $\Lambda_x^n (A') \subset W$.

Soit $m \in \mathbb{N}$ tel que $m > 3n_0 + 2$. Alors $m = 3n+r$ avec $0 \leq r \leq 2$ et $n > n_0$. Comme $\Lambda_x \circ V_x = V_x \circ \Lambda_x$ on a $(\Lambda_x \circ V_x)^m (B) = (\Lambda_x \circ V_x)^{3n} \circ (\Lambda_x \circ V_x)^r (B)$. Donc $(\Lambda_x \circ V_x)^m (B) \subset (\Lambda_x \circ V_x)^{3n} (H)$ pour tout $x \in A$. Or $(\Lambda_x \circ V_x)^{3n} (H) = [\Lambda_x^n \circ (\Lambda_x \circ V_x^3)^n \circ \Lambda_x^n] (H)$. Comme $n > n_0$, on a :

$$(\Lambda_x \circ V_x)^{3n} (H) \subset [\Lambda_x^n \circ (\Lambda_x \circ V_x^3)^n] (A') \quad \text{(d'après (2))}$$

$$\subset \Lambda_x^n (A') \quad \text{(d'après (1))}$$

$$\subset W \quad \text{(d'après (3))}.$$

Donc, pour tout $x \in A$, pour $m > 3n_0 + 2$:

$$N_x^{-m} (B) \subset W.$$

Donc, pour tout $(x,y) \in A \times B$, pour $m > 3n_0 + 2$:

$$\mathcal{N}^{-m} (x,y) \in \{x\} \times W \text{ car } \mathcal{N}^{-m} (x,y) = (x, N_x^{-m} (y)).$$

Soit alors $\mathcal{K}$ un compact de $\mathbb{C}^n = \mathbb{C}^\ell \times \mathbb{C}^p$. On peut recouvrir $\mathcal{K}$ par un nombre fini d'ouverts du type $A \times B$. Donc, il existe un entier m_0, tel que si $m > m_0$, $\mathcal{N}^{-m} (x,y) \in \{x\} \times W$ pour tout $(x,y) \in \mathcal{K}$, d'où le résultat.

COROLLAIRE 4.5.3 - *Soit* $\Omega = \mathbb{C}^\ell \times O_{\mathbb{C}^p}$ *une variété linéaire de* $\mathbb{C}^n$. *Soit* $\mathcal{N}$ *une forme réduite d'une application* $\mathcal{F}$ *analytique au voisinage de* Ω, Ω *étant une variété fixe et répulsive pour* $\mathcal{F}$. *Alors pour tout compact* $\mathcal{K}$ *de* $\mathbb{C}^\ell \times \mathbb{C}^p$ *et pour tout voisinage* $\mathcal{W}$ *de* Ω, *il existe un entier positif* k_0, *tel que si* $k > k_0$, $\mathcal{N}^{-k} (\mathcal{K}) \subset \mathcal{W}$.

Démonstration - L'ensemble $\Pi(\mathcal{K})$ est compact, et il existe $\varepsilon>0$, tel que $(x,y)\in\mathcal{W}$ pour $x \in \Pi(\mathcal{K})$ et $\|y\|<\varepsilon$. D'autre part, il existe k_0 tel que $\|\mathcal{N}^{-k}(x,y)-(x,0)\|<\varepsilon$ pour $(x,y) \in \mathcal{K}$ et $k>k_0$. Comme $\Pi(\mathcal{N}^{-k}(x,y)) = (x,0)$ on a bien $\mathcal{N}^{-k}(x,y)\in \mathcal{K}$ pour $(x,y) \in \mathcal{W}$ et $k>k_0$.

Remarquons que si $\mathcal{V}$ est un ensemble et si $\mathcal{F} : \mathcal{V} \longrightarrow \mathcal{V}$ est injective, alors $\mathcal{F}|_{\mathcal{W}}$ est bijective où $\mathcal{W} = \bigcap_{m\geq 1} \mathcal{F}^m(\mathcal{V})$. Ceci permet de définir $\mathcal{F}^{-m}(z)$ pour tout $z \in \mathcal{W}$ et tout $m \geq 1$.

THEOREME 4.5.4 - Soit $\Omega = \mathbb{C}^{\ell}\times 0_{\mathbb{C}^p}$ *une variété linéaire de* $\mathbb{C}^n$ *et soit* $\mathcal{V}$ *un ouvert de* $\mathbb{C}^n$ *contenant* Ω. *Soit* $\mathcal{F} : \mathcal{V} \longrightarrow \mathcal{V}$ *une application analytique admettant* Ω *comme variété fixe et répulsive. Alors* $\mathcal{F}$ *admet une forme réduite et, pour toute forme réduite* $\mathcal{N}$ **de** $\mathcal{F}$, *il existe une application entière* $\mathcal{P} : \mathbb{C}^n \longrightarrow \mathcal{V}$ *telle que* $\mathcal{P}(z) = z$, $\mathcal{P}'(z) \in GL(\mathbb{C}^n)$ *pour tout* $z \in \Omega$ *et vérifiant* $\mathcal{F}o\mathcal{P}=\mathcal{P}o\mathcal{N}$. *De plus,* $\mathcal{P}(\mathbb{C}^n)$ *est l'ensemble des* $z \in \mathcal{V}$ *vérifiant la propriété suivante :*

(1) Il existe une suite convergente (z_m) *d'éléments de* $\mathbb{C}^n$ *telle que* $\lim_{m\to+\infty} z_m \in \Omega$ *et telle que*

$z=\mathcal{F}^m(z_m)$ *pour tout* $m \geq 1$.

Si $\mathcal{F}'(z) \in GL(\mathbb{C}^n)$ pour tout $z\in\mathcal{V}$, alors $\mathcal{P}'(z) \in GL(\mathbb{C}^n)$ pour tout $z\in\mathbb{C}^n$. De plus, si $\mathcal{F}$ est injective sur $\mathcal{V}$, $\mathcal{P}$ est injective sur $\mathbb{C}^n$ et, dans ce cas, $\mathcal{P}(\mathbb{C}^n)$ est l'ensemble des $z \in \bigcap_{m\geq 1} \mathcal{F}^m(\mathcal{V})$ tels que la suite $(\mathcal{F}^{-m}(z))_{m\geq 1}$ converge vers un élément de Ω.

Démonstration - Il résulte du théorème 4.4.5 que $\mathcal{F}$ possède une forme réduite $\mathcal{N}$. Il existe donc un germe d'application analytique $\mathcal{P}$ au voisinage de Ω, laissant fixe Ω telle que $\mathcal{P}'(z) \in GL(\mathbb{C}^n)$ pour tout $z\in\Omega$ et telle que $\mathcal{F}o\mathcal{P}o\mathcal{N}^{-1} = \mathcal{P}$ au sens des germes. Soit $\mathcal{U}$ un voisinage connexe de Ω tel que $\mathcal{P}$ soit analytique sur $\mathcal{U} \cup \mathcal{N}^{-1}(\mathcal{U})$ et tel que $\mathcal{P}(\mathcal{U} \cup \mathcal{N}^{-1}(\mathcal{U})) \subset \mathcal{V}$.

On a donc pour tout $z \in \mathcal{U}$, $(\mathcal{F}o\mathcal{P}o\mathcal{N}^{-1})(z) = \mathcal{P}(z)$. Posons, pour $m > 1$, $\mathcal{U}_m=\mathcal{N}^{m-1}(\mathcal{U})$ et $\mathcal{P}_m=\mathcal{F}^m \circ \mathcal{P} \circ \mathcal{N}^{-m}$. $\mathcal{P}_m$ est définie et analytique sur l'ouvert $\mathcal{U}_m$.

Soit d'autre part une suite croissante de compacts $(K_p)_{p\geq 1}$ d'intérieur connexe tels que $\bigcup_{p\geq 1} \mathring{K}_p=\mathbb{C}^n$ et $K_1 \subset \mathcal{U}$. D'après le lemme 4.5.2, pour tout compact K_p, il existe un entier m_p tel que si $m>m_p$, alors $\mathcal{N}^{-m+1}(K_p) \subset \mathcal{U}$, c'est-à-dire $K_p \subset \mathcal{U}_m$. On construit ainsi une suite (m_p) que l'on peut supposer croissante. De plus, pour tout $p \geq 1$, si $m > m_p$ on a :

$$\begin{aligned}\mathcal{P}_m(z) &= \mathcal{F}^{m-1} \circ (\mathcal{F}o\mathcal{P}o\mathcal{N}^{-1}) \circ \mathcal{N}^{-m+1}(z)\\ &= \mathcal{F}^{m-1} \circ \mathcal{P}o\mathcal{N}^{-m+1}(z)\\ &= \mathcal{P}_{m-1}(z) \quad (z \in K_p).\end{aligned}$$

Donc, pour tout $p \geq 1$ si $m > m_p$, on a $\mathcal{P}_m(z) = \mathcal{P}_{m_p}(z)$ $(z \in K_p)$. Considérons la suite $(\mathcal{P}_{m_p})_{p\geq 1}$. Pour tout $p \geq 1$ on a : $\mathcal{P}_{m_p}|_{\overset{\circ}{K}_1} = \mathcal{P}$, et $\mathcal{P}_{m_p}$ est le prolongement analytique de $\mathcal{P}$ à l'ouvert $\overset{\circ}{K}p$. On en déduit l'existence d'une application analytique définie sur $\mathbb{C}^n$, qui est le prolongement de $\mathcal{P}$ à $\mathbb{C}^n$ tout entier. Notons le également $\mathcal{P}$. On a donc sur $\mathbb{C}^n$: $\mathcal{F} \circ \mathcal{P} = \mathcal{P} \circ \mathcal{N}$ et donc $\mathcal{F}^m \circ \mathcal{P} = \mathcal{P} \circ \mathcal{N}^m$ pour tout $m \geq 1$.

Etudions $\mathcal{P}(\mathbb{C}^n)$: Soit $u \in \mathcal{P}(\mathbb{C}^n)$; alors $u = \mathcal{P}(x,y)$ où $(x,y) \in \mathbb{C}^n$. Posons $u_m = \mathcal{N}^{-m}(x,y)$; alors $\lim_{m\to+\infty} \mathcal{N}^{-m}(x,y) = (x,0)$, d'après le lemme 4.5.2, et $\lim_{m\to+\infty} \mathcal{P}(u_m) = (x,0)$. O

$$u = \mathcal{P} \circ \mathcal{N}^m(u_m) = \mathcal{F}^m \circ \mathcal{P}(u_m) = \mathcal{F}^m(z_m) \text{ et } \lim_{m\to\infty} z_m = (x,0), \text{ donc } \lim_{m\to+\infty} z_m \in \Omega.$$

Réciproquement si, $z = \mathcal{F}^m(z_m)$ pour tout $m \geq 1$ et $\lim_{m\to+\infty} z_m \in \Omega$, il existe un entier m_0 tel que, pour $m \geq m_0$ on ait $z_m \in \mathcal{P}(\mathcal{U})$. Posons $z_{m0} = \mathcal{P}(u)$ et $u_m = \mathcal{N}^m(u)$.

Alors $z_{m0} = \mathcal{P} \circ \mathcal{N}^{-m0}(u_{m0})$ et donc $z = \mathcal{F}^{m0}(z_{m0}) = \mathcal{F}^{m0} \circ \mathcal{P} \circ \mathcal{N}^{-m0}(u_{m0}) = \mathcal{P}(u_{m0})$. Donc $z \in \mathcal{P}(\mathbb{C}^n)$.

Supposons que $\mathcal{F}'(z) \in GL(\mathbb{C}^n)$ pour tout $z \in \mathcal{V}$. Comme $\mathcal{P}'(z) \in GL(\mathbb{C}^n)$ pour tout $z \in \Omega$, il existe un ouvert $\mathcal{O}_1$ contenant Ω sur lequel $\mathcal{P}'(z) \in GL(\mathbb{C}^n)$. Soit maintenant z quelconque dans $\mathbb{C}^n$; d'après le corollaire 4.5.3, il existe un entier k, tel que, si $k > k_0$, $\mathcal{N}^{-k}(z) \in \mathcal{O}_1$. Donc pour $k > k_0$, on peut écrire

$$\mathcal{P}'(z) = (\mathcal{F}^k)'[\mathcal{P} \circ \mathcal{N}^{-k}(z)] \circ \mathcal{P}'[\mathcal{N}^{-k}(z)] \circ (\mathcal{N}^{-k})'(z)$$

et on en déduit que $\mathcal{P}'(z) \in GL(\mathbb{C}^n)$ pour tout $z \in \mathbb{C}^n$. Supposons $\mathcal{F}$ injective sur $\mathcal{V}$, alors $\mathcal{F}_{|\mathcal{W}}$ est bijective (où $\mathcal{W} = \bigcap_{m\geq 1} \mathcal{F}^m(\mathcal{V})$) et donc $\mathcal{P}(\mathbb{C}^n)$ est l'ensemble des $z \in \mathcal{W}$ tels que $(\mathcal{F}^{-m}(z))$ converge vers un élément de Ω. De plus, pour tout $m \geq 1$ et pour tout $z \in \mathbb{C}^n$ $\mathcal{F}^{-m} \circ \mathcal{P}(z) = \mathcal{P} \circ \mathcal{N}^{-m}(z)$. D'autre part, comme $\mathcal{P}'(z) \in GL(\mathbb{C}^n)$ pour tout $z \in \Omega$, il existe un ouvert $\mathcal{O}_2$, contenant Ω, tel que $\mathcal{P}$ soit injective sur $\mathcal{O}_2$. Soit maintenant z_1 et $z_2 \in \mathbb{C}^n$ tels que $\mathcal{P}(z_1) = \mathcal{P}(z_2)$. Alors $\mathcal{F}^{-m} \circ \mathcal{P}(z_1) = \mathcal{F}^{-m} \circ \mathcal{P}(z_2)$, donc $\mathcal{P} \circ \mathcal{N}^{-m}(z_1) = \mathcal{P} \circ \mathcal{N}^{-m}(z_2)$ pour tout $m \geq 1$. Or, il existe un entier m_0 tel que si $m \geq m_0$, $\mathcal{N}^{-m}(z_1) \in \mathcal{O}_2$ et $\mathcal{N}^{-m}(z_2) \in \mathcal{O}_2$, donc $\mathcal{N}^{-m}(z_1) = \mathcal{N}^{-m}(z_2)$ et $z_1 = z_2$: $\mathcal{P}$ est donc injective sur $\mathbb{C}^n$. Ceci achève la démonstration.

Au chapitre IV , nous avons établi l'existence d'une forme réduite pour les germes d'applications analytiques au voisinage d'une variété linéaire de $\mathbb{C}^n$ fixe et répulsive. Cette forme réduite $\mathcal{N}$ est une application $\mathcal{N}$: $(x,y) \longmapsto (x,N(x,y))$ où pour tout $x \in \mathbb{C}^\ell$, l'application $\mathcal{N}_x : y \longrightarrow \mathcal{N}_{(x,y)}$, vérifie $N_x = L_x \circ U_x = U_x \circ L_x$ avec L_x linéaire et diagonalisable et U_x unipotente pour tout $x \in \mathbb{C}^\ell$,. Il est naturel de chercher à obtenir un résultat un peu plus fort, à savoir l'existence d'une forme normale, c'est-à-dire une forme réduite $\mathcal{N}$ où L_x ne dépend pas de x et est diagonale. Le problème se réduit à trouver une application entière $x \longrightarrow S_x$ de $\mathbb{C}^\ell$ dans $GL(\mathbb{C}^p)$ telle que $S_x^{-1} \circ L_x \circ S_x$ soit diagonale pour tout $x \in \mathbb{C}^\ell$, en s'appuyant sur le fait que les valeurs propres de L_x sont constantes.

On sait obtenir la forme normale désirée si $\ell=1$ et aussi dans le cas où $n \leq 4$ (voir [7], chapitre IV). Par contre, nous n'avons pas pu obtenir jusqu'à présent la solution dans le cas général. Ces questions seront discutées dans un article ultérieur.

BIBLIOGRAPHIE

---oOo---

[1] R. ARENS, "Dense inverse limit rings". Michigan Math. J.5 (1958), 169-182.

[2] L. BIEBERBACH, "Beispiel zwier ganzen funktionen zweier komplexen Variablen, welche eine schlichte volumtreue Abbildung des R^4 auf einem Teil seiner selbst vermitten" S.B. Preuss. Akad. Wiss. (1933), 476-479.

[3] H. CARTAN, "Théorie élémentaire des fonctions analytiques d'une ou plusieurs variables complexes". Sixième édition. Hermann, Paris, 1961.

[4] L. CHAMBADAL, J.L. OVAERT, "Algèbre linéaire et algèbre tensorielle". Dunod, Paris, 1968.

[5] J.B. CONWAY, "Functions of one complex variable". Graduate texts in Mathematics n°11. Springer Verlag. 1973.

[6] D. COUTY, "Sur l'image d'une fonction entière de deux variables complexes". Proc. Edinburgh Math. Soc. (1984), 27, 327-331.

[7] D. COUTY, "Formes réduites et normales des automorphismes analytiques de $\mathbb{C}^n$ à variété linéaire fixe et répulsive". (Thèse-Bordeaux, mars 1987).

[8] P.G. DIXON and J. ESTERLE, "Michael's problem and the Poincaré-Fatou-Bieberbach phenomenon". Bull. A.M.S. 15(2) (1986), 127-187.

[9] J. ESTERLE, "Problème de Michael et fonctions entières de plusieurs variables complexes". Proc. Conf. Analyse complexe en plusieurs variables, Toulouse, 1983, Lecture Notes in Math, vol. 1094, Springer-Verlag, Berlin and New-york (1984), 65-85.

[10] J. ESTERLE, communication privée.

[11] P. FATOU, "Sur certaines fonctions uniformes de deux variables". C.R. Acad. Sci. Paris 175 (1922), 1030-1033.

[12] P. GRIFFITHS and J. HARRIS, "Principles of algebraic geometry". Wiley. Inter-Science Publication. New-York.

[13] L. HORMANDER, "An introduction to complex analysis in several variables". North-Holland. Amsterdam. London. Paris. 1973

[14] J.E. HUMPHREYS, "Linear algebraic groups". Graduate texts in Mathematics n° 21. Springer-Verlag. 1975.

[15] K. KODAIRA, "Holomorphic mappings of polydiscs into compact complex manifolds,J. Differential geometry 6 (1971), 33-46.

[16] S. LATTES, Sur les formes réduites des transformations ponctuelles à deux variables, C.R. Acad. Sci. Paris 152 (1911), 1566-1569.

[17] L. LEAU, "Etude sur les équations fonctionnelles à une ou plusieurs variables", Ann. Fac. Sci. Toulouse 11 (1897), 1-110.

[18] J. MARTINET et J.P. RAMIS, "Classification analytique des équations différentielles non linéaires résonnantes du premier ordre". Ann. Sci. Ecole Norm. Sup. 16 (1983), 571-621.

[19] Y. NISHIMURA, "Automorphismes analytiques admettant des sous-varétés de points fixes attractives dans la direction transversale". J. Math. Kyoto Univ. 23 (1983), 289-299.

[20] E. PICARD, "Sur une classe de transcendantes nouvelles", Acta Math. 18 (1894), 133-154.

[21] H. POINCARE, Sur une classe nouvelle de transcendantes uniformes". J. de Math. 6 (1890), 313-365.

[22] L. REICH, "Das typenproblem bei-formal biholomorphen abbildungen mit anziehendem fixpunkt", Math. Ann. 179 (1969), 227-250.

[23] L. REICH, "Normalformen biholomorphen Abbildungen mit anziehendem Fixpunkt". Math. Ann. 180 (1969), 233-255.

[24] A. SADULLAEV, "On Fatou's example", Mat. Zametki 6 (1969), 437-441 (Russian).

[25] L. STEHLE, "Plongement du disque dans $\mathbb{C}^2$. Séminaire P. Lelong (Analyse). 1970. Springer Lecture Notes 275, 119-130.

[26] S. STERNBERG, "Local contraction and a theorem of Poincaré". Amer. J. Math. 79 (1957), 809-823.

[27] V.I. ARNOLD, Chapitres supplémentaires à la théorie des équations différentielles ordinaires. Editions Mir, Moscou 1980.

[28] M. CHAPERON : "Sur certains diffeomorphismes normalement hyperboliques", C.R.A.S. Paris 281 (1975), 695-698.

[29] M. CHAPERON : "Sur certains diffeomorphismes normalement hyperboliques", C.R.A.S. Paris 281 (1975), 731-733.

[30] M. CHAPERON : "Invariant manifolds and a preparation lemma for local holomorphic flows and actions, "preprint non publié.

[31] M. CHAPERON : Geometrie différentielle et singularités de systèmes dynamiques, Asterisque 138-139 (1986), S.M.F.

[32] M. CHAPERON : C^k conjugacy of holomorphic flows near a singularity, Publications Mathématiques de l'IHES, 64 (1987) 143-183.

[33] F. TAKENS : "Partially hyperbolic fixed points", Topology, 10 (1971), 133-147.

[34] V.V. LYCHIAGIN : "Local classification of non-linear first order partial differentiel equations" Usp. Math. Nowk 30-1 (1975), 101-171, Russian Math. Surveys 30-1 (1975) 105-175.

[35] T. OSHIMA : "Singularities in contact geometry and degenerate pseudo-differential equations" J Fac. Sc. Univ. Tokyo 21(1974), 43-83.

[36] G. ROBINSON : "A global approximation theorem for hamiltonian systems", Global Analysis, Proc. Symp. Pure Math 14 (1970) 233-243, A.M.S.

232: P.C. Schuur, Asymptotic Analysis of Soliton Problems. VIII, ages. 1986.

1233: Stability Problems for Stochastic Models. Proceedings, Edited by V.V. Kalashnikov, B. Penkov and V.M. Zolotarev. VI, ages. 1986.

1234: Combinatoire énumérative. Proceedings, 1985. Edité par belle et P. Leroux. XIV, 387 pages. 1986.

1235: Séminaire de Théorie du Potentiel, Paris, No. 8. Directeurs: relot, G. Choquet et J. Deny. Rédacteurs: F. Hirsch et G. bodzki. III, 209 pages. 1987.

1236: Stochastic Partial Differential Equations and Applications. edings, 1985. Edited by G. Da Prato and L. Tubaro. V, 257 s. 1987.

1237: Rational Approximation and its Applications in Mathematics hysics. Proceedings, 1985. Edited by J. Gilewicz, M. Pindor and emaszko. XII, 350 pages. 1987.

1238: M. Holz, K.-P. Podewski and K. Steffens, Injective Choice ions. VI, 183 pages. 1987.

1239: P. Vojta, Diophantine Approximations and Value Distribu- heory. X, 132 pages. 1987.

1240: Number Theory, New York 1984–85. Seminar. Edited by Chudnovsky, G.V. Chudnovsky, H. Cohn and M.B. Nathanson. 4 pages. 1987.

1241: L. Gårding, Singularities in Linear Wave Propagation. III, ages. 1987.

1242: Functional Analysis II, with Contributions by J. Hoffmann- nsen et al. Edited by S. Kurepa, H. Kraljević and D. Butković. VII, ages. 1987.

1243: Non Commutative Harmonic Analysis and Lie Groups. edings, 1985. Edited by J. Carmona, P. Delorme and M. Vergne. 9 pages. 1987.

1244: W. Müller, Manifolds with Cusps of Rank One. XI, 158 s. 1987.

1245: S. Rallis, L-Functions and the Oscillator Representation. 239 pages. 1987.

1246: Hodge Theory. Proceedings, 1985. Edited by E. Cattani, F. n, A. Kaplan and F. Puerta. VII, 175 pages. 1987.

1247: Séminaire de Probabilités XXI. Proceedings. Edité par J. a, P.A. Meyer et M. Yor. IV, 579 pages. 1987.

1248: Nonlinear Semigroups, Partial Differential Equations and tors. Proceedings, 1985. Edited by T.L. Gill and W.W. Zachary. 35 pages. 1987.

1249: I. van den Berg, Nonstandard Asymptotic Analysis. IX, 187 s. 1987.

1250: Stochastic Processes – Mathematics and Physics II. edings 1985. Edited by S. Albeverio, Ph. Blanchard and L. Streit. 59 pages. 1987.

1251: Differential Geometric Methods in Mathematical Physics. edings, 1985. Edited by P.L. García and A. Pérez-Rendón. VII, pages. 1987.

1252: T. Kaise, Représentations de Weil et GL_2 Algèbres de on et GL_n. VII, 203 pages. 1987.

1253: J. Fischer, An Approach to the Selberg Trace Formula via elberg Zeta-Function. III, 184 pages. 1987.

1254: S. Gelbart, I. Piatetski-Shapiro, S. Rallis. Explicit Construc- of Automorphic L-Functions. VI, 152 pages. 1987.

1255: Differential Geometry and Differential Equations. Proceed- 1985. Edited by C. Gu, M. Berger and R.L. Bryant. XII, 243 s. 1987.

1256: Pseudo-Differential Operators. Proceedings, 1986. Edited O. Cordes, B. Gramsch and H. Widom. X, 479 pages. 1987.

1257: X. Wang, On the C*-Algebras of Foliations in the Plane. V, pages. 1987.

1258: J. Weidmann, Spectral Theory of Ordinary Differential ators. VI, 303 pages. 1987.

Vol. 1259: F. Cano Torres, Desingularization Strategies for Three-Dimensional Vector Fields. IX, 189 pages. 1987.

Vol. 1260: N.H. Pavel, Nonlinear Evolution Operators and Semigroups. VI, 285 pages. 1987.

Vol. 1261: H. Abels, Finite Presentability of S-Arithmetic Groups. Compact Presentability of Solvable Groups. VI, 178 pages. 1987.

Vol. 1262: E. Hlawka (Hrsg.), Zahlentheoretische Analysis II. Seminar, 1984–86. V, 158 Seiten. 1987.

Vol. 1263: V.L. Hansen (Ed.), Differential Geometry. Proceedings, 1985. XI, 288 pages. 1987.

Vol. 1264: Wu Wen-tsün, Rational Homotopy Type. VIII, 219 pages. 1987.

Vol. 1265: W. Van Assche, Asymptotics for Orthogonal Polynomials. VI, 201 pages. 1987.

Vol. 1266: F. Ghione, C. Peskine, E. Sernesi (Eds.), Space Curves. Proceedings, 1985. VI, 272 pages. 1987.

Vol. 1267: J. Lindenstrauss, V.D. Milman (Eds.), Geometrical Aspects of Functional Analysis. Seminar. VII, 212 pages. 1987.

Vol. 1268: S.G. Krantz (Ed.), Complex Analysis. Seminar, 1986. VII, 195 pages. 1987.

Vol. 1269: M. Shiota, Nash Manifolds. VI, 223 pages. 1987.

Vol. 1270: C. Carasso, P.-A. Raviart, D. Serre (Eds.), Nonlinear Hyperbolic Problems. Proceedings, 1986. XV, 341 pages. 1987.

Vol. 1271: A.M. Cohen, W.H. Hesselink, W.L.J. van der Kallen, J.R. Strooker (Eds.), Algebraic Groups Utrecht 1986. Proceedings. XII, 284 pages. 1987.

Vol. 1272: M.S. Livšic, L.L. Waksman, Commuting Nonselfadjoint Operators in Hilbert Space. III, 115 pages. 1987.

Vol. 1273: G.-M. Greuel, G. Trautmann (Eds.), Singularities, Representation of Algebras, and Vector Bundles. Proceedings, 1985. XIV, 383 pages. 1987.

Vol. 1274: N. C. Phillips, Equivariant K-Theory and Freeness of Group Actions on C*-Algebras. VIII, 371 pages. 1987.

Vol. 1275: C.A. Berenstein (Ed.), Complex Analysis I. Proceedings, 1985–86. XV, 331 pages. 1987.

Vol. 1276: C.A. Berenstein (Ed.), Complex Analysis II. Proceedings, 1985–86. IX, 320 pages. 1987.

Vol. 1277: C.A. Berenstein (Ed.), Complex Analysis III. Proceedings, 1985–86. X, 350 pages. 1987.

Vol. 1278: S.S. Koh (Ed.), Invariant Theory. Proceedings, 1985. V, 102 pages. 1987.

Vol. 1279: D. Ieşan, Saint-Venant's Problem. VIII, 162 Seiten. 1987.

Vol. 1280: E. Neher, Jordan Triple Systems by the Grid Approach. XII, 193 pages. 1987.

Vol. 1281: O.H. Kegel, F. Menegazzo, G. Zacher (Eds.), Group Theory. Proceedings, 1986. VII, 179 pages. 1987.

Vol. 1282: D.E. Handelman, Positive Polynomials, Convex Integral Polytopes, and a Random Walk Problem. XI, 136 pages. 1987.

Vol. 1283: S. Mardešić, J. Segal (Eds.), Geometric Topology and Shape Theory. Proceedings, 1986. V, 261 pages. 1987.

Vol. 1284: B.H. Matzat, Konstruktive Galoistheorie. X, 286 pages. 1987.

Vol. 1285: I.W. Knowles, Y. Saitō (Eds.), Differential Equations and Mathematical Physics. Proceedings, 1986. XVI, 499 pages. 1987.

Vol. 1286: H.R. Miller, D.C. Ravenel (Eds.), Algebraic Topology. Proceedings, 1986. VII, 341 pages. 1987.

Vol. 1287: E.B. Saff (Ed.), Approximation Theory, Tampa. Proceedings, 1985–1986. V, 228 pages. 1987.

Vol. 1288: Yu. L. Rodin, Generalized Analytic Functions on Riemann Surfaces. V, 128 pages, 1987.

Vol. 1289: Yu. I. Manin (Ed.), K-Theory, Arithmetic and Geometry. Seminar, 1984–1986. V, 399 pages. 1987.

Vol. 1290: G. Wüstholz (Ed.), Diophantine Approximation and Transcendence Theory. Seminar, 1985. V, 243 pages. 1987.

Vol. 1291: C. Mœglin, M.-F. Vignéras, J.-L. Waldspurger, Correspondances de Howe sur un Corps p-adique. VII, 163 pages. 1987

Vol. 1292: J.T. Baldwin (Ed.), Classification Theory. Proceedings, 1985. VI. 500 pages. 1987.

Vol. 1293: W. Ebeling, The Monodromy Groups of Isolated Singularities of Complete Intersections. XIV, 153 pages. 1987.

Vol. 1294: M. Queffélec, Substitution Dynamical Systems – Spectral Analysis. XIII, 240 pages. 1987.

Vol. 1295: P. Lelong, P. Dolbeault, H. Skoda (Réd.), Séminaire d'Analyse P. Lelong – P. Dolbeault – H. Skoda. Seminar, 1985/1986. VII, 283 pages. 1987.

Vol. 1296: M.-P. Malliavin (Ed.), Séminaire d'Algèbre Paul Dubreil et Marie-Paule Malliavin. Proceedings, 1986. IV, 324 pages. 1987.

Vol. 1297: Zhu Y.-l., Guo B.-y. (Eds.), Numerical Methods for Partial Differential Equations. Proceedings. XI, 244 pages. 1987.

Vol. 1298: J. Aguadé, R. Kane (Eds.), Algebraic Topology, Barcelona 1986. Proceedings. X, 255 pages. 1987.

Vol. 1299: S. Watanabe, Yu.V. Prokhorov (Eds.), Probability Theory and Mathematical Statistics. Proceedings, 1986. VIII, 589 pages. 1988.

Vol. 1300: G.B. Seligman, Constructions of Lie Algebras and their Modules. VI, 190 pages. 1988.

Vol. 1301: N. Schappacher, Periods of Hecke Characters. XV, 160 pages. 1988.

Vol. 1302: M. Cwikel, J. Peetre, Y. Sagher, H. Wallin (Eds.), Function Spaces and Applications. Proceedings, 1986. VI, 445 pages. 1988.

Vol. 1303: L. Accardi, W. von Waldenfels (Eds.), Quantum Probability and Applications III. Proceedings, 1987. VI, 373 pages. 1988.

Vol. 1304: F.Q. Gouvêa, Arithmetic of p-adic Modular Forms. VIII, 121 pages. 1988.

Vol. 1305: D.S. Lubinsky, E.B. Saff, Strong Asymptotics for Extremal Polynomials Associated with Weights on $\mathbb{R}$. VII, 153 pages. 1988.

Vol. 1306: S.S. Chern (Ed.), Partial Differential Equations. Proceedings, 1986. VI, 294 pages. 1988.

Vol. 1307: T. Murai, A Real Variable Method for the Cauchy Transform, and Analytic Capacity. VIII, 133 pages. 1988.

Vol. 1308: P. Imkeller, Two-Parameter Martingales and Their Quadratic Variation. IV, 177 pages. 1988.

Vol. 1309: B. Fiedler, Global Bifurcation of Periodic Solutions with Symmetry. VIII, 144 pages. 1988.

Vol. 1310: O.A. Laudal, G. Pfister, Local Moduli and Singularities. V, 117 pages. 1988.

Vol. 1311: A. Holme, R. Speiser (Eds.), Algebraic Geometry, Sundance 1986. Proceedings. VI, 320 pages. 1988.

Vol. 1312: N.A. Shirokov, Analytic Functions Smooth up to the Boundary. III, 213 pages. 1988.

Vol. 1313: F. Colonius, Optimal Periodic Control. VI, 177 pages. 1988.

Vol. 1314: A. Futaki, Kähler-Einstein Metrics and Integral Invariants. IV, 140 pages. 1988.

Vol. 1315: R.A. McCoy, I. Ntantu, Topological Properties of Spaces of Continuous Functions. IV, 124 pages. 1988.

Vol. 1316: H. Korezlioglu, A.S. Ustunel (Eds.), Stochastic Analysis and Related Topics. Proceedings, 1986. V, 371 pages. 1988.

Vol. 1317: J. Lindenstrauss, V.D. Milman (Eds.), Geometric Aspects of Functional Analysis. Seminar, 1986–87. VII, 289 pages. 1988.

Vol. 1318: Y. Felix (Ed.), Algebraic Topology – Rational Homotopy. Proceedings, 1986. VIII, 245 pages. 1988

Vol. 1319: M. Vuorinen, Conformal Geometry and Quasiregular Mappings. XIX, 209 pages. 1988.

Vol. 1320: H. Jürgensen, G. Lallement, H.J. Weinert (Eds.), Se groups, Theory and Applications. Proceedings, 1986. X, 416 pag 1988.

Vol. 1321: J. Azéma, P.A. Meyer, M. Yor (Eds.), Séminaire Probabilités XXII. Proceedings. IV, 600 pages. 1988.

Vol. 1322: M. Métivier, S. Watanabe (Eds.), Stochastic Analy Proceedings, 1987. VII, 197 pages. 1988.

Vol. 1323: D.R. Anderson, H.J. Munkholm, Boundedly Contro Topology. XII, 309 pages. 1988.

Vol. 1324: F. Cardoso, D.G. de Figueiredo, R. Iório, O. Lopes (Ed Partial Differential Equations. Proceedings, 1986. VIII, 433 pag 1988.

Vol. 1325: A. Truman, I.M. Davies (Eds.), Stochastic Mechanics Stochastic Processes. Proceedings, 1986. V, 220 pages. 1988.

Vol. 1326: P.S. Landweber (Ed.), Elliptic Curves and Modular Form Algebraic Topology. Proceedings, 1986. V, 224 pages. 1988.

Vol. 1327: W. Bruns, U. Vetter, Determinantal Rings. VII,236 pag 1988.

Vol. 1328: J.L. Bueso, P. Jara, B. Torrecillas (Eds.), Ring The Proceedings, 1986. IX, 331 pages. 1988.

Vol. 1329: M. Alfaro, J.S. Dehesa, F.J. Marcellan, J.L. Rubio Francia, J. Vinuesa (Eds.): Orthogonal Polynomials and their Appl tions. Proceedings, 1986. XV, 334 pages. 1988.

Vol. 1330: A. Ambrosetti, F. Gori, R. Lucchetti (Eds.), Mathema Economics. Montecatini Terme 1986. Seminar. VII, 137 pages. 19

Vol. 1331: R. Bamón, R. Labarca, J. Palis Jr. (Eds.), Dynam Systems, Valparaiso 1986. Proceedings. VI, 250 pages. 1988.

Vol. 1332: E. Odell, H. Rosenthal (Eds.), Functional Analysis. ceedings, 1986–87. V, 202 pages. 1988.

Vol. 1333: A.S. Kechris, D.A. Martin, J.R. Steel (Eds.), Cabal Sem 81–85. Proceedings, 1981–85. V, 224 pages. 1988.

Vol. 1334: Yu.G. Borisovich, Yu. E. Gliklikh (Eds.), Global Anal – Studies and Applications III. V, 331 pages. 1988.

Vol. 1335: F. Guillén, V. Navarro Aznar, P. Pascual-Gainza, F. Pue Hyperrésolutions cubiques et descente cohomologique. XII, pages. 1988.

Vol. 1336: B. Helffer, Semi-Classical Analysis for the Schrödir Operator and Applications. V, 107 pages. 1988.

Vol. 1337: E. Sernesi (Ed.), Theory of Moduli. Seminar, 1985. VIII, pages. 1988.

Vol. 1338: A.B. Mingarelli, S.G. Halvorsen, Non-Oscillation Dom of Differential Equations with Two Parameters. XI, 109 pages. 1988

Vol. 1339: T. Sunada (Ed.), Geometry and Analysis of Manifo Procedings, 1987. IX, 277 pages. 1988.

Vol. 1340: S. Hildebrandt, D.S. Kinderlehrer, M. Miranda (Ec Calculus of Variations and Partial Differential Equations. Proceedir 1986. IX, 301 pages. 1988.

Vol. 1341: M. Dauge, Elliptic Boundary Value Problems on Co Domains. VIII, 259 pages. 1988.

Vol. 1342: J.C. Alexander (Ed.), Dynamical Systems. Proceedi 1986–87. VIII, 726 pages. 1988.

Vol. 1343: H. Ulrich, Fixed Point Theory of Parametrized Equiva Maps. VII, 147 pages. 1988.

Vol. 1344: J. Král, J. Lukeš, J. Netuka, J. Veselý (Eds.), Pote Theory – Surveys and Problems. Proceedings, 1987. VIII, 271 pa 1988.

Vol. 1345: X. Gomez-Mont, J. Seade, A. Verjovski (Eds.), Holomor Dynamics. Proceedings, 1986. VII, 321 pages. 1988.

Vol. 1346: O. Ya. Viro (Ed.), Topology and Geometry – R Seminar. XI, 581 pages. 1988.

Vol. 1347: C. Preston, Iterates of Piecewise Monotone Mapping an Interval. V, 166 pages. 1988.

Vol. 1348: F. Borceux (Ed.), Categorical Algebra and its Applicati Proceedings, 1987. VIII, 375 pages. 1988.

Vol. 1349: E. Novak, Deterministic and Stochastic Error Bounc Numerical Analysis. V, 113 pages. 1988.